Polymers
Properties and Applications

1

B. Rånby J. F. Rabek

ESR Spectroscopy in Polymer Research

Springer-Verlag
Berlin Heidelberg New York 1977

Dr. B. Rånby, Professor of Polymer Technology
Dr. J. F. Rabek, Assistant Professor of Polymer Technology

Department of Polymer Technology, The Royal Institute of Technology
(Technical University of Stockholm), Teknikringen, Stockholm, Sweden

This volume continues the series *Chemie, Physik und Technologie der Kunststoffe in Einzeldarstellungen*, which is now entitled *Polymers/ Properties and Applications*.

With 356 Figures

ISBN 3-540-08151-8 Springer-Verlag Berlin Heidelberg New York

ISBN 0-387-08151-8 Springer-Verlag New York Heidelberg Berlin

Library of Congress Cataloging in Publication Data
Rånby, Bengt G., ESR spectroscopy in polymer research. (Polymers; v. 1) Bibliography: p. Includes index. 1. Electron paramegnetic resonance spectroscopy. 2. Polymers and polymerization. I. Rabek, J. F., joint author. II. Title. III. Series. QD96.E4R36 547'.84'028 77-5392

Typesetting and printing: Schwetzinger Verlagsdruckerei GmbH, 6830 Schwetzingen. Bookbinding: Konrad Triltsch, Graphischer Betrieb, 8700 Würzburg

2152/3020 543210

This book is dedicated with affection to our wives and families who, during the time of its preparation, had to sacrifice our company during innumerable evenings, weekends and holidays.

Bengt Rånby   Jan F. Rabek

Preface

Thirty years ago, Zavoisky, in Moscow (USSR), reported the first successful experimental observations of the ESR phenomenon. Its application to polymer problems began about 20 years ago. ESR belongs to the most specific and useful methods in the study of polymer reactions.

The main purpose of this book is to collect the present available information on the applications of electron spin resonance (ESR) spectroscopy in polymer research. The book has been written both for those who want an introduction to this field, and for those who are already familiar with ESR and are interested in application to polymers. Therefore, the fundamental principles of ESR spectroscopy are first outlined, the experimental methods including computer applications are described in more detail, and the main emphasis is on the application of ESR methods to polymer problems. Many results obtained are only briefly treated for lack of space.

The authors hope that this book will provide a useful source of information by giving a coherent treatment and extensive references to original papers, reviews, and discussions in monographs and books. In this way we hope to encourage polymer chemists, organic chemists, biochemists, physicists, and material scientists to apply ESR methods to their research problems.

Although we have carefully reviewed the literature, including a search of Chemical Abstracts for 1960–1975, it has not been possible to include all published papers on the subject matter treated. We apologize for inevitable omissions. We thank hundreds of scientists who have responded to our request and sent us reprints of their publications. We also invite those who have more material to offer to send it or inform us about additional references. We would also appreciate the readers' remarks, and new material gathered will be considered for subsequent new editions.

Acknowledgement

The original work from this department on ESR applications to polymers referred to in this book has largely been sponsored and supported by The Swedish Technical Research Council, The Swedish Board for Technical Development, and the Swedish Polymer Research Foundation. This help is gratefully acknowledged.

Stockholm, May 1977   Bengt Rånby and Jan F. Rabek

Table of Contents

Chapter 1

Generation of Free Radicals

1.1. Free Radicals, Biradicals, and Radical Ions

Most reactions initiated by thermal energy, light, γ-radiation, and mechanical forces involve the formation of free radicals. In such reactions a two-electron chemical bond is cleaved either symmetrically or unsymmetrically:

$$A:B \longrightarrow A\cdot \ + \ B\cdot \tag{1.1}$$

$$A:B \longrightarrow A^+ \ + \ :B^- \tag{1.2}$$

In reaction (1.1) free radicals are formed, whereas in reaction (1.2) ions are formed.

A "free radical" is defined as an atom, a group of atoms, or a molecule in a certain state containing one unpaired electron which occupies an outer orbital, e.g. atomic hydrogen (H·), hydroxyl (HO·), or methyl ($H_3C\cdot$) radicals. Free radicals are usually very reactive an account of the strong tendency of their unpaired electrons to interact with other electrons and form electron pairs (chemical bond).

Some common molecules contain an odd number of electrons in their normal state e.g. nitrous oxide (NO), nitrogen dioxide (NO_2), or chlorine dioxide (ClO_2), and are, according to the above definition, free radicals although they may be stable and longlived.

A "biradical" is a species containing two unpaired electrons in outer orbitals, e.g. methylene radical ($H_2C:$) or oxygen (O_2) in its ground state.

A "radical ion" is a free radical with positive or negative charge, e.g. protonated amine radical ($H_3N^{\cdot +}$) or naphthalene anion ($C_{10}H_8^{\cdot -}$).

Some of the transition metal ions have unpaired electrons in their inner orbitals and have, therefore, some properties in common with free radicals.

1.2. Basic Properties of Free Radicals

An unpaired electron, as in a free radical, has magnetic properties due to the intrinsic angular momentum which is known as "the electron spin". The unpaired electron is a magnetic dipole which is assigned to a spinning motion of the electric charge. In a magnetic field, the electron spin has a tendency to align itself in the direction of the applied field by making a precession motion around an axis in the direction of the field. This is the basis for paramagnetic properties of matter. Very few stable molecules are paramagnetic because electrons have a tendency to form pairs with opposing (antiparallel) spins. The effect of this interaction is that the magnetic dipoles of the electrons cancel each other out. Only when a molecule contains an uneven number of electrons in the outer orbitals, e.g. NO, NO_2, and ClO_2, or contains two electrons with parallel spins in its lowest energy state, e.g. molecular oxygen (O_2), does it have a magnetic moment. The theory of free radicals is presented in several publications[96, 231, 769, 892, 894, 1435, 1710, 1769].

Free radicals can be detected by measuring "Magnetic Susceptibility" and "Electron Paramagnetic Resonance (EPR)" = "Electron Spin Resonance (ESR)".

Recently, "Electron Spin Echo (ESE)"[790, 1147, 1156, 1495, 1497, 1498] and "Chemically Induced Dynamic Nuclear Polarization (CIDNP)"[141, 150−152, 162, 677, 681, 1351, 2384],

1

have been applied to the study of free radicals in polymers and related problems. These methods will not be further discussed here, as both measured effects, ESE and CIDNP, do not belong to ESR spectroscopy.

1.3. Free Radical Reactions

Typical free radical reactions are chain reactions which occur in three steps:
    1. "The initiation step" is a radical formation process, e.g. the radicals are formed in pairs by homolytic cleavage of a two-electron bond (1.1).
    2. "The propagation step" is a transfer reaction of free radicals in which the site of free radical is changed. There are four types of propagation reactions:
        1. *"Atom transfer reactions"*, e.g. abstraction of hydrogen by a free radical:

$$A\cdot + RH \longrightarrow AH + R\cdot \tag{1.3}$$

        2. *"Addition reactions"*, involving the addition of free radical to a double bond:

$$A\cdot + C{=}C \longrightarrow A{-}C{-}C\cdot \tag{1.4}$$

This type of reaction is the basis for the free radical polymerization and is known as "the chain propagation step".
        3. *"Fragmentation reactions"*. A typical example of these reactions is known as the "β-scission", in which an unpaired electron in a molecule splits a bond in β position and produces a free radical and a molecule containing a double bond:

$$R{:}C{-}C\cdot \longrightarrow R\cdot + C{=}C \tag{1.5}$$

    4. *"Rearrangement reactions"*, in which a free radical changes position in a molecule, giving are two types of termination reactions:

$$\begin{array}{ccc} & R & \\ & | & \\ R{-}\overset{\textbf{.}}{C}{-}CH_2{-}R & \longrightarrow & R{-}\overset{}{C}{-}\overset{\textbf{.}}{C}H_2 \\ | & & | \\ R & & R \end{array} \tag{1.6}$$

    3. *"Termination reactions"* which occur in all systems where free radicals are present. There are two types of termination reactions:
        1. *"Combination"* of two radicals:

$$R\cdot + R\cdot \longrightarrow R{-}R \tag{1.7}$$

        2. *"Disproportionation"* e.g. involving the transfer of hydrogen:

$$R\cdot + RCH_2{-}\overset{\textbf{.}}{C}HR \longrightarrow R{-}H + RCH{=}CHR \tag{1.8}$$

Both (1.7) and (1.8) are common termination reactions in free radical polymerization.

1.4. Interaction of Electromagnetic Radiation with Matter

When electromagnetic radiation passes through matter its intensity decreases, primarily as a result of scattering and energy absorption by some of the irradiated molecules. The radiation chemist may use two types of electromagnetic radiation:
    1. Low energy photons (visible and ultraviolet radiation).
    2. High energy photons (X-ray and gamma radiation).

1.4.1. Interaction of Low Energy Photons with Matter

These processes are investigated in "photochemical research"[255, 256, 381, 485, 492, 687, 761, 762, 1077, 1295, 1298, 1335, 1445, 1623, 1814, 1928, 2014, 2316, 2317, 2396, 2399].

Molecules in their ground state, i.e. their lowest energy state, may absorb low-energy photons and form "excited singlet states" ($S_1, S_2, S_3 \ldots$). In a singlet state the spins of the electrons are paired. The main photochemical reactions occur from the lowest excited singlet state (S_1). The very fast rate of internal conversion (10^{11} to 10^{14} s^{-1}) from high excited states ($S_2, S_3 \ldots$) to the lowest excited singlet state (S_1) makes photochemical reactions involving higher states unlikely (Fig. 1.1).

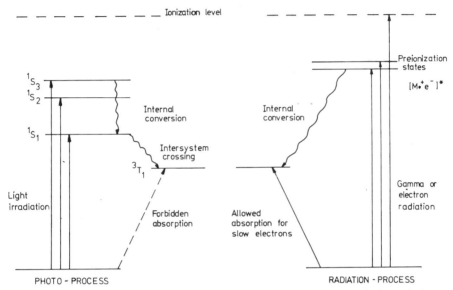

Fig. 1.1. Modified diagram for the most important processes involving electronically excited states and preionization states

The lowest excited triplet state (T_1) is mainly formed by a radiationless transition (intersystem crossing) from the lowest excited singlet state (S_1). The formation of a triplet state by direct absorption of a photon is a forbidden transition. In the triplet state the spins of electrons are unpaired. Higher triplet states ($T_2, T_3 \ldots$) may only be formed when a molecule in its lowest triplet state (T_1) absorbs a new photon.

A necessary condition for a photochemical reaction is the absorption by a molecule of a light quantum of sufficient energy. A photochemical reaction usually occurs in two stages:

1. "The primary photochemical reaction" which is directly due to the absorbed light quantum and involves electronically excited states (singlet or triplet).
2. "Secondary (dark) reaction" of the various chemical species as molecules, ions with free radicals, which are produced by the primary photochemical reactions.

The bond dissociation energies are listed in Table 1.1, whereas the energies of light quanta (photons) of different wavelengths are shown in Table 1.2.

It is seen that bond energies in common polymers are in the energy range of low-energy photons. When a polymer molecule absorbs a photon, it is usually expected that a particular bond may dissociate with the formation of free radicals. The majority of polymers contain only C–C, C–H, C–O, C–N, and C–Cl bonds. They are therefore not expected to absorb photons of wavelength longer than 1900 Å. The fact that free radicals are formed after irradiation of many polymers

3

Table 1.1. Approximate bond dissociation energies of various chemical bonds

Chemical bonds	Bond dissociation energy (kcal mol^{-1})
C–H (primary)	99
C–H (tertiary)	85
C–H (allylic)	77
C–C	83
C=C	145
C≡C	191
C–Cl	78
C–N	82
C=N	153
C≡N	191
C–O	93
C=O	186
C–Si	78
Si–H	76
N–O	37
–O–O–	66

Table 1.2. Energy of light

Wavelength (A)	Energy (kcal mol^{-1})
10.000	28.6
7.000	40.8
6.200	46.1
6.000	47.6
5.800	49.3
5.300	53.9
5.000	57.1
4.700	60.8
4.200	68.1
4.000	71.4
3.660	78.0
3.530	81.0
3.250	88.0
3.000	95.3
2.890	99.0
2.537	112.7
2.000	142.9

1 eV = 23.060 kcal mol^{-1} = 8.067.49 cm^{-1} = 1.239.5 nm = 123.950 A

with UV light, even with λ > 3000 Å in vacuum, indicates that some kinds of chromophores are present in these polymers. Small amounts of low-molecular impurities may also be responsible for the absorption of light quanta and act as photosensitizers[1 777, 1 779, 1 782–1 785, 1 814].

1.4.2. Interaction of High Energy Photons with Matter

These processes are investigated in "radiation chemistry"[106, 241, 250, 359, 430−432, 437, 505, 566, 844, 907, 980, 1224, 1233, 1599, 2062, 2102, 2410, 2414, 2434]. There are three ways in which high-energy photons can transfer their energy to matter:

1. "A photoelectric process", which involves complete absorption in a single step of a high-energy photon ($h\nu$) by the sample. In this process electrons may subsequently be rejected from the systems with an energy (E) given by the equation:

$$E = h\nu - E_e \qquad (1.9)$$

where E_e is the electron binding energy. The ejected electrons have an angular distribution which is energy-dependent. At low photon energies, the electrons are ejected mainly at 90° to the path of the incident high-energy photons. When the photon energy rises, ejection in the forward direction occurs with increased probability.

2. "Compton scattering" occurs when the energy of the incident photons increases above 10 keV. This process involves an elastic collision of high-energy photon ($h\nu$) with an electron (Fig. 1.2).

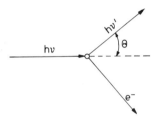

Fig. 1.2. The Compton scattering process

The energy of the scattered photon ($h\nu'$) is represented by the equation:

$$h\nu' = \frac{h\nu}{1 + h\nu/mc^2} (1 - \cos \Theta) \qquad (1.10)$$

where m is the stationary electron mass, c is the velocity of light, and Θ is the angle between the incident and scattered photons.

3. "Pair production" occurs when the high-energy photons ($h\nu$) have an energy equal to at least 1 MeV. The energy is annihilated in the field of a nucleus with high atomic number to produce an electron-positron pair. Since the polymers usually contain only atoms of low atomic numbers, this process is of little importance in the radiation chemistry of polymers.

1.5. Interaction of Charged Particles with Matter

These processes also belong to "radiation chemistry"[180, 469, 553, 762, 1223, 1224, 1407, 1824, 2177, 2410, 2414]. Of interest are mainly two types of charged particles:

1. Electrons with a negative charge,
2. Protons with a positive charge.

"Electrons" are more heavily charged particles than protons. They can lose their energy by means of a transfer of energy to molecular electrons. When the amount of energy transferred to a molecular electron is lower than its lowest ionisation potential, it may still be sufficiently large

5

to displace the electron from its ground state to "a preionization state". In other words, excited molecules are created through direct radiation-chemical interaction:

$$M \longrightarrow\!\!\!\!\!\!\!\!\sim\!\!\!\!\sim\!\!\!\!\sim\!\!\!\!\rightarrow [M^{\cdot+}e^-]^* \tag{1.11}$$

These states are similar to the corresponding excited states produced by absorption of a light quantum in the same medium absorption (Fig. 1.1):

$$M + h\nu \longrightarrow M^* \tag{1.12}$$

It may be assumed that the subsequent reactivity of such excited molecules (M*) is the same whether the excitation is brought about by a photochemical or a radiation-chemical act. This assumption is supported by experimental evidence. In many chemical compounds, including polymers, the products of photolysis are at least quantitatively similar to the products of radiolysis. The difference between the products of photolysis and those of radiolysis may be accounted for by the formation in the latter case of a charge separation and energy transfer to suitable molecule (D) of lower ionization potential. Figure 1.3 summarizes the various processes leading to charge separation and excited-state formation in the radiation chemistry of liquids and solids[2414, 1415].

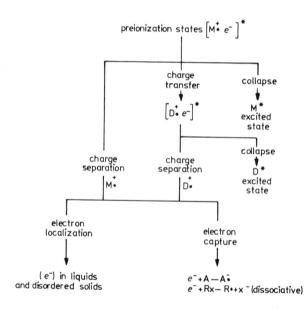

Fig. 1.3. Schematic representation of processes leading to charge separation in the condensed state[2414]

When the incident particle is a slow electron which has not enough energy to excite the preionization states of irradiated molecule, it may still be capable of exciting low-lying states differing in multiplicity from the ground state (i.e. triplet state) through a forbidden transition (Fig. 1.1):

$$M + \text{slow electron} \longrightarrow\!\!\!\!\!\!\!\!\sim\!\!\!\!\sim\!\!\!\!\sim\!\!\!\!\rightarrow M^* \text{ (triplet)} \tag{1.13}$$

Such triplet states do not usually possess enough energy to dissociate by breaking chemical bonds. But since these states possess two unpaired electrons, the chemical behavior of triplet states is analogous to that of biradicals.

The triplet states may moreover produce chemical rearrangements. The number of excitations to triplet state caused by slow electrons represent as much as 15% of the total number of ionizations. All electrons ejected from the molecules of the irradiated sample lose their kinetic energy by ionizing and exciting other particles along their path. In a substance which is incapable of capturing slow electrons, the only reaction is the neutralization of positive ions. However, if the substance contains traces of impurities having high electron affinity, the impurities may capture the majority of slow electrons and various important reactions may be observed, e.g. energy transfer processes.

High energy electrons produce not only excited states but also ionized states in atoms and molecules. In this way positive radical ions ("free radical cations") are produced which are formed when electrons leave their orbital, e.g.:

$$R:R + e^- \longrightarrow\!\!\!\!\sim\!\!\!\!\sim\!\!\!\!\sim\!\!\!\!\longrightarrow R \cdot R^+ + 2\,e^- \qquad\qquad (1.14)$$

"Protons" — positively charged particles in the energy range below 10 MeV — interact with matter mainly electrostatically. These interactions depend upon the distance of approach to the particular molecular electron, the impact parameter, and the velocity with which the encounter takes place. The application of positively charged particles in radiation chemistry of polymers has not yet aroused much interest.

1.6. Physical Sources of Free Radicals

1.6.1. Ionizing Radiation

There are three main categories of ionizing radiation devices developed and used in practice[435, 567, 1055]):
1. Charged particle accelators.
2. Radioisotope sources.
3. Nuclear reactors.

1.6.1.1. Charged Particle Accelerators

There are a few types of these devices:
1. "Van der Graaff accelerators" are of different constructions. The 3 MeV (million electron volts) Van der Graaff accelerator is especially useful for radiation chemistry. This device produces electrons of variable energy from less than 1 to 3 MeV with beam currents from about 10 μA to a maximum of 1 mA[1381]).
2. "Medium energy electron accelerators" produce an electron beam in the wide range (Table 1.3), with an output of electron of the order of several miliampers when driven by a klystron power source and a radio frequency power feedback is applied.
This type of accelerator operates in a pulsed manner at the rate of several hundred pulses per second and a pulse duration from nsec to msec.
3. "X-ray generators". There are several commercial types of low-energy electromagnetic radiation sources. They produce X-rays in the energy range of 40–300 eV.

1.6.1.2. Radioisotope Sources

There are two commercial sources of the ionizing radiation for research in the field of radiation chemistry: cobalt-60 and cesium-137[180]).
1. Cobalt-60 emits γ-radiation with the half-life of the order of 5.2 years and the average energy of 1.25 MeV.
2. Cesium-137 also emits γ-radiation with the energy of 0.66 MeV and the half-life of about 30 years.

Table 1.3. Medium energy electron accelerators (Table submitted by Instrument AB Scanditronix, Uppsala, Sweden)

Feature \ Accelerator type	Microtron	Racetrack microtron	Linear accelerator	Synchrotron	Betatron
Diagram 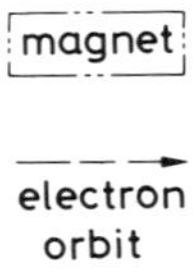	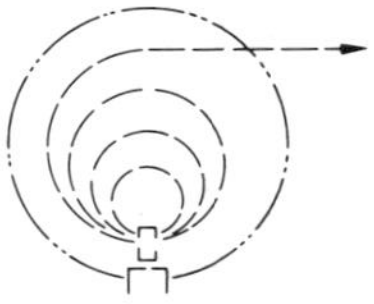	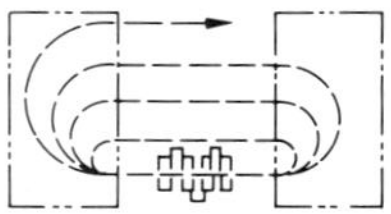	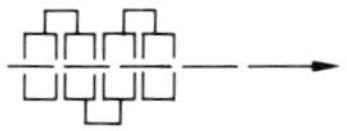	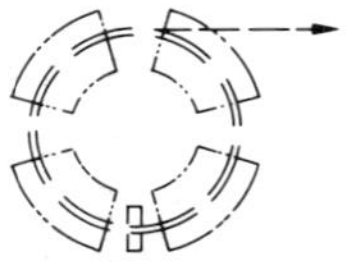	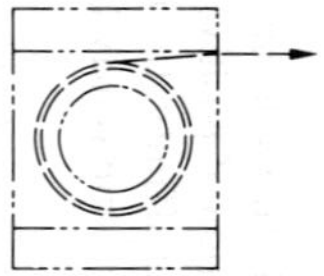
Energy source		3 GHz	r.f. power	200 MHz	induction 50 Hz
Orbital magnetic field	0.1 T	1 T	0	1 T	0.5 T
Injection and preacceleration		Pulsed electron gun 50 keV		microtron. linac betatron action	pulsed electron gun 50 keV
Energy gain per revolution	0.5–1 MeV	1–10 MeV	0.2–0.8 MeV/cavity	1–100 keV	50 eV
Max energy built	40 MeV	20 MeV	20 GeV	6 GeV	340 MeV
Limitation on energy	magnetic field homogeneity	cost	cost	cost radiation loss	radiation loss cost
Practical energy range	2–20 MeV	5–100 MeV	2–100 MeV	10–100 MeV	5–50 MeV

8

Table 1.3. (continued)

Feature \ Accelerator type	Microtron	Racetrack microtron	Linear accelerator	Synchrotron	Betatron
Energy spread	50 keV	300 keV	0.2–2 MeV	50 keV	50 keV
Energy variability	fair	good	good	excellent	excellent
Absolute energy definition	excellent	good	fair	fair	fair
Acceleration time	200 nsec	100 nsec	10 nsec	10 msec	10 msec
Extracted beam power — pulse		1 MW		100 kW	5 kW
Extracted beam power — average		1 kW		20 W	1 W
Extracted electrons per pulse		$2 \cdot 10^{12}$ ($2 \cdot 10^8$ per micropulse)		10^{11}	$5 \cdot 10^9$
Dose rate at 1 meter — electrons		$10^6 \frac{\text{rad}}{\text{min}}$		$5 \cdot 10^4 \frac{\text{rad}}{\text{min}}$	$2 \cdot 10^3 \frac{\text{rad}}{\text{min}}$
		$5 \cdot 10^3 \frac{\text{rad}}{\text{min}}$		$10^3 \frac{\text{rad}}{\text{min}}$	$4 \cdot 10^2 \frac{\text{rad}}{\text{min}}$

Note: The table is in a way a fiction as the devices can not be compared in a straight forward way. Different devices behave in a different way at different energies.

1. Generation of Free Radicals

3. Spent fuel elements from nuclear reactors have also been used as radiation sources. They change their radiation characteristics with time, on account of the different decay rates of the various fission products.

1.6.1.3. Nuclear Reactors

They provide a very intense mixed radiation of gamma quanta and neutrons. For that reason nuclear reactors have no wide application in the radiation chemistry.

1.6.2. Light Irradiation

To initiate photochemical reactions electromagnetic radiation in the range 2000–3000 Å is usually applied. Emission spectra and the intensity of emitted light radiation depend on the construction of light sources. Extensive description of the characteristics of a variety of commercial lamps used in photochemical research is given in several publications[381, 864, 1176, 1179, 1814].
Experimental techniques in vacuum ultraviolet (1000–2000 Å) are seldom applied in polymer research and require special light sources[381, 1279, 1450].

1.6.3. Other Physical Sources for the Production of Free Radicals

Other sources for the production of free radicals include silent electric discharge, radio frequency discharge, and field ionization[358, 489, 1452], and mechanical forces accompanying milling, stretching, breakage, etc. may bring forward the formation of free radicals.

1.7. Chemical Sources for the Production of Free Radicals

There are hundreds of different chemical reactions used to produce free radicals for research. One of the most widely applied in ESR research is the redox reaction obtained by mixing aqueous hydrogen peroxide and titanium trichloride (See Chapter 4.6.3). The hydroxyl radicals (HO·) are the main species formed which react further, e.g. by abstraction:

$$Ti^{3+} + H_2O_2 \longrightarrow Ti^{4+} + HO· + HO^- \tag{1.15}$$

$$HO· + RH \longrightarrow R· + H_2O \tag{1.16}$$

Another example is the formation of free radical anions (negative radial ions) by one-electron transfer reduction of an organic compound, e.g. formation of a colored naphthalene ion in the reaction of naphthalene with sodium (see Chapter 4.7.1):

(1.17)

Chapter 2

Principles of ESR Spectroscopy

2.1. Introduction

This chapter presents the background, nomenclature, and theory of ESR spectroscopy. It is written as a necessary introduction for readers who are not familiar with the ESR techniques and ESR literature. The basic interpretation of ESR spectra requires understanding of quantum mechanics and the use of matrix, vector, and tensor calculations. We have attempted to present this material with a minimum of mathematical treatment. Polymer chemists who are interested in carrying out experiments in ESR spectroscopy are advised to improve their knowledge by reading some of the review papers[20, 87, 105, 147, 291, 333, 448, 554, 556, 585, 595, 648, 676, 691, 713, 716, 765, 768, 791, 798, 833, 880, 881, 906, 921, 962, 972, 1073, 1074, 1085, 1124, 1128, 1129, 1163, 1175, 1277, 1299, 1331, 1408, 1422, 1448, 1458, 1526, 1575, 1595, 1805–1807, 1810, 1844, 1911, 1917, 1918, 1921, 2048, 2070, 2092, 2335, 2404, 2430, 2456) and monographs[33, 40, 84, 89, 107, 109, 217, 234, 394, 472, 625, 758–760, 886, 894, 898, 1300, 1324, 1387, 1446, 1447, 1457, 1552, 1617, 1643, 1660, 1690, 1760, 1761, 1763, 1770, 1910, 1919, 1920, 1924, 1950, 2024, 2046, 2066, 2067, 2103, 2340, 2405, 2417, 2489)].

Examples of hundreds of ESR spectra of different organic free radicals and radical ions are collectes in Atlas of Electron Spin Resonance Spectra[222].

Application of ESR spectroscopy in polymer chemistry has also been reviewed by several authors[157, 290–292, 300, 301, 333, 349, 366, 374, 383, 666, 669, 671, 672, 675, 685, 802, 961, 1025, 1026, 1065, 1456, 1499, 1603, 1640, 2337, 2370, 2375)].

2.2. Interaction Between the Electron Spin and Internal Magnetic Field

The electrons in atoms and molecules form pairs. For each electron in a certain orbital with the spin quantum number

$$M_S = -\frac{1}{2},$$

there is another electron in the same orbital with the spin quantum number

$$M_S = +\frac{1}{2}.$$

The paired electrons do not give an ESR signal. An unpaired electron has no other electron as partner in the same orbital, and for that reason it gives an ESR signal.

The electron is characterized by its rest mass m = (9.109558 ± 0.000054) x 10^{-28} g, charge e = (4.803250 ± 0.000021) x 10^{-10} esu[a], intrinsic angular momentum, and its magnetic moment.

---

a) All data taken from paper[2167].

2. Principles of ESR Spectroscopy

"The intrinsic angular momentum of the electron" is known as the spin vector with the symbol (**S**)[b]. The values of its component in any direction are $\pm\frac{1}{2}\hbar$, where

$$\hbar = \frac{h}{2\pi} = (1.0545919 \pm 0.0000080) \times 10^{-27} \text{ erg s}$$

and h is Planck's constant h = (6.626196 ± 0.0000050) × 10⁻²⁷ erg s.

"The magnetic moment vector for the unbonded electron" (μ_e)[c] is related to the spin vector (**S**) by the following equation:

$$\mu_e = -g_e\beta_e\mathbf{S} \quad (\text{erg G}^{-1})^{c)} \tag{2.1}$$

where g is dimensionless constant called "the Lande g-factor" or "electron free-spin g factor" $g_e = 2.0023192778 \pm 0.0000000062$, β_e is "the Bohr magneton"

$$\beta_e = \frac{eh}{2mc} = (9.274096 \pm 0.000065) \times 10^{-21} \text{ erg G}^{-1} \tag{2.2}$$

where c is velocity of light

$$c = (2.997925 \pm 0.000010) \times 10^{10} \text{ cm s}^{-1}.$$

The minus in eq. 2.1 indicates that the vector of the magnetic moment (μ_e) of the electron is in opposite direction to the spin vector (**S**).

When an electron is in a magnetic field (H), its magnetic moment vector (μ_e) has two possible orientations (Fig. 2.1), which correspond to the two possible orientations of the spin vector (**S**); parallel and antiparallel to the applied magnetic field **H**.

The angle (Θ) between the magnetic moment vector (μ_e) of an electron and the direction of the magnetic field (**H**) can have only two possible values: 35° 15′ and 144° 45′. The orientation in

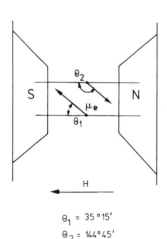

$\Theta_1 = 35°15'$

$\Theta_2 = 144°45'$

Fig. 2.1. The two orientations of the electron magnetic moment (μ_e) with respect to external magnetic field (H)

---

[b] Thickly written letters mean the description in vector language.
[c] G = gauss.

which the magnetic moment is antiparallel to the field is the orientation of maximum energy. The electron energy (E) in the magnetic field (H) is given by the equation:

$$E = -\mu_e H \ (erg) \tag{2.3}$$

where μ_e is the electron magnetic moment (vector) (erg G^{-1}) and H* is the magnetic field (vector) (G). Substituting eq. (2.1) into eq. (2.3) we obtain:

$$E = g_e \beta_e SH \ (erg) \tag{2.4}$$

The spin quantum number (M_S) of one electron is 1/2: the spin vector (S) in a magnetic field can have one of the two possible values: +1/2 and −1/2. It follows that the energy of an electron in a magnetic field (H) is either

$$E_1 = +\frac{1}{2} g_e \beta_e H \ (erg) \tag{2.5}$$

or

$$E_2 = -\frac{1}{2} g_e \beta_e H \ (erg) \tag{2.6}$$

The increase of the number of energy levels of an electron by a magnetic field in called "the Zeeman splitting of the electron" (Fig. 2.2).

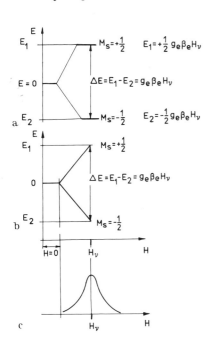

Fig. 2.2. Energy level diagrams showing Zeeman splitting for a free electron:
a. In constant external magnetic field (H = const) and variable frequency (ν),
b. In variable external magnetic field and constant frequency (ν = const), H_ν is the magnetic field at which the resonance conditions occur, c. Absorption line

The energy difference between the two states of the electron is given by:

$$\Delta E = E_1 - E_2 = g_e \beta_e H \ (erg) \tag{2.7}$$

The measurement of the energy difference (ΔE) is the basis of electron paramagnetic resonance experiments. Application of an oscillating radio frequency field perpendicular to magnetic field

13

(H), whose quanta (hν) have an energy equal to $g_e\beta_e H$, induce transition between the two states of the electron:

$$\Delta E = E_1 - E_2 = h\nu = g_e\beta_e H \text{ (erg)} \tag{2.8}$$

where ν is the frequency of the radio waves (cycles s^{-1}). Microwaves have wavelengths from a millimeter to a meter. A commonly used X-band spectrometer operates at about 3 cm wavelength corresponding to a frequency of about 9500 Mc s^{-1}, which requires a magnetic field (H) about 3400 G. The shorter wavelengths of the Q-band spectrometer have a frequency of about 35000 Mc s^{-1}, and the magnetic field (H) of 12500 G. The Q-band spectrometer is commonly used for studies at very low temperatures.

Only electromagnetic waves with the frequency

$$\nu = \frac{g_e\beta_e H}{h}$$

contain the right amount of energy to produce transition between the two energy states of the electrons. This coincidence of the energy of the microwave quantum (hν) and the energy difference between the two states of the electron (ΔE) is called resonance.

2.3. Spin Relaxation

The ESR emission or absorption can only be detected when there is a population difference between the two spin levels. The transition between the two energy levels

$$E_1 \left(M_S = +\frac{1}{2} \right) \text{ and } E_2 \left(M_S = -\frac{1}{2} \right)$$

can occur by either emission on absorption. The rates of both processes are proportional to:

1. The population of the state considered,
2. The microwave energy density,
3. The square of the transition matrix element.

The microwaves of the oscillating magnetic field (hν) induce two types of transition:

1. From the upper state (E_1) to the lower state (E_2) (the emission process)
2. From the lower state (E_2) to the upper state (E_1) (the absorption process)

When the unpaired electron population is equally divided between the upper $N(E_1)$ population and the lower $N(E_2)$ population, the emission of microwave quanta is higher than the absorption of these quanta, and no observable signal is obtained. When the population of the unpaired electrons is higher in the state (E_2) than in the state (E_1) a radio frequency absorption is observed. These arguments also apply to the electron and nuclear spins (see Chapter 2.6). The ratio of the two populations in thermal equilibrium is determined by the Boltzmann equation:

$$\frac{N(E_2)}{N(E_1)} = e^{\frac{E_1 - E_2}{kT}} = e^{\frac{\Delta E}{kT}} = e^{\frac{g_e\beta_e H}{kT}} \tag{2.9}$$

where k is the Boltzmann constant k = 1,38044 x 10^{-16} erg degree^{-1}.

The ESR signal recorded in an ESR spectrometer increases with increasing ratio $N(E_2)/N(E_1)$, because the absorption more and more predominates over the emission. The signal intensity is also increased when the temperature decreases.

The population difference Δn is expressed by the equation:

$$\Delta n = \frac{N(E_1) - N(E_2)}{2\,kT} = \frac{N g_e \beta_e H}{2\,kT} \tag{2.10}$$

where N is the total number of unpaired electrons in both populations: $N = N(E_1) + N(E_2)$. The signal intensity is directly proportional to the quantity Δn, which varies inversely to the absolute temperature. For that reason the intensity of the absorption signal at liquid nitrogen temperature (77 K) is approximately four times higher than at room temperature (300 K).

The probability of the transition between the energy levels E_1 and E_2 is proportional to the microwave energy density. With increasing incident microwave energy, the ESR signal intensity increases to a maximum. Further increase of power decreases and broadens the signal up to a moment when the signal completely disappears. This is the result of a saturation process, where the population $N_1(E_1)$ is equal to $N_2(E_2)$. The return of the system to equilibrium is called a "relaxation process". Such processes are exponential and characterized by "relaxation times".

One of the processes which restores the equilibrium ratio of $N(E_1)$ and $N(E_2)$ consists of "the spin-lattice" interaction with the relaxation time (T_1). This process involves thermal equilibrium of the spin system with the vibration of the crystal lattice. Another important process which influences the equilibrium ratio is "the spin-spin interaction" with the relaxation time (T_2)[51]. It involves the decay to random condition of the spin phase coherence in the microwave magnetic field, which oscillates in the xy plane. The calculation of the relaxation times T_1 and T_2 is necessary for the study of transition probabilities between the two spin levels, and it presents a rather difficult problem[95, 404, 977, 989]. Spin-lattice and spin-spin relaxation times also provide a means of distinguishing the molecular movements in the crystalline phase from those in the amorphous regions of polymers[724, 841, 1964].

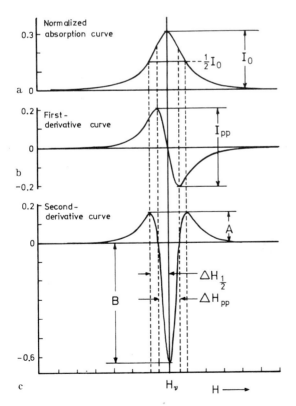

Fig. 2.3. Lorentzian line shape:
a. Absorption spectrum, b. First-derivative spectrum, c. Second-derivative spectrum[2405]

15

2.4. The Shape of ESR Resonance Lines

The magnetic interaction between the electron spins and the internal magnetic fields produces typical energy absorption spectra (Figs. 2.3 a and 2.4 a). Contrary to the principles of NMR

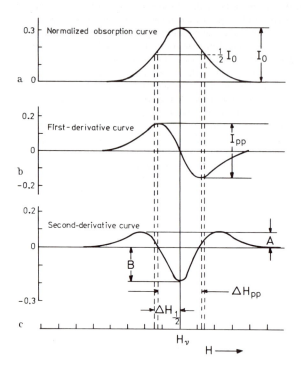

Fig. 2.4. Gaussian line shape: a. Absorption spectrum, b. First-derivative spectrum, c. Second-derivative spectrum[2405])

spectrometers which usually record directly the absorption spectra, the ESR spectrometers draw at once "the first derivative of the energy absorption curves" (Figs. 2.3 b and 2.4 b).

This curve is obtained by varying of the magnetic field and measuring the corresponding variation of energy absorption. It is often of considerable interest to obtain "the second derivative of the energy absorption spectra" by tuning the amplifier to double modulation frequency (Figs. 2.3 c and 2.4 c). The shape of the second-derivative recordings is very sensitive to the nature of the absorption line.

2.4.1. Line Shape

There are two important types of line shape in the ESR absorption curves: "Lorentzian line shape" and "Gaussian line shape". The main difference between the two types of line shape is that the outer wings of the Lonrentzian line are much longer and drop more slowly than those of the Gaussian on (Figs. 2.3 and Fig. 2.4). Measurable experimental parameters characterizing the two types of spectral lines are given in Table 2.1.

In several cases the shape of an absorption line is a mixture of the two types[51]. The mathematical description of the exact shape of an absorption line is very complicated. There are two methods which simplify this treatment and distinguish the two types:

    1. "The slope of the derivative curve", presented in Figure 2.5[1357].

For a Gaussian curve the ratio of slopes $a:b = 2.2$ and for a Lorentzian curve $a:b = 4$.

    2. "The normalization plot" is shown in Figure 2.6[986].

Table 2.1. Properties of Lorentzian and Gaussian Lines[2405]

	Lorentzian shape	Gaussian shape
Half-width at half-height	$\Delta H_{1/2}$	$H_{1/2}$
Peak-to-peak width	$H_{pp} = \dfrac{2}{\sqrt{3}}\, \Delta H_{1/2}$	$H_{pp} = \left(\dfrac{2}{\ln 2}\right)^{1/2} \Delta H_{1/2}$
Peak amplitude	$I_0 = \dfrac{1}{\pi \Delta H_{1/2}}$	$I_0 = \left(\dfrac{\ln 2}{\pi}\right)^{1/2} \dfrac{1}{\Delta H_{1/2}}$
Equation for absorption line	$I = I_0 \dfrac{(\Delta H_{1/2})^2}{(\Delta H_{1/2})^2 + (H - H_r)^2}$	$I = I_0 \exp\left[\dfrac{(-\ln 2)(H - H_r)^2}{(\Delta H_{1/2})^2}\right]$
Peak-to-peak amplitude	$I_{pp} = \dfrac{3\sqrt{3}}{4\pi}\, \dfrac{1}{(\Delta H_{1/2})^2}$	$I_{pp} = \left(\dfrac{2}{\ln 2}\right)^{1/2} \Delta H_{1/2}$
Equation for first derivative of absorption line	$I' = -I_0 \dfrac{2(\Delta H_{1/2})^2 (H - H_r)}{[(\Delta H_{1/2})^2 + (H - H_r)^2]^2}$	$I' = -I_0 \dfrac{2(\ln 2)(H - H_r)}{(\Delta H_{1/2})^2} \exp\left[\dfrac{(-\ln 2)(H - H_r)^2}{(\Delta H_{1/2})^2}\right]$
Peak amplitude of positive lobe	$A = I_0 \left(\dfrac{1}{2(\Delta H_{1/2})^2}\right)$	$A = I_0 \left(\dfrac{4\,e^{-3/2} \ln 2}{(\Delta H_{1/2})^2}\right)$
Peak amplitude of negative lobe	$B = -I_0 \left(\dfrac{2}{(\Delta H_{1/2})^2}\right)$	$B = -I_0 \left(\dfrac{2 \ln 2}{(\Delta H_{1/2})^2}\right)$

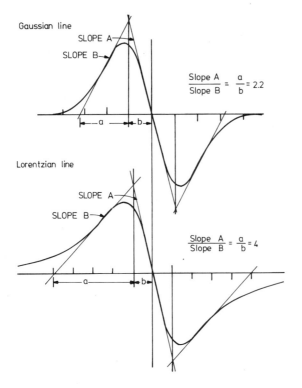

Fig. 2.5. The slope method for identification of Lorentzian and Gaussian derivative curves[1357)]

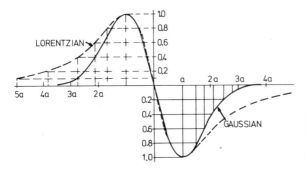

Fig. 2.6. The normalization technique for identification of Lorentzian and Gaussian derivative curves[986)]

In this method, the base line is divided into units of m, where m is the point on the abscissa where the curve reaches a maximum, and the ordinates are normalized to unity.

2.4.2. Line Width

ESR lines measured under specified conditions have a finite "width". There are two usual ways of expressing line width:

1. "The half width at half-height ($\Delta H_{1/2}$) of the absorption line (Figs. 2.3 and 2.4 and Table 2.1).
2. "Peak-to-peak width" (ΔHpp)-the full width between the extremes of the first-derivative curve (Figs. 2.3 and 2.4 and Table 2.1).

The width is influenced not only by interaction between the electron spin and the external magnetic field applied, but also by interaction of the electron spin with the environment inside the sample. This fact allows to obtain information about the spin environment from the line width. Typical environment effects are, for example, electron spin exchange between molecules and crystal lattice vibration. ESR lines can be broadened homogeneously when all spins have the same environment (Fig. 2.7a) and broadened inhomogeneously when spins are in different environments (Fig. 2.7b).

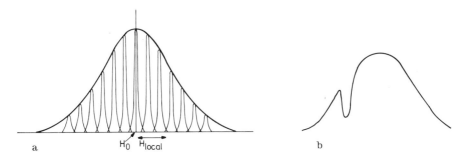

a $\quad$ H_0 H_{local} $\qquad$ b

Fig. 2.7. Hole burning in inhomogenously broadened line: a. Initial line, b. Localized saturation develops and appears as a hole

In the last case a selective saturation of some particular population of spins, for which the intensity becomes zero, is observed. The resonance line shows typical "holes", and the process is called "burning a hole"[230]. Most of the recorded ESR spectra are inhomogeneously broadened. This effect may be caused by irregularities in the external magnetic field and also by heterogeneities in the sample. Most powdered samples, for example, show anisotropies in the dipolar interaction which alters the inhomogeneous broadening. Similar effects are observed when there are impurities in the sample or when a single crystal contains lattice defects. In solution inhomogeneously broadened lines are due to the viscosity of the liquid phase.

The width of homogeneously broadened lines is determined by the equation:

$$g_e \beta_e \Delta H_{1/2} \geqslant \frac{\hbar}{2} \text{ or } \Delta E \Delta t \geqslant \frac{\hbar}{2} \qquad (2.11)$$

where $\Delta H_{1/2}$ is the line width (G), Δt is the lifetime of spin state (sec), ΔE is the difference in energy between the two electron levels. It is seen from Eq. 2.11 that increase of the lifetime of a state results in a decrease of line width. When a sufficient amount of microwave power is added to the sample the lifetime of the spin state decreases and a broadening of the absorption line (Fig. 2.8) is produced.

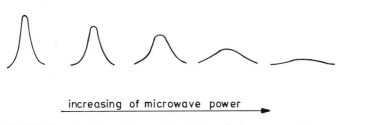

increasing of microwave power

Fig. 2.8. The saturation and power broadening of a homogenous line

19

This process is known as "saturation broadening". In the recording of a spectrum a wide line is much more difficult to detect than a narrow line of the same intensity.

The width of lines is dependent on the symmetry of the radicals examined[1441, 1442]. The reduction in the symmetry of the radical leads to weakening of the spin-orbital coupling and to an increase in the spin-lattice relaxation time, and consequently to narrowing of the lines.

2.4.3. Intensity

The intensity (I) of the absorption line as a function of the outer magnetic field (H) for Lorentzian and Gaussian line shape is given by the equations:

$$I = \frac{I_0}{T_2^2(H-H_r)^2 + 1} \quad \text{Lorentzian shape} \tag{2.12}$$

$$I = I_0 e^{-b(H-H_r)^2 T_2^2} \quad \text{Gaussian shape} \tag{2.13}$$

where I_0 is the intensity of the absorption line at its center, H_r is the field of resonance at the line center, T_2 is the spin-spin relaxation time, b is a constant.

The integrated intensity of the ESR signal represents the total energy absorbed by the sample at resonance conditions. This intensity is expressed by the total area under the resonance curve and it can be used for the determination of the concentration of free radicals in the sample as the number of unpaired spins per gram or per milliliter.

2.5. g-Value

Various ESR spectrometers use different frequencies and various types of samples require different frequencies for the experiment. For that reason the position of ESR lines cannot be indicated by the frequencies. The line position are instead denoted in terms of "g-value", which is expressed as a function of microwave frequency (ν) and outer magnetic field (H) at resonance. From Eq. 2.8 we obtain the conditions for the resonance of an electron spin:

$$g_e = \frac{h\nu}{\beta_e H} \tag{2.14}$$

As mentioned before the g_e value of a free electron is 2.002322, whereas the g-values for free radicals in atoms, molecules, and crystals depend on their electronic structure. A compilation of g-values published up to March 1964 have been collected by Fischer in Landolt-Börnstein Tables[667] (a new edition will be published in 1977) and in other publications[559, 2179, 2180].

The application of a double cavity in the ESR apparatus simplifies the measurement of g-values. Special standard compounds with precisely determined g-value are used as reference (see Chapter 3.3). These standard compounds are inserted into the double cavity and their spectra are recorded by a dual-channel recorder. The separation of the centers of the two spectra represents the magnetic field difference at the two positions of the sample. One of the standard samples is replaced by a sample to be examined and the field difference (ΔH) (corrected for the standard sample field difference) between the centers of the ESR spectra is measured. The g-value of the unknown sample (g_u) can be calculated from the equation:

$$g_u = g_s - \frac{\Delta H}{H} g_s \tag{2.15}$$

where g_s is the g-value of the known standard sample and H is the external magnetic field. Detailed descriptions of measurements and necessary corrections have been given[1149, 1909, 1930].

For various reasons, studies of the ESR spectra of randomly oriented radicals in a solid are likely to provide a less detailed picture of their structure and electron distribution than studies of radicals in single crystals or in solution. There are many systems, e.g. polymers and frozen aqueous systems at low temperature, where studies of single crystals or in solution are not possible so that polycrystalline or amorphous matrices must be used. For that reason, studies of randomly oriented radicals are still the most common in ESR spectroscopy of polymers.

In single crystals the g-value is dependent on the direction of the external magnetic field in relation to the crystallographic or molecular axes, e.g. x, y, and z. Consequently in polycrystalline samples (powdered samples) a set of g_1, g_2 and g_3 values (Fig. 2.9) is necessary to define the anisotropy of the g-value, which can be expressed by the following equation[1164]:

$$g^2 = g_1^2 \sin^2\Theta \cos^2\Phi + g_2^2 \sin^2\Theta \sin^2\Phi + g_3^3 \cos^2\Theta \qquad (2.16)$$

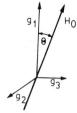

Fig. 2.9. Relations between the directions associated with g_1, g_2, g_3 and H_0 [109]

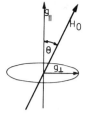

Fig. 2.10. Relations between the directions associated with $g_\parallel, g_\perp$ and H_0 [109]

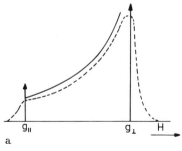

a

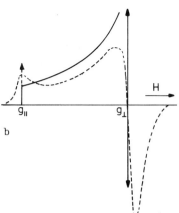

b

Fig. 2.11. Axial symmetry $g_\parallel \neq g_\perp$: a. Absorption spectrum, b. First-derivative spectrum[109]

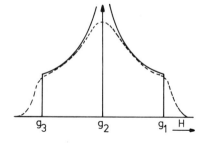

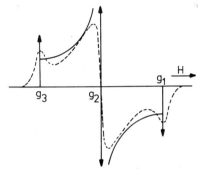

Fig. 2.12. An asymmetric g-tensor $g_1 < g_2 < g_3$: a. Absorption spectrum, b. First-derivative spectrum[109]

21

where Φ and Θ are the polar angles of the field direction with respect to the coordinate system of the principal directions of the g-tensor.

With H_0 in the g_1g_3 plane the observed g-factor varies sinusoidally with Θ, passing through the value of g_2. Similarly for all angles Φ of H_0 with respect to the g_1g_3 plane there is a value of Θ for which the observed g-factor is equal to g_2. For crystals having axial symmetry, expressions $g_{\parallel}$ (crystal axis parallel to H_0) and $g_{\perp}$ (crystal axis perpendicular to H_0) are commonly used (Fig. 2.10).

g-Value varies sinusoidally from $g_{\parallel}$ to $g_{\perp}$ with variation in Θ. The anisotropy of the g-value causes an asymmetry of the ESR absorption line (Figs. 2.11 and 2.12)[1024, 1034, 1164, 1343, 1895].

The idealized shapes shown in continous lines arise because there is a whole plane of orientations with the magnetic field perpendicular to the symmetry axis, but only one orientation with the magnetic field in the parallel position. For the nonsymmetrical system there is similarly a plane of orientations for which the g-value corresponds to the intermediate value g_2, but only one orientation each for g_1 and g_3. More realistic line-shapes are obtained by including broadening corresponding to random Gaussian distribution of neighboring dipoles, which are mainly the nuclei of neighboring molecules. The resulting shapes for the absorption and first derivative curves are shown as dotted lines in Figures 2.11 and 2.12. The interpretation of ESR spectra of free radicals in the crystalline solid state requires a detailed understanding of the nature of anisotropic interactions.

For free radicals in solution the g-value is shifted and measured as an average value[229, 1018, 1164, 1536]. The deviation of the g-value of an organic radical from that of a free electron gives significant information. The g-value depends upon the type of free radical, e.g. the g-value for a peroxy radical $RO_2 \cdot$ is higher than that for an alkyl radical $R \cdot$.

2.6. Interaction Between the Electron and the Nuclear Spins

The magnetic interaction between the electron and the nuclear spins in the same molecule gives an ESR spectrum with a number of lines instead of one single line. This type of interaction is called *"hyperfine coupling"* and produces ESR spectra with *"hyperfine splitting"*[2179, 2180]. By analyzing the number of lines, their separation, and their relative intensities it is usually possible to determine the type and number of nuclear spins which interact with the electron, i.e. the structure of the free radical. The line separation depends on the interaction between the electron spin and each nuclear spin, and on the magnetic moments of the involved nuclei. From the number of absorption lines in the hyperfine ESR spectrum and their relative intensities, it is possible to identify the number of equivalent nuclei in the molecule and to determine the structure.

"The spin angular moment vector of a nucleus" $(I)^{a)}$ and its magnetic moment vector (μ_N) is given by the equation:

$$\mu_N = \frac{g_N \beta_N I}{\hbar} \tag{2.17}$$

where g_N is "the nuclear g-factor" which is characteristic for each type of nucleus, the g_N for proton is $g_N = 5.585564 \pm 0.000017^{b)}$ and β_N "the nuclear magneton" given by equation:

$$\beta_N = \frac{e\hbar}{2\,m_p c} = (5.050951 \pm 0.000050) \times 10^{-24} \text{ erg G}^{-1} \tag{2.18}$$

where m_p is proton mass, $m_p = (1.672614 \pm 0.000011) \times 10^{-24}$ g and c is velocity of light, $c = (2.997925 \pm 0.000010) \times 10^{10}$ cm s^{-1}.

---

a) Thickly written letters mean the description in vector language.
b) All data taken from paper[2167].

Table 2.2. Nuclear spins for different nuclei

^{1}H	1/2	^{31}P	1/2	^{67}Zn	5/2
^{2}H	1	^{33}S	3/2	^{75}As	3/2
^{6}Li	1	^{35}Cl	3/2	^{77}Se	1/2
^{7}Li	3/2	^{37}Cl	3/2	^{79}Br	3/2
^{9}Be	3/2	^{39}K	3/2	^{81}Br	3/2
^{10}B	3	^{43}Ca	7/2	^{83}Kr	9/2
^{11}B	3/2	^{45}Sc	7/2	^{85}Rb	5/2
^{13}C	1/2	^{47}Ti	5/2	^{87}Rb	3/2
^{14}N	1	^{49}Ti	7/2	^{95}Mo	5/2
^{15}N	1/2	^{51}V	7/2	^{97}Mo	5/2
^{17}O	5/2	^{53}Cr	3/2	^{107}Ag	1/2
^{19}F	1/2	^{55}Mn	5/2	^{109}Ag	1/2
^{23}Na	3/2	^{57}Fe	1/2	^{127}I	5/2
^{25}Mg	5/2	^{59}Co	7/2	^{129}Xe	1/2
^{27}Al	5/2	^{61}Ni	3/2	^{131}Xe	3/2
^{29}Si	1/2	^{63}Cu	3/2	^{133}Cs	7/2
		^{65}Cu	3/2	^{207}Pb	1/2

The nuclear spins of many nuclei are listed in Table 2.2.

The spin angular moments (I) in the external magnetic field have the restricted orientations: $I, I-1, I-2 \ldots -I$. In this way, the nuclei have a limited number of spin quantum numbers $M_I = 2I + 1$, i.e. they split an electron magnetic level into $2I + 1$ sublevels.

The magnetic moment of the electron (μ_e) is about thousand times larger than the magnetic moment of the nuclei (μ_N). The applied external magnetic field (H) is normally many times larger than the local magnetic field at the electron produced by the nuclei. In a strong external field, the magnetic interaction is influenced by the interaction between the electron and the nucleus, the so-called "contact interaction". This interaction is determined by the "isotropic hyperfine coupling constant" (A) (measured in Hertz = cycles s^{-1}).

There are two other kinds of hyperfine interactions[721]:

1. *"Dipolar hyperfine interaction"* which is caused by the presence of a nucleus near the electron but not just in the immediate vicinity of the electron.

2. *"Fermi contact interaction"* which is referred as electrostatic interaction of electron and nucleus. It is proportional to the electron density at the nucleus.

The last two kinds of hyperfine interaction will not be further discussed here. They are well-treated in the basic monographs on ESR spectroscopy.

2.6.1. Hyperfine Structure Spectrum due to a Single Proton

In the external magnetic field, the magnetic moment of the proton has two possible orientations $\pm 1/2$, which affect the magnetic energy levels of an electron (E_1 and E_2) and split them into four sublevels (E_I, E_{II}, E_{III} and E_{IV}). This process is called "the Zeeman hyperfine splitting" for proton (Figs. 2.13a and 2.13b).

First-order splitting of the lines occurs when the hyperfine coupling constant energy (hA) is very much smaller than the electron energy ($g_e\beta_e H$). The transition between four sublevels E_I, E_{II}, E_{III} and E_{IV} gives the "hyperfine spectrum", but there are certain selection rules which determine the possible transitions between various levels. The allowed transitions are (Figs. 2.13a and 2.13b):

1. Between energetic levels E_I and E_{IV} and levels E_{II} and E_{III}, because they require the absorption or emission of energy and change in angular momentum by the electron only.

2. Between energetic levels E_I and E_{II} and levels E_{III} and E_{IV}, because they require the absorption or emission of energy and change in angular momentum by the nucleus only.

23

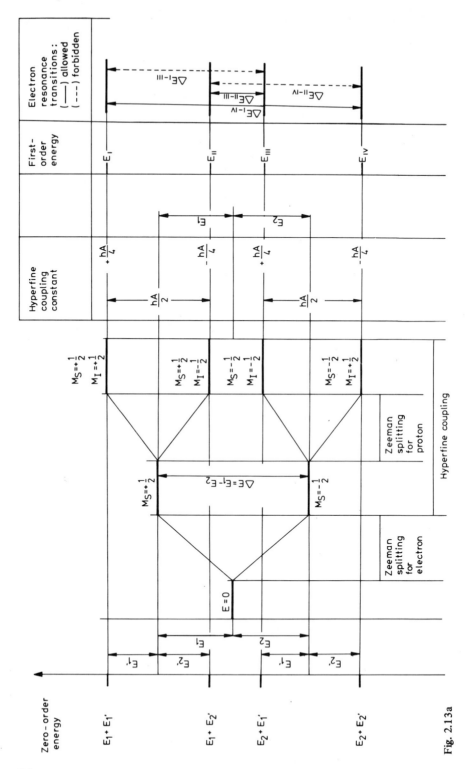

Fig. 2.13a

Transition between energetic levels E_I and E_{III} and levels E_{II} and E_{IV} are not allowed because they require a simultanous absorption or emission of energy by both the electron and the nucleus. In conclusion the selection rules give:

1. The allowed transitions for: $\Delta M_S = \pm 1$ and $\Delta M_I = 0$,
2. The forbidden transitions for: $\Delta M_S = \pm 1$ and $\Delta M_I = \pm 1$.

At a constant external magnetic field (Fig. 2.13a), the transition from level E_{IV} to level E_I occurs at a higher frequency than the transition from level E_{III} to level E_{II}. The difference between the two frequencies is called "the hyperfine frequency" ($\Delta\nu$) and is equal to A.

At a constant frequency (Fig. 2.13b), the transition from level E_I to level E_{IV} occurs at a lower magnetic field than the transition from level E_{II} to level E_{III}.

In the commercial spectrometers, the microwave frequency is maintained constant but the magnetic field is varied. In this case the energy ($\Delta E = h\nu$) of the transition between energy levels $E_I - E_{IV}$ and $E_{II} - E_{III}$ must be equal:

$$h\nu_{I-IV} = h\nu_{II-III} = h\nu \tag{2.19}$$

In a constant magnetic field the transition in energy differs by the isotropic hyperfine coupling constant (A):

$$g\beta H_{I-IV} - \frac{hA}{2} = g\beta H_{II-III} + \frac{hA}{2} \tag{2.20}$$

Fig. 2.13a. Energy levels of the proton (hydrogen atom) (system: $S = 1/2$, $I = 1/2$) in constant external magnetic field (H = const) and variable frequency (ν). Description of energy:

Electron Zeeman splitting energy:

$$E_1 = +\frac{1}{2} g_e \beta_e H$$

$$E_2 = -\frac{1}{2} g_e \beta_e H$$

Nucleus Zeeman splitting energy:

$$E_{1'} = +\frac{1}{2} g_N \beta_N H$$

$$E_{2'} = -\frac{1}{2} g_N \beta_N H$$

Zero-order energy:

$$E_1 + E_{1'} = +\frac{1}{2} g_e \beta_e H + \frac{1}{2} g_N \beta_N H$$

$$E_1 + E_{2'} = +\frac{1}{2} g_e \beta_e H - \frac{1}{2} g_N \beta_N H$$

$$E_2 + E_{1'} = -\frac{1}{2} g_e \beta_e H + \frac{1}{2} g_N \beta_N H$$

$$E_2 + E_{2'} = -\frac{1}{2} g_e \beta_e H - \frac{1}{2} g_N \beta_N H$$

First-order energy (which consider contact interaction between the electron and nucleus, which is determined by hyperfine coupling constant):

$$E_I = E_1 + \frac{hA}{4}$$

$$E_{II} = E_1 - \frac{hA}{4}$$

$$E_{III} = E_2 + \frac{hA}{4}$$

$$E_{IV} = E_2 - \frac{hA}{4}$$

The energies for allowed transition are:

$$\Delta E_{I-IV} = E_I - E_{IV} = g_e \beta_e H + \frac{hA}{2}$$

$$\Delta E_{II-III} = E_{II} - E_{III} = g_e \beta_e H - \frac{hA}{2}$$

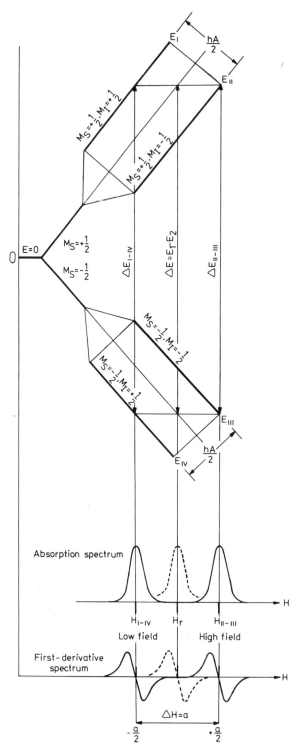

Absorption spectrum

H_{I-IV} H_r H_{II-III}

Low field High field

First-derivative spectrum

$\Delta H = a$

$-\dfrac{a}{2}$ $+\dfrac{a}{2}$

Fig. 2.13b. Energy levels of the proton (hydrogen atom) (system: $S = 1/2$, $I = 1/2$) in variable external magnetic field and constant frequency (ν = const).

Electron resonance transitions will occur at the resonant field:

$$H_{I-IV} = \frac{h\nu}{g_e\beta_e} - \frac{a}{2}$$

$$H_{II-III} = \frac{h\nu}{g_e\beta_e} + \frac{a}{2}$$

where a is the hyperfine splitting constant (G)

$$H_{I-IV} - H_{II-III} = \frac{hA}{g\beta} \qquad (2.21)$$

$$\Delta H = \frac{hA}{g\beta} = a \qquad (2.22)$$

The value $\Delta H = a$ is called *"hyperfine splitting constant"* and expresses in gauss (G) the separation between the two hyperfine lines of the spectrum (Fig. 2.13b).

In the case when there is no nuclear interaction (A = 0), the transition for the electron (S = 1/2) occurs at one resonance field (H_r) (Fig. 2.2):

$$H_r = \frac{h\nu_0}{g_e\beta_e} \qquad (2.23)$$

When the magnetic moment of an unpaired electron (S = 1/2) is affected by the magnetic field of a nucleus with I = 1/2 (proton), the transition occurs at two resonance fields (Fig. 2.13b):

$$\text{for } M_I = +1/2, \quad H_{I-IV} = \frac{h\nu_0}{g_e\beta_e} - \frac{a}{2} \qquad (2.24)$$

$$\text{for } M_I = -1/2, \quad H_{II-III} = \frac{h\nu_0}{g_e\beta_e} + \frac{a}{2} \qquad (2.25)$$

For the system with S = 1/2 and I = 1 (the $^2_1$H-deuterium), the transition occurs at three resonance fields (Figs. 2.14a and 2.14b):

$$\text{for } M_I = +1, \quad H_{I-VI} = \frac{h\nu}{g_e\beta_e} - a \qquad (2.26)$$

$$\text{for } M_I = 0, \quad H_{II-V} = \frac{h\nu}{g_e\beta_e} \qquad (2.27)$$

$$\text{for } M_I = -1, \quad H_{III-IV} = \frac{h\nu}{g_e\beta_e} + a \qquad (2.28)$$

Generally speaking, for a system where a single nucleus interacts with one unpaired electron, 2 I + 1 lines of equal intensity are obtained. They are separated by the constant hyperfine splitting (a). Using the hyperfine splitting constant (a) we can express the hyperfine coupling frequency ($\Delta\nu$) as:

$$\Delta\nu = \frac{g_e\beta_e a}{h} \qquad (2.29)$$

for g = 2

$$\Delta\nu = 2.8\,a \quad (Mc\ s^{-1}) \qquad (2.30)$$

2.6.2. Spectra with Hyperfine Structure due to Equivalent Protons

In the majority of free radicals the unpaired electron interacts with a number of magnetic nuclei, producing spectra with several lines. When a free radical interacts with n equivalent nuclei of spin I, the number of levels for each M_S value is 2 nI + 1.

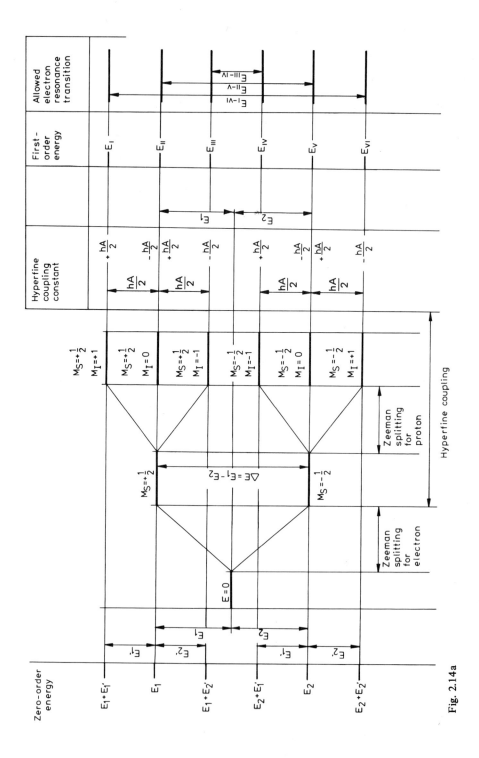

Fig. 2.14a

28

When one unpaired electron interacts with two equivalent protons (with I = 1/2), the analysis of the resulting spectra may be considered as follows: It follows from the interaction between the unpaired electron and the first proton that the M_S = +1/2 and M_S = −1/2 levels must be split by $\frac{hA}{2}$ (Figs. 2.15a and 2.15b).

Interaction with the second proton splits each level again by $\frac{hA}{2}$. The intermediate levels (M_I = 0) coincide and the relative probability for the transitions between the two levels becomes a factor 2 (Fig. 2.15b).

For three equivalent protons (with I = 1/2) the method of spectral analysis is analogous. The repetition of splitting gives four levels, both for the M_S = +1/2 and M_S = −1/2. The coincidence of the intermediate levels gives a threefold relative probability for the transitions between these levels (Figs. 2.16a and 2.16b).

The interaction of the unpaired electron with n equivalent protons results in n + 1 lines whose relative intensities are proportional to the coefficients of the binomial expansion of $(1 + x)^n$ (Fig. 2.17).

Summing up, it is seen that when we have a single set of equivalent protons, having only one hyperfine coupling constant A, the energy levels for hyperfine interaction are obtained by repetitive equal splitting of the hyperfine levels M_S = +1/2 and M_S = −1/2. The maximum possible number of lines (when there is no second-order splitting, see Chapter 2.6.4) is determined by $\Pi_i(2 n_i I_i + 1)$ where Π_i indicates a product over all values of i, n_i is the number of equivalent nuclei with spin I_i. It is important to point out that the overall shape of the spectra is influenced by the ratio of splitting to line width (Fig. 2.18).

Fig. 2.14a. Energy levels of the deuterium atom (system: S = 1/2, I = 1) in constant external magnetic field (H = const) and variable frequency (ν).

Description of energy:
Electron Zeeman splitting energy:

$E_1 = +1/2\, g_e\beta_e H$

$E_2 = -1/2\, g_e\beta_e H$

Nucleus Zeeman splitting energy:

$E_{1'} = +1\, g_N\beta_N H$

$E_{2'} = 0\, g_N\beta_N H = 0$

$E_{3'} = -1\, g_N\beta_n H$

Zero-order energy:

$E_1 + E_{1'} = +1/2\, g_e\beta_e H + g_N\beta_N H$

$E_1 + E_{2'} = +1/2\, g_e\beta_e H$

$E_1 + E_{3'} = +1/2\, g_e\beta_e H - g_N\beta_N H$

$E_2 + E_{1'} = -1/2\, g_e\beta_e H + g_N\beta_N H$

$E_2 + E_{2'} = -1/2\, g_e\beta_e H$

$E_2 + E_{3'} = -1/2\, g_e\beta_e H - g_N\beta_N H$

First-order energy:

$E_I = E_1 + \frac{hA}{2}$

$E_{II} = E_1$

$E_{III} = E_1 - \frac{hA}{2}$

$E_{IV} = E_2 - \frac{hA}{2}$

$E_V = E_2$

$E_{VI} = E_2 - \frac{hA}{2}$

The energies of allowed transitions are:

$E_{I-VI} = E_I - E_{VI} = g_e\beta_e H + hA$

$E_{II-III} = E_{II} - E_{III} = g_e\beta_e H$

$E_{III-IV} = E_{III} - E_{IV} = g_e\beta_e H - hA$

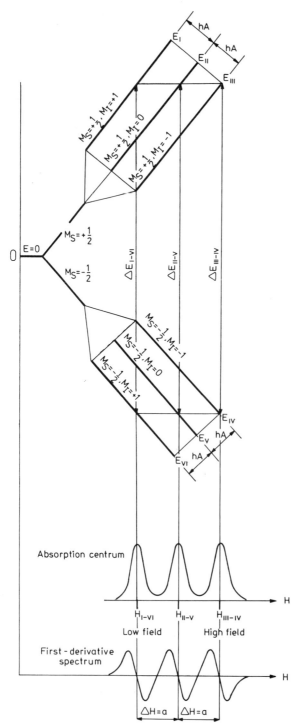

Absorption centrum

First - derivative spectrum

Fig. 2.14b. Energy levels of the deuterium atom (system: $S = 1/2$, $I = 1$) in variable external magnetic field and constant frequency (ν = const). Electron resonance transition will occur at the resonant field:

$$H_{I-VI} = \frac{h\nu}{g_e\beta_e} - a$$

$$H_{II-V} = \frac{h\nu}{g_e\beta_e}$$

$$H_{III-IV} = \frac{h\nu}{g_e\beta_e} + a$$

where a is the hyperfine splitting constant (G).

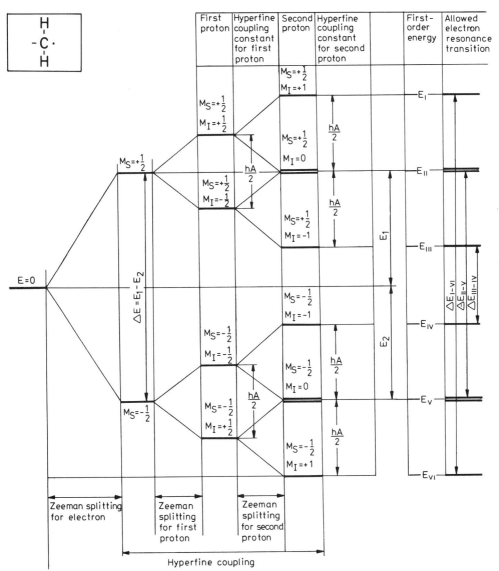

Fig. 2.15a. Energy levels of the free radical with two equivalent protons (System: S = 1/2, I = 1/2, I = 1/2) in constant external magnetic field (H = const) and variable frequency (ν)

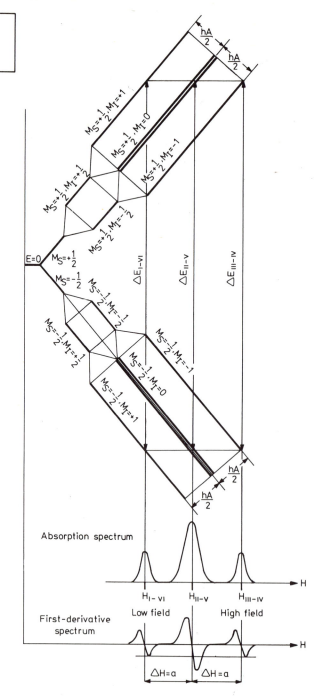

Fig. 2.15 b. Energy levels of the free radical with two equivalent protons (System: S = 1/2, I = 1/2, I = 1/2) in variable external magnetic field and constant frequency (ν = const).
Peak at frequency H_{II-V} in absorption spectrum is twice as intense as peaks at frequency H_{I-VI} and H_{III-IV}

32

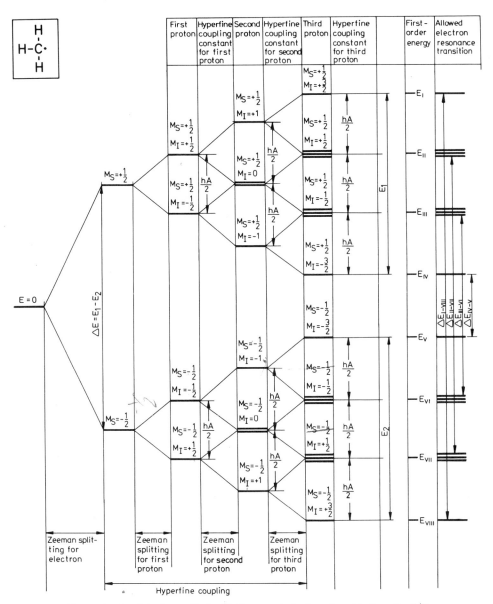

Fig. 2.16a. Energy levels of the free radical with three equivalent protons (System: S = 1/2, I = 1/2, I = 1/2, I = 1/2) in constant external magnetic field (H = const) and variable frequency (ν)

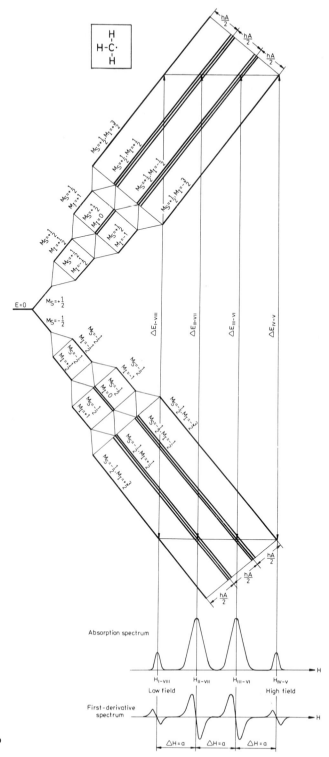

Fig. 2.16b

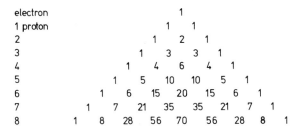

```
electron                        1
1 proton                     1     1
2                          1    2    1
3                        1    3    3    1
4                      1    4    6    4    1
5                    1    5   10   10    5    1
6                  1    6   15   20   15    6    1
7                1    7   21   35   35   21    7    1
8              1    8   28   56   70   56   28    8    1
```

Fig. 2.17. Pascal's triangle

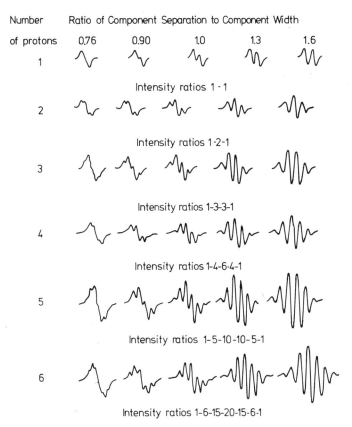

Fig. 2.18. Theoretical hyperfine structure curves for a Gaussian derivative curves[1762]

Fig. 2.16b. Energy levels of free radical with three equivalent protons (System: $S = 1/2$, $I = 1/2$, $I = 1/2$, $I = 1/2$) in variable external magnetic field and constant frequency (ν = const)
Peaks at frequency H_{II-VII} and H_{III-VI} are three times as intense as peaks at frequency H_{I-VII} and H_{IV-V}

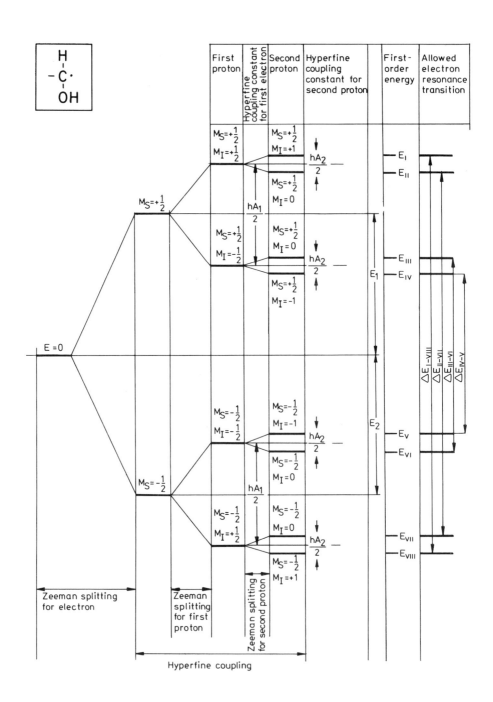

Fig. 2.19a. Energy levels of the free radical with two nonequivalent protons (System: S = 1/2, I = 1/2, I = 1/2) in constant external magnetic field (H = const) and variable frequency (ν)

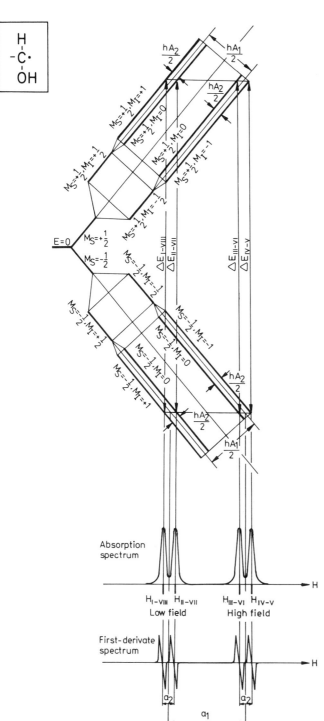

Fig. 2.19b. Energy levels of the free radical with two nonequivalent protons (System: $S = 1/2$, $I = 1/2$, $I = 1/2$) in variable external magnetic field and constant frequency (ν = const)

2.6.3. Hyperfine Structure Spectra due to Nonequivalent Protons

When the free radical structure contains two nonequivalent protons, having different hyperfine coupling constants (A_1 and A_2, where $A_1 \gg A_2$), the energy level diagram may be constructed in the samy way as before. The first splitting $\frac{hA}{2}1$ is due to the proton with hyperfine splitting constant A_1. The second splitting is due to that with $\frac{hA}{2}2$ (Figs. 2.19a and 2.19b).

Summing up, it is seen that when we have two nonequivalent sets of n and m protons respectively, the maximum number of lines in the ESR spectrum is $(n+1)(m+1)$. The positions of lines in the spectrum are a function of the hyperfine splitting constants a_n and a_m (a_i). The position of the lines at the field H_k in the spectrum is given by:

$$H_k = H_c - \Sigma a_i M_i \tag{2.31}$$

where H_c is the magnetic field at the center of the spectrum, M_i is the sum of the M_I values for the protons of the i-th set.

When: $a_n \gg a_m$ the obtained spectrum is well resolved.

$\quad a_n > a_m$ but a_m is sufficiently large the crossing of line groups is observed. The analysis of such spectra is much more complicated.

$\quad a_n \approx a_m$ (the difference is small) some of the lines are overlapping and the spectrum consists of a lower number of resolved lines than expected.

The separation of the energy levels is determined by $\Pi_i(n_i I_i + 1)^n$ where Π_i indicates a product over all values of i, and n_i is the number of equivalent nuclei with spin I_i.

The positions of the lines in the isotropic spectrum should be symmetric about a central position. Asymmetry of line positions may result from:
1. Anisotropy of the g-tensor of the radical
2. Superposition of two spectra with different g-values
3. Anisotropy of the hyperfine interaction tensor
4. Second order splitting when the hyperfine splittings are large
5. Variation in spectral linewidth which arises from a slow tumbling rate of the radical.
6. A factor due to the spectrometer.

In many cases, in the presence of an odd number of nonequivalent nuclei with $I = 1/2$, the spectrum has no apparent center line[663, 1797]. The sum of the hyperfine splitting constants for all nuclei with $I = 1/2$, is equal to the separation in gauss between the outermost lines, which may be very weak and are therefore missed. When the line resolution is poor or when there appears to be more lines than expected, it is useful to carry out a computer simulation of the spectrum based on assumed hyperfine splitting constants and line-width[624, 1202, 2074].

2.6.4. Second Order Splittings of Lines

When hyperfine coupling energy (hA) is very large in comparison with the electron energy ($g_e \beta_e H$) or when a small magnetic field is used, additional splitting of some line may occur, which is called "second-order splitting" (Fig. 2.20).

The detailed treatment of these phenomena is not discussed here, because it requires the application of perturbation theory.

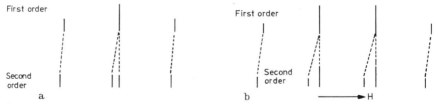

Fig. 2.20. Second-order splitting: a. for a $1:2:1$ triplet arising from two equivalent nuclei, $I = 1/2$, b. for a $1:3:3:1$ quartet from three equivalent nuclei, $I = 1/2$

Chapter 3

Experimental Instrumentation of Electron Spin Resonance

3.1. ESR Spectrometers

There are several commercial ESR spectrometers operating at different frequencies produced today, e.g. Varian (USA) E-Line spectrometer, Jeol (Japan) JES-3 BX and JES-PE-3 X spectrometers, and Bruker (Germany) B-ER-414, B-ER-418 and B-ER-420 spectrometers. The commercial ESR spectrometers operate at a fixed microwave frequency and scan the ESR spectra by linear variation of the magnetic field. A typical ESR spectrometer has the following main components (Fig. 3.1).

1. The source of microwave radiation is a klystron producing microwave oscillation in a small frequency range. ESR spectrometers have been constructed for the following bands:

Band	S	X	K	Q	E	
Approx. frequency	3	9	24	35	70	GHz
Approx. wave-length	90	30	12	8	4	mm
Approx. field for g = 2	1.1	3.3	8.5	12.5	25	kG

There are differences between the European and American names for the bands. To avoid misunderstandings the frequency should always be given. The large majority of ESR spectrometers are designed for a frequency around 9.5 GHz (X-band). The most common second frequency is 35 GHz (Q-band). Some properties of ESR spectra are frequency-dependent and it is advantageous to study these spectra at different frequencies. The frequency of monochromatic radiation is determined by the voltage applied to the klystron. Stabilization of the frequency is made by an automatic frequency control (AFC) system which works on the voltage. The power of the klystrons used in ESR spectrometers is usually a few hundred milliwatts. The heat generated by the klystron is removed by circulating water. The life-times of modern klystrons exceed 7000 hours.

2. The magnet is a source of static magnetic field. It should have a large field region, a wide air gap, and good field homogeneity. For g = 2 a field of 3.4 kG is necessary for resonance of free electron spins at 9.5 GHz (X-band) and a field of 12.5 kG at 35 GHz (Q-band). The field has to be very stable and kept within ± 10 mG. The stability of the magnet is often obtained by regulation with a Hall crystal in the magnet gap. The crystal gives a voltage proportional to the field and any change can be corrected with a feed-back circuit to the current.

3. The cavity system contains the sample. The quality factor Q of a cavity measures its power to store microwave energy:

$$Q = \frac{2 \text{ (maximum microwave energy stored in the cavity)}}{\text{energy lost per cycle}} \qquad (3.1)$$

When sample and sample holder are introduced in the cavity, the Q-value may change appreciably. Since the signal is proportional to Q, it is necessary to consider the best way to introduce the sample without decreasing Q too much. Cavities can be designed in different shapes and sizes for various purposes. The most common cavity is the TE_{102} rectangular cavity used for large samples and especially for liquid samples. Another cavity design is a cylinder type TE_{011} useful

39

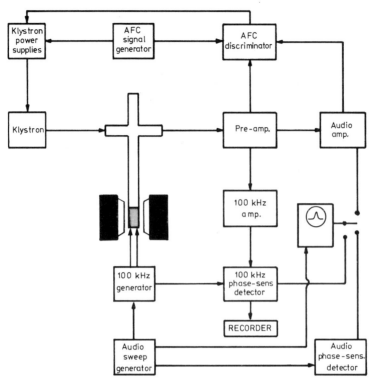

Fig. 3.1. Block diagram of a 100 kHz ESR spectrometer with AFC and phase sensitive detection[109]

for gaseous systems and liquid samples in capillaries. The microwave energy enters and leaves the sample cavity by the same hole, called the iris.

4. The modulation system at the commonly used frequency of 100 kHz consists of small Helmholtz coils, one on each side of the cavity along the axis of the static field.

5. The detection is usually a semiconducting silica crystal for measuring the microwave power absorbed by the specimen. A super-heterodyne electronic system is applied for sensitive detection.

6. The recording system allows the display of the phase-detected signal on an oscilloscope for quick observation or on a paper-chart for permanent record. It may also be stored in a computer for further data handling, such as the integration or the multi-scan time averaging. A computer is also convenient for storing time-dependent signals which change so rapidly that mechanical recorders cannot follow.

There are certain requirements for a good ESR spectrometer. It should have the maximal sensitivity and stability and it should have a good resolution. A spectrometer to be used for many different types of experiments should also be highly versatile.

The sensitivity should allow detection of very small numbers of unpaired electrons. The sensitivity of good commercial ESR spectrometers in the X-band amounts of 5×10^{10} spin G^{-1} for 100 kHz modulation and $5 - 15 \times 10^{11}$ spin G^{-1} for 80 kHz, in the K-band 3×10^{10} spin G^{-1} for 100 kHz modulation, and in the Q-band 6×10^9 spin G^{-1} for 100 kHz modulation.

The stability should allow the recording of linear and reproducible spectra. The resolution should allow the separation of lines which are very close to each other. The versatility should allow for large changes of the spectrometer parameters and provide for many different accessories to be added for special experiments. For further details of the construction and the operation of ESR spectrometers the reader should consult original papers, reviews, and books, e.g.[33, 986, 1760, 1919, 2417].

40

For study of rapid chemical reactions occurring in a few milliseconds, a rapid scan ESR spectrometer is used[1654, 2052, 2053]. Block diagram of a modified ESR spectrometer for this purpose is shown in Figure 3.2.

With the single shot saw tooth generator, the duration of the scan is raised in the range from 10^{-3} to one second. The signal, amplified by a power amplifier, is fed to a Helmholtz coil. The sweep of the two-element synchroscope is synchronized with the original single shot. The release of the camera shutter is also synchronized with the single shot. One of the advantages of the rapid scan method is its ability to investigate phenomena occurring rapidly as a single event, e.g. the decay of electrons trapped in an irradiated solid matrix.

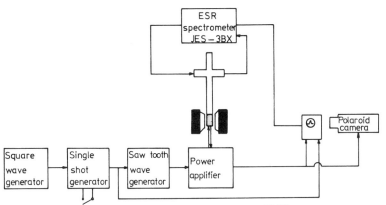

Fig. 3.2. Block diagram of a rapid scan ESR spectrometer[2053]

3.2. Field Sweep Calibration

In order to measure line widths and hyperfine splittings, an accurate calibration of the field is necessary. This is easily made by using a standard sample in which the line positions have been carefully determined. The standard sample can be introduced in the dual cavity and the spectra recorded simultaneously. Various compounds containing stable unpaired electrons are used as standards[42]:

1. Mn^{2+} in SrO powder characterized by 6 lines, 1.6 G wide, totally over 420 G, splittings 84 ± 0.2 G[1853].
2. Peroxylamine disulphate ion in solution characterized by 3 lines 250 mG wide, outer splitting 26.18 G, splittings 13.1 ± 0.004 G[635].
3. Tetracyanoethylene anion radical (TCNE) in solution (Varian standard sample) characterized by 9 lines, 100 mG wide, outer splitting 1.6 G, splittings 1.57 G[1820].
4. Wurster's blue perchlorate characterized by 39 lines listed for half spectrum, total spectrum 86 G, accuracy ± 0.01 G[2103].

Direct measurements and statistical analysis of line width is given in paper[1931].

3.3. Measurements of Spin Concentration

The number of spins (per gram, per milliliter, or per mm length of sample) is proportional to the area under the absorption curve. The derivative signal has to be integrated to obtain this area. This is equivalent to the first moment of the derivative signal. The integration is made graphically,

41

electronically, or digitally with a computer. The accuracy of all methods decreases when the signals are noisy or have long wings, and when the base line drifts or is affected by a background signal. These problems can most easily be overcome by using a digital computer.

A method for graphical calculation of spin concentration is to divide the spectrum into constant intervals and measure the amplitude in each interval. These amplitudes are cumulatively added and each partial sum written down. The last value should be zero, otherwise it has to be divided by the number of intervals and cumulatively substracted from the partial sums. The second integral is then obtained by adding all the partial sums[2432].

The intensity of the ESR spectra is influenced by several experimental factors such as:
1. The overall spectrometer gain
2. The microwave frequency
3. The modulation amplitude at the sample
4. The concentration of free radicals in the sample
5. The g-factor of the sample
6. The transition probability
7. The filling factor of the sample
8. The sample temperature

Since many factors influence the intensity of the ESR signal, the absolute determination of free radical concentration involves many errors and requires many corrections, and it is carried out very rarely.

The usual method for measuring the concentration of free radicals in an unknown sample is to compare its ESR signal with that of a sample containing a known quantity of free radicals[33, 2103].

A standard sample should have the following properties:
1. Line width and line shape of its ESR spectrum similar to that of the unknown sample.
2. Its number of spins should be similar to that of the unknown sample.
3. Physical shape and dielectric loss of the reference sample similar to those of the unknown sample.
4. Stability both with time and temperature. The number of spins, the line width, and the g-value should remain constant.
5. A short relaxation time (T_1) to avoid easy saturation of the signal.

Since line width and spin concentration vary over several order of magnitude it may be necessary to prepare a set of standard samples. Various secondary standard samples are used[33, 42]:
1. Synthetic ruby crystal[2015]. It can be placed in the cavity permitting simultaneous observation of standard and unknown samples.
2. Charred dextrose[973]. This standard sample does not change properties between 4 K and 500 K.
3. Powdered coal (pitch) diluted with KCl. Varian supplies two such standards (peak-to-peak line width 1.7 G, g = 2.0028) of different concentrations (1×10^{13} and 3×10^{15} spins cm^{-1}) with each spectrometer.
4. Mn^{2+} in MgO[1123, 2234, 2405], CaO[1925] and $CaCO_3$[1015].

The secondary standards are calibrated against less stable samples (primary standards) with spin concentration that are determined by independent methods:
1. Freshly recrystallized DPPH (α,α'-diphenyl-β-picrylhydrazyl)[985, 986, 1234, 2483].
2. Fresh, deep blue $CuSO_4 \cdot 5\ H_2O$ crystal[986, 1015, 1141].
3. Fresh $MnSO_4 \cdot H_2O$[2015].
4. $K_2NO(SO_3)_2$ in aqueous solution[1063].
5. O_2 gas[634, 2186, 2406].

Recalibrations with one type of primary standard in the same laboratory are reproducible within 10% while absolute accuracy may not be better than 50%. Relative comparison with a secondary standard can be done with an accuracy of 5−10%.

A number of factors influence the measurements and can lead to systematic errors if not accounted for:
1. Filling factors of the samples should be identical. The ratio of radiofrequency field and modulation field is not linear, so a linear sample cannot easily be compared with another of a

different diameter or with a point sample. The samples should be carefully positioned in the maximum radiofrequency field.

2. Materials with a high dielectric constant such as water and quartz may distort the micro-wave field.

3. Materials with high dielectric loss change the cavity characteristic.

4. The scanned width should be sufficiently large.

5. The signals should not be partially saturated.

6. The sweep rate should not distort the spectrum.

7. g-Value, resonance frequency and temperature should be identical.

8. The spin quantum number and (for $S > 1/2$) the transition should be the same (for Mn^{2+} $I = 5/2$)[1727].

The dual cavity allows the use of samples with similar filling factors and position in the radio-frequency field[1171]. The change of samples is also simple. Differences in the two signal measuring channels are eliminated by exchange of samples. A careful analysis of the errors in spin concentration determination has been given in papers[405, 2022].

3.4. Measurements of ESR Spectra at Different Temperatures

Low temperatures are used to a large extent in ESR measurements, because they provide a great deal of information about the spin system. Apart from the temperature dependence of g-factor and hyperfine interaction, some of the advantages of low temperature are[1913]:

1. An increase of spectrometer sensitivity, because of the increase of the thermal equilibrium population difference between the Zeeman levels.

2. The spin lattice relaxation time increases and yields information about the coupling between the spin system and the lattice.

3. Several double resonance methods, like the ENDOR technique, require reasonably long relaxation times and correspondingly low temperature.

4. Similarly, most methods of optically detected magnetic resonance (ODMR) require slow relaxation rates.

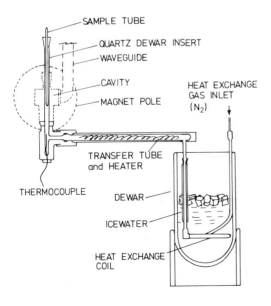

Fig. 3.3. Gas flow apparatus for varying and controlling the temperature of the sample[33, 1745]

3. Experimental Instrumentation of Electron Resonance

Measurements of ESR spectra at different temperatures (77–300 K) are made by applying a temperature variation and control device (Fig. 3.3)[1032].

This sample arrangement is a heat exchanger. Dry nitrogen is passed through a copper coil immersed in liquid nitrogen and conducted to the quartz Dewar insert through a Vycor transfer tube, which contains a heater. The cooling gas passes around the quartz sample tube and is discharged through the top of the Dewar in the cavity. Temperature control is achieved by varying the flow rate of the cooling gas and controlling the heater current. Temperature control may be obtained using thermocouple or platinum resistors. The Varian E-257 and E-268 Variable Temperature Units are examples of devices using this method.

Dangerous phenomena of explosions have been observed involving irradiated Dewars of liquid nitrogen with various atmospheric impurities including oxygen and water vapor from the ambient air and solid catalysts[1849].

For measurements of ESR spectra in the temperature range 4.2–77 K it is necessary to use a special low temperature continuous flow cryostat (Fig. 3.4).

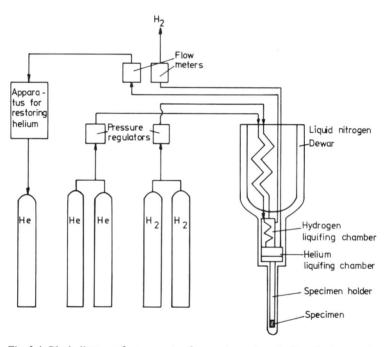

Fig. 3.4. Block diagram of an apparatus for varying and controlling the temperature lower than 77 K[1032]

The sample is cooled by a controlled flow of cold helium gas. Temperature measurements in this region is conventionally achieved by monitoring the resistivity of a carbon resistor. Such cryostats are commercially available, e.g. Leybold-Heraeus ESR-Verdampferkryostat, or Oxford Instruments ESR-9 Continuous Flow Cryostat. These kinds of experiments are difficult to perform on account of the necessity of using liquid helium which is not always available and not easy to handle[950,1461].

For measurements of ESR spectra in the temperature range 1.2–4.2 K the sample has to be immersed in the bath of liquid helium. Lowering the vapor pressure yields a corresponding decrease of the bath temperature. Various cryostats for 1.2–4.2 K[120, 1187, 1912, 1926, 1939, 2002, 2211, 2338] and below 1 K[643] temperature range are described in original papers.

3.5. Computer Application into ESR Spectroscopy

Application of different computer programs to evaluate ESR experiments have been reported in several papers[334, 396, 626, 656, 674, 686, 830, 832, 1061, 1162, 1318, 1322, 1342, 1521, 1594, 1654, 1757, 1758, 1825, 2101, 2172, 2320, 2442].

A program for evaluation of ESR spectra with unresolved hyperfine structure must be calculated for:

1. Various line numbers, intensity distributions, and ratios of component width to splitting on the assumptions that the components are symmetrical and give the same width and splitting (this is for many free radicals and ion-radicals in the liquid state).

2. Anisotropic g-factors (for free radicals and ion-radicals in solids, biopolymers, etc.).

3. The assumption that the unresolved components differ in width (for free radicals and ion-radicals in glass matrices, amorphous polymers, etc.).

Varian Associates (Palo Alto, California, USA) have prepared the EPR/ADAPTS system written in E-BASIC computer language. It is easy for an inexperienced computer operator to use for solving his specific ESR problems. Operator errors in the entering of data are reported so that they can be corrected immediately. An ESR application software package is also included in the system; it contains a set of 21 programs designed to perform many of the routine tasks demanded by the majority of ESR experiments. The tapes may be divided into four specific categories: data acquisition, data display, data reduction and special functions. These programs are all sorted as files on a 123-K disk, and are loaded into a CPU memory. All loading is done by the use of DASMR assembly language routines when speed is essential to obtain the required data throughput. A description of these programs is given in Appendix (p. 57–60).

Lebedev et al.[1323] have published the Atlas of Electron Spin Resonance Spectra which contains theoretically computer calculated multicomponent symmetrical spectra.

3.6. Preparation of Samples for ESR Experiments

3.6.1. Gases

There are three types of gases which can be studied by ESR[318]:

1. Paramagnetic molecules: O_2, NO, NO_2, ClO_2
2. Reactive atoms: H, N, O
3. Free radicals in a gas phase.

The ESR study of gases is carried out for two main reasons:

1. The theory of coupling in molecular species.
2. Gas phase chemistry involving reactive intermediate radicals.

The ESR spectra of paramagnetic gas molecules are generally more complicated (see Chapter 6) than those for solids and liquids, because the magnetic field does not uncouple the strong interaction between the magnetic moment of the unpaired electron and the angular momentum of the rotating molecule[40, 393, 1005, 1704, 2186]. The main experimental problem is to produce a sufficient concentration of free radicals in the gas to obtain a good signal-to-noise ratio without losing resolution[1438, 2087]. Measurements at a pressure of 10–20 mm Hg (10^{17}–10^{18} molecules cm^{-3}) sometimes give line broadening. Pressures of 0.1–1 mm Hg (10^{15}–10^{16} molecules cm^{-3}) give spectra with better resolution. For that reason it is necessary to have a gas-handling system which permits pressure adjustment to optimize signals and resolutions. Two methods are commonly used:

1. Differential pumping to remove excess molecules.
2. Dilution with inert diamagnetic gases such as helium or argon.

Atomic gases are produced by several different methods:

1. Electrical discharge.
2. Electrodeless discharge.
3. Pyrolysis, e.g. by high-temperature arc or flames.

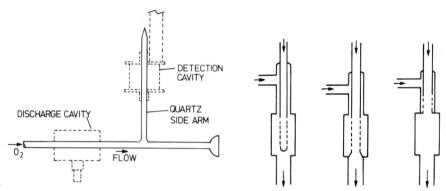

Fig. 3.5. Arrangement of flow apparatus for studying microwave discharge products[393]

Fig. 3.6. Different types of quartz flow mixing cells[393]

On account of the short lifetime of these radicals the production devices should be arranged close to the cavity system (Fig. 3.5).

In order to investigate reactions which occur in gas phase, special flow devices are placed in the microwave cavity (Fig. 3.6).

3.6.2. Liquids

The studies of free radicals in liquids are devoted to the following problems:
1. Radicals generated by radiolysis and photolysis of liquids.
2. Negative and positive radical ions produced in solution.
3. Intermediate species in chemical reactions.

Radicals generated by radiolysis are obtained by two main techniques:
1. Radicals generated in a radiation source and then pumped through the resonant cavity (Fig. 3.7).
2. Radicals formed in the cavity by direct introduction of high energy electrons into the cavity (Fig. 3.8).

From an experimental point of view, both methods are complicated.

Radicals generated by photolysis are obtained by direct irradiation of the sample in the cavity equipped with a quartz window in one wall (Fig. 3.9) or by irradiating the sample outside the cavity.

Far UV-photolysis ($\lambda < 1800$ A) of liquid samples in a rigid phase at 77 K requires specially designed apparatus for measurement of ESR spectra[2418]. This instrument gives spectra resolved up to six different times after the UV flash (Fig. 3.10).

The sample is placed in a commercial ESR spectrometer and irradiated with flashes from a xenon tube[563]. At a fixed magnetic field the spectrometer output describes the radical decay curve. After each flash the output is monitored by six different sampling channels and recorded by a multi-channel recorder.

A practical method for measuring triplet states by ESR has been described in detail[834]. The intensity of the $\Delta m = 2$ spin resonance for a given number of randomly oriented triplets is computed as a function of zero-field splitting and line width. Then it is compared with the intensity of a standard sample with known yield.

A photolytic continuous-flow method operating at room temperature makes possible the observation of radicals formed by the addition of photochemically generated primary radicals to different monomers[2035, 2042].

Negative free radical ions are produced by chemical methods, e.g. by mixing an alkali metal with an aromatic compound in a polar solvent such as 1,2-dimethyloxyethylene (DME), tetra-

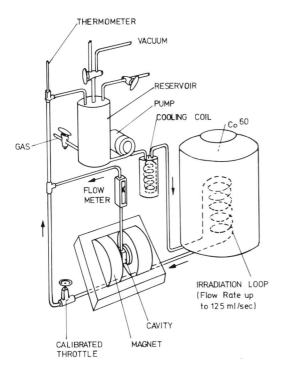

Fig. 3.7. Continuous-flow apparatus for measuring radicals produced in liquids by irradiation with ^{60}Co γ-rays[33)]

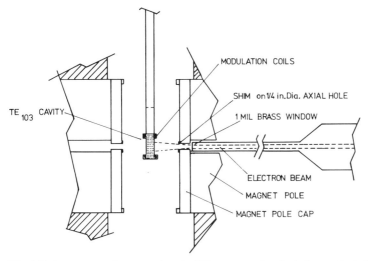

Fig. 3.8. Arrangement for measuring the ESR spectra of radicals formed by 2.8 MeV electron irradiation of a sample in the cavity[652)]

hydrofuran (THF), or dioxane in high vacuum. A variety of different arrangements have been developed; one of them is shown in Figure 3.11.

Negative free radical ions may also be produced by an electrolytic method. This method is sometimes preferred because the ESR spectra obtained are not complicated by the presence of alkali metals[408)]. Different types of electrolytic cells are described in the references[243, 911)]. Positive

47

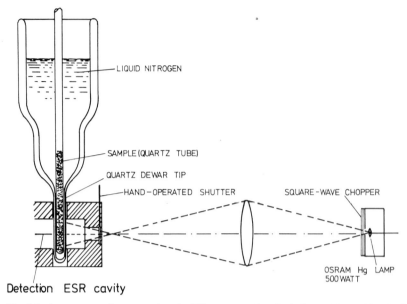

LIQUID NITROGEN

SAMPLE (QUARTZ TUBE)

QUARTZ DEWAR TIP

HAND–OPERATED SHUTTER

SQUARE–WAVE CHOPPER

OSRAM Hg LAMP
500 WATT

Detection ESR cavity

Fig. 3.9. Arrangement for measuring the ESR spectra of radicals formed by UV-irradiation of a sample in the cavity[33]

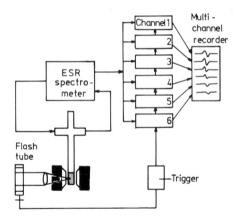

Channel 1

Multi-channel recorder

2

3

ESR spectro-meter

4

5

6

Flash tube

Trigger

Fig. 3.10. Block diagram of the multiscan ESR spectrometer[2418]

free radical ions can be produced in solution, e.g. by dissolving aromatic hydrocarbons in concentrated sulphuric acid[243].

Intermediate species in chemical reactions are observed mainly in flow systems[426]. These systems are widely used for the investigation of redox reactions, e.g. to initiate polymerization (see Chapter 4.6.3). Figure 3.12 shows a typical arrangement for such experiments.

Containers L_1 and L_2 hold the redox reagents and the monomer in solution. These solutions are pumped, or flow by gravity, from the containers to the mixing cell which is placed in the ESR spectrometer cavity. The flow rate is regulated and measured in the flow meter. In order to obtain good results the mixing cell should be correctly designed. At present modifications of the Dixon-Norman mixing cell are widely used (Fig. 3.13)[556, 557, 2393].

This cell consists of two tubes with a flat cell on top. The inner tube contains a spray head, from which one solution (e.g. the reductant) is injected into the second solution (e.g. the oxidant). Free radicals are immediately formed after mixing the two solutions in the ESR spectrometer

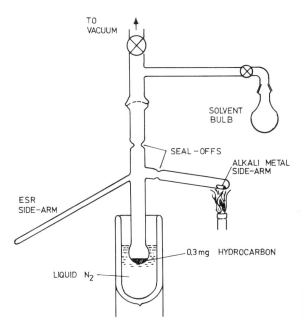

Fig. 3.11. Arrangement for preparation of anion radicals by reduction with an alkali metal[243]

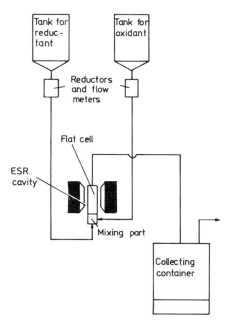

Fig. 3.12. Arrangement for the production of a high concentration of free radicals in a flow system[1808]

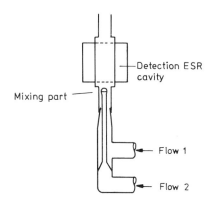

Fig. 3.13. The Dixon-Norman type mixing cell for the ESR study of free radicals formed in a flow system

49

cavity. The flow rate of the solvents should experimentally be selected, and the rates between 1 and 10 cm^3 s^{-1} are habitually used.

The flow system method is mainly limited by the solvent used, water being the most frequent choice. A second limitation is due to the necessary presence of the redox system components in the reaction mixture. The presence of these components may be the source of secondary reactions complicating the overall kinetic analysis and may reduce the attainable spectroscopic resolution[494, 1455, 2043]. This method has also been developed for the study of short-lived free radicals formed in UV-irradiated solutions[1382, 1383, 2035, 2042].

3.6.3. Solids

Free radicals in solids are produced by photolysis, radiolysis, pyrolysis, and electrical discharge.

The generation of free radicals by photolysis and radiolysis can be directly made in the cavity or the radicals can be transferred into the cavity after their production outside. In the latter method the sample manipulation is relatively simple and straight-forward down to liquid nitrogen temperature. Transfer operations are more difficult at liquid helium temperature. In most experiments with radicals down to 77 K, the radicals are generated externally and transferred to the cavity for measuring the ESR spectra. At lower temperatures it is usually better to irradiate the sample in the cavity to minimize manipulations.

Samples which are solids at room temperature, such as crystals, polymers, or glasses, are readily irradiated as it is shown on Figure 3.8. Many compounds which are liquid at room temperature are prepared in the form of small beads by freezing drops of the sample at liquid nitrogen temperature. Several different arrangements for direct irradiating of samples in the cavity and outside the cavity are described in the literature[33, 1518, 2379].

Measurements of the orientation dependence of the ESR spectra of single crystals require the application of a movable amount with cross leaders for fine adjustment of the sample position in the cavity[1032].

Inserting the sample directly into liquid nitrogen frequently causes problems, because the nitrogen gas bubbles produce an excessive noise in the recording of spectrum. This effect may be suppressed by blowing helium gas into the liquid nitrogen.

3.6.4. Stabilization of Free Radicals

The life-times of many organic free radicals and ionic radical intermediates are too short for observation by ESR spectroscopy. For that reason it is very important to stabilize free radicals and radical ions[165]. The following methods are applied:

1. Stabilization of free radicals by cooling to low temperatures, e.g. liquid nitrogen $(-196\,^\circ C = 77$ K)[1501].

2. Stabilization of free radicals on synthetic zeolites[1615]. Synthetic zeolites seem to be good stabilizing matrices for free radicals at relatively high temperatures. Application of this method in radiolysis is limited as the radiation induces formation of paramagnetic centers in the zeolites[10, 288, 1231, 1611, 2068, 2313, 2354, 2427].

3. Stabilization of free radicals and ionic radicals in organic glass matrices. Organic glass matrices are widely applied for stabilization of ionic radical intermediates (see Chapter 4.2.2). It has been found that free electrons are trapped only in amorphous frozen (glassy) states but never in crystalline states[2030].

4. Stabilization of free radicals in organic crystalline matrices, e.g. alcohols. Methanol and other alcohol glasses are frequently used as trapping matrices in radiation and photochemical studies. Irradiation effects in frozen alcohols have been reported in several papers[34, 111, 416, 417, 499–501, 1138, 1514, 1995, 2093]. Generally, electrons and hydrogen abstraction radicals RĊHOH of the alcohols are formed by ionizing radiation. In the case of methanol this may be explained in the following mechanism[111, 499, 500]:

$$CH_3OH \xrightarrow{\quad\wedge\wedge\wedge\quad} CH_3OH^{\cdot+} + e^- \qquad (3.2)$$

$$CH_3OH^{\cdot +} + CH_3OH \longrightarrow \begin{cases} \dot{C}H_2OH + CH_3OH_2^+ & (3.3) \\[2ex] CH_3O\cdot + CH_3OH_2^+ & (3.4) \end{cases}$$

It has been found from application of spin trapping methods that methoxy and hydroxymethyl radicals are formed in about equal amounts[1418, 1896]. Methoxy radicals are not observed in X-irradiated frozen methanol matrices because they probably give rise to hydroxymethyl radicals:

$$CH_3O\cdot + CH_3OH \longrightarrow \dot{C}H_2OH + CH_3OH \qquad (3.5)$$

During UV-irradiation the hydroxymethyl radicals are photolyzed to formyl radicals[34, 1054]:

$$\dot{C}H_2OH \xrightarrow{h\nu} \dot{C}HO + H_2 \qquad (3.6)$$

These results show that application of alcohol glasses as trapping matrices should receive considerable attention.

3.6.5. Apparatus for High Pressure Annealing of Samples

The simultaneous measurements of ESR spectra of free radicals at high pressure and at different temperatures directly in the cavity resonator of an ESR spectrometer are difficult from experimental point of view. This problem can be solved by the following method[2104]. Samples are irradiated at low temperature and the starting concentration of free radicals is determined by ESR spectroscopy. Then the sample is inserted into the press (Fig. 3.14) and heated to the necessary temperature.

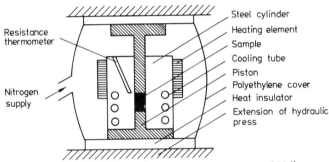

Fig. 3.14. Arrangement for high-pressure sample annealing[2104]

After annealing at certain pressure and temperature, the sample is cooled down, the pressure is read, and the concentration measured by ESR at lower temperature. If the sample is transferred into the press and then again into the ESR spectrometer at a temperature at which the decay is negligible, and if the heating rate is sufficiently fast, the rate constant of free radical decay may be determined with high accuracy from the concentration difference and the annealing time.

3.6.6. Loading Apparatus for ESR Stretching Experiments

The radical formation during the tensile deformation (constant stress, constant strain, or constant elongation) is made in the specially designed loading apparatus, which is placed in the cavity assembly of the ESR spectrometer (Fig. 3.15)[1556, 1558].

51

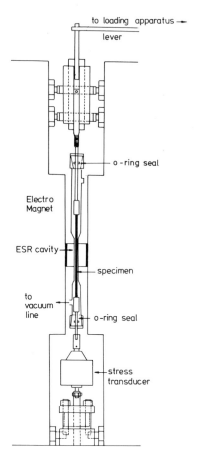

Fig. 3.15. Loading apparatus for ESR stretching experiments[1556, 1558)

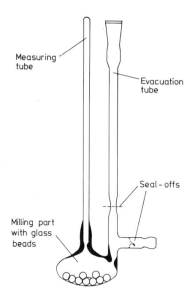

Fig. 3.16. Glass grinding mill used for evacuation, grinding and measuring ESR spectra of polymers[1747)

The ends of the specimen are held in brass chucks on either side of the microwave cavity. At the lower end a load transmitting brass rod is connected to a stress transducer. At the upper end a second rod is attached to a lever connected to a motor-driven loading framework, as is shown in Figure 3.15. To stretch fibers in vacuo or in a stream of nitrogen, the specimen is placed in a glass tube, of which the part in the microwave cavity is made of vitrous quartz.

3.6.7. Milling Devices for Grinding Samples

Several different constructions of milling devices are described in literature[369, 578, 1747, 1878, 2057). A simple vibration mill is shown in Figure 3.16.

The vibration frequency is 50–60 cycles min^{-1} and the milling time 4–6 h. In order to trap the free radicals formed, the milling process is carried out at liquid nitrogen temperature (77 K). After milling, the glass ampoule is turned 180° to an upright position and inserted in the ESR cavity for measurement. For this purpose the ampoule is equipped with a special quartz finger.

3.6.8. Preparation of Samples Containing Free Radicals by Sawing Technique

In this method polymer blocks are cut in liquid nitrogen with a metal saw and the saw dust of polymer produced by this sawing is transferred in liquid nitrogen to a quartz tube and then inserted into the microwave cavity. Before the ESR measurements the liquid nitrogen is removed by evacuation[1879, 2057].

3.7. Special Techniques for ESR Spectroscopy

During the last few years new techniques for ESR spectroscopy have been developed, as:

1. "Electron double resonance" (ELDOR) in which the magnetic resonance of the specimen is affected by a second resonance frequency. It, therefore, leads to higher resolution of hyperfine structure than might otherwise be obtained. This technique has been applied to study relaxation mechanisms[88, 91, 621, 894, 990].

2. "Electron-nuclear double resonance" (ENDOR) is the detection of changes in a monitored ESR signal, due to the induction of nuclear resonance of nuclei coupled to the paramagnetic center. The detailed description of ENDOR theory and application has been given in several publications[11, 88, 91, 503, 534, 645, 714, 715, 720, 798, 881, 894, 900, 988, 992, 993, 1067, 1115, 2405]. ENDOR corresponds to ordinary NMR applied to nuclei close to unpaired electrons, where electrons serve as a means of observing the nuclear resonance. Block diagram of ENDOR spectrometer is shown in Figure 3.17.

The radio frequency (rf) generator having power (200 W) and NMR radio frequency (2–30 MHz), and large power output is connected by ferrite-core transformers to the side coils of the special constructed cavity (Fig. 3.18).

Readers can find the basic description of ENDOR spectrometers in references[37, 502, 829, 1409].

A partial saturation of some particular electronic transitions by a microwave frequency produces an ESR signal of a given amplitude. When the second field having an NMR radiofrequency is slowly altered, a change takes place in the spin population of the electronic levels at values corresponding to the resonance of a given type of nuclei. This change affects the amplitude of the original ESR signal, and it is the magnitude of this "difference in amplitudes" which is observed as function of the rf frequency. This effect is commonly known as ENDOR. At two frequencies of the rf generator (ν_1 and ν_2), the recorder plots two separated peaks (Fig. 3.19). This type of spectrum is called "ENDOR spectrum". The difference $\nu_2 - \nu_1$ is numerically equal to the hyperfine coupling constant (A). The ENDOR method is especially used when the hyperfine lines are not well-resolved in the ESR spectrum, and when the identity of an interacting nucleus is to be established by measuring of the nuclear g-factor.

When the unpaired electron interacts with more than one proton, or with more nuclear spin systems, the ENDOR lines appear at frequencies corresponding to the hyperfine coupling constants, no matter how many equivalent spins are present in each spin system.

The differences between ESR and ENDOR spectra are shown in an example of the triphenyl-methyl radical (Fig. 3.20).

In this radical three proton spin systems are found which correspond to the ortho-, para- and metha-protons:

Each ortho and metha system contains 6 protons, the para system contains 3 protons. The total number of ESR hyperfine spectral lines is:

$$N = (2 \times (6 + 1) \times 1/2)^2(2 \times (3 + 1) \times 1/2) = 196$$

In the spectrum shown in Figure 3.20a, about 100 lines are resolved. It is evident that the determination of the coupling constant for such complex spectrum is rather difficult.

The three coupling constants are easily determined from the ENDOR spectrum (Fig. 3.20b), but from this spectrum it is not evident how to assign the coupling constants. For that reason

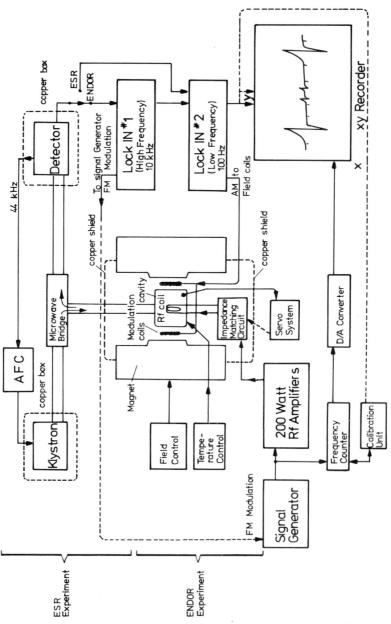

Fig. 3.17. Block diagram of an X-band high power ENDOR spectrometer[502)]

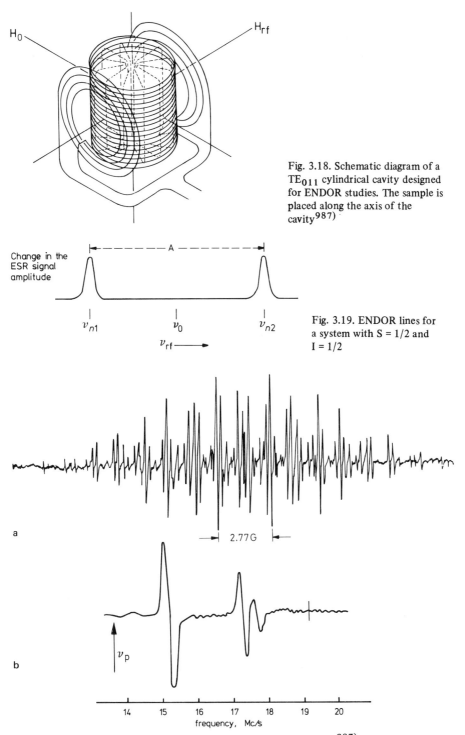

Fig. 3.18. Schematic diagram of a TE$_{011}$ cylindrical cavity designed for ENDOR studies. The sample is placed along the axis of the cavity[987]

Fig. 3.19. ENDOR lines for a system with S = 1/2 and I = 1/2

Fig. 3.20. The ESR and ENDOR spectra of the triphenylmethyl radical[987]

55

the ESR spectrum is more helpful. For example, the highest coupling constant corresponds to the 3 protons, because the corresponding splitting 2.78 ± 0.02 G gives a quartet (3 + 1 lines) in the ESR spectrum, indicating that 3 protons are involved. The septet corresponding to the 6 ortho-protons with a coupling constant 2.55 ± 0.03 G is also observed. The coupling 1.156 G caused by the 6 metha-protons is not directly observable in the ESR spectrum, but is very clear in the ENDOR spectrum. It is seen that ENDOR is valuable for accurate determination of hyperfine coupling constants of complex radicals when the spectra are compared with ESR. ENDOR spectroscopy is more useful than ESR spectroscopy in two main respects. One is that the spectra are simpler, as each group of equivalent protons produces only one single ENDOR line regardless of the number of equivalent protons in the group. The second adventage is that the resolution is more improved.

3. "Optical detection of magnetic resonance" (ODMR). The basic idea of this method is that radiofrequency transitions which occur in the course of the magnetic resonance can be monitored by the changes which they cause in the polarization, intensity, or spectral distribution of either emitted or absorbed light[503, 975, 1530]. Thus the ODMR combines the high selectivity of optical spectroscopy with the sensitivity of ESR spectroscopy. This method has not yet been applied to the study of free radicals in polymers.

Appendix

MSAV: Multiscan Average

MSAV provides the user with a means to acquire and sum together several scans of EPR data in order to increase the signal-to-noise ratio. Assuming the spectrometer noise is random, it will add as the square root of the number of scans, i.e. $\sqrt{N}$, while the EPR signal (not random) will add directly as the number of scans, N. Thus, an improvement in the signal-to-noise ratio equal to the square root of the number of scans can be achieved by multiscan averaging.

MSAVST: Multiscan Average Using Standard Parameters

MSAVST operates in the same manner as MSAV. All necessary input parameters are retrieved from a standard parameter file, STPARA, and the acquisition of data is begun. The purpose of this program is to maintain a standard set of multiscan averages in a file called STPARA. Hence, any future efforts in averaging spectra can be accomplished without having to answer the questions asked by loading the program MSAV.

MSAVRP: Multiscan Average Repeating Parameters in Data File

MSAVRP operates similarly to the MSAVST program described above, but reuses the parameters in the data file rather than the parameters in STPARA. The data file parameters are generally those used to obtain the most recent MSAV. For example, if the operator completes a MSAV, changes samples, and wishes to do an identical experiment, typing MSAVRP will initiate a multiscan average without the necessity of reentering the earlier input parameters.

RESUME: Resume Multiscan Average

This program allows the user to add scans to the data file if necessary. If the user, for instance, has performed a 40-scan MSAV on a dilute sample and found that the resultant signal-to-noise ratio was only fair, 40 additional scans may be added to improve this ratio by $\sqrt{2}$ with the RESUME option.

Data Reduction

Each of the five data reduction programs reads the data file created by the program and generates a plot file. The plot file may be displayed on the EPR oscilloscope, the E-80A EPR recorder, the optional 611 CRT, or it may be stored for future reference.

DERIVT: Derivative

DERIVT takes the derivative of the data file and stores the result in the plot file. The numerical derivative can be taken over as little as three or as many as 23 data points. The numerical derivative of normal EPR data is displayed as a second derivative.

SMOOTH: Smooth with Variable Length Convolution Function

Information in the data file may be processed with this program by using a variable number of values in a convolution function to smooth the data, with the result being stored in the plot file. The operator specifies the number of points, between 5, . . ., 23, used to calculate the smoothed value at each point in the data file.

INT1: First Integral

INT1 calculates the first integral (i.e. the absorption spectrum) of the standard EPR spectrum (1st derivative) emanating from the spectrometer.

3. Experimental Instrumentation of Electron Spin Resonance

INT2SC: Second Integral/Spin Concentration

This program calculates the second integral (intensity) of the normal EPR spectrum, prints a value for the intensity on the Teletype, stores the result in the plot file, and asks if a spin concentration calculation is desired. If not, the program terminates.

When the spin concentration calculation is desired, the user must supply a sample of known concentration (e.g. Varian strong pitch). The computer then calculates the known sample's second integral and compares this value to the second integral computed for the unknown sample. Spin concentration for the unknown is then printed out.

STICK: Stick Plot of Experimental Spectrum

STICK integrates each line in the standard *experimental* EPR spectrum and produces a set of lines positioned at line maxima with height proportional to the integral of the line. A list of line positions in Gauss and T_2-1's is provided as a user option. The purpose of the program is to allow the user to generate a theoretical stick plot of the EPR spectrum and compare this to the experimental stick plot generated by the computer, thereby giving a better feel for the accuracy of the proposed molecular structure.

Data Display Programs

OUTD: Output Data File

OUTD outputs the data file to the EPR recorder, EPR scope, or the 611 CRT storage scope (optional). OUTD provides both horizontal and vertical scaling.

OUTP: Output Plot File

OUTP outputs the plot file to the EPR recorder, EPR scope, or the 611 CRT storage scope (optional). OUTP provides both horizontal and vertical scaling.

Special Function Programs

EPRSIM: Simulate EPR Spectrum from Theoretical Parameters

EPRSIM calculates and stores in the data file an EPR spectrum using a first order approximation to determine the line positions. It produces a spectrum by first generating a list of all line positions and intensities, and then calculating the spectral shape over the region requested. The number of EPR lines generated is dependent upon the number of nonequivalent nuclei, the number of equivalent nuclei within each nonequivalent set, the nuclear spin, and the magnitude of the hyperfine splitting constant associated with each set of nonequivalent nuclei.

The line intensities related to a given set of nonequivalent nuclei are calculated as coefficients of the binomial expansion $(a + b)^n$. Final intensity of the simulated spectrum is obtained by taking a superposition of the various sets of inequivalent nuclei. Line positions are calculated in field units from the number of lines generated, the magnitude of the hyperfine splitting constant, and the magnitude of the magnetic field can range (in Gauss).

The final spectrum is determined by incorporation of a line shape function, either Lorentzian or Gaussian or some combination of the two, whose character is determined by an input parameter corresponding to the peak-to-peak width of the EPR line. A line shape is generated for each of the calculated EPR resonances. When two or more EPR lines overlap, a superposition of the intensities at each value of the magnetic field is calculated and saved in the data file.

ADDEL: Add Two Spectral Files

ADDFL adds two spectral files together and saves the result in a sum file. One application of this program might be to add an experimental spectrum to its computed counterpart but with each spectrum of opposite phase. If the result in the sum file is a straight line then the theoretical spectrum is exactly identical to the experimental spectrum; if a wiggly line appears, the theoretical spectrum does not exactly represent its experimental counterpart.

MULFL: Multiply Spectral File by Constant

Data values in a particular file are multiplied by a constant and the result saved in a new file. For example, multiplying the data file by the constant −1, will invert the phase of the EPR spectrum.

LISTFL: List Parameters from Spectral File

This program lists all of the parameters in the file header of a particular program. The user may also request a search of the data file to find and print the maximum and minimum values. Thus, if the program MSAV were in the data file, the operator would be able to list (print) all of the existing input parameters and their values and search for the maximum and minimum values of the multiscan averaged spectrum residing in the data file.

LINW: Quadratic Curve Fit of EPR Linewidths of a Spectrum

LINW fits a quadratic equation to the nuclear spin (independent variable) and EPR linewidth, T_2^{-1}, data. The pair of parameters, M_I, T_2^{-1}, is input for each of the EPR hyperfine lines in the spectrum. The program calculates the coefficients of the quadratic equation:

$$T_2^{-1} = A + Bm_I + Cm_I^2$$

LDPARA: Load Standard Parameter File (STPARA).

The header information from the data file is copied into the STPARA file. Thus, the last set of parameters used can be saved and reused whenever MSAVST is executed, i.e. MSAVST will use those parameters in the STPARA file to run a multiscan average.

Software Drivers

The four software drivers constitute a set of assembly language programs generally used to interact with various hardware on the EPR spectrometer in the fastest possible time.

EPRRCD: EPR Recorder Interface

This subroutine provides a general interface between the ADAPTS computer and the E-80A EPR recorder via the interface cable. Computed spectra are plotted on the EPR recorder by using EPRRCD.

EPRSCP: EPR Oscilloscope Interface

Provides an interface between the ADAPTS computer and the EPR E-200 refresh oscilloscope via the interface cable. Computed spectra are displayed on the EPR scope by using EPRSCP.

59

CRT: 611 CRT Storage Oscilloscope Interface

This subroutine provides a general interface between the ADAPTS computer and a Model 70-6400 CRT (optional hardware). Computed spectra are displayed on the CRT via this program.

PROCIN: Process Control Interpreter

This subroutine allows the user to mix different types of I/0 commands at a reasonably fast rate so that control applications requiring input and output of data on a point-by-point basis may be performed. In addition, control pulses may be generated or sense lines checked. The user is able to specify all of the input and output data channels and control sense lines in any order desired.

The 21 programs described above constitute the software package available with the EPR/ADAPTS spectrometer system at the present time. Future EPR software developments will be made available to EPR/ADAPTS owners upon completion.

Chapter 4

ESR Study of Polymerization Processes

4.1. Homogeneous Chain-growth Polymerization

Chain-growth polymerization occurs by free radical or ionic mechanism. The growth of a single polymer chain (a macromolecule) is due to the propagation of one kinetic chain reaction. Every free-radical chain reaction requires a separate initiation step in which a radical species is generated in the reaction mixture.

Radical polymerizations can be divided into two general types according to the manner in which the initial radical species is formed:

1. Homolytic cleavage of a covalent bond in the monomer by energy absorption (radiation, photo, thermal, or ultrasonic initiation)

$$AB \xrightarrow{\text{+ energy}} A\cdot + B\cdot \tag{4.1}$$

2. Unpaired electron transfer to the monomer from initiator (or photoinitiator) fragments, formed by dissociation, e.g.:

$$R-R \xrightarrow{\text{+energy}} 2\,R\cdot \tag{4.2}$$

$$R\cdot + CH_2{=}CHX \longrightarrow R-CH_2-\overset{\cdot}{C}HX \tag{4.3}$$

In radiation initiation, the energy absorbed by the monomer is often higher than that required for the excitation of the molecule and dissociation of a covalent bond, and sufficient for the ionization of the molecule. An electron is ejected and the ionization of the monomer molecule produces a radical-cation containing one unpaired electron and a positive charge (see Chapter 1.4):

$$AB \xrightarrow{\gamma} AB^{\cdot+} + e^- \tag{4.4}$$

Such ion-radicals are often unstable and may dissociate into a free radical and a cation:

$$AB^{\cdot+} \longrightarrow A^+ + B\cdot \tag{4.5}$$

But this dissociative process can also occur in the same step in which the electron (e^-) is ejected from the irradiated monomer:

$$AB \xrightarrow{\gamma} A^+ + B\cdot + e^- \tag{4.6}$$

61

When the electron does not have an excessive energy it may be attracted back to the cation, and a second free radical is produced:

$$A^+ + e^- \longrightarrow A\cdot \tag{4.7}$$

Otherwise, the ejected electron is eventually trapped by another monomer molecule and either forms a radical-anion:

$$AB + e^- \longrightarrow AB\cdot^- \tag{4.8}$$

or causes molecular dissociation to one radical and one anionic species:

$$AB + e^- \longrightarrow A\cdot + B^- \tag{4.9}$$

In conclusion, the ejection of an electron from a molecule by ionizing radiation can produce two free radicals ($A\cdot$ and $B\cdot$) and one cation (A^+) and one anion (B^-).

Radiation-initiated polymerizations of vinyl monomers are generally, but not always, free radical reactions. The effect of free-radical inhibitors and the results of kinetic studies indicate that the polymerization of most vinyl monomers in the liquid state proceeds by a free radical mechanism[430]. There is, however, good evidence that the mechanism of initiation and propagation is ionic for some vinyl and diene monomers in solution at very low temperatures or in the solid state[2413].

An application of ESR spectroscopy for the study of polymerization mechanism has been for the first time made by Bresler et al.[297-299]. These authors have measured the concentration of free radicals formed during polymerization of vinyl monomers such as methyl methacrylate, methyl acrylate, and vinyl acetate.

ESR spectra of monomers polymerizing in bulk and highly viscous systems (e.g. precipitation polymerization) are not well resolved[1195, 1196]. On the other hand, during polymerization in nonviscous media (e.g. polymerization in solution) it is difficult to detect free radicals by the ESR method because the concentration of free radicals is very low. If sufficiently high concentration is reached, the ESR spectra obtained during polymerization in nonviscous media are well resolved and easy to interprete. The ESR detection and identification of free radicals during polymerization under various conditions have been reviewed in detail by Fischer[672, 685] and Takakura and Rånby[2136].

4.2. Ionizing Polymerization

4.2.1. Solid-state Polymerization

In this method monomers in crystalline or glassy solid state are exposed to a continuous dose of γ-radiation from a ^{60}Co source or to an electron beam from a Van

Table 4.1. Polymerization of Olefin monomers in the solid state initiated by γ-radiation[1404]

Olefin monomer	Monomer M.P., °C	Irradiation temp., °C	Tendency to polymerize
Styrene	−30	−51 to −78	−
α-Methylstyrene	−23	−78 to −196	−
2,4-Dimethylstyrene	−	−80	+
Vinyl acetate	−159	−196	+
Vinyl carbazole	+64	6−20	+
Vinyl chloride	−138	−138	−
1,3-Butadiene	−	−196	+
N-Vinylphthalimide	−	−	−
Isobutene	−141	−80, −196	+
Acrylic acid	+12	−	+
Methacrylic acid[b]	+16	0	+
Acrylamide	+85	0−60	+
Acrylonitrile	−83	−83 to −135	+
Methyl methacrylate[c]	−50	−78, −196	−
Octadecyl methacrylate	+17	−10, +30	+

de Graaff generator or a linear accelerator. For investigations of polymerization in a nonsteady state, pulse radiolysis has been applied. A very high dose is fed to the system in a very short time ($10^{-6}-10^{-9}$ s). Examples of solid state polymerization are shown in Table 4.1.

The exact mechanism of the solid-state polymerization is unknown. Whether the polymerization occurs as a free-radical, or an ionic reaction, or both cannot be determined directly because of the wide variety of reactive species which can be formed by ionizing radiation.

It is difficult to prepare a single crystal of a polymer which is sufficiently large to permit observation of ESR after irradiation. The determination of the nature of free radicals trapped in a vinyl polymer either by polymerization or by radiation damage, is made by the study of an irradiated single crystal of the corresponding monomer. However, only few ESR studies of irradiated single crystals of vinyl monomers have been carried out[27, 137, 774, 1628, 1993, 2323]. This is partly because several monomers are readily polymerized during evaporation of the solvent from their solutions in the preparation of single crystals. Another difficulty is that most vinyl monomers are liquid at room temperature (see Table 4.1). The final difficulty in the preparation of single crystals of most common vinyl monomers is the low temperature required for this purpose.

In the crystalline monomers with a high degree of symmetry an "excition absorption" of irradiation energy may be observed[186, 599]. This means that absorbed energy in the form of mobile "exciton" may migrate over a long distance in crystalline monomer (10^2-10^6 Å). Such a migrating exciton may be trapped by the imperfections where free radicals are then formed.

4.2.1.1. Olefin Monomers

ESR studying of radiation induced polymerization of ethylene (m.p. -103.6 K) have been reported in a few papers[607, 756, 757]. ESR spectrum of γ-irradiated ethylene monomer adsorbed on silica gel at 77 K is shown in Figure 4.1.

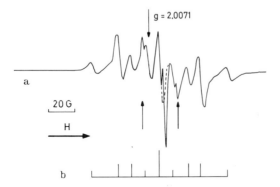

Fig. 4.1. a. ESR spectrum of γ-irradiated ethylene adsorbed on silica gel at 77 K and measured at 133 K, b. Stick plot for the radical $-CH_2\dot{C}H_2$[607]

The overall ESR signal is due to four different radicals. The first component due to a radiation defect in the silica gel is a sharp line observed at $g = 2.0007$. If the signal from silica gel is substracted, the remaining spectrum corresponds to the dotted line in the center of Figure 4.1 a. The predominant component of the spectrum is assigned to the radical $-CH_2\dot{C}H_2$, for which the coupling constant are $A_\alpha = 22$ G and $A_\beta = 32$ G. A stick plot with these coupling constants is shown in Figure 4.1 b. The third component is attributed to ethyl radical $CH_3-\dot{C}H_2$. The contribution of this species to the spectrum is small and almost camouflaged by the main spectrum. Two sharp lines with a splitting constant of 26 G, marked in Figure 4.1 by arrows, are the central part of the strongest quartet of the ethyl radical. Other lines which originate from this radical are considerably lower in amplitude and are not observed. Although the fourth radical is not shown in Figure 4.1 a, a strong line centered at $g = 2.0071$ with a linewidth of 3.3 G is observed for irradiated polycrystalline ethylene at 77 K.

This line is too unstable to allow its observation even at 77 K one hour after the irradiation. Because of its instability, this signal line is presumed to originate from electrons which are ejected from ethylene molecules and loosely trapped on the surface of silica gel.

The overall mechanism of radical formation from ethylene is given as follows[607]:

$$CH_2=CH_2 \xrightarrow{\gamma} (CH_2=CH_2)^{+} + e^{-} \qquad (4.10)$$

$$(CH_2=CH_2)^{+} + n\ CH_2=CH_2 \longrightarrow \cdot CH_2-CH_2-(CH_2-CH_2)_{n-1}-CH_2CH_2^{+} \qquad (4.11)$$

$$CH_2=CH_2 + H\cdot \longrightarrow CH_3-\dot{C}H_2 \qquad (4.12)$$

Electrons ejected in the reaction (4.10) are trapped and the polymerization proceeds from the cation radicals by a cationic mechanism, a neutral radical being located at one end of each growing chain. The ethyl radical is formed by addition of a hydrogen atom which for example is present on silica gel, to the double bond of the ethylene molecule (4.12). The ethyl radical is not expected to initiate a polymerization reaction at the low temperature (77 K). Silica gel may contribute to the stabilization of the cation radical by trapping the released electrons from the ethylene molecules, promoting in this way the growth of polymer chains.

Solid state polymerization of isobutene (m.p. 128 K) initiated with γ-irradiation gives the ESR spectrum presented in Figure 4.2[2298, 2300].

Fig. 4.2. ESR spectrum of γ-irradiated isobutene in dark at 77 K and measured at 77 K[2300]

It consist of an 8-line spectrum with the separation of about 21 G. The hyperfine structure of the spectrum may be attributed to radicals of the type $R-CH_2-\dot{C}(CH_3)_2$.

ESR studies of free radicals formed in γ-irradiated solid tetrafluoroethylene (m.p. 130.5 K) at 77 K[324, 325] and 4.2 K[1550] and adsorbed on silica gel[1945] have also been reported.

A xenon matrix isolation technique has also been applied to ESR studies of the primary processes in γ-irradiated ethylene[1606] and tetrafluoroethylene[1607]. The main radical species which was observed in xenon-ethylene matrix was the ethyl radical (Fig. 4.3).

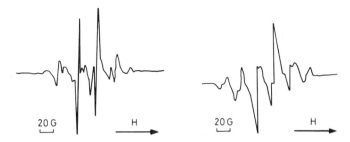

Fig. 4.3. ESR spectra of γ-irradiated: a. solid ethylene at 77 K, b. ethylene (0.57 mol %) in xenon matrix at 77 K[1606]

The concentration of ethyl radicals in a xenon matrix decreased to zero in 1–2 days even at 77 K (Figs. 4.4b and 4.4c) and after their disappearance a broad spectrum was obtained (Figs. 4.4c and 4.5a), which may be assigned to a polymer (oligomer) radical[1606].

65

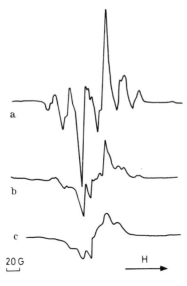

Fig. 4.4. ESR spectra of: a. γ-irradiated ethylene (4.7 mol %) in xenon matrix at 77 K, b. after 1 day, c. after 2 days[1606)]

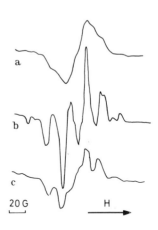

Fig. 4.5. ESR spectra of: a. γ-irradiated ethylene (2.0 mol %) in xenon matrix at 77 K, b. after light irradiation (λ > 3000 A), c. after 1 day storage at 77 K in dark[1606)]

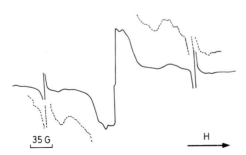

Fig. 4.6. ESR spectrum of γ-irradiated solid tetrafluoroethylene at 77 K[1607)]

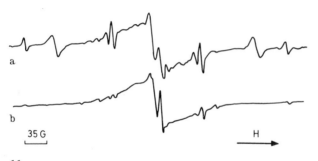

Fig. 4.7. ESR spectra of γ-irradiated tetrafluoroethylene: a. (1.2 mol %), b. (19 mol %) in xenon matrices at 77 K[1607)]

It is well known that pure ethylene is not polymerized in the solid state by γ-irradiation at 77 K. During γ-irradiation of ethylene in xenon matrix, an energy transfer from xenon to ethylene monomer may take place. The mobility of ethylene radicals in a xenon-matrix may be higher than in pure solid ethylene. The polymer radical was found to isomerize upon UV-irradiation ($\lambda > 3000$ Å) and the isomerized radical species has a well-resolved six-line spectrum (Fig. 4.5 b), which can reversibly be transformed after storage in liquid nitrogen into the former broad spectrum (Fig. 4.5 c).

ESR spectrum of γ-irradiated pure tetrafluoroethylene consists of a broad component at the centre and "wing peaks" (Fig. 4.6)[1607].
This spectrum was assigned to $-CF_2-CF_2$ polymer radical. On the other hand, ESR spectrum of γ-irradiated tetrafluoroethylene in xenon matrix (Fig. 4.7) shows several fragment radicals, e.g. $\cdot CF_3$.
This result suggests that the scission of carbon-carbon bond takes place by γ-irradiation of tetrafluoroethylene in xenon matrices.

4.2.1.2. Dienes

Butadiene can be polymerized by radical, cationic, and anionic mechanisms[476].
ESR studies of ionizing radiation-induced polymerization of 1,3-butadiene show that cation radicals (4.1) are precursors of allyl type radicals (4.2) which are formed by the following mechanism:

$$CH_2=CH-CH=CH_2 \xrightarrow{\gamma} [CH_2-CH=CH-CH_2]^{\cdot+} + e^- \qquad (4.13)$$
$$(4.1)$$

$$[CH_2-CH=CH-CH_2]^{\cdot+} + CH_2=CH-CH=CH_2 \longrightarrow CH_2=CH-\overset{\cdot}{C}H-CH_2-CH_2-CH=CH-CH_2^{+}$$
$$(4.2) \qquad (4.14)$$

A typical ESR spectrum of γ-irradiated 1,3-butadiene single crystal at 77 K is shown in Figure 4.8.

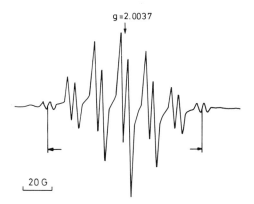

g=2.0037

20 G

Fig. 4.8. ESR spectrum of butadiene single crystal after γ-irradiation at 77 K[1395]

67

The signal has been attributed to allyl free radical (4.2)[1395, 1973]. Detailed analysis of the hyperfine coupling anisotropy of this ESR spectrum is interpreted as trans-conformation of the free radical (4.2):

$$\text{(4.3)}$$

H
|
H₂C   C   CH₂
\ / \\ / \
Ċ   C
| |
H   H

$$(4.3)$$

ESR studies of γ-irradiated 1,3-butadiene adsorbed on silica gel at 77 K also indicate that cation radicals may be precursors of further free radical formation:

$$S \xrightarrow{\gamma} S^* \tag{4.15}$$

$$S^* + CH_2{=}CH{-}CH{=}CH_2 \longrightarrow S\cdots[CH_2{-}CH{=}CH{-}CH_2]^{\cdot+} + e^-$$

$$S\cdots[CH_2{-}CH{=}CH{-}CH_2]^{\cdot+} + CH_2{=}CH{-}CH{=}CH_2 \longrightarrow \tag{4.16}$$

$$\longrightarrow S\cdots CH_2{=}CH{-}\dot{C}H{-}CH_2{-}CH_2{-}CH{=}CH{-}CH_2^+$$

Where S^* represents an active (electron-accepting) centre formed on the gel (see Chap. 3.6.4.).

The mechanism of γ-irradiation initiated polymerization of 1,3-butadiene in CCl_4 crystalline matrix at 77 K occurs also via the formation of butadiene cation radicals[1974, 1975].

It has been shown[113, 932, 2139, 2140] that free radicals formed during electron irradiation of other diene monomers as isoprene, 1,3-pentadiene, and 2,3-dimethyl-1,3-butadiene are quite different from those produced by chemical free radical initiation or UV-irradiation (Table 4.2).

The ESR spectrum of isoprene irradiated with electrons at 77 K is shown in Figure 4.9. The spectrum consists of seven lines with a hyperfine coupling constant of 14 G and has been attributed to free radical of the allylic type (4.4):

CH₃
|
−CH₂−C—CH=CH₂   (4.4)
.

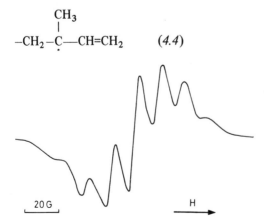

20 G   H

Fig. 4.9. ESR spectrum of isoprene electron-irradiated at 77 K and measured at 77 K[2139]

Table 4.2. The structure of trapped free radicals and the types of reactions of conjugated dienes (assigned from ESR spectra)[2139].

Reaction method	isoprene	pentadiene	dimethyl-butadiene
	$\overset{\overset{\textstyle CH_3}{\mid}}{CH_2=C-CH=CH_2}$ 1 2 3 4	$\overset{\overset{\textstyle CH_3}{\mid}}{CH_2=CH-CH=CH}$ 1 2 3 4	$\overset{\overset{\textstyle CH_3 \; CH_3}{\mid \quad \mid}}{CH_2=C\!-\!\!-\!C=CH_2}$ 1 2
Electron-irradiation	R-addition type (1-carbon) $\overset{\overset{\textstyle CH_3}{\mid}}{RCH_2-\underset{\cdot}{C}-CH=CH_2}$	R-addition type (1-carbon) $\overset{\overset{\textstyle CH_3}{\mid}}{RCH_2-\underset{\cdot}{CH}-CH=CH}$	R-addition type (1-carbon) $\overset{\overset{\textstyle CH_3 \; CH_3}{\mid \quad \mid}}{RCH_2-\underset{\cdot}{C}\!-\!\!-\!C=CH_2}$
Reaction with HO·	abstraction of H· atom $\overset{\overset{\textstyle \cdot CH_2}{\mid}}{CH_2=C-CH=CH_2}$	addition of HO· (3-position) $\overset{\overset{\textstyle CH_3}{\mid}}{CH_2=CH-CH-\underset{\cdot}{CH}}$ $\overset{\mid}{OH}$	no radical observed
Reaction with CH₃·	abstraction of H· atom $\overset{\overset{\textstyle \cdot CH_2}{\mid}}{CH_2=C-CH=CH_2}$	complicated spectra not assigned	addition of CH₃· (1-position) $\overset{\overset{\textstyle CH_3 \; CH_3}{\mid \quad \mid}}{CH_3CH_2-\underset{\cdot}{C}\!-\!\!-\!C=CH_2}$
Ultraviolet-irradiation	loss of H· atom $\overset{\overset{\textstyle \cdot CH_2}{\mid}}{CH_2=C-CH=CH_2}$	addition of R· (3-position) $\overset{\overset{\textstyle CH_3}{\mid}}{CH_2=CH-CH-\underset{\cdot}{CH}}$ $\overset{\mid}{R}$ (B) cyclic structure	loss of H· atom $\overset{\overset{\textstyle \cdot CH_2 \; CH_3}{\mid \quad \mid}}{CH_2=C\!-\!\!-\!C=CH_2}$

Free chemical radical initiation

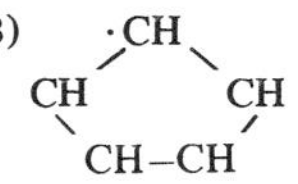

69

The process of the formation of allylic free radicals is probably an ion-molecule reaction involving a cation-radical as a reaction intermediate:

$$CH_2=\overset{\overset{\displaystyle CH_3}{|}}{C}-CH=CH_2 \xrightarrow{\sim\!\sim\!\sim} [CH_2=\overset{\overset{\displaystyle CH_3}{|}}{C}-CH=CH_2]^{\cdot+} + e^- \tag{4.17}$$

$$[CH_2=\overset{\overset{\displaystyle CH_3}{|}}{C}-CH=CH_2]^{\cdot+} + M \longrightarrow {}^+M-CH_2-\overset{\overset{\displaystyle CH_3}{|}}{\underline{C}}\underline{-CH-CH_2} \tag{4.18}$$

$$^+M-CH_2-\overset{\overset{\displaystyle CH_3}{|}}{\underline{C}}\underline{-CH-CH_2} + nM \longrightarrow {}^+M_{n+1}-CH_2-\overset{\overset{\displaystyle CH_3}{|}}{\underline{C}}\underline{-CH-CH_2} \tag{4.19}$$

Similarily the ESR spectra of electron-irradiated 1,3-pentadiene (Fig. 4.10a) and 2,3-dimethyl-1,3-butadiene (Fig. 4.11 a) in crystalline state are different from those obtained in photo-initiated free radical polymerization (Figs. 4.10b) and 4.11b).

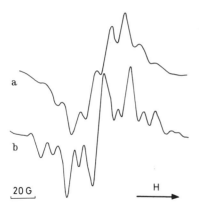

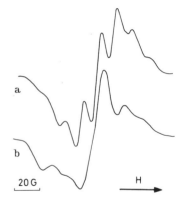

Fig. 4.10. a. ESR spectrum of 1,3-pentadiene electron-irradiated at 77 K and measured at 77 K. b. ESR spectrum of 1,3-pentadiene-tert-butyl-hydroperoxide solution, photo-irradiated at 77 K and measured at 77 K[2139)

Fig. 4.11. a. ESR spectrum of 2,3-dimethyl-1,3-butadiene electron-irradiated at 77 K and measured at 77 K, b. ESR spectrum of 2,3-dimethyl-1,3-butadiene-tert-butyl-hydroperoxide solution, photo-irradiated at 77 K and measured at 77 K[2139)

For 1,3-pentadiene the following ion-radical initiation mechanism has been proposed. An ion-radical is primarily formed by the loss of one π-electron from a monomer molecule by the ionizing radiation. This molecule-ion is then added to a monomer and in this way initiates the cationic polymerization from its ionic end:

$$CH_2=CH-CH=\overset{\overset{\displaystyle CH_3}{|}}{CH} \xrightarrow{\sim\!\sim\!\sim} [CH_2=CH-CH=\overset{\overset{\displaystyle CH_3}{|}}{CH}]^{\cdot+} + e^- \tag{4.20}$$

$$[CH_2=CH-CH=CH]^+ + M \longrightarrow {}^+M-CH_2-\underset{\underline{\quad\quad}}{CH}-\overset{\displaystyle CH_3}{\overset{|}{\underline{CH}}}-\underset{\underline{\quad}}{CH} \qquad (4.21)$$

(with CH₃ groups on the second and fourth carbons)

The radical end of the parent ion-radical, which becomes isolated from the ionic and by the ionic propagation, is thought to be unreactive at 77 K (i.e. of low reactivity compared with the cationic chain ends) and trapped in the solid monomer matrix. The observed radicals are inactive radical residues involved in the ionic propagation process, though their ESR spectra are the same as those expected for a propagating free-radical end in radical polymerization. Primary ion-radicals are not observed by ESR measurements, because ion-radicals are unstable and reactive even at 77 K.

ESR study of polymerization of butadiene in urea canal complexes (clathrates) presents a separate problem. A detailed review of the polymerization of different monomers in clathrates has been published by Rabek[1776, 1778]. The ESR spectrum of γ-irradiated 1,3-butadiene in urea clathrates at 77 K consists of 7 lines (Fig. 4.12) with a hyperfine splitting constant of 15 G[1637].

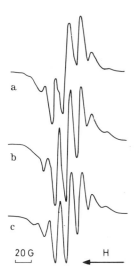

a

b

c

20 G   H

Fig. 4.12. ESR spectra of 1,3-butadiene-urea complexes γ-irradiated at 77 K: a. measured at 77 K immediately after irradiation, b. measured at 77 K after warming to 130 K. c. Computed ESR spectrum of monomer radical[1637]

This spectrum was assigned to monomer radical of the allylic type (4.5)

$$\cdot CH_2-CH=CH-CH_3 \qquad (4.5)$$

which may be produced by hydrogen addition to 1,3-butadiene. Upon warming to room-temperature the spectrum changes into 6 x 2-line spectrum (Fig. 4.13a) ($a_1 = 14$ G and $a_2 = 4$ G), which was attributed to the propagating radical (4.6) occluded in the urea canals.

$$\cdot CH_2-CH=C_\alpha H-C_\beta H_2-R \qquad (4.6)$$

71

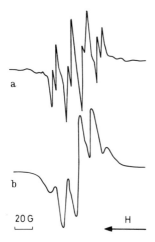

20 G → H

Fig. 4.13. ESR spectra of
1,3-butadiene-urea com-
plexes γ-irradiated at 77 K
after warming to room temperature: a. Room tem-
perature ESR spectrum,
b. ESR spectrum measured
at 77 K[1637]

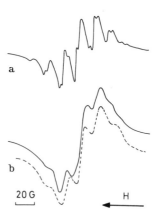

20 G → H

Fig. 4.14. ESR spectra of
1,3-butadiene-urea complexes
γ-irradiated at 77 K after
warming to 400 K: a. Room
temperature ESR spectrum,
b. ESR spectrum measured
at 77 K. Dotted lines indicate
the computed ESR spec-
trum[1637]

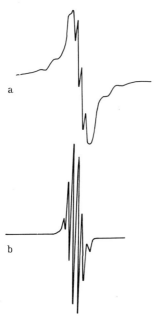

Fig. 4.15. ESR spectra of polycrystalline cyclopentadiene
γ-irradiated at 77 K and successively annealed: a. at 89 K,
and b. at 158 K[642]

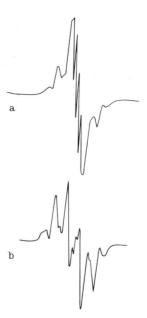

Fig. 4.16. ESR spectra of glassy
cyclopentadiene γ-irradiated
at 77 K and successively annealed: a. at 89 K, and b. at
158 K[642]

On cooling back to 77 K, the 6 x 2-line spectrum is transformed into a 5-line spectrum (Fig. 4.13b) (a = 14 G).

This spectral change is reversible with temperature and it is interpreted in terms of the oscillation of the propagating radical around the C_α–C_β bond. On further warming to 400 K the propagating radical is converted to the "chain" allyl type radical (4.7), the room temperature ESR spectrum of which consists of 7 x 2-lines ($a_1 = 14$ G, $a_2 = 4$ G) (Fig. 4.14).

$$R-CH_2-CH=CH-\overset{\cdot}{C}H-CH_2-R \qquad (4.7)$$

This radical conversion is explained by a hydrogen transfer reaction within the propagating radical through the radical chain and/or between the propagating radical and a neighboring polymer chain. Irradiated 1,3-butadiene-1,1,4,4-d_4 (deuteriated butadiene) in urea canals was also examined and the results confirm the above-mentioned identifications[1637].

4.2.1.3. Cyclic Olefines

ESR studies of γ-irradiated cyclopentadiene at 77 K as polycrystalline (Fig. 4.15) and glassy (Fig. 4.16) samples show the formation of two different radicals.

One of these radicals was identified as a radical of the type (4.8) with five equivalent protons (Fig. 4.15) formed by hydrogen abstraction and the second to a cation radical (4.9) (Fig. 4.16) formed by electron transfer[55, 642, 1375, 1646]. Based on ESR studies of γ-radiation-induced polymerization of cyclic olefins the following mechanisms have been proposed for:

1. crystalline samples:

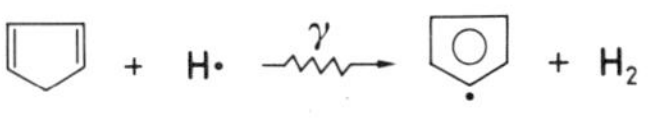

(4.22)

(4.23)

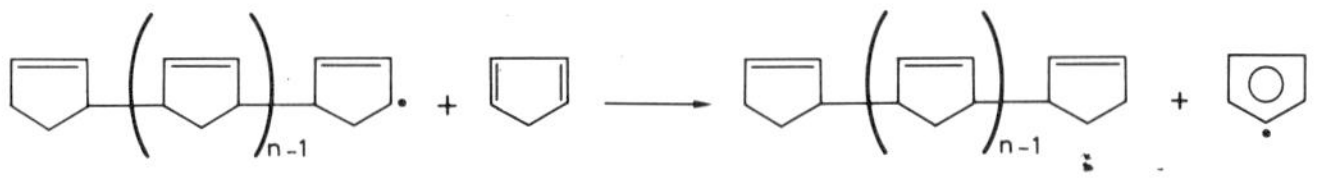

2. glassy samples:

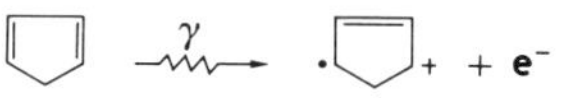

(4.24)

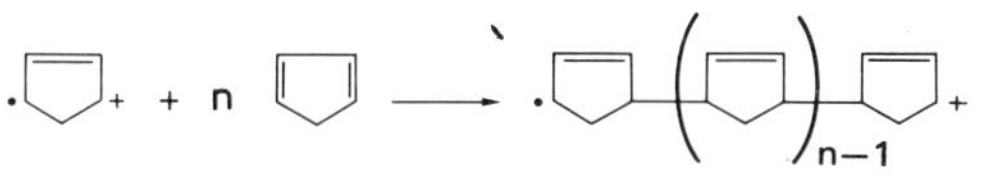

(4.25)

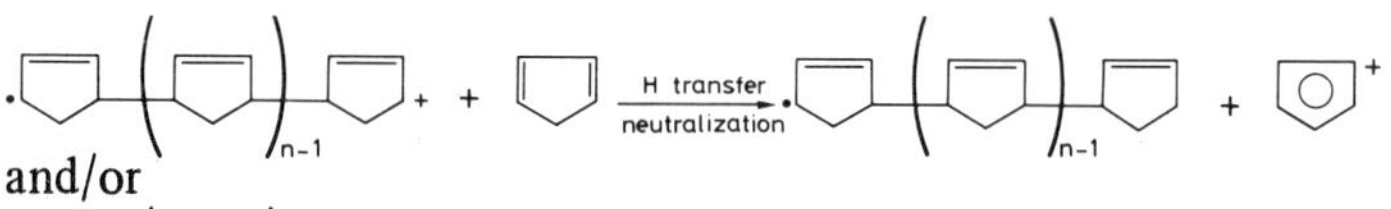

(4.26)

and/or

(4.27)

73

4.2.1.4. Vinyl Monomers

4.2.1.4.1. Acrylic and Methacrylic Acide Monomers. Many workers using ESR spectroscopy have studied the polymerization of acrylic acid[707, 1301, 1993], methacrylic acid[134, 137, 418, 641, 1882–1886], their amides[25, 175, 415, 637, 865, 1527, 1835, 2323, 2325, 2328], and metal salts[251–254, 482, 1301, 1528, 1628] irradiated with γ-radiation in solid-state. Ohnishi et al.[1993] identified the free radicals formed during electron beam or γ-irradiation of a single crystal of acrylic acid at 77 K and made a detailed study of the hyperfine structure of the spectrum, which is shown in Figure 4.17a.

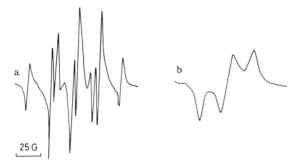

Fig. 4.17. ESR spectra of acrylic acid electron-irradiated at 77 K: a. measured at 77 K, b. after warming to 195 K[1933]

This double quartet spectrum was attributed to radical (*4.10*):

$$CH_3-\overset{\cdot}{C}H-COOH \qquad (4.10)$$

After annealing the sample at 150–200 K the double quartet spectrum decayed to a triplet attributed to a propagation radical (*4.11*):

$$R-CH_2-\overset{\cdot}{C}H-COOH \qquad (4.11)$$

where R is a chain containing an indefinite number of units[707]. Above 250 K an irreversible reaction is observed.

It has been shown[1883–1885] that the free radicals trapped in γ-irradiated polycrystalline methacrylic acid have the same structure as those trapped in solid poly(methacrylic acid) when irradiated with ionizing radiation. This free radical was attributed to propagating radical (*4.12*):

$$R-C_\beta H_2-\overset{\cdot}{C}_\alpha-COOH \qquad (4.12)$$
$$|$$
$$CH_3$$

The lifetime and the ESR spectrum of this radical are different, depending on the matrix which surrounds the radicals. The radicals trapped in the poly(methacrylic acid) are fairly stable even at room-temperature, while the radicals trapped in the crystalline methacrylic acid monomer are stable only at low temperatures and can-

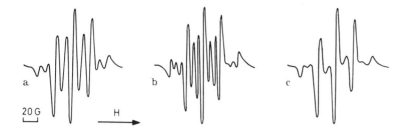

Fig. 4.18. ESR spectra of methacrylic acid γ-irradiated at 77 K: a. measured at 113 K (in phase II), b. measured at 149 K (in phase I), c. ESR spectrum of poly(methacrylic acid) at 298 K[1883]

not be observed near the melting point (200 K). The ESR spectrum of the radicals in the monomer matrix and in the polymer matrix are shown in Figure 4.18. The ESR spectrum of the radicals in the monomer matrix changes with temperature. These radicals have a 9-line (5 + 4) structure in phase II (below the transition point) (Fig. 4.18a) and a 13-line (5 + 8) spectrum in phase I (above the transition) (Fig. 4.18b). It has been shown that the radicals in the monomer matrix have an asymmetrical conformation. The two β-protons of the propagating radical (4.12) have unequal coupling constants, giving the conformational angle of the half-filled p-orbital to the C–H bond as 55° and 65°, respectively. These conformations may depend on the length of the radical chain, the temperature, and the environment.

4.2.1.4.2. Acrylamides. During γ-irradiation of a single crystal of acrylamide at 77 K the free radical (4.13) is only formed and stable up to 148 K[26, 27].

$$CH_3-\overset{\cdot}{C}H-CO-NH_2 \qquad (4.13)$$

The radical (4.13) disappears by addition of monomer molecule and a new radical (4.14) is then formed:

$$CH_3-\underset{\underset{CONH_2}{|}}{CH}-CH_2-\underset{\underset{CONH_2}{|}}{\overset{\cdot}{C}H} \qquad (4.14)$$

which retains a definite direction with respect to the crystallographic axes up to 243 K. Above this temperature the ESR spectrum indicates that polymerization does actually occur and the radicals then exhibit random orientations. Additional information on hyperfine structure of free radicals (4.13) have been obtained from ESR spectra of irradiated polycrystalline acrylamide[415].

A detailed ESR study of a γ-irradiated single crystal of methacrylamide in the temperature range from 77 K to 288 K have been presented in paper[2323]. The number and the spacings of the spectral lines change according to the temperature of irradiation and not to the temperature of measurements. ESR study of γ-irradiated polycrystalline methacrylamide[2325] show that the free radicals stable at room-temperature begin to change their conformation near the melting point.

ESR studies of polymerization of p-vinylbenzamide in solid-state initiated thermally[1710] and cyclopolymerization of N-methyldimethacrylamide[1169] and N-n-propyldimethylacrylamide[1168] initiated with γ-irradiation have also been made.

4.2.1.4.3. Methyl Methacrylate. Some monomers such as methyl acrylate and methyl methacrylate are difficult to polymerize in the crystalline state[429]. The importance of crystal imperfections as reaction sites has been pointed out[135, 138].

ESR spectrum of methyl methacrylate in polycrystalline state γ-irradiated at 77 K in the dark is shown in Figure 4.19.

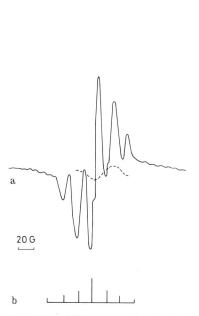

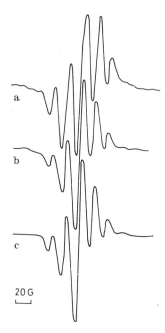

Fig. 4.19. a. ESR spectra of methyl methacrylate γ-irradiated at 77 K, b. Stick plot for the radical (4.15)[2131]

Fig. 4.20. ESR spectra of methyl methacrylate γ-irradiated at 77 K. Photo-bleached with light of wave length: a. > 4000 A, b. > 3000 A and c. > 2800 A[774]

The spectrum is composed of a seven-line component, a broad singlet, and an additional outer group of weak spectral lines[193, 297, 774, 1181, 2131]. The seven line spectrum with a coupling constant of 22 G with the intensity ratio of 1:6:10:20:-10:6:1 was assigned to the initiating radical (4.15) which is formed by the addition of a hydrogen atom to the monomer:

$$
\begin{matrix}
& CH_3 & & & CH_3 \\
& | & & & | \\
CH_2=C & & \xrightarrow[+H\cdot]{\gamma} & CH_3-C\cdot & \\
& | & & & | \\
& COOCH_3 & & & COOCH_3
\end{matrix}
\qquad (4.15) \qquad\qquad (4.28)
$$

76

The spectrum is changed by photo-illumination of the irradiated sample as shown in Figure 4.20.

By illuminating it with light in the wavelength region above 4000 Å, the singlet decays rapidly and the intensity of the outer spectrum with fine structure decreases gradually. The singlet may be attributed to trapped electrons. The outer spectrum, which disappears completely, is assigned to a radical pair[774, 2131] (see Chapter 4.2.1.4.8.). These radical pairs absorb energy and are excited by illumination of light above 3000 Å; then the radicals disappear by combining with each other.

The main spectrum was found to change gradually from the seven-line spectrum to a five-line spectrum by the photo-illumination[2131]. These experimental results indicate that the initiation radicals can be converted to propagating radicals by photo-excitation at 77 K.

The temperature dependence of the ESR spectrum of a photo-bleached sample is shown in Figure 4.21.

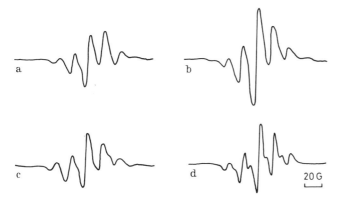

Fig. 4.21. ESR spectra of methyl methacrylate γ-irradiated at 77 K after photo-bleaching: a. 103 K, b. 128 K, c. 168 K and d. 213 K[774]

The intensity of the spectrum decreases with increasing temperature in the region 103 K to 312 K. On the other hand, the seven-line spectrum changes to a nine-line spectrum during the temperature elevation. The nin-line spectrum becomes sharper at 213 K. This means that the propagation reaction takes place slowly during the increase of temperature. In the course of temperature increase of nonbleached samples from 77 K, it was observed that the radical pairs disappear at 213 K. On the other hand, the concentration of initiating radicals was observed to be increased by photo-bleaching of trapped electrons. It seems that new initiating radicals are produced by the neutralization. An ESR study of γ-irradiated polymerization of methyl methacrylate in mineral oils has been presented in paper[1424].

The γ-initiated polymerization of octadecyl methacrylate in solid state has also been extensively studied[252].

4.2.1.4.4. Acrylonitrile and Methacrylonitrile. Radiation-induced solid-state polymerization of these monomers was studied by ESR spectroscopy[153, 154, 193, 194, 196, 461, 772, 1066, 1529, 1657]. ESR spectra of acrylonitrile recorded at 88 K and

Fig. 4.22. ESR spectra of acrylonitrile γ-irradiated at 77 K in the dark and measured: a. at 88 K, b. at 177 K[194]

a   25 G

b

177 K after γ-irradiation in polycrystalline state at 77 K in the dark are shown in Figure 4.22.

When the sample temperature is increased, the observed spectrum is irreversibly transformed. Two out of the five lines progressively disappear. The three-line spectrum, with intensity $1:2:1$ was, attributed to the propagating radical (4.16):

$$R-CH_2-\overset{\cdot}{C}H \quad (4.16)$$
$$\underset{CN}{|}$$

where R is a chain containing an indefinite number of units.

The ESR spectrum of γ-irradiated methacrylonitrile[1529] is a seven-line spectrum which has been attributed to radical (4.17) formed by addition of a hydrogen atom to the vinyl group.

$$CH_3-\overset{\cdot}{C}-CH_3 \quad (4.17)$$
$$\underset{CN}{|}$$

By warming the irradiated monomer to about 223 K the spectrum is changed to a 5(+4)-line spectrum which corresponds to a propagating radical with a very short chain. However, post-polymerization to high molecular weight does not actually take place.

4.2.1.4.5. Vinyl Acetate. Free radicals produced by γ-irradiation of vinyl acetate in the solid-state have been studied by ESR[153, 935]. An oriented solid vinyl acetate sample was prepared for ESR measurements by solidification of vinyl acetate around an aluminium rod at 77 K or 156 K and then γ-irradiated at 77 K. The oriented crystalline solid prepared at 156 K gives an ESR spectrum consisting of a singlet (Fig. 4.23) and additional absorption lines, especially when the irradiation was made in the dark at 77 K (Fig. 4.24).

After irradiation with visible light for 10 min the ESR spectra of oriented crystalline solid vinyl acetate exhibit a distinct hyperfine structure (Fig. 4.25) and a definite anisotropy when the sample is rotated in the magnetic field.

These spectra consist of a quartet with the coupling constant 12.5 G, an anisotropic triplet (coupling constant 33 G*) and four equivalent lines (doublet of doublet with coupling constants 20.8 G and 143 G*) corresponding to the radicals (4.18), (4.19) and (4.20) respectively:

$$CH_3\overset{\cdot}{C}OO, \quad (4.18) \qquad CH_2=CHCOO\overset{\cdot}{C}H_2, \quad (4.19) \qquad CH_2=\overset{\cdot}{C}H \quad (4.20)$$

---

*) Corrected by the authors of this book.

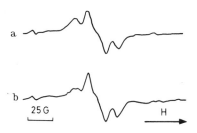

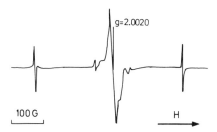

Fig. 4.23. a. ESR spectrum of γ-irradiated crystalline vinyl acetate at 77 K, b. Spectrum obtained when the sample tube was rotated about its vertical axis by 90° from the position where spectrum a was observed[935]

Fig. 4.24. ESR spectrum of γ-irradiated crystalline vinyl acetate in the dark at 77 K[935]

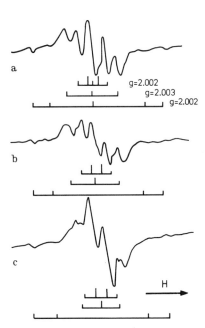

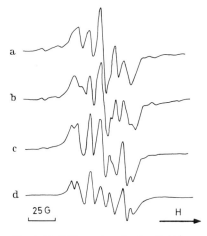

Fig. 4.25. ESR spectra of γ-irradiated crystalline vinyl acetate at 77 K and next 10 min irradiated with visible light. Orientation angle: a. 0°, b. 45°, c. 90°[935]

Fig. 4.26. ESR spectra of γ-irradiated crystalline vinyl acetate at 77 K at orientation angle 0° and next irradiated with visible light for: a. 10 min, b. 60 min, c. 180 min and d. 300 min[935]

Prolonged irradiation with visible light for 300 minutes yields an apparent five-line spectrum, consisting of a pair of triplets (with coupling constants 10 G and 20 G) (Fig. 4.26) which was attributed to the radicals (4.21) and (4.19) respectively.

$\cdot CH_2 CHO$ (4.21)

The ESR spectrum of γ-irradiated glassy solid samples of vinyl acetate prepared at 77 K consists of four strong lines (Fig. 4.27) which are confirmed to be isotropic and a weak doublet splitted by 143 G[2)] corresponding to radical (*4.20*).

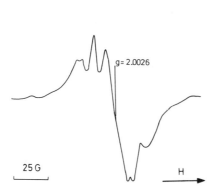

Fig. 4.27. ESR spectrum of γ-irradiated glassy vinyl acetate at 77 K in the dark[935)]

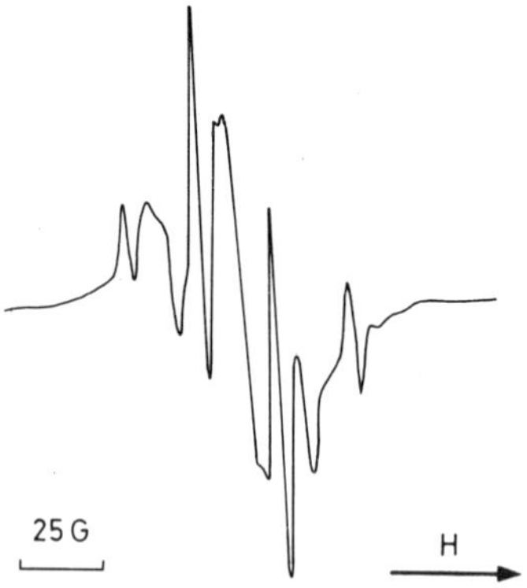

Fig. 4.28. ESR spectrum of γ-irradiated glassy vinyl acetate at 77 K and next 60 min irradiated with visible light[935)]

Irradiation of glassy solid vinyl acetate with visible light for 30 minutes gives a sharp quartet (Fig. 4.28) with a splitting constant 23 G and assigned to methyl radical (*4.22*) produced by further photolysis of radical (*4.18*).

·CH$_3$ (*4.22*)

A broad triplet with a coupling constant 22 G becomes observable on elevation of the temperature to 113 K and remains unchanged up to 173 K. This triplet was assigned to radical (*4.19*). The ESR studies indicate that both the crystalline and the glassy vinyl acetate produce the same radicals on irradiation.

4.2.1.4.6. Styrene. Solid-state polymerization of styrene initiated with γ-irradiation in polycrystalline form[2260)] gives a broad ESR spectrum as shown in Figure 4.29 a. After warming up the sample to about 221 K a five-line spectrum with a coupling constant of about 16 G is observed (Fig. 4.29 b). This spectrum was attributed to free radicals (*4.23*) formed by hydrogen addition to the vinyl group.

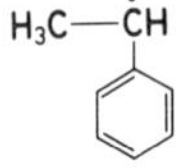

H$_3$C—ĊH   (*4.23*)

The coupling constant of the methyl protons is as low as about 16 G, and that of the α-proton has also about the same value due to the delocalization of the free spin. Therefore a five-line spectrum is expected.

4.2.1.4.7. N-Vinylcarbazole. When single crystal of N-vinylcarbazole is γ-irradiated at 77 K, the ESR spectrum observed before warming consists of three peaks (Fig. 4.30a)[118].

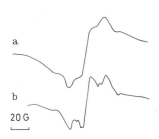

Fig. 4.29. ESR spectra of γ-irradiated styrene at 77 K and measured: a. at 77 K and b. at 221 K[2260]

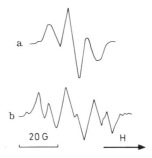

Fig. 4.30. ESR spectra of γ-irradiated single crystal N-vinylcarbazole at 77 K and measured: a. at 77 K, and b. at 291 K[118]

This spectrum has been attributed to a radical-cation with the unpaired spin associated mainly with the nitrogen atom (*4.24*).

$$(4.24) \qquad (4.25)$$

Above 90 K polymerization initiated by the cation occurs and the spectrum changes into four lines (1:3:3:1) at certain orientations and into six lines (1:1:2:2:1:1) at others (Fig. 4.30b). This spectrum may be assigned to an alkyl radical (*4.25*), trapped in the polymer[108, 118]. Unfortunately the ESR spectra give no information about the polymerization, since the unpaired spin is left on the nonactive end of the chain (*4.26*):

$$(4.26)$$

The radical is presumably trapped at the chain end after termination has occurred at the growing end by charge neutralization or proton transfer.

4.2.1.4.8. Radical Pairs in γ-Irradiated Vinyl Monomers. Pairwise trapping of radicals has been reported for many organic substances[148, 328, 584, 745, 1255, 1256, 1258, 1319]. The theory of ESR spectra of radical pairs is discussed in several papers[774, 793, 923, 1255–1257, 1319, 1449, 2209, 2421, 2516].

The ESR spectra of radical pairs in γ-irradiated vinyl monomers in polycrystalline or single crystal state consist of series of outer lines. Radical pairs are observed as $\Delta M_S = 1$ and $\Delta M_S = 2$ transitions.

As mentioned in Chapter 4.2.1.4.3. γ-irradiation of polycrystalline methyl methacrylate gives a spectrum with seven broad lines and additional outer groups of spectral lines forming the pairs marked (B_1, B_2) and (B_3, B_4) (Fig. 4.31).

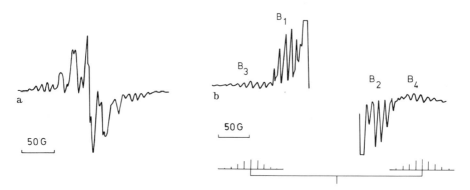

Fig. 4.31. a. ESR spectra of γ-irradiated single crystal of methyl methacrylate at 77 K, b. spectrum with the same orientation and at higher gain[774]

A pair of radicals gives a spectrum consisting of two groups of 13 lines with a hyperfine coupling constant of 11.2 G. The distance between the radicals forming the pairs is between 5.45 and 6.3 Å[774] or less than 6.67 Å[2131].

γ-irradiation of a single crystal of methyl acrylate showed a spectrum consisting of five main lines with a splitting of 22 G[772, 774] (Fig. 4.32).
Beside the central five-line spectrum there are four groups of lines forming the pairs (B_1, B_2) and (B_3, B_4). The mean value of the hyperfine coupling constant is 10.8 G. The distance between the radicals in one of the pairs was calculated to be 5.9 Å.

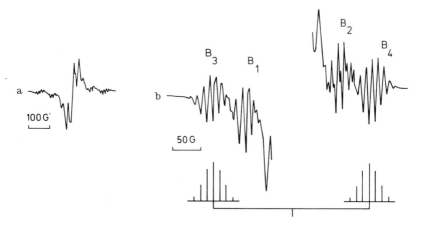

Fig. 4.32. a. ESR spectra of γ-irradiated single crystal of methyl acrylate at 77 K, b. spectrum with the same orientation and at higher gain[774]

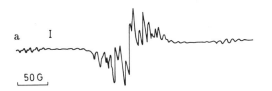
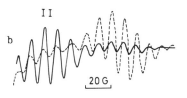

Fig. 4.33. a. ESR spectra of γ-irradiated single crystal and methacrylonitrile at 77 K and measured at 77 K, b. Dotted line — right wing of spectrum (a) at higher gain, solid line — after 25 min visible light irradiation (4750 A)[772]

γ-irradiation of partially crystalline[1529] and a single crystal[772] of methacrylonitrile gives the formation of radical pairs. In Figure 4.33 the ESR spectrum of methacrylonitrile single crystal γ-irradiated at 77 K is shown.

The complex central part of this spectrum (Fig. 4.33) was not analyzed, but on the wings there are two strongly anisotropic groups of lines assigned to radical pairs. The hyperfine coupling constant within each group was measured and found to be 10.8 ± 2 G, and the maximum splitting between the outermost groups was 351 G, which corresponds to a radical distance r < 5.4 Å. When the crystal was illuminated with red light (>4750 Å), the radical pair (I) was slowly bleached and a new pair (II) appeared with its spectrum placed closer to the central spectrum, as illustrated in Figure 4.33b. The maximum splitting of pair II, 207 G corresponds to r < 6.45 Å. Further rotation of a bleached crystal showed that yet another radical pair (III) was present with a maximum splitting of 228 G, which gives a calculated separation r < 6.25 Å between the radicals. For the pairs II and III, six lines were observed on the high-field side of the central line in the radical pair spectra as shown in Figure 4.34.

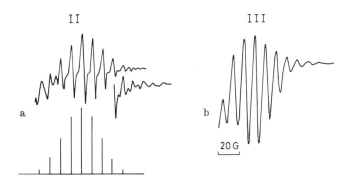

Fig. 4.34. ESR spectra from: a. radical pairs II, b. radical pairs III in a bleached methacrylonitrile single crystal measured at 77 K[772]

A detailed analysis of these ESR spectra is given in the original paper[772].

The ESR spectra of an irradiated acrolein single crystal are very anisotropic and for most orientations exhibit an asymmetric profile[774]. A polycrystalline sample gave an asymmetric poorly resolved spectrum. The $\Delta M_S = 2$ transition observed in a polycrystalline sample at 77 K consists of four lines with a separation of 9.8 G.

4.2.1.5. Cyclic Monomers

ESR spectra of 3,3-bis(chloromethyl)oxethane γ-irradiated at 77 K (Fig. 4.35) consists of a triplet (coupling constant 24 G) which was superimposed by a doublet spectrum (coupling constant 15 G)[2254].

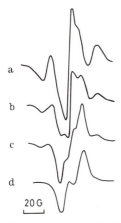

a

b

c

d

20 G

Fig. 4.35. Change of ESR spectra of electron irradiated 3,3-bis(chloromethyl)oxetane with the increasing temperature: measured at: a. 77 K, b. 117 K, c. 184 K, d. 252 K[2254]

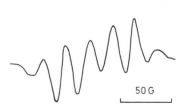

50 G

Fig. 4.36. ESR spectrum of γ-irradiated n-butyl chloride glass at 77 K[2458, 2462]

As the temperature was raised, the triplet spectrum disappeared at about 203 K and only the doublet spectrum was observed, which disappeared at about 288 K. Detailed analysis of the ESR spectra shows that polymerization occurs via an ion radical mechanism with participation of cation-radicals (*4.27–4.30*):

$$ClH_2C-C\overset{\displaystyle CH_2Cl}{\underset{\displaystyle O}{\bigsqcup}} \xrightarrow{\gamma} ClH_2C-C\overset{\displaystyle CH_2Cl}{\underset{\displaystyle O^{\cdot+}}{\bigsqcup}} + e^- \qquad (4.27) \qquad\qquad (4.29)$$

$$ClH_2C-C\overset{\displaystyle CH_2Cl}{\underset{\displaystyle O^{\cdot+}}{\bigsqcup}} \longrightarrow ClH_2C-\underset{\displaystyle \underset{\displaystyle CH_2-O^+}{\overset{\displaystyle CH_2Cl}{|}}}{\overset{\displaystyle }{C}}-CH_2 \; \text{ or } \; ClH_2C-\underset{\displaystyle \underset{\displaystyle CH_2-O^{\cdot}}{\overset{\displaystyle CH_2Cl}{|}}}{\overset{\displaystyle }{C}}-CH_2^+ \qquad (4.30)$$

$$(4.28) \qquad\qquad\qquad (4.29)$$

$$CIH_2-\underset{\underset{CH_2-O\cdot}{|}}{\overset{\overset{CH_2Cl}{|}}{C}}-CH_2^+ \longrightarrow CIH_2C-\underset{\underset{\underset{\cdot}{CH-OH}}{|}}{\overset{\overset{CH_2Cl}{|}}{C}}-CH_2^+ \qquad (4.30) \qquad\qquad\qquad (4.31)$$

The ESR spectra of γ-irradiated single crystal of tetraoxane at 77 K show the formation of two types of radicals: cation-radical (4.31) and free radicals (4.32–4.35)[1873]:

$$\begin{array}{cc}
CH_2-O-CH_2 & CH_2-O-CH_2 \\
| \quad\quad\; | & | \quad\quad\; | \\
O \quad\quad O^{\cdot+} & O \quad\quad O \\
| \quad\quad\; | & | \quad\quad\; | \\
CH_2-O-CH_2 & CH_2-O-\underset{\cdot}{C}H
\end{array} \qquad -CH_2-\overset{\cdot}{O} \quad -\overset{\cdot}{C}H_2 \quad -\overset{\cdot}{C}H-$$

$$(4.31) \qquad\qquad (4.32) \qquad\qquad\qquad (4.33) \qquad (4.34) \qquad (4.35)$$

The ESR spectra of all radicals (4.31)–(4.35) are complicated to interprete because they depend on the position of the single crystal of tetroxane in the magnetic field.

4.2.2. Polymerization in Organic Glass Matrices

4.2.2.1. ESR Studies of Ionic Processes in γ-Irradiated Organic Glasses

During γ-irradiation of some frozen organic glasses radical cations and electrons (anionic intermediates) are formed. Electrons are directly trapped in the frozen glassy state at 77 K[858, 1906, 2350, 2410] and 4.2 K[2033, 2461].

The following organic glasses have been examined:

1. *n-Butylchloride*[119, 2027, 2458, 2462]. After γ irradiation at 77 K n-butyl-chloride shows a six-line spectrum (Fig. 4.36) due to n-butyl radicals. In this glassy state, the liberated electrons are captured by n-butylchloride molecules and the anionic intermediates are stabilized as Cl^-. Therefore, primary cationic intermediates, i.e. cation radicals of n-butylchloride, have a long lifetime, and they can migrate in n-butylchloride glass probably by positive charge transfer.

2. *3-Methylpentane*[616, 726, 996, 1007, 1012, 1118, 1119, 1369, 1371, 1373, 1997, 1999, 2289, 2292, 2295, 2303, 2458]. The ESR spectrum of γ-irradiated 3-methylpentane in glassy state is shown in Figure 4.37[1007].

It is composed of a broad six-line spectrum, which was attributed to 3-methyl-pentyl radicals and a central sharp spectrum assigned to the trapped electrons. The later spectrum is characterized by g = 2.0025 ± 0.0003 and the width 3.7 G. Trapped electrons are readily bleached out by room lights and the signal shape changes as shown in Figure 4.37 (*dotted line*). Trapped electrons rapidly disappear even at 77 K (Fig. 4.38).

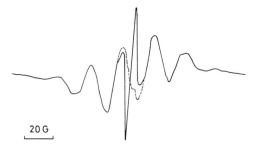

Fig. 4.37. ESR spectra of γ-irradiated 3-methylpentane glass at
77 K (solid line) and after photo-
bleaching by visible light (dotted
line)[1007]

20 G

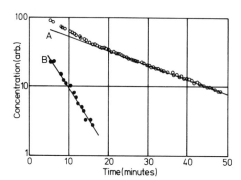

Fig. 4.38. The decay of the intensity of
the sharp ESR line of the trapped electrons
in γ-irradiated 3-methyl pentane glass at
77 K: ○ − observed value and ● − after
substraction of the value corresponding
to the stright line A from the observed
values. Line A represents the slowly disappearing electrons and line B represents
the fast disappearing electrons[2458]

It is seen from Figure 4.38 that this disappearance is described by the superposition
of two first-order reactions[1007, 2458]. The first-order decays indicate that the electrons recombine with their counterpart cations.

Trapped electrons may be classified into two groups[1007]:

1. Those which are stabilized in shallow traps and decay rapidly.

2. Those which are trapped deeply and decay slowly. The recombination of the
slowly disappearing electrons with positive charges is interpreted as being brought
about mainly by the migration of the positive charge, while the recombination of the
fast disappearing electrons should be caused by the migration of the electrons in the
3-methylpentane glass.

No evidence for cation radicals of 3-methylpentane has been obtained by ESR
spectroscopy.

3. *2-Methyltetrahydrofuran*[497, 498, 885, 996, 1630, 1887, 1999, 2031, 2032, 2458, 2462].
During γ-irradiation of 2-methyltetrahydrofuran at 77 K primary cation radicals
(*4.36*) and free electrons are formed.

$$\text{(diagram)} \quad \xrightarrow{\gamma} \quad \left(\text{(diagram)}\right)^{\overset{\cdot}{+}} +e^- \qquad (4.36) \qquad\qquad (4.32)$$

The ESR spectrum of γ-irradiated 2-methyltetrahydrofuran at 77 K is shown in
Figure 4.39.

86

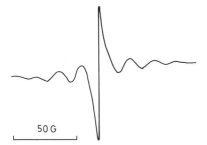

50 G

Fig. 4.39. ESR spectrum of γ-irradiated 2-methyl-tetrahydrofuran glass at 77 K[2458, 2462]

It is composed of a sharp single line due to trapped electrons and a broad seven-line spectrum due to free radicals formed by intramolecular hydrogen transfer in the primary cation radicals (4.33), and also by intermolecular proton transfer between the primary cation radicals and neighboring 2-methyltetrahydrofuran (4.34):

$$(4.33)$$

$$(4.34)$$

Cationic intermediates are stabilized in the vicinity of their native site, and only the electrons migrate in the glass until they are trapped. The decay of trapped electrons at 77 K in the dark is very slow and this fact suggests that cation intermediates are trapped through the reactions (4.33) and (4.34), while the electrons are also trapped by the electric dipole of 2-methyltetrahydrofuran molecules and the charge recombination reaction is suppressed[2458]. Thermal decay of trapped electrons is reasonably well understood as being due to neutralization by positively charged entities which are not identified at present. The first-order decay of trapped electrons suggests that each electron reacts with a predestined partner. According to Smith and Pieroni[2031], trapped electrons and free radicals or radical ions are distributed inhomogeneously, with local concentrations of these species 11 ± 5 times higher than the bulk concentration estimated on the assumption of homogeneous distribution. If this "spur model" of initial distribution is correct, neutralization reactions apparently occur in the spur.

The thermal decay of trapped electrons in 2-methyltetrahydrofuran has also been studied by means of rapid-scan measurements[1630]. The decay curve of trapped electrons was found to be a superposition of two reactions having different first-order rate constant (Fig. 4.40).

These rate constants were determined to be 2.2×10^{-1} s^{-1} and 4.5×10^{-2} s^{-1} at 93 K for the fast and the slow reactions respectively. The fast and slow decay reactions of trapped electrons were ascribed to two kinds of positive entities in the spurs[1630]. The mechanism of the electron trapping was also studied at 4.2 K[2033, 2461], and under very high doses of irradiation[1998]. Application of ENDOR spectroscopy

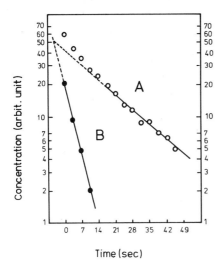

Fig. 4.40. The decay of the trapped electrons at 93.1 K in γ-irradiated 2-methyl-tetrahydrofuran glass at 77 K: o -- observed value and • − remainder after subtraction of the corresponding to the straight line A[1630]

may elucidate the details of the trapping environment. Matrix ENDOR lines or protons associated with trapped electrons in γ-irradiated glassy matrices of 2-methyltetrahydrofuran have been observed and interpreted[900].

It is important to point out that carbon dioxide is an efficient electron scavenger in γ-irradiated and photoionized organic glasses[1062, 1370, 1373, 2289, 2290, 2291].

4.2.2.2. ESR Studies of Ionic Polymerization in Organic Glass Matrices

In the last decade several ESR studies have been devoted to the radiation-induced polymerization of monomer-solvent mixture which form glasses at low temperatures and have been reviewed in detail by Hayashi and Yoshida[883, 2458]. The ionic radical formation of the following monomers have been studied: nitroethylene[884, 1656, 2293, 2301, 2445, 2458], isobutene[884, 1656, 2264, 2300], dienes[1395, 1965, 1971, 1974, 1975], vinyl ethers[883, 1008, 1009, 2458], acrylic monomer[771, 777, 1027], acrylonitrile[238], styrene[883, 1023, 2260, 2299, 2458, 2462] and α-methylstyrene[2457, 2462].

When an organic glass matrix (S) (see Chap. 4.2.2.1.) is γ-irradiated, the primary reaction is the ionization of organic glass with the formation of electrons and radical cations (Fig. 4.41).

Most of the electrons and radical cations recombine immediately, but some of them may react further. If the organic glass, e.g. 2-methyltetrahydrofuran or n-butyl chloride contain a small amount of monomer (M), the secondary reaction of electrons with monomer gives radical anions which may initiate anionic polymerization. The radical cation of organic glass can undergo positive charge transfer or proton transfer to monomer and in both cases may initiate a cationic polymerization (Fig. 4.41).

In 3-methylpentane glass, the anionic reaction and/or the cationic reaction may occur, depending on the nature of the monomer. Considering the properties of organic glasses, we can assume that 2-methyltetrahydrofuran is useful for studies of the an-

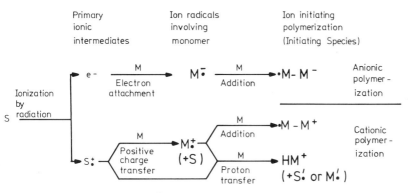

Fig. 4.41. Primary reactions occurring in γ-irradiated organic glasses[2458]

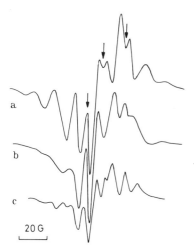

Fig. 4.42. ESR spectra of 2-methyltetrahydro-furan glasses containing nitroethylene with concentration of: a. 1.5 mole %, b. with 28 mole %, c. sample b after warming to 123 K, Arrows indicate the spectrum due to added nitro-ethylene[2458]

ionic reaction of the dissolved monomers, whereas n-butyl chloride glass would be useful for the cationic reactions.

4.2.2.2.1. Nitroethylene.

γ-Irradiation of a small amount of nitroethylene (1.5 mole %) in 2-methyltetrahydrofuran glass gives an ESR spectrum as shown in Figure 4.42a[885, 2293, 2301].

The comparison between this signal and the signal of pure 2-methyltetrahydrofuran glass, shows that the solute nitroethylene captures the electrons, eliminating the sharp electron spectrum. New anionic intermediates are formed, which are paramagnetic and give a new ESR spectrum, indicated in Figure 4.42 by arrows.

γ-Irradiation of a larger amount of nitroethylene (28 mole %) in 2-methyltetra-hydrofuran glass, gives an ESR spectrum as shown in Figure 4.42b. When the temperature is increased, the seven-line spectrum, due to the free radicals formed from 2-methyltetrahydrofuran, disappears, leaving a new spectrum. This spectrum has the hyperfine structure due to three protons with the hyperfine constant of 16 G and a nitrogen nucleus with the coupling constant 8 G. The hyperfine spectrum suggests that it is due to the anion radicals of nitroethylene. The anion radicals disappear when the temperature of the glass is increased to 133 K. The spectrum of the anion radicals is bleached by visible light by photo-detachment of electrons from the anion radicals. Because of the low concentration of nitroethylene and its low mobility in the glass matrices, the degree of polymerization will be low, about $210^{2301)}$. Another mechanism for the initiation process can be proposed. Free electrons are formed primarily from solvent molecules by irradiation (4.32) and are then readily captured by nitroethylene molecules, giving anion radicals of nitroethylene which are primary intermediates of the monomer (4.35):

$$CH_2=CH + e^- \longrightarrow \left(CH_2=CH \atop NO_2\right)^{\cdot-} \quad (4.37) \qquad (4.35)$$

The reaction (4.35) is favored by the strong electron affinity of nitroethylene. The carbanion (4.38) may be formed by an ion-molecule reaction between the anion radicals (4.37) and nitroethylene molecules (4.36). The nitroethylene is successively added. The polymerization proceeds by anionic propagation $(4.37)^{2458)}$.

$$\left(CH_2=CH \atop NO_2\right)^{\cdot-} + CH_2=CH \atop NO_2 \longrightarrow \cdot CH_2-CH-CH_2-CH^- \atop NO_2 \quad NO_2 \qquad (4.38) \qquad (4.36)$$

$$-CH_2-CH^- + CH_2=CH \longrightarrow -CH_2-CH-CH_2-CH^- \atop NO_2 \quad NO_2 \qquad NO_2 \qquad NO_2 \qquad (4.37)$$

ESR studies have not given evidence of reaction (4.36) which is supported by mass-spectroscopy$^{2446)}$.

4.2.2.2.2. Isobutene. Isobutene γ-irradiated in 3-methylpentane glass at 77 K in the dark gives the ESR spectra presented in Figure 4.43.
The spectrum consist of an eight-line component with the hyperfine separation of about 23 G, a 15-line component with 11.5 G separation and a sharp singlet spectrum. With a decreasing fraction of isobutene in the mixture, the 15-line spectrum reduces its intensity relative to that of the eight-line spectrum. The singlet spectrum is bleached by visible light (Fig. 4.43b *dotted line*). These spectra were attributed to cation-radicals $(4.40)^{2300)}$, which can be formed by the following mechanism: When pure 3-methylpentane (MP) is γ-irradiated, the ejected electrons recombine readily with their parent cations (4.39) because they are trapped unstably (4.38). When isobutene is added to 3-methylpentane, the parent cations transfer their positive charge to isobutene (IB) (4.39).
In further reaction (4.40) cation-radicals (4.40) are formed:

$$MP \xrightarrow{\gamma} (MP)^{\cdot+} + e^- \quad (4.39) \qquad (4.38)$$

$$(MP)^{\cdot+} + IB \longrightarrow MP + (IB)^{\cdot+} \qquad (4.39)$$

90

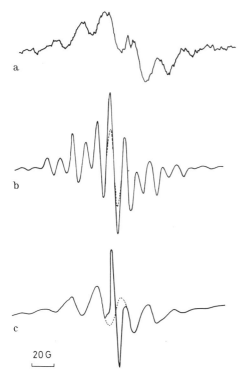

a

b

c

20 G

Fig. 4.43. ESR spectrum of γ-irradiated isobutene solid in the dark
a. at 77 K and measured at 77 K,
b. ESR spectrum of γ-irradiated
3-methylpentane and isobutene
(2:1) at 77 K before bleaching
(solid line) and after bleaching
with visible light (dotted line),
c. ESR spectrum of γ-irradiated
3-methylpentane and isobutene
(71:1) at 77 K[2300)

$$(IB)^{+} + IB \longrightarrow \cdot C(CH_3)_2 - CH_2 - R^{+} \qquad (4.40) \qquad\qquad (4.40)$$

Polymerization proceeds from the cation radicals (4.40) by cationic mechanism.

4.2.2.2.3. 1,3-Butadiene. 1,3-Butadiene γ-irradiated in butyl chloride (BuCl) glass at
77 K gives ESR spectrum shown in Figure 4.44[1965, 1974)].

This spectrum has been attributed to butadiene cation-radicals (4.41) which are responsible for the cationic polymerization mechanism (4.44):

$$BuCl \xrightarrow{\gamma} BuCl^{+} + e^{-} \qquad\qquad (4.41)$$

$$BuCl + e^{-} \longrightarrow \dot{C}H_2 - CH_2 - CH_2 - CH_3 + Cl^{-} \qquad\qquad (4.42)$$

$$BuCl^{+} + CH_2{=}CH{-}CH{=}CH_2 \longrightarrow BuCl + (CH_2{=}CH{-}CH{=}CH_2)^{+} \qquad (4.41) \ (4.43)$$

$$(CH_2{=}CH{-}CH{=}CH_2)^{+} + CH_2{=}CH{-}CH{=}CH_2 \longrightarrow CH_2{=}CH{-}\dot{C}H{-}CH_2{-}CH_2{-}CH{=}CH{-}\overset{+}{C}H_2$$

$$(4.42) \ (4.44)$$

The successive addition of butadiene monomers to the cationic end produces the
polymer of butadiene, yielding an allyl-type radical (4.42) at the other end of the
polymer chain.

20 G   20 G

Fig. 4.44. ESR spectrum of γ-irradiated n-butyl chloride glass containing 1,3-butadiene (with concentration of 2.4 mol %) at 77 K
and measured at 100 K[1974]

Fig. 4.45. ESR spectrum of γ-irradiated 2-methyltetrahydrofuran
glass containing 1,3-butadiene
(3.0 mol %) in the dark at 77 K
and measured at 77 K[1974]

The ESR spectrum of 1,3-butadiene γ-irradiated in the 2-methyltetrahydrofuran glass at 77 K is shown in Figure 4.45[1965, 1974].

This spectrum has been assigned to the butadiene anion radical (*4.43*) which can be formed by reaction (4.45):

$$\text{(structure)} \xrightarrow{\gamma} \left(\text{(structure)}\right)^{+} + e^{-} \tag{4.32}$$

$$CH_2=CH-CH=CH_2 + e^{-} \longrightarrow (CH_2=CH-CH=CH_2)^{-} \quad (4.43) \tag{4.45}$$

Butadiene anion radical (*4.43*) is responsible for an anionic polymerization mechanism (4.46):

$$(CH_2=CH-CH=CH_2)^{-} + CH_2=CH-CH=CH_2 \longrightarrow CH_2=CH-\dot{C}H-CH_2-CH_2-CH=CH-\bar{C}H$$
$$(4.44) \qquad (4.46)$$

In the reaction (4.46) the allyl-type radical (*4.44*) is formed.

The ESR spectrum of 1,3-butadiene γ-irradiated in 3-methylpentane (MP) glass at 77 K is shown in Figure 4.46[1974].

20 G

Fig. 4.46. ESR spectrum of γ-irradiated
3-methylpentane glass containing 1,3-
butadiene-1,1-4,4-d$_4$ (1.7 mole %) in the
dark at 77 K and measured at 77 K[1974]

This spectrum was attributed to the cation-radical (*4.41*) formed by the following mechanism:

$$MP \xrightarrow{\gamma} MP^{\cdot+} + e^{-} \tag{4.38}$$

$$CH_2=CH-CH=CH_2 + MP^{\cdot+} \xrightarrow{\prime} (CH_2=CH-CH=CH_2)^{\cdot+} + MP \quad (4.41) \tag{4.47}$$

Butadiene cation-radicals (*4.41*) add to monomers, yielding the allyl-type radical (*4.42*) in reaction (4.44). It is seen that the type of propagation, cationically (4.44) or anionically (4.46), depends on the type of organic glass matrix in which the reaction occurs under γ-irradiation.

4.2.2.2.4. Vinyl Ethers. Methyl vinyl ether γ-irradiated in 3-methylhexane glass at 77 K gives ESR spectrum shown in Figure 4.47[1009].

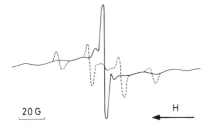

Fig. 4.47. ESR spectrum of -irradiated 3-methylhexane glass containing methyl vinyl ether (9.1 mole %) at 77 K. Solid line presents spectrum after 10 min after irradiation and broken line after photo-bleaching of the trapped electrons[1009]

This spectrum contains a sextet assigned to 3-methylhexyl radicals[2290]. A broad central spectrum is due to cation-radicals of methyl vinyl ether, and a sharp singlet spectrum to trapped electrons, similar to the case of γ-irradiation of methyl vinyl ether in 3-methylpentane glass at 77 K (Fig. 4.48)[1008].

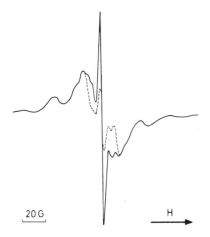

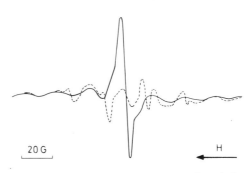

Fig. 4.48. ESR spectrum of γ-irradiated 3-methylpentane glass containing methyl vinyl ether (8 mole %) at 77 K. Solid line presents spectrum 10 min after irradiation and dotted line after 1 hr of infrared irradiation[1009]

Fig. 4.49. ESR spectrum of γ-irradiated 2-methyltetrahydrofuran glass containing methyl vinyl ether (8.0 mole %) at 77 K. Solid line presents spectrum 10 min after irradiation, dotted line after photo-bleaching of the trapped electrons[1009]

Methyl vinyl ether γ-irradiated in 2-methyltetrahydrofuran glass at 77 K gives ESR spectrum shown in Figure 4.49[1009].

The septet-line spectrum and a sharp singlet spectrum have been attributed to free radicals formed by intramolecular hydrogen transfer in the primary 2-methyltetrahydrofuran cation-radicals *(4.36)* (see reaction 4.33) and trapped electrons, respectively. There are no cation-radicals of methyl vinyl ether formed in this glass matrix.

The yield of trapped electrons increases remarkably with increasing concentration of methyl vinyl ether in 3-methylhexane glass (a nonpolar solvent), while the increase is insignificant in 2-methyltetrahydrofurane glass (a polar solvent). The rate of photobleaching of the trapped electrons with visible light decreases with the addition of methyl vinyl ether to 3-methylhexane, while it increases in 2-methyltetrahydrofuran. The bleaching of trapped electrons results in the formation of methyl radicals by a dissociative electron capture process:

$$CH_3OCH=CH_2 + e^- \longrightarrow \cdot CH_3 + CH_2=CHO^- \qquad (4.48)$$

The dissociative electron capture process of methyl vinyl ether occurs when the electrons are liberated from their physical traps by visible light. This is interpreted to mean that the cross section of methyl vinyl ether for physical electron trapping is larger than that for dissociative electron capture. Under the visible light illumination, the electrons are repeatedly detrapped and trapped and finally captured by methyl vinyl ether to give methyl radicals[1009].

Similar ESR studies have been made with n-butyl vinyl ether γ-irradiated in 3-methylpentane glass matrix at 77 K[1008, 2458].

4.2.2.2.5. Acrylic Acid. An interesting ESR study has been made of the reaction of electrons with acrylic acid in glass matrices having no polar proton such as 2-methyltetrahydrofuran, triethylamine, and 3-methylhexane, and in solvents having polar protons such as alcohols, alkaline ice (8 N NaOH-frozen), and propionic acid[1027].

ESR spectra of acrylic acid γ-irradiated in the frozen glasses 2-methyl-tetrahydrofuran (Fig. 4.50) and triethylamine (Fig. 4.51) show formation of a three-line spectrum which was attributed to the anion radical *(4.45)*.

$$[CH_2=CHCOOH]^- \qquad (4.45)$$

It is assumed that the acrylic acid anion radical *(4.45)* is a delocalized π radical having the resonance structure *(4.46)*:

$$CH_2=CH-\overset{\cdot}{C}\begin{array}{c} O^- \\ \diagup \\ \diagdown \\ OH \end{array} \rightleftharpoons \overset{\cdot}{C}H_2-CH=CH\begin{array}{c} O^- \\ \diagup \\ \diagdown \\ OH \end{array} \qquad (4.46)$$

where the unpaired electron mainly occupies the p orbital of the C atoms in the end COOH or CH_2 groups. The 1:2:1 three-line feature of the spectrum with the coupling 8.5 G indicates that it is due to a species having two equivalent protons. The outer two lines of the three-line spectrum exhibit familiar line shape arising from the hyperfine anisotropy of α-proton couplings. Figure 4.50d shows the simulated spectrum for the $-\dot{C}H_2 \pi$ radical assuming the typical anisotropy of α proton coupling with spin a density of 0.5 on the radical carbon atom.

A frozen methanol glass containing 5% water and ca. 2 mol % acrylic acid γ-irradiated at 77 K did not give the ESR signal of the anion radical *(4.45* or *4.46)* or of

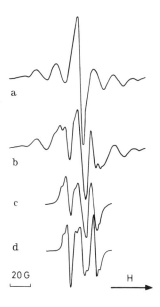

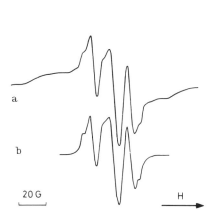

Fig. 4.50. ESR spectra of γ-irradiated: a. 2-methyltetrahydrofuran glass at 77 K, b. 2-methyltetrahydrofuran glass containing acrylic acid (2 mole %) at 77 K, c. Difference spectrum obtained by substracting the spectrum a from spectrum b, d. Computed spectrum[1027]

Fig. 4.51. ESR spectra of γ-irradiated: a. triethylamine glass containing acrylic acid (4 mole %), b. Difference spectrum obtained by substracting the spectrum of the solvent from spectrum a[1027]

the trapped electron. Instead of these the superimposed spectra attributed to hydrogen addition radical (4.47) and to solvent radical were found as shown in Figure 4.52a.

$$CH_3-\dot{C}H-COOH \qquad (4.47)$$

The ESR spectrum obtained from γ-irradiated alkaline ice (8 N NaOH-frozen) containing ca. 3 mol % acrylic acid is shown in Figure 4.52b. The spectrum is essentially the same as that of γ-irradiated pure acrylic acid, indicating an efficient yield of hydrogen addition radical. This spectrum was attributed to the dissociated (basic) form of free radical (4.48):

$$CH_3-\dot{C}H-COO^- \qquad (4.48)$$

Rapid protonation in alcohol and alkaline ice matrices may be interpreted by two alternative mechanism: direct proton transfer from surrounding solvent molecules (4.49 and 4.51) or proton transfer from the protonated cation of the solvent (4.50 and 4.52).

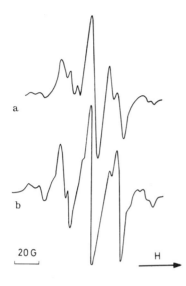

a

b

20 G

H

Fig. 4.52. ESR spectra of γ-irradiated at 77 K: a. frozen methanol solution containing water (5 mole %) and acrylic acid (2 mole %), b. a frozen alkaline (8 N NaOH) aqueous solution containing acrylic acid (3 mole %)[1027]

In the alcohol matrices:

$$[CH_2=CH-COOH]^- + ROH \longrightarrow CH_3-\dot{C}H-COOH + RO^- \qquad (4.49)$$

$$[CH_2=CH-COOH]^- + H^+ \text{ from } ROH_2^+ \longrightarrow CH_3-\dot{C}H-COOH + ROH \qquad (4.50)$$

In the alkaline ice matrices:

$$[CH_2=CH-COO]^{2-} + H_2O \longrightarrow CH_3-\dot{C}H-COO^- + OH^- \qquad (4.51)$$

$$[CH_2=CH-COO]^{2-} + H^+ \text{ from } H_2O^+ \text{ or } H_3O^+ \longrightarrow CH_3-\dot{C}H-COO^- \qquad (4.52)$$

It is important to point out that in addition to the species formed in the γ-irradiated alkaline ice matrices at 77 K, electrons produced by irradiation are also trapped[1655].

4.2.2.2.6. Methyl Methacrylate. γ-Irradiation at 77 K in the dark of organic glasses of 2-methyltetrahydrofuran (Fig. 4.33) and 3-methylpentane (Fig. 4.54) containing methyl methacrylate gives an ESR spectrum containing a three-line main component with a coupling constant of 11.0 ± 0.2 G, overlapping the spectrum of pure glass radicals[777].

20 G

Fig. 4.53. ESR spectrum of γ-irradiated 2-methyltetrahydrofuran glass containing methyl methacrylate (1 mole %) at 77 K in the dark and measured at 77 K. The dotted line presents spectrum obtained after photo-bleaching[777]

96

Fig. 4.54. ESR spectrum of γ-irradiated 3-methyl-pentane glass containing methyl methacrylate (0.04 mole %) at 77 K. The dotted line presents spectrum obtained after photo-bleaching[777]

The main ESR spectrum has been attributed to radical anions of methyl methacrylate (*4.49*) formed by trapping of electrons in the glass:

$$\left[\begin{array}{c} CH_3 \\ | \\ CH_2=C \\ | \\ COOCH_3 \end{array}\right]^{\cdot-} \qquad (4.49)$$

The radical anion (*4.49*) in the 2-methyltetrahydrofuran glass is bleached by visible light and methyl radicals derived from the methoxy group are produced. In the 3-methylpentane glass the radical anion (*4.49*) is bleached by infrared light and no methyl radicals were observed.

ESR studies of anion radicals of the following acrylic monomers: methyl acrylate, ethyl acrylate, allyl methacrylate, acrolein, acrylamide, acrylic acid, crotonic acid, and methyl crotonate in organic glasses have also been described[771].

4.2.2.2.7. Acrylonitrile. ESR study shows that anion radicals are involved in the initiation process of the radiation-induced anionic polymerization of acrylonitrile in 2-methyltetrahydrofuran[238]. ESR spectra of γ-irradiated acrylonitrile with different content of monomer in 2-methyltetrahydrofurane are shown in Figure 4.55.

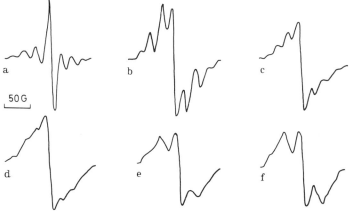

Fig. 4.55. ESR spectra of γ-irradiated 2-methyltetrahydrofuran glass: a. pure and containing acrylonitryle at concentration: b. 5 mole %, c. 12 mole %, d. 20 mole %, e. 44 mole %, f. 60 mole %[238]

97

These ESR spectra may be attributed to radical anion (4.50)

$$\left[\begin{array}{c} CH_2{=}CH \\ | \\ CN \end{array} \right]^{\cdot -} \qquad (4.50)$$

4.2.2.2.8. Styrene and α-Methylstyrene. ESR spectra of small amounts of styrene[1023], [2260] and α-methylstyrene[1023, 2457] in 2-methyltetrahydrofuran glass γ-irradiated at 77 K are shown in Figures 4.56 and 4.57, respectively.

The ESR spectra for styrene and α-methylstyrene in 2-methyltetrahydrofurane glass show a broad singlet spectrum (width 33 G) superimposed on a seven-line spectrum assigned to free radicals formed by intramolecular hydrogen transfer in the primary cation radicals of 2-methyltetrahydrofuran (4.33). No sharp singlet spectrum due to trapped electrons is observed. The broad singlet spectrum is bleached by visible light.

ESR spectra of small amounts of styrene and α-methylstyrene in n-butyl chloride glass γ-irradiated at 77 K are shown in Figures 4.58 and 4.59 respectively[1023]. The comparison of the spectra obtained in n-butyl chloride glass and 2-methyltetrahydrofurane matrices shows that the type of ionic radical formation mechanism is

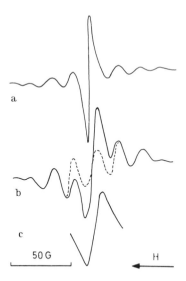

Fig. 4.56. ESR spectra of γ-irradiated 2-methyltetrahydrofuran glass at 77 K and measured at 77 K: a. pure, b. containing styrene (0.42 mole %) (solid line) and after photo-bleaching (dotted line), c. styrene anion radicals[1023]

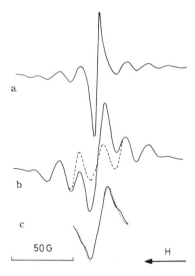

Fig. 4.57. ESR spectra of γ-irradiated 2-methyltetrahydrofuran glass at 77 K and measured at 77 K: a. pure, b. containing α-methylstyrene (5.31 mole %) (solid line) and after photo-bleaching (dotted line), c. α-methylstyrene anion radicals (solid line — observed and dotted line calculated)[1023]

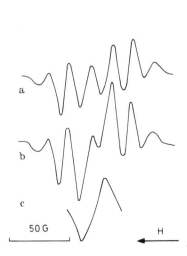

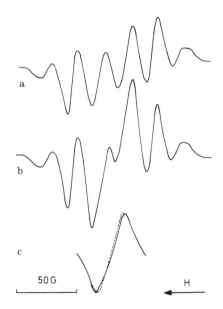

Fig. 4.58. ESR spectra of γ-irradiated n-butyl chloride at 77 K and measured at 77 K: a. pure, b. containing styrene (0.15 mole %), c. styrene cation radicals[1023]

Fig. 4.59. ESR spectra of γ-irradiated n-butyl chloride at 77 K and measured at 77 K: a. pure, b. containing α-methyl styrene (0.69 mole %), c. α-methyl styrene cation radicals (solid line-observed and dotted line calculated)[1023]

dependent on the type of glass matrix in which the γ-irradiation of monomer was made[1023, 2458, 2462, 2465]. The spectra obtained in 2-methyltetrahydrofuran glass matrix were attributed to styrene (4.51) or γ-methylstyrene radical anion formed by an electron capture reaction (4.53), whereas the ESR spectra in n-butyl chloride glass matrix were attributed to radical cations (4.52) formed by positive hole transfer from the glass matrix to the solute (4.54):

$$H_2C=CH + e^- \longrightarrow \left[H_2C=CH \right]^{\cdot-} \qquad (4.51) \qquad\qquad (4.53)$$

$$H_2C=CH + (n-BuCl)^{\cdot+} \longrightarrow n-BuCl + \left[H_2C=CH \right]^{\cdot+} \qquad (4.52) \qquad\qquad (4.54)$$

The ESR study suggests that the cation-radicals (4.52) after an addition of monomer form carbonium ions (4.53) as follows:

$$\left[H_2C=CH \right]^{\cdot+} + H_2C=CH \longrightarrow \overset{\cdot}{C}H-CH_2-CH_2-CH^+ \qquad (4.53) \qquad\qquad (4.55)$$

99

Anion radicals (*4.51*) formed in 2-methyltetrahydrofuran[1010, 1023, 2260, 2465] cannot initiate anionic polymerization, whereas cation-radicals (*4.52*) formed in n-butyl chloride[1023] may react with monomer to (*4.53*) and give cationic polymerization at one chain end.

4.2.3. Ionizing Polymerization-Miscellaneous Problems

ESR spectroscopy has also been applied to studies of radiation-induced polymerization of formaldehyde[420], acetaldehyde[419], maleimide[1071], ethylene oxide[1609], and vinyl fluoride, vinylidene fluoride and vinyl chloride[1608] in xenon matrices, methyl methacrylate in sulphur dioxide[1106], N-vinylpyrrolidone[1545], vinylsiloxanes[1186], acrylamide-water solid systems[30], and cyclopolymerization of acrylic and methacrylic anhydrides[1021].

ESR studies of copolymerization of ethylene with vinyl acetate[655] and also styrene with methyl methacrylate in the polymer matrix at high pressure[247] have been reported.

4.3. Photopolymerization

Without photoinitiators or photosensitizers added, the ESR signals obtained in direct photopolymerization of pure monomers are generally too weak for identification of radical species.

In "photosensitized polymerization" various organic or inorganic substances, which are decomposed readily fo form free radicals under UV irradiation, are added to monomers. Various types of photosensitizers for photopolymerization are reviewed by Oster and Yang[1676] and Rabek[1775, 1779]. The mechanism involved in photosensitized reactions is inadequately known and cannot, therefore, be used for

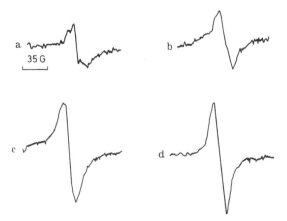

Fig. 4.60. ESR spectra of photolyzed hydrogen peroxide: a. 30% H_2O_2 at 262 K, b. 30% H_2O_2 plus 10 ml/l H_2SO_4 at 261 K, c. 45% H_2O_2 at 240 K, d. 45% H_2O_2 plus 10 ml/l H_2SO_4 at 237 K[1382]

a classification of these initiators. In practice it is convenient to group the sensiti reactions according to the chemical nature of the sensitizers.

4.3.1. Photosensitized by Hydrogen Peroxide

Hydroxyl radicals (HO·) are mainly associated with γ-radiolysis of aqueous solutions. Chemically, hydroxyl radicals can be readily generated from hydrogen peroxide by direct UV-irradiation[1382, 2044, 2127]. ESR spectra from photolyzed hydrogen peroxide are shown in Figure 4.60.

ESR spectra of hydroxyl (HO·) and hydroperoxy (HOO·) radicals may only be observed in frozen glassy state[995, 1794, 2333] and their parameters are listed in Table 4.3.

Hydroxyl radicals react with organic substrates, e.g. alcohols[766, 1382, 1383] and a variety of carbon-centered radicals. Hydroxyl radicals are capable of initiating free radical polymerization of different monomers. Kuwata et al.[1280] have applied ESR spectroscopy to study steady-state photopolymerization of vinyl acetate and methyl methacrylate using hydrogene peroxide as initiator, whereas Fischer and Hefter[899] have investigated reaction kinetics of free radicals formed from photolysis of H_2O_2 with acrylonitrile in methanol.

4.3.2. Photosensitized by Hydroperoxides and Peroxides

The tert-butyl peroxy radical (4.54) (g = 2.0137) is formed during the photolysis of tert-butyl hydroperoxide (Bu^tOOH)[456, 1003, 1405, 2178].

$$Bu^tOOH \xrightarrow{+h\nu} Bu^tO\cdot + \cdot OH \tag{4.56}$$

$$Bu^tO\cdot + Bu^tOOH \longrightarrow Bu^tOO\cdot + Bu^tOH \qquad (4.54) \tag{4.57}$$

Photolysis of tert-butyl hydroperoxide on silica gel at 77 K produces an asymmetric singlet, unchanged after prolonged irradiation (Fig. 4.61).

This ESR spectrum has been attributed to tert-butyl peroxy radical (4.54). Photolysis of liquid cumene hydroperoxide and isopropyl hydroperoxide also produces peroxy radicals[218, 1003, 1325, 2178].

Photolysis of tert-butyl peroxide (Bu^tOOBu^t) gives formation of butoxy radicals (4.55)[1003, 1405, 2100]:

$$Bu^tOOBu^t \xrightarrow{+h\nu} 2 Bu^tO\cdot \qquad (4.55) \tag{4.58}$$

which further decompose to methyl radicals and acetone:

$$Bu^tO\cdot \longrightarrow CH_3COCH_3 + \cdot CH_3 \tag{4.59}$$

101

Table 4.3. ESR parameters of oxygen-centered radicals in the solid state[1166]

Radical	Medium	Temperature (K)	g tensor				Hyperfine tensor (G)			
			g_x	g_y	g_z	g_{aT}	A_x	A_y	A_z	A_{iso}
$H\dot{O}$	Ice	77	2.005	2.009	2.06	2.025	-26 ± 4	-44 ± 2	0 ± 6	-26.4
	Gas	–	–	–	–	–	–	–	–	-26.7
$HO\dot{O}$	H_2O_2 glass	77	2.0353	2.0086	2.0042	2.0160	-14.0	-3.57	-15.7	-11
$DO\dot{O}$	D_2O_2 glass	77	2.0344	2.0086	2.0031	2.0154	-2.14	-0.54	-2.32	-1.7

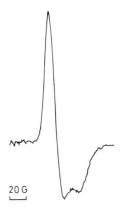

20 G

Fig. 4.61. ESR spectrum
of photolyzed t-butyl hydroperoxide on activated silica
gel at 77 K[1405]

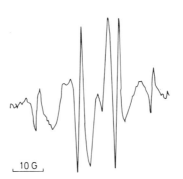

10 G

Fig. 4.62. ESR spectrum of photolyzed
t-butyl peroxide on activated silica gel
at 77 K[1405]

This reaction is still more complicated by secondary reactions:

$$Bu^tO\cdot + CH_3COCH_3 \longrightarrow CH_3CO-\dot{C}H_2 + Bu^tOH \qquad (4.56) \qquad\qquad (4.60)$$

$$\cdot CH_3 + CH_3COCH_3 \longrightarrow CH_3CO-\dot{C}H_2 + CH_4 \qquad\qquad (4.61)$$

Photolysis of tert-butyl peroxide on silica gel at 77 K gives an ESR spectrum consisting of a quartet (approximate intensity ratio 1:3:3:1) superimposed on a triplet (Fig. 4.62).

The quartet with splitting of 23.8 G has been attributed to methyl radical, whereas the triplet with splitting of about 20 G to the radical (4.56)[1405].

ESR spectroscopy has been applied to the study of:
polymerization of dienes[2139, 2140] and acrylonitrile[140, 141, 143, 456, 1003] photo-sensitized by tert-butyl peroxide,
zinc alkyl-peroxides (or hydroperoxides) initiator system for polymerization[1565, 1854],
benzoyl peroxide interpreted as photopolymerization of methyl methacrylate[1181, 1182, 1752, 2055, 2056],
styrene[294, 295],
and copolymerization of maleic anhydride with dimethylbutadiene[2078].

The ESR spectrum of radical species formed by UV-irradiation of methyl methacrylate at 77 K with 0.1 wt. % of benzoyl peroxide shows the superposition of a septet, an asymmetric singlet, and a doublet from hydrogen atoms (Fig. 4.63)[1181].
The asymmetric singlet and the septet spectrum are assigned as due to a benzoyl peroxy radical (4.57) and a methyl methacrylate parent radical (4.58), respectively. These radicals may be formed by the following reactions:

$$C_6H_5CO-O-O-OCC_6H_5 \xrightarrow{+h\nu} 2\,C_6H_5CO-O\cdot \qquad (4.57) \qquad\qquad (4.62)$$

103

$$CH_2=\underset{\underset{COOCH_3}{|}}{\overset{\overset{CH_3}{|}}{C}} \quad \xrightarrow{+h\nu} \quad CH_2=\underset{\underset{COOCH_2}{|}}{\overset{\overset{CH_3}{|}}{C}} \quad + H\cdot \qquad (4.63)$$

$$H\cdot + CH_2=\underset{\underset{COOCH_3}{|}}{\overset{\overset{CH_3}{|}}{C}} \quad \longrightarrow \quad CH_3-\underset{\underset{COOCH_3}{|}}{\overset{\overset{CH_3}{|}}{C}}\cdot \qquad (4.58)$$

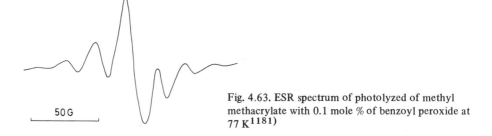

50 G

Fig. 4.63. ESR spectrum of photolyzed of methyl methacrylate with 0.1 mole % of benzoyl peroxide at 77 K[1181]

After warming to higher temperatures, the observed ESR spectrum changes and a new pattern, which contains a septet and a quintet, is formed. The latter signal might be due to methyl methacrylate to which a benzoyl peroxy radical is added (4.59)[1181, 2056].

$$C_6H_5COO\cdot + CH_2=\underset{\underset{COOCH_3}{|}}{\overset{\overset{CH_3}{|}}{C}} \quad \longrightarrow \quad C_6H_5COO-CH_2-\underset{\underset{COOCH_3}{|}}{\overset{\overset{CH_3}{|}}{C}}\cdot \qquad (4.59) \quad (4.64)$$

Other authors[13, 1006, 2125] have reported the formation of radical (4.60) analogous in structure to radical (4.59).

$$X-CH_2-\underset{\underset{COOCH_3}{|}}{\overset{\overset{CH_3}{|}}{C}}\cdot \qquad (4.60)$$

At higher polymerization temperature a poorly resolved four-line spectrum can be observed. This spectrum was attributed to radical (4.61) formed by a disproportionation reaction (4.65) and further hydrogen atom abstraction from the end group (4.66):

104

$$
\begin{array}{ccccc}
& \text{CH}_3 && \text{CH}_3 && \text{CH}_3 && \text{CH}_2 \\
& | && | && | && \| \\
-\text{CH}_2-\text{C} \cdot & & + & -\text{CH}_2-\text{C} \cdot & \longrightarrow & -\text{CH}_2-\text{CH} & + & -\text{CH}_2-\text{C} \\
& | && | && | && | \\
& \text{COOCH}_3 && \text{COOCH}_3 && \text{COOCH}_3 && \text{COOCH}_3
\end{array}
$$

$$(4.65)$$

$$
\begin{array}{ccccc}
& \text{CH}_2 && & \text{CH}_2 \\
& \| && & \| \\
\text{R} \cdot + -\text{CH}_2-\text{C} & \longrightarrow & \text{RH} + & -\overset{\cdot}{\text{C}}\text{H}-\text{C} \\
& | && & | \\
& \text{COOCH}_3 && & \text{COOCH}_3
\end{array}
$$

$$(4.61) \qquad (4.66)$$

During the photopolymerization of methyl methacrylate peroxyradicals may also be formed[1753].

4.3.3. Photosensitized by Diacyl Peroxides and Tert-Butyl Peresters

The direct photolysis of diacyl peroxides (4.67) and tert-butyl peresters (4.68) has been used for selective generation and ESR studies of a wide variety of alkyl radicals[1019, 1167]:

$$
\begin{array}{cc}
\text{O} & \text{O} \\
\| & \| \\
\text{RC}-\text{O}-\text{O}-\text{CR} & \xrightarrow{+h\nu} 2\,\text{R} \cdot + 2\,\text{CO}_2
\end{array}
$$

$$(4.67)$$

$$
\begin{array}{c}
\text{O} \\
\| \\
\text{RC}-\text{O}-\text{O}-\text{Bu}^t \xrightarrow{+h\nu} \text{R} \cdot + \text{CO}_2 + \text{Bu}^t\text{O} \cdot
\end{array}
$$

$$(4.68)$$

ESR studies of photolytic decomposition of di-acyl peroxides and peresters in the presence of vinyl monomers can provide information on the structure and conformation of carboxy radical adducts to monomers, the decarboxylation of carboxy radicals of different structural types, and the competition between addition to a C=C group and hydrogen abstraction.

4.3.4. Photosensitized by Dialkyl Peroxydicarbonates

ESR studies of photolysis of dialkyl peroxydicarbonates have shown the formation of carboxy radicals[606]:

$$
\begin{array}{ccc}
\text{O} & \text{O} & \text{O} \\
\| & \| & \| \\
\text{RO}-\text{C}-\text{O}-\text{O}-\text{C}-\text{OR} & \xrightarrow{+h\nu} & 2\,\text{RO}-\text{C}-\text{O} \cdot
\end{array}
$$

$$(4.69)$$

ESR studies of photopolymerization of perfluorobutadiene with $\text{CF}_3-\text{OO}-\text{CF}_3$ and diisopropyl peroxydicarbonate as initiators have been described in paper[2236].

105

4.3.5. Photosensitized by Benzophenone

The observation of formation of the triplet state of benzophenone by ESR spectroscopy is only possible in the solid-state (single crystal) at 4.2 K[1951]. UV-irradiation of benzophenone in solution, e.g. in isopropyl alcohol, leads to ketyl radicals (*4.62*) and dimethyl-hydroxy methyl radicals (*4.63*)[2143]:

$$(C_6H_5)_2C=O + (CH_3)_2CH-OH \xrightarrow{\;+h\nu\;} (C_6H_5)_2\dot{C}-OH + (CH_3)_2\dot{C}-OH \qquad (4.70)$$

$$(4.62) \qquad\qquad (4.63)$$

The benzophenone ketyl radicals (*4.62*) have a short lifetime (80 ms) and for that reason it is difficult to determine their ESR spectra[1546]. Iwakura et al.[1022] found that benzophenone ketyl radicals are stably trapped in cellulose triacetate film after UV irradiation of the film containing benzophenone. Using this method the ESR spectra of various substituted benzophenone ketyl radicals were determined (Fig. 4.64).

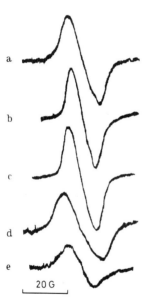

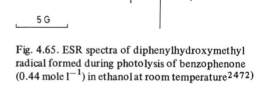

Fig. 4.64. ESR spectra of substituted benzophenone ketyl radicals trapped in cellulose triacetate at room temperature: a. benzophenone, b. 4-methylbenzophenone, c. 4-methoxybenzophenone, d. 4-benzoylbenzophenone, e. 4-phenylbenzophenone[1022]

Fig. 4.65. ESR spectra of diphenylhydroxymethyl radical formed during photolysis of benzophenone (0.44 mole l^{-1}) in ethanol at room temperature[2472]

A symmetrical broad singlet spectrum was obtained at temperatures higher than 253 K. The radicals trapped in the film can initiate polymerization of methyl methacrylate to high molecular weight.

Yoshida and Warashina[2472], using a special UV-irradiation method[1382], have obtained ESR spectra of ketyl radicals (4.62) with well-resolved hyperfine structure consisting of 150 lines (partly overlapping) (Fig. 4.65).

The structure of this ESR spectrum is described as containing four equivalent protons in ortho-position of the rings with a hyperfine coupling constant of 3.1 G, four protons at meta-position with a coupling constant of 1.2 G, two protons a para-position of the rings with a coupling constant of 3.6 G, and one proton in the hydroxyl group with a constant of 2.9 G.

Benzophenone is a very effective sensitizer for photopolymerization, photografting, and photodegradation[1779, 1783]. Until now, there are no detailed ESR studies of the photoinitiation process of polymerization by benzophenone published in the literature.

4.3.6. Quinones

Formation of semiquinone anions and semiquinone radicals during reduction or oxidation processes between quinones and hydroquinones, have been observed by ESR[92, 463, 525, 801, 846, 1245, 2023, 2073, 2075, 23–2357, 2422, 2460] and ENDOR[92].

The semiquinone anion radical was observed by photolysis of benzoquinone[2460] and duroquinone[801]. When the solution of p-benzoquinone in ethanol is photolyzed, the ESR signal is observed as shown in Figure 4.66a.

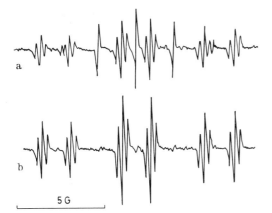

a

b

5 G

Fig. 4.66. ESR spectra observed during photolysis of p-benzoquinone solution in ethanol: a. Spectrum due to benzosemiquinone radical (triplet-doublet-triplet) and due to benzosemiquinone anion (quintet) formed in neutral solution. b. Spectrum due to benzosemiquinone radical in acidic solution containing 30 mmole of acetic acid[2460]

The signal is composed of a quintet spectrum and that of a doublet of triplets of small triplets. The former is attributed to the p-benzosemiquinone anion radical (4.64) with four equivalent protons, while the latter is due to the p-benzosemiquinone radical (4.65). When the solution is made acidic by adding a small amount of acetic acid, almost all of the semiquinone anions are transformed into the semiquinone radicals through protonation:

$$(4.71)$$

(4.64)   (4.65)

and only a weak spectrum of the semiquinone anion radical remains (Fig. 4.66b). The mechanism of the formation of semiquinone anion and semiquinone radicals is given in the original papers.

ESR spectra of radicals formed during photolysis of chloranil in solutions are reported in several papers[847, 1101, 2429, 2463]. The ESR spectra observed in solutions in alcohols are identified as due to the neutral semiquinone radical (Fig. 4.67a)[2463].

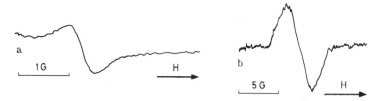

Fig. 4.67. ESR spectra observed during photolysis of 1 mmol of chloranile at room temperature: a. in isopropanol, b. in benzene[2463]

The peak-to-peak line width is 0.55 G. ESR spectrum of chloranil photolyzed in benzene solution is shown in Figure 4.67b. The spectrum has the width 3.4 G without resolved hyperfine structure, and is much wider than the spectrum of the neutral semiquinone radical. These results suggest that the excited chloranil molecule under UV light may abstract hydrogen from a benzene ring.

Quinones have a very important role in inhibiting polymerization processes and also as photosensitizers of polymer photodegradation and photooxidation[1814]. But these problems have not yet been examined by ESR spectroscopy.

4.3.7. Photosensitized by Azo-Compounds

Detailed ESR studies of the free radicals formed during photolysis of several azo-compounds have been made, e.g. azo-bis(isobutyronitrile)[110, 221, 326, 341, 1320, 1384, 1535, 1721, 2042, 2397, 2398, 2493] and 1,1'-azobis(1-cyanocyclopentane)[2038].

During the photolysis of azo-bis(isobutyronitrile)(AIBN), 2-cyano-2-propyl radicals (4.66) are formed by the following mechanism:

$$\underset{\underset{\text{CN}}{|}}{\overset{\overset{\text{CH}_3}{|}}{\text{H}_3\text{C}-\text{C}}}-\text{N}=\text{N}-\underset{\underset{\text{CN}}{|}}{\overset{\overset{\text{CH}_3}{|}}{\text{C}}}-\text{CH}_3 \xrightarrow{+h\nu} 2\,\text{CH}_3-\underset{\underset{\text{CN}}{|}}{\overset{\overset{\text{CH}_3}{|}}{\text{C}\cdot}} + \text{N}_2 \qquad (4.66) \qquad (4.72)$$

The formation of the radical (4.66) has also been observed during the thermal decomposition of AIBN[278, 341].

The ESR spectrum of radical (4.66) formed in polycrystalline state at 77 K shows a septet with satellites (Fig. 4.68).

A detailed interpretation of its hyperfine structure is given in[1535].

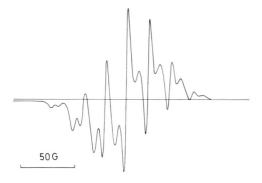

Fig. 4.68. ESR spectrum of photolysed azo-bis(isobutyronitrile) in
polycrystalline state at 77 K[1535]

50 G

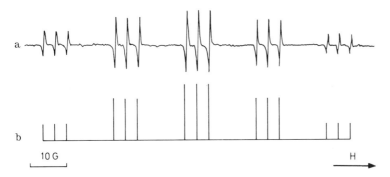

a

b

10 G   H

Fig. 4.69. a. ESR spectrum of photolyzed azo-bis(isobutyronitrile) in toluene at 298 K, b. stick
plot of spectrum a[2042]

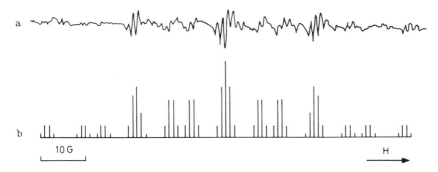

a

b

10 G   H

Fig. 4.70. a. ESR spectrum of photolyzed azo-bis(isobutyronitrile) in methyl methacrylate at
313 K, b. stick plot for a quartet (1:3:3:1) of doublets (1:1) of quartets (1:3:3:1) consistent
with a radical (4.67)[2042]

The photolysis of AIBN in toluene at 298 K gives a completely different ESR
spectrum (Fig. 4.69)[2042] assigned to the radical (4.66) in solution.

AIBN is a very effective sensitizer for photopolymerization of methacrylic
acid[142] and its esters[711, 1610, 2029, 2036, 2042]. For each ester, two ESR spectra
(Fig. 470) were observed, one from the radical (4.66) and the other from radicals
formed by addition and having the general formula (4.67):

109

$$R- \left[\begin{matrix} CH_3 \\ | \\ CH_2-C \\ | \\ COOY \end{matrix} \right]_n \begin{matrix} CH_3 \\ | \\ -CH_2-C\cdot \\ | \\ COOY \end{matrix} \qquad (4.67)$$

where $n \geqslant 0$ and Y are methyl, ethyl, i-butyl, or n-butyl groups.

Bullock et al.[344] measured the formation and the decay of free radicals trapped in poly(methyl methacrylate) gels which were produced either by direct UV-irradiation (2537 Å) of pure monomer or by photodecomposition (3650 Å) of AIBN dissolved in pure monomer. ESR spectra showed two types of chain end radicals (4.68) and (4.69) and two types of allyl-radicals (4.70) and (4.71). It was proposed that the allyl radicals can be produced both by photodegradation of the polymer (4.73)–(4.74) and by reactions of other radicals (e.g. alkyl radicals) with the unsaturated end of polymer chains (4.75):

$$-CH_2-\underset{\underset{COOCH_3}{|}}{\overset{\overset{CH_3}{|}}{C}}-----CH_2-\underset{\underset{COOCH_3}{|}}{\overset{\overset{CH_3}{|}}{C}}- \quad \xrightarrow{+h\nu} \quad -CH_2-\underset{\underset{COOCH_3}{|}}{\overset{\overset{CH_3}{|}}{C}}\cdot \quad + \cdot CH_2-\underset{\underset{COOCH_3}{|}}{\overset{\overset{CH_3}{|}}{C}}- \quad (4.73)$$

(4.68)   (4.69)

$$-CH_2-\underset{\underset{COOCH_3}{|}}{\overset{\overset{CH_3}{|}}{C}}----\dot{C}H_2 \longrightarrow -\dot{C}H-\underset{}{\overset{\overset{CH_3}{|}}{C}}=CH_2 + CH_3COOH \qquad (4.74)$$

(4.69)   (4.70)

$$R\cdot + -CH_2-\underset{\underset{COOCH_3}{|}}{\overset{\overset{CH_3}{|}}{C}}=CH_2 \longrightarrow -\dot{C}H-\underset{\underset{COOCH_3}{|}}{\overset{\overset{CH_3}{|}}{C}}=CH_2 + RH \quad (4.71) \qquad (4.75)$$

Decay curves for radicals obtained in the above experiments are shown in Figure 4.71. Initially the propagating radical concentration falls rapidly and after prolonged heating it falls asymptotically. The allyl radical concentration, however, shows a short but sharp drop and thereafter it increases noticeably before decaying slowly during prolonged heating time. In the AIBN-initiated samples the initial spin concentrations are ten to thirty times lower than those of the directly photopolymerized samples. The propagating radical concentration decreases in about the same way in both cases. After an initial decrease, the rise in allyl radical concentration for the AIBN-initiated system (Fig. 4.71 b) is much more marked than for the corresponding curve in Figure 4.71 a. In the directly photopolymerized samples the initial presence of allyl radicals can be accounted for in terms of photo-induced main chain scission and subsequent disproportionation of radical (4.69) into methyl formate and allyl radical (4.70) (reaction 4.74). This type of reaction cannot account for the allyl radicals in AIBN-initiated samples, since the radiation of 3600 Å has no effect on poly(methyl metha-

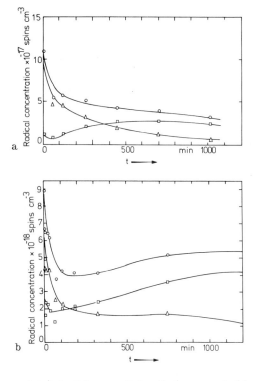

Fig. 4.71. Decay curves for radicals: a. in directly photopolymerized methyl methacrylate at 353.5 K, b. in azobis(isobutyronitrile) sensitized photopolymerization of methyl methacrylate at 363 K: o – total radical, Δ propagating radical and □ allyl radical concentration[344]

crylate). In this case allyl radicals are probably formed by attack of another radical at a terminal double bond of a dead chain (4.75). Here R· may be a propagating radical or an AIBN fragment. Free allyl radicals (4.70) and (4.71) differ in structure, but trapped in a polymer matrix they cannot be differentiated by ESR. The low spin density on the central carbon atom, together with the large line widths, prevents the methyl protons in radical (4.70) from showing resolved hyerfine splittings.

ESR studies have also been made on methyl methacrylate photopolymerization initiated with 4,4'-azo(bis(4-cyanopentanoic acid))[2037].

4.3.8. Photosensitized by Metal Halides

Below 123 K ferric chloride photosensitized initiation has been used to initiate polymerization of vinyl monomers in rigid glasses of alcohols[72, 73, 870–873]. When the temperature of photolyzed glasses was increased, free radicals derived from alcohols (4.72)[930] reacted with vinyl monomers to initiate polymerization:

$$CH_3OH + FeCl_3 \xrightarrow{+h\nu} FeCl_2 + HCl + \dot{C}H_2OH \qquad (4.72) \qquad \qquad (4.76)$$

$$\dot{C}H_2OH + CH_2{=}\underset{\underset{COOR}{|}}{\overset{\overset{R}{|}}{C}} \longrightarrow HO{-}CH_2{-}CH_2{-}\underset{\underset{COOR}{|}}{\overset{\overset{R}{|}}{C}}\cdot \qquad (4.73) \qquad \qquad (4.77)$$

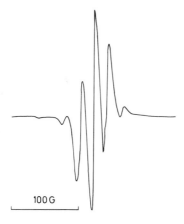

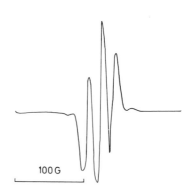

Fig. 4.72. ESR spectrum of propagating radical formed during ferric chloride sensitized photopolymerization of methyl methacrylate in methanol glass at 103 K and measured at 103 K[871]

Fig. 4.73. ESR spectrum of propagating radical formed during ferric chloride sensitized photopolymerization of methacrylamide in methanol glass at 113 K and measured at 113 K[871]

Typical ESR spectra of propagating radicals of methyl methacrylate and acrylamide are shown in Figures 4.72, and 4.73, respectively.

It was found that one type of propagating radical (4.73) is formed in all cases of monomers investigated. Propagating radicals derived from methyl methacrylate were found to exist in two conformations, and radicals derived from methylacrylate in one conformation. The ESR spectra generated by the propagating radical in reaction 4.77 are dependent on rotation at the C_α–C_β bond and on interaction of H_α, H_{β_1} and H_{β_2} (nonequivalent protons) with the unpaired electron on C_α.

ESR spectroscopy has also been applied for the study of photopolymerization of vinyl monomers sensitized with $ZnCl_2$, $AlCl_3$, $AlBr_3$, $SnCl_4$, $TiCl_4$, $TiBr_4$, and VCl_4[753, 755, 938, 998, 1420, 1421, 2518] and for photosensitized polymerization of N-vinylcarbazole with sodium chloroaurate ($NaAuCl_4 \cdot 2 H_2O$)[81].

4.3.9. Photosensitized by Metal Acetylacetonates

ESR spectroscopy has been used to study photopolymerization of vinyl monomers (e.g. methyl methacrylate) sensitized with metal acetylacetonates[1068, 1069]. $Mn(acac)_3$ and $Co(acac)_3$ have a great sensitizing effect. Since sensitization is high in the wavelength region of charge transfer absorption and low in the ligand region, photoreduction of Co(III) to Co(II) is suggested to occur during the formation of initiating radicals. ESR spectroscopy shows that the initiation is due to acetylacetonyl radicals from the chelate. The central metal ion is simultaneously reduced.

Bartoň and Horanská[159] have reported that some metal acetyl acetonates, e.g. copper(II)-acetylacetonate in carbon tetrachloride, can themselves form radical intermediates which may initiate the polymerization of vinyl monomers, e.g. styrene.

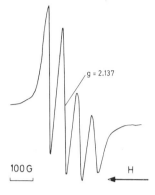

g = 2.137

100 G

H

Fig. 4.74. ESR spectrum of copper(II) acetylacetonate in carbon tetrachloride (10^{-3} mole l^{-1}) at 318 K[159]

Figure 4.74 shows the ESR spectrum of a solution of copper(II)-acetylacetonate in CCl_4.

The hyperfine structure of the ESR spectrum (an equal quartet) is due to the interaction of the unpaired electron of Cu^{2+} with the nucleus having the spin I = 3/2.

4.4. Thermal Polymerization

ESR studies of growing polymer chain radicals formed by thermal initiation are made difficult by the high reactivity of free radicals at the usually required elevated temperatures. This leads to low steady-state concentration of radicals. There are a few methods in which growing polymer chain radicals can be stabilized:
1. Application of a polymer matrix for studies of growing chain radicals at high pressure.
2. Stabilization of free radicals on the surface of synthetic zeolites.
3. Spin-trapping method (see Chap. 11.3).

4.4.1. Application of a Polymer Matrix for Studies of Growing Chain Radicals at High Pressure

A new method which involves trapping the growing polymer chain radicals in polymer matrix by thermal decomposition of benzoyl peroxide at high pressure has been described by Czechoslovakian scientists in several papers[2106, 2108, 2109, 2113].

The principle of this method is the stabilizing effect of the high pressure on the radicals generated in the polymer matrix of poly(methyl methacrylate) or polystyrene, by thermal decomposition of benzoyl peroxide. The decrease of the polymer viscosity due to the increased temperature is compensated by the pressure effect on the polymer system, e.g. in poly(methyl methacrylate). At the pressure 10,000 atm the free radicals are relatively stable with a lifetime which makes them measurable by ESR even at 423 K[2105]. The study of the mechanism of the free-radical generation revealed that the polymer radicals can easily be trapped when the polymer matrix used contains, in addition to benzoyl peroxide, a small amount of a vinyl monomer. In such cases, polymerization of the vinyl monomer takes place and the lifetime of propagating radicals is prolonged by the effect of high pressure even more than that of the initiator radicals.

4.4.2. Stabilization of Free Radicals on the Surface of Synthetic Zeolites

The growing polymer chain radicals during thermal polymerization can be stabilized by adsorption on synthetic zeolites (synthetic faujasite of the NaX type)[1864, 1866, 1867].

ESR spectra of radicals obtained for n-butyl methacrylate and methacrylic acid by this method are shown in Figure 4.75.

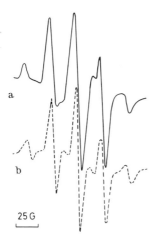

a

b

25 G

Fig. 4.75. ESR spectra of growing radicals of: a. n-butyl methacrylate and b. methacrylic acid, stabilized by absorption on synthetic fajausite, measured at 363 K[1867]

This spectrum was attributed to the propagating radical (*4.74*):

$$\begin{array}{c} CH_3 \\ | \\ -CH_2-C\cdot \\ | \\ COOX \end{array} \qquad (4.74)$$

where X = H or alkyl.

The stabilization of growing macroradicals on zeolite is very useful for studying reactions of the stabilized radicals (Fig. 4.76).

The reaction of stabilized alkyl radicals with oxygen at low temperatures leads to the formation of peroxy radicals with a characteristic asymmetric spectrum. At 313 K the asymmetric ESR spectrum changes simultaneously into a symmetric singlet. In

114

Fig. 4.76. ESR spectrum of n-butyl methacrylate polymer radicals stabilized on synthetic fajau-site and their change during the transformation to peroxy radicals in air and during reverse reaction in vacuum[1867]

vacuum, radicals represented by both the asymmetric and the symmetric singlet spectrum, change to the original alkyl radicals.

4.4.3. Popcorn Polymerization

Free radical initiation of styrene-divinylbenzene mixtures in a narrow range of compositions gives proliferous polymerization to produce what is commonly called "popcorn polymers". This occurs in a range of low concentrations of divinylbenzene, for which the exact position depends on polymerization conditions such as temperature, initiator concentration, and diluent. During proliferous polymerization intensive ESR signals are obtained, due to high concentrations of growing polymer radicals[285-287, 560, 1492, 1576]. These free radicals are partly growing chain ends with low termination rate and partly formed by chain scission during the swelling of the styrene-divinyl-benzene copolymer in styrene. The radicals may be trapped in the gel phase for many days. ESR signals are also obtained in the polymerization of technical divinyl-benzene leading at high conversion to glassy polymers, indicating in this case also the presence of radicals in high concentrations[711].

4.4.4. Thermal Polymerization-Miscellaneous Problems

ESR spectroscopy has also been applied to studies of thermal polymerization of methyl methacrylate and vinyl acetate[301, 302], styrene[294] in the presence of benzoyl peroxide, radical polymerization of methyl methacrylate in the presence of H_2SO_4[15], polymerization of: vinyl monomers with N-chlorosuccinimide[1165], dibenzofulvene initiated by AIBN[706], maleic anhydride by AIBN or benzoyl peroxide[1564], cyclo-polymerization of acrylic anhydride by AIBN[1459] copolymerization of vinyl mono-mers[689, 903], retardation of free radical polymerization of styrene by nitrosocom-pounds[2094, 2095, 2311, 2312], azobenzenes[1264], aromatic hydrazyles[279], inhibition of vinyl acetate polymerization by quinones[2007], determination of the dissociation constants of polymerization initiators[954], estimation of rate constants of chain growth in the thermal polymerization in high viscous media of polyacrylate esters[1195, 1196], calculation of Q-e values[689, 903].

4.5. Polymerization Initiated by Mechano-radicals

Free radicals formed during mechanical degradation of polymers by cutting, milling, grinding, etc. (see Chap. 8.6) are named "mechano-radicals". In the presence of monomer they can initiate a polymerization reaction. There are two general methods of such polymerizations[2057]:

1. The "post-contact method" in which monomer is added to a sample with mechano-radicals which have been produced previously.
2. The "simultaneous fracture", method in which monomer (in gas or liquid form) is added simultaneously with the mechanical treatment of the polymer at low temperatures.

ESR spectra of free radicals observed in the post-contact method for grinded (milled) poly(tetrafluoroethylene) and methyl methacrylate are shown in Figure 4.77.

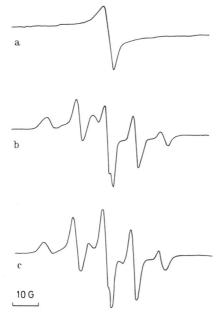

a

b

c

10 G

Fig. 4.77. Changes of ESR spectra after the contact of methyl methacrylate with grinded poly-(tetrafluoroethylene) at 77 K: a. no heat treatment, b. after 1 min heat treatment at 273 K, c. after 5 min heat treatment at 273 K. All spectra were measured at 77 K[2057]

This quartet-quintet spectrum has been attributed to a characteristic propagation radical (4.75):

$$-CH_2-\underset{\underset{COOCH_3}{|}}{\overset{\overset{CH_3}{|}}{C}}\cdot \qquad (4.75)$$

The spectrum observed after the simultaneous fracture of poly(tetrafluoroethylene) in the presence of methyl methacrylate at 77 K is also a quartet-quintet and may be

116

assigned to the same propagation radical (4.75). No trace of a free poly(tetrafluoroethylene) radical has been observed.

Similar ESR studies have also been reported for the polymerization of ethylene initiated by poly(tetrafluoroethylene) mechano-radicals[2057].

Russian scientists have also reported that free radicals formed by mechanical grinding of frozen solutions (80–150 K) of different polymers immersed in styrene, methyl methacrylate, acrylic acid and other monomers are added to double bonds of the monomers[577].

4.6. Polymerization Initiated by Various Free Radical Initiation Systems

4.6.1. Hydroperoxide-SO$_2$ System

The reaction of hydroperoxides with sulphur dioxide is a convenient radical source for initiation polymerization of vinyl monomers[1439]. Using ESR with flow system the nature of radical intermediates has been investigated in the reaction between RO$_2$H (R = H, Bu^t, PhCMe$_2$) and sulphur dioxide in several solvent media at 293 K[692, 693].

The details of the ESR spectra (Fig. 4.78) varied with R and with the solvent.

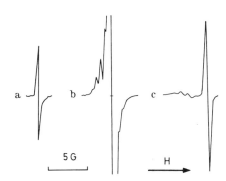

Fig. 4.78. ESR spectra of free radicals obtained in the systems: a. HOSO$_2$ from H$_2$O$_2$–SO$_2$–H$_2$O, b. Bu^{t}OSO$_2$ and HOSO$_2$ from Bu^{t}OOH–SO$_2$–MeOH, c. MeSO$_2$ and HOSO$_2$ from Bu^{t}OOH–SO$_2$–H$_2$O[692]

In every case the main signal (with R = H, only a single line Fig. 4.78) consists of a singlet with g = 2.0033 ± 0.0002 which was assigned to the radical 4.76 formed by the reactions:

$$RO_2H + SO_2 \longrightarrow RO\cdot + HO\dot{S}O_2 \qquad (4.76) \qquad\qquad (4.78)$$

$$RO_2H + SO_2 \longrightarrow HO\cdot + RO\dot{S}O_2 \qquad (4.77) \qquad\qquad (4.79)$$

$$HO\cdot + SO_2 \longrightarrow HO\dot{S}O_2 \qquad\qquad\qquad\qquad\qquad (4.80)$$

In addition to the singlet the spectrum recorded during the reaction of t-butyl hydroperoxide in methanol contained a multiplet, which was attributed to radical 4.78 (Fig. 4.78b):

117

$Bu^{t}O\dot{S}O_2$ (4.78)

A corresponding seven-line multiplet was also observed for the reaction of $PhCMe_2OOH$ in methanol, where free radicals (4.77) are formed according to the reaction:

$$RO\cdot + SO_2 \longrightarrow RO\dot{S}O_2 \quad (4.77) \tag{4.81}$$

In water used as solvent the spectra are different. Both organic hydroperoxides now produce a quartet centered on $g = 2.0055 \pm 0.0002$ and attributed to radical (4.79) (Fig. 4.78c) which can be formed by the reactions:

$$Mc_3CO\cdot \longrightarrow Me_2CO + \dot{M}e \tag{4.82}$$

$$\dot{M}e + SO_2 \longrightarrow Me\dot{S}O_2 \tag{4.83}$$

$$Me_3CO\cdot + SO_2 \longrightarrow Me_3CO\dot{S}O_2 \tag{4.84}$$

$$Me_3CO\dot{S}O_2 \longrightarrow Me_2CO + Me\dot{S}O_2 \quad (4.79) \tag{4.85}$$

It is not clear at present why the formation of the radicals (4.79), rather than radicals (4.77), is favored in water. No radicals could be detected when the reaction of t-butyl hydroperoxide with sulphur dioxide is conducted in ether or acrylonitrile, but the addition of a few percent of water or methanol is sufficient to obtain the radicals.

Ivin et al.[693-695, 1021] have measured ESR spectra of radicals obtained from t-butyl hydroperoxide and sulphur dioxide in the presence of different vinyl monomers and other unsaturated compounds. The characteristics of the spectra which appear in the presence of various compounds are summarized in Table 4.4 and a few typical spectra are shown in Figure 4.79.

Most of the ESR spectra are attributed to monomer radicals with the structure $(4.80-4.82)$:

$RCH_2\dot{C}HX$, (4.80) $RCH_2\dot{C}XY$, (4.81) $RCHX\dot{C}HY$ (4.82)

In some cases there is a second weaker set of lines corresponding to a radical with slightly different coupling constants. For both styrene and vinyl acetate the appearance of the stronger set of lines correlates with the disappearance of radical (4.76) (by addition to monomer), while the weaker set correlates with the disappearance of radical (4.77) (also by addition to monomer) which is the more reactive radical. The two sets of lines are thus attributed to radicals $(4.83-4.85)$:

$HOSO_2M\cdot$, (4.83) $Bu^{t}OSO_2M\cdot$, (4.84) $Bu^{t}OM\cdot$ (4.85)

were M is monomer.

ESR studies of the sulphite radical ion addition to monomers have also been reported[1618].

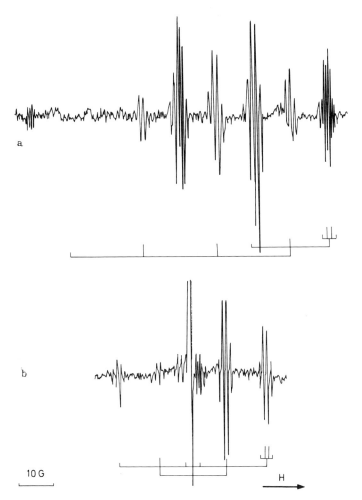

Fig. 4.79. ESR spectra obtained from t-butyl hydroperoxide and sulphur dioxide in the presence of different monomers: a. vinyl acetate, b. methyl methacrylate[692]

4.6.2. Tetraphenylborate-Organic Acid Systems

It was found that the binary system of tetraphenylborate salts and organic acids, e.g. trichloroacetic acid, may serve as radical initiator of the vinyl polymerization[1899]. ESR studies of the reaction of the initiation system show that triphenylboron is produced first in reaction (4.86) and next decomposes into phenyl-radicals (4.86) and diborophenyl radicals (4.87) (reaction 4.87):

$$[B(C_6H_5)_4]^- + CCl_3COOH \longrightarrow B(C_6H_5)_3 + C_6H_6 + CCl_3COO^- \qquad (4.86)$$

$$2\ B(C_6H_5)_3 \longrightarrow \dot{C}_6H_5 + \cdot B_2(C_6H_5)_5 \qquad (4.87)$$
$$\qquad (4.86) \qquad\quad (4.87)$$

119

Table 4.4. Radicals obtained from t-butyl hydroperoxide and sulphur dioxides in the presence of additives[695]

Additive M	Radical	Type of spectrum $A \times B \times C$ (α σ γ)	A (α)	B (β)	C (γ)	g
CH₂=CHX						
Acrylic acid	$RCH_2\dot{B}CH^ACOOH$	2×3	20.4	17.2		2.0037
	$RM_n\cdot$					
Methyl acrylate	$RCH_2\dot{B}CH^ACOOCH_3^C$	2×3	20.4	23.2		2.0036
	$RCH_2\dot{B}CH^ACOOCH_3^C$	$2 \times 3 \times 4$	20.4	16.7	1.5	2.0039
Ethyl acrylate	$RCH_2\dot{B}CH^ACOOCH_2^CCH_3$	$2 \times 3 \times 3$	20.4	16.8	1.4	2.0038
Acrylamide	$RCH_2\dot{B}CH^ACON^CH_2^C$	$2 \times 3 \times 5$	20.0	16.8	2	2.0033
Methyl vinyl ketone	$RCH_2\dot{B}CH^ACOCH_3^C$	$2 \times 3 \times 4$	19.1	15.9	1.0	2.0045
Acrylonitrile	$RCH_2\dot{B}CH^ACN^C$	$2 \times 3 \times 3$	20.4	14.5	3.4	2.0033
Vinyl acetate	$RCH_2\dot{B}CH^AOCOCH_3^C$	$2 \times 3 \times 4$	20.4	12.9	1.4	2.0033
Vinyl propionate	$RCH_2\dot{B}CH^AOCOCH_2^CCH_3$	$2 \times 3 \times 3$	19.5	13.2	1.5	2.0033
Vinyl isobutyrate	$RCH_2\dot{B}CH^AOCOCH^C(CH_3)_2$	$2 \times 3 \times 2$	19.6	12.0	1.0	2.0033
Vinyl chloride	$RCH_2\dot{B}CH^ACl^C$	$2 \times 3 \times 4$	20.8	14.8	2.8	2.0062
Styrene	$RCH_2\dot{B}CH^AC_6H_5^C$	$2 \times 3 \times 3$	15.32	12.12	5.00 (o)	2.0029
		$\times 3$			1.76 (m)	
		$\times 2$		10.6	5.92 (p)	
Hex-1-ene	$RCH_2\dot{B}CH^A(CH_2Pr^n)\dot{S}O_2$	2×5	4.7	2.1		2.0055
Allyl alcohol	$RCH_2\dot{B}CH^A(CH_2OH)\dot{S}O_2$	2×5	4.8	1.6		2.0055
CH₂=CXY						
Methacrylic acid	$RCH_2\dot{B}\dot{C}(CH_3^A)COOH$	4×3	22.6	12.0		2.0035
Methyl methacrylate	$RCH_2\dot{B}\dot{C}(CH_3^A)COOCH_3^C$	$4 \times 3 \times 4$	22.4	12.0	1.4	2.0037
	$RCH_2\dot{B}\dot{C}(CH_3^A)COOCH_3^C$	$4 \times 3 \times 4$	22.4	11.3	1.4	2.0037

Monomer	Radical	Splitting				g-value
Ethyl methacrylate	$RCH_2^B\dot{C}(CH_3^A)COOCH_2^C CH_3$	$4 \times 3 \times 3$	22.4	12.0	1.5	2.0036
Methacrylamide	$RCH_2^B\dot{C}(CH_3^A)CON^C H_2^C$	$4 \times 3 \times 5$	24.4	12.8	2	2.0034
Methacrylonitrile	$RCH_2^B\dot{C}(CH_3^A)CN^C$	$4 \times 3 \times 3$	21.2	11.2	3.2	2.0031
		$4 \times 3 \times 3$	21.8	10.8	3.2	
Itaconic acid	$RCH_2^B\dot{C}(CH_2^A COOH)COOH$	3×3	13.6	12.5		2.0036
Dimethyl itaconate	$RCH_2^B\dot{C}(CH_2^A COOCH_3)COOCH_3^C$	$3 \times 3 \times 4$	13.6	13.2	1.4	2.0036
Isopropenyl acetate	$RCH_2^B\dot{C}(CH_3^A)OCOCH_3^C$	$4 \times 3 \times 4$	22.8	11.6	0.5	2.0033
α-Methylstyrene	$RCH_2^B\dot{C}(CH_3^A)C_6H_5^C$	$4 \times 3 \times 3$ $\times 3$ $\times 2$	16.52	9.40	4.80 (o) 1.64 (m) 5.40 (p)	2.0029
Isobutene	$RCH_2^B\dot{C}(CH_3^A)_2$	7×3	23.0	11.5		2.0033
	$RCH_2C(CH_3)_2SO_2$	1				2.005
CHX=CHY						
Maleic acid	$RCH^B(COOH)\dot{C}H^A COOH$	2×2	20.4	6.4		2.0037
Crotonic acid	$RCH^B(CH_3^C)\dot{C}H^A COOH$	$2 \times 2 \times 4$	20.4	6.0	1.0	2.0033
Cyclopentene	$CH_2^A CH_2^A CH^B=CH^C \dot{C}H^B$	$5 \times 3 \times 2$	22.5	14.4	2.8	2.0030
	$RM\dot{S}O_2$					2.0055
Cyclohexene	$RM\dot{S}O_2$	3	3.6			2.0055
Cycloheptene	$RM\dot{S}O_2$	$2 \times 2 \times 3$	5.4	4.2	2.4	2.0054

Primary radicals for comparison

	$HO\dot{S}O_2$	1	0.28			2.0033
	$(CH_3^A)_3CO\dot{S}O_2$	10	0.9			2.0034
	$CH_3\dot{S}O_2$	4				2.0055

121

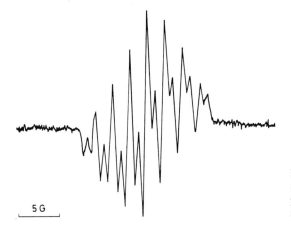

Fig. 4.80. ESR spectrum of free radicals formed in the system tetraphenylboratetrichloroacetic acid at room temperature[1899])

Figure 4.80 shows the ESR spectrum of radical (*4.87*) obtained from reactions (4.86–4.87) in ethanol solution.

The radical (*4.87*) is very stable in ethanol, but unstable in other solvents such as acetonitrile, methyl methacrylate, and dimethylformamide in which its ESR signal could not be observed.

4.6.3. Redox Systems

The most widely used initiation redox system for free radical generation consists of Ti^{3+}–H_2O_2 acting in aqueous solution at $pH < 2$[125–128, 168, 453, 481, 555, 556, 662, 670, 697, 1051, 1460, 1616, 1744, 2004, 2072, 2133, 2134, 2470]). This redox system was first used by Dixon and Norman[555)] for generating HO· and HOO· radicals in the study of intermediates formed during oxidation of alcohols. Reactions occurring in this redox system are:

$$Ti^{3+} + H_2O_2 \longrightarrow Ti^{4+} + HO· + HO^- \tag{4.88}$$

$$HO· + H_2O_2 \longrightarrow H_2O + HOO· \tag{4.89}$$

When aqueous solutions of $TiCl_3$ and H_2O_2 are mixed in the resonance cavity ("flow system for ESR studies", see Chapter 3.6.2) of the ESR spectrometer a narrow singlet spectrum with line width 1 G has been observed, which was attributed to the HO· radical[168, 555)]. In the presence of an alcohol, this spectrum is replaced by a more complex spectrum of alcohol radicals[556)].

Several authors[1744, 2004, 2134)] have shown that this redox system gives two ESR signals which vary in intensity depending mainly on the acidity, ratio of H_2O_2 to $TiCl_3$ (Fig. 4.81), flow rate and temperature.

The concentration of recorded radicals is a function of the flow rate (Fig. 4.82). The plot shows a reciprocal peak heigth of major peaks vs. reciprocal flow rate. At 293 K the concentration of major peak radicals (assigned to HO·) shows a maximum

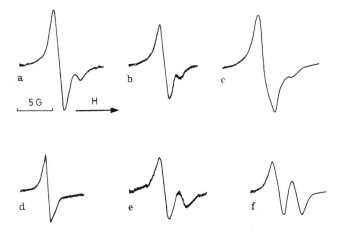

Fig. 4.81. ESR spectra from titanium salts and hydrogen peroxide in aqueous solution: a. TiCl$_3$-(0.01 mole l^{-1})-H$_2$O$_2$ (0.1 mole l^{-1}) at 301 K, b. same concentrations as in (a) at 331 K, c. TiCl$_3$ from anhydrous salt, oxygen excluded at 301 K, d. TiCl$_3$(0.003 mole l^{-1})-H$_2$O$_2$-(0.1 mole l^{-1}) at 301 K, e. TiCl$_3$(0.03 mole l^{-1})-H$_2$O$_2$(0.1 mole l^{-1}) at 301 K, f. TiCl$_3$-(0.09 mole l^{-1})-H$_2$O$_2$(0.1 mole l^{-1}) at 301 K[2004]

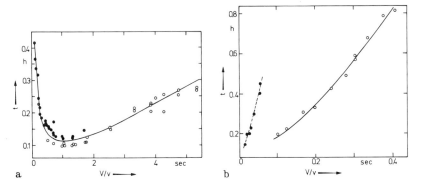

Fig. 4.82. Concentration of hydroxyl radicals as a function of flow rate at: a. 293 K, ● (V = 0.623 cm^3 s^{-1}), ○ (V = 3.5 cm^3 s^{-1}), b. 231 K, ● (V = 0.10 cm^3 s^{-1}), ○ (0.623 cm^{-3} s^{-1})[2004]

as the flow rate is varied, while at 331 K the concentration increases monotoniously with increasing flow rate[2004].

At high H$_2$O$_2$: Ti^{3+} ratio, the following reactions may occur to an appreciable extent[453, 2004, 2134]:

$$Ti^{3+} + H_2O_2 \longrightarrow Ti^{4+} + HO\cdot + HO^- \qquad (4.88)$$

$$HO\cdot + HO\cdot \longrightarrow H_2O_2 \qquad (4.90)$$

$$Ti^{3+} + HO\cdot \longrightarrow Ti^{4+} + OH^- \qquad (4.91)$$

123

$$HO \cdot + H_2O_2 \longrightarrow H_2O + HOO \cdot \tag{489}$$

$$HOO \cdot + HOO \cdot \longrightarrow H_2O_2 + O_2 \tag{4.92}$$

$$HOO \cdot + HO \cdot \longrightarrow H_2O + O_2 \tag{4.93}$$

$$Ti^{3+} + HOO \cdot + H^+ \longrightarrow Ti^{4+} + H_2O_2 \tag{4.94}$$

$$HOO \cdot + H_2O_2 \longrightarrow H_2O + O_2 + HO \cdot \tag{4.95}$$

$$HOO \cdot \rightleftharpoons O_2{}^- + H^+ \tag{4.96}$$

Takakura and Rånby[2134] have shown that the addition of H_2SO_4 to the reacting solution has a very pronounced effect on the relative intensity of the two ESR signals (4.83).

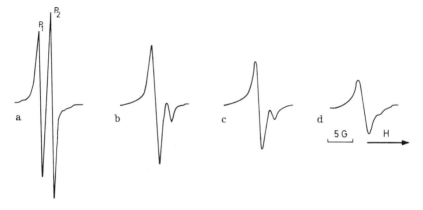

Fig. 4.83. Effect of H_2SO_4 concentration on ESR spectra obtained from reaction between $TiCl_3$ (0.007 mole l^{-1}) and H_2O_2 (0.15 mole l^{-1}) in aqueous solution: a. without H_2SO_4 at pH 1.6, b. with H_2SO_4 (0.011 mole l^{-1}) at ph 1.4, c. with H_2SO_4 (0.044 mole l^{-1}) at pH 1.1, d. with H_2SO_4 (0.264 mole l^{-1}) at pH 0.5, P_1 and P_2 refer to peak 1 and peak 2, respectively[2134]

The principal peak (P_1) is at low magnetic field (g = 2.0128) and the minor peak (P_2) at high field (g = 2.0114). The assignment of the main peak to HO· radicals is questioned in the following arguments: The HO· radicals are highly reactive and are, therefore, expected to be too shortlived to reach a steady-state concentration large enough for ESR observation. Several authors[453, 670, 697, 2004, 2134] have interpreted the two peaks (P_1 and P_2) as being due to HOO· and HO· radicals, respectively, both coordinated with Ti(IV) ions or a Ti(IV)–H_2O_2 complex. This assignment of the ESR spectra to the formation of complexes between HO· and HOO· and Ti(IV)–H_2O_2 is further based on the marked intensity enhancement of the colored Ti(IV)–H_2O_2 complex and the observed Ti hyperfine structure.

According to Fischer's[670] suggestion, the main complex radical species can be Ti–O–O·$^{3+}$ formed by the following reactions:

124

$$Ti^{4+} + H_2O_2 \longrightarrow Ti\underset{O}{\overset{O^{2+}}{<}} | \quad + 2\,H^+ \tag{4.97}$$

$$HO\cdot + Ti\underset{O}{\overset{O^{2+}}{<}} | \quad \rightarrow Ti-O-O^{\cdot 3+} + HO^- \tag{4.98}$$

A more detailed discussion of the kinetics of $Ti-O-O^{\cdot 3+}$ formation was given by Florin[697]. In strongly acidic solution (pH $<$ 1.4) the formation of complexes of HO· and HOO· with other species in the solution should be predominant.

The reaction in the $Ti^{3+}-H_2O_2$ redox system has been studied in the presence of ethylenediamine tetraacetic acid (EDTA) as a chelating agent[734, 735, 905, 2134]. When a $TiCl_3$ solution, containing an equimolar concentration of EDTA, is mixed with H_2O_2 solution in a flow system, the ESR signals due to the two peaks shown in Figure 4.82 disappear, and the reaction mixture is almost discolored. This suggests that the titanium chloride chelate with EDTA, in which EDTA is tightly bound to the Ti coordination sphere, cannot any longer coordinate with the resulting HO· and HOO· radicals. $Ti^{3+}-H_2O_2$ system containing EDTA can initiate the formation of vinyl monomer radicals, e.g. those from vinyl acetate[2134]. This is an additional proof for the existence of HO· and HOO· as free radicals in the system. These radicals are also detectable by ESR spectroscopy under experimental conditions.

Yoshida and Rånby[2468] have found that when 1,3,5-trioxane was added to the $Ti^{3+}-H_2O_2$ reactants, the spectra of the primary HO· radicals disappear and spectra related to trioxane appear as shown in Figure 4.84, tentatively assigned to a hydrogen-bonded radical.

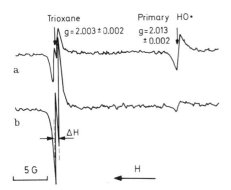

Fig. 4.84. ESR spectra from HO· radicals in a redox system containing trioxane at 296 K. Trioxane concentration: a. (0.083 mole l^{-1}), b. (0.016 mole l^{-1})[2468]

With a sufficient amount of trioxane added, the HO· spectra are completely converted to the trioxane spectra (Fig. 4.84 b). The ESR spectra related to trioxane radicals are doublets with a very small splitting. Their g-value is 2.003 ± 0.002, while g = 2.013 ± 0.002 was recorded for the primary HO· radicals.

Another substantial evidence for the formation of HO· and HOO· complexes with $TiCl_4$ has been given by Takakura and Rånby[2134] and others[125, 127, 128]. They use the well-known redox system $Fe^{2+}-H_2O_2$ in water solution at pH < 2 which does not give an ESR signal, even though HO· and HOO· radicals are known to be formed[556, 905]. Addition of a small amount of $TiCl_4$ to the $Fe^{2+}-H_2O_2$ system gives two intense ESR signals (Fig. 4.85) identical with those observed for the $TiCl_3-H_2O_2$ system (Fig. 4.83).

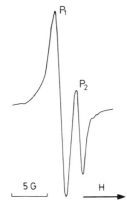

Fig. 4.85. ESR spectrum of free radicals obtained from reaction between $FeCl_2$ (0.007 mole l^{-1}) and H_2O_2 (0.15 mole l^{-1}) in the presence of $TiCl_4$ (0.007 mole l^{-1}) and H_2SO_4 (0.022 mole l^{-1})[2134]

These results may be interpreted as showing that HO· and HOO· radicals generated from the reaction of Fe^{2+} with H_2O_2 are associated with Ti(IV) ions, giving the radical type complexes assigned to the two ESR peaks.

Similar results were obtained with another redox system $Ce^{4+}-H_2O_2$ in aqueous solution at pH < 2[555, 1744, 1972]. In this redox system only HOO· radicals are formed by the following reaction:

$$Ce^{4+} + H_2O_2 \longrightarrow Ce^{3+} + HOO· + H^+ \tag{4.99}$$

Addition of small amounts of $TiCl_4$ to this system gives the characteristic one-line ESR spectrum assigned to a complex $HOO-Ti(IV)-H_2O$[125, 2134]. Addition of other metal ions such as ZrO^{2+}, Hf^{4+}, Th^{4+} and UO_2^{2+} [128] produces the formation of ESR spectra similar to those obtained by addition of very small amounts of $TiCl_4$. It appears that some of the transition metal ions in high and stable oxidation states form complexes with free radical species in oxidation-reduction systems generating characteristic ESR spectra.

Another initiation redox system is the $Ti^{3+}-NH_2OH$ system, where amino radicals are formed[481, 511]:

$$Ti^{3+} + NH_2OH \longrightarrow Ti^{4+} + H_2N· + HO^- \tag{4.100}$$

Amino radicals ($H_2N·$) are not observed by ESR spectroscopy, probably because they react immediately with Ti^{3+} ions:

$$Ti^{3+} + ·NH_2 \longrightarrow Ti^{4+} + NH_2^- \tag{4.101}$$

giving finally a paramagnetic complex.

126

The redox system $Ti^{3+}-(CH_3)_3COOH$ at pH < 2 in aqueous solution produces free methyl radicals in a two-step reaction[555, 556]:

$$Ti^{3+} + (CH_3)_3COOH \longrightarrow Ti^{4+} + HO^- + (CH_3)_3C-O\cdot \qquad (4.102)$$

$$(CH_3)_3C-O\cdot \longrightarrow CH_3-CO-CH_3 + \cdot CH_3 \qquad (4.103)$$

The formation of methyl radicals in these conditions is concluded from the ESR observation of a quartet spectrum of narrow lines with a separation of about 23 G.

Radical species of HO· and HOO· are also formed in the Fenton reagent system $Fe(II)-EDTA-H_2O_2$ [1970].

4.6.3.1. Polymerization Initiated by Redox Systems

Several reviews of this field have been published[672, 685, 1807–1809, 2136]. The conditions for the reactions occurring in ESR flow system cell (see Chapter 3.6.2) are quite different from those of conventional polymerization during a steady-state process, e.g. the free radical concentration is much higher in the flow cell. The hyperfine structure of the ESR spectra observed by the flow technique provides, however, straight-forward information about the structure, concentration, reactivity, and even the steric conformation of the transient radicals involved, particularly for the initial stage of the polymerization.

Mixing of a monomer with the redox system leads to the formation of new radicals by addition:

$$R\cdot + M \longrightarrow RM\cdot \qquad (4.104)$$

where R· is a primary radical from the redox system, M the monomer, and RM· a monomer radical which reacts further:

$$RM\cdot + nM \longrightarrow RM_{n+1}\cdot = P_n\cdot \qquad (4.105)$$

where P· is a growing polymer radical.

To obtain monomer and polymer radicals with concentrations large enough for ESR analysis ($\geqslant 10^{-6}$ M), the initiation rate must be much faster than in conventional polymerizations where radical concentrations are about 10^{-8} M. With the initiating aqueous system $Ti^{3+}-H_2O_2$, the rate of initiation in flow experiments is estimated to be $\geqslant 10^{-2}$ mole liter^{-1} sec^{-1}.

4.6.3.1.1. Acrylic and Methacrylic Acid Monomers. Application of the flow method to investigate radical polymerization of acrylic and methacrylic acid monomers was first made by Fischer[481, 659–662, 678].

The redox polymerization is initiated by addition of HO·, ·NH$_2$·, ·CH$_3$ and ·CH$_2$OH radicals to the monomer. The generation of ·CH$_2$OH radicals for addition to the monomers was achieved with a large excess of methanol in the $Ti^{3+}-H_2O_2$

127

system[481, 665]. After addition of the initiating radicals to monomers, well-resolved ESR spectra assigned to the monomer radicals have been obtained. Only "head radicals" of the structure $R-CH_2-CX_1X_2$ and no "tail radicals" $R-CX_1X_2-\dot{C}H_2$ were detected in the ESR spectra. This result agrees with the general concept that the methylene group of $CH_2=CX_1X_2$ is normally more reactive toward free radical addition than the substituted carbon. Furthermore, head radicals have more resonance stabilization. For some monomers, e.g. acrylic and methacrylic acid, well-defined ESR spectra attributed to growing polymer radicals were also observed when the monomer concentration was increased above certain levels. For acrylic acid the spectrum of polymer radicals was dominant already at a monomer concentration of 1.5×10^{-1} mole 1^{-1} [662].

The coupling constant for various acrylic and methacrylic radicals obtained are summarized in Table 4.5.

Table 4.5. Coupling constants for radicals from acrylic and methacrylic acid[2136]

Radical	Coupling Constants (G)				References
	$a_{H\alpha}$	$a_{H\beta}$	a_{HCH_3}	a_N	

HO$-$CH$_2$$-$$\dot{C}$H
    |
    COOH
	20.45	27.58	–	–	662)

CH$_3$$-CH_2$$-$$\dot{C}$H
    |
    COOH
	20.17	23.78	–	–	678)

NH$_2$$-CH_2$$-$$\dot{C}$H
    |
    COOH
	21.17	25.03	–	3.40	481)

HOCH$_2$$-CH_2$$-$$\dot{C}$H
    |
    COOH
	20.23	22.81	–	–	481)

HO$-$CH$_2$$-$$\dot{C}CH_3$
    |
    COOH
	–	19.98	23.03	–	662)

HO$-$ CH$_2$$-CH-CH_2$$-$$\dot{C}$H
     | |
     COOH   COOH
	22.62
	or
	20.67
		21.34
		
		22.06
			–
			
			–
				–
				
				–
					662)
					
					678)

CH$_3$$-CH_2$$-$$\dot{C}CH_3$
    |
    COOH
	–	15.37	21.83	–	

NH$_2$$-CH_2$$-$$\dot{C}CH_3$
    |
    COOH
	–	16.98	24.19	4.96	481)

HO$-$CH$_2$$-$$\dot{C}CH_3$
    |
    COOH
	–	14.45	22.27	–	481)

HO$-$(CH$_2$$-CCH_3$)$_3$$-CH_2$$-$$\dot{C}CH_3$
     | |
     COOH   COOH
	–	11.04
		13.75
			22.45	–	662)

The variation in the β-coupling constant (a_{H_β}) of $R-\overset{.}{C}H_2-CX_1X_2$ radicals with different R groups (Table 4.5) has been interpreted as largely being caused by steric hinderance to bond rotation of the R group attached to the β-carbon (C_β) atom[664,678]. This is in agreement with the theory of H_β-coupling related to the conformation as described by Eq. (4.106)[904, 2123]:

$$a_{H_\beta} = B_{H_\beta}\rho \cos^2\Theta \tag{4.106}$$

where B_{H_β} is a constant, ρ is the spin density on the α-carbon (C_α) and Θ is the angle between the axis of the $2\,p_z$ orbital of the unpaired electron and the direction of the $C_\beta-H$ bond, both projected on a plane perpendicular to the $C_\alpha-C_\beta$ bond direction (Fig. 4.86).

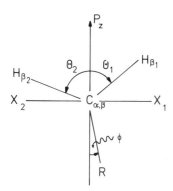

Fig. 4.86. Steric conformation of free radicals of the type:

$$R-C_\beta H_2-\overset{\overset{\displaystyle X_1}{|}}{\underset{\underset{\displaystyle X_2}{|}}{C_\alpha}}\cdot$$

with the $C_\alpha-C_\beta$-axis perpendicular to the paper plane[2136]

If the two β-protons are equivalent, the a_{H_β}-values are given by Eq. (4.107)[662]:

$$a_{H_\beta} = B_{H_\beta}\rho\,\frac{1}{4}(3 - 2\cos^2\Phi) = B_H\rho\,\overline{\cos^2\Theta} \tag{4.107}$$

where Φ is the angle of free rotation between the projection of the $C_\alpha-R$ bond and the axis of the $2\,p_z$ orbital, representing the average position of the substituent R as shown in Figure 4.86, B_{H_β} is 58.6 G[652], and $\overline{\cos^2\Theta}$ is a mean value for all angles attained.

With increasing size of the substituent R, in the order $HO\cdot$, $\cdot CH_3$, $\cdot NH_2$ and $\cdot CH_2OH$ radicals, the a_{H_β} values tend in most cases to decrease. This indicates that the angle of free rotation Φ decreases. The smaller a_{H_β} values for polymer radicals, e.g. acrylic acid, compared with monomer radicals can be explained in a similar way as an effect of much larger size of the group (i.e. the chain) attached to the β-carbon (C_β) atom. From the observed spectra, however, it is difficult to distinguish between dimer, trimer, and larger radicals, because the expected variation in the a_{H_β} values with the degree of polymerization is not significant.

Methacrylic acid has an ESR spectrum with well-resolved hyperfine structure consisting of 16 lines (Fig. 4.87a) when observed during redox polymerization in a flow system[661, 662].

129

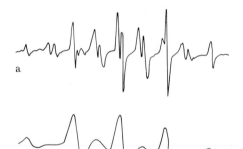

a

b

25 G

H

Fig. 4.87. ESR spectra of polymerization radicals of methacrylic acid: a. measured at 320 K in aqueous solution, b. at 234 K with solid methacrylic acid irradiated at 77 K with 1 MeV electrons[661]

This spectrum is assigned to the growing polymer radicals. According to this interpretation, the two β-protons are not equivalent, since they show two different coupling constant (Table 4.5). In this conformation, the angles Θ_1 and Θ_2 for β_1- and β_2-protons are 63.1° and 56°4, respectively, indicating that the R group of long chains in radical (4.88) is "locked" in relation to the α-carbon, i.e. has a hindered rotation giving two different Θ averages.

$$R-CH_2-\underset{\underset{COOH}{|}}{\overset{\overset{CH_3}{|}}{C}}\cdot \qquad (4.88)$$

Moreover, this interpretation leads to the conclusion, that the so-called "5 + 4-line" spectrum which has been reported for irradiated (1 MeV) methacrylic monomers in glass state (Fig. 4.87) is indeed caused by polymer radicals (4.88).

On the basis of the steric structure and internal rotation of propagating radicals derived from ESR spectra, Fischer[664, 685] discussed the relative occurrence of isotactic and syndiotactic units in the polymer.

A kinetic study of redox polymerization of acrylic acid initiated by methyl radicals ($\cdot CH_3$) in aqueous solutions in the flow system has been made by Fischer[668]. The rate constants obtained under flow system conditions with a rate of initiation $R_i = 1.83 \times 10^{-2}$ mole 1^{-1} sec^{-1} were for propagation $k_p = 0.64 \times 10^5$ 1 sec mole^{-1} and for termination $k_t = 1.52 \times 10^8$ 1 sec mole^{-1}, both of which are about one order of magnitude higher than for conventional steady-state solution polymerization. This was interpreted as a reason for the much shorter chain length of the propagation radicals in the ESR flow experiments.

4.6.3.1.2. Acrylamides. A flow system was applied to obtain high resolution ESR spectra of free radicals produced by the addition of hydroxyl radicals to acrylamide and methacrylamide in aqueous solution[340]. A comparison of the hyperfine coupling constants of the β-protons with those observed for other acrylic monomers leads to the conclusion that the radicals (*4.89*) and (*4.90*) observed contain only one monomer unit:

$$
\underset{\underset{\text{CONH}_2}{|}}{\overset{\overset{\text{H}}{|}}{\text{HO–CH}_2\text{–C·}}} \qquad (4.89) \qquad \underset{\underset{\text{CONH}_2}{|}}{\overset{\overset{\text{CH}_3}{|}}{\text{HO–CH}_2\text{–C·}}} \qquad (4.90)
$$

The two protons in the amide group were found to be magnetically inequivalent what can be explained in two different ways:

1. A difference in NH bond length would be due to a σ–π exchange interaction to induce different spin densities at the amide protons, and

2. The two protons may be slightly out of the molecular NCO plane and to a different extent.

It is important to point out that free radicals from acrylamide and methacrylamide derived in various ways and trapped in a variety of solid matrices give poorly resolved ESR spectra[27, 193, 425, 1929], as compared to the flow method[340].

4.6.3.1.3. Acrylonitrile. Using the TiCl_3–H_2O_2 or TiCl_3–H_2O_2–CH_3OH radical generating method in flow system a few authors have reported ESR studies of acrylonitrile[662, 2039, 2136]. Only the ESR spectrum from the radicals (*4.91*) was observed at low concentration of acrylonitrile.

$$
\underset{\underset{\text{CN}}{|}}{\overset{\overset{\text{H}}{|}}{\text{HO–CH}_2\text{–C·}}} \qquad (4.91)
$$

However, as the acrylonitrile concentration was progressively raised, a new ESR absorption appeared and was assigned to the polymer radicals (*4.92*):

$$
\text{HO}\!\left[\underset{\underset{\text{CN}}{|}}{\overset{\overset{\text{H}}{|}}{\text{CH}_2\text{–C}}}\right]_n\!\!\underset{\underset{\text{CN}}{|}}{\overset{\overset{\text{H}}{|}}{\text{CH}_2\text{–C·}}} \qquad (4.92)
$$

with $n \geqslant 1$.

The coupling constants for radicals obtained in various redox initiation systems are shown in Table 4.6.

4.6.3.1.4. Vinyl Esters. The application of the flow method to investigate radical polymerization of vinyl esters was first made by Rånby and coworkers[2070, 2133, 2137]. The ESR spectra from vinyl acetate radicals (*4.93*) produced in the system of TiCl_3–H_2O_2 are shown in Figure 4.88.

Table 4.6. Coupling constants for radicals from acrylonitrile[2136)]

	Coupling Constants (G)				References
Radical	a_{H_α}	a_{H_β}	a_{HCH_3}	a_N	
HO–CH$_2$–ĊH 　　｜ 　　CN	20.10	28.15	–	3.53	662)
CH$_3$–CH$_2$–ĊH 　　　｜ 　　　CN	20.10	25.19	–	3.53	678)
NH$_2$–CH$_2$–ĊH 　　　｜ 　　　CN	20.45	23.75	–	3.25	481)
HOCH$_2$–CH$_2$–ĊH 　　　　｜ 　　　　CN	20.08	22.89	–	3.44	481)

$$\begin{array}{c} H \\ | \\ HO-CH_2-C \cdot \\ | \\ OCOCH_3 \end{array} \qquad (4.93)$$

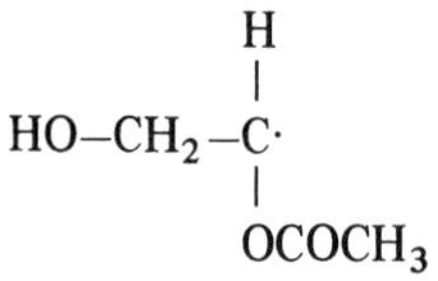

The spectrum shown in Figure 4.88 a, was obtained at an ordinary monomer concentration of 5×10^{-2} mole l^{-1} and can be described as a doublet of triplets of narrow quartets (g = 2.0031). It is assigned to the monomer radical (4.93). The spectrum in Figure 4.88 b shows half of the spectrum obtained at saturated vinyl acetate concentration, i.e. about 0.3 mole l^{-1}. The weak signals superimposed on the spectrum of vinyl acetate monomer radicals (4.93) are considered to be due to growing polymer radicals. The weak signals denoted as P_1 and P_2 in Figure 4.88 a are interpreted as being due to a residue of the initiating system (HOO· and HO· complex forms as described in Chap. 4.6.3).

The ESR spectra for the following vinyl esters monomers in the following initating systems are shown:

TiCl$_3$–H$_2$O$_2$ for isopropenyl acetate (Fig. 4.89) and vinyl butyrate (Fig. 4.90).

TiCl$_3$–NH$_2$OH for isopropenyl acetate (Fig. 4.91) and vinyl propionate (Fig. 4.92), at monomer concentrations below 0.1 mole l^{-1}.

Isopropenyl acetate gave the spectrum (Fig. 4.89) of a quartet of triplets. Each line is further split into a narrow quartet (barely visible) assigned to the expected very weak coupling with the three protons in the acetate group. Vinyl propionate (Fig. 4.92), vinyl butyrate (Fig. 4.90), and isopropenyl acetate (Fig. 4.91) gave almost the same spectra as vinyl acetate (Fig. 4.88), except for the narrow triplets arising from the two protons next to the carbonyl in the ester group. The ESR spectrum from vinyl crotonate was rather weak and the splitting from the ester protons could not be resolved.

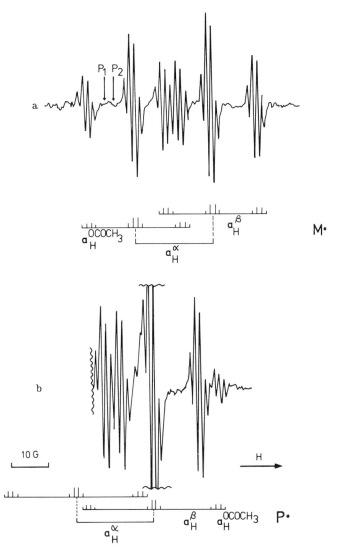

Fig. 4.88. ESR spectra from vinyl acetate in the system $TiCl_3-H_2O_2$: a. Monomer radicals: vinyl acetate concentration (0.055 mole l^{-1}), P_1 and P_2 indicate the position of peaks 1 and 2 respectively, which appear in the absence of monomer and are assigned to HOO· and HO· complexes, b. Monomer and polymer radicals, vinyl acetate concentration (0.3 mole l^{-1}). The stick spectrum shows the hyperfine lines for the radicals attributed to the polymer radicals. $TiCl_3$ (0.007 mole l^{-1}) H_2O_2 (0.11 mole l^{-1}) and H_2SO_4 (0.022 mole l^{-1})[2137]

The assignments of the observed spectra to vinyl ester monomer radicals are verified by using ·NH_2 radicals as initiator. The ESR spectra from these vinyl ester radicals (Fig. 4.91 and 4.92) show the triplet coupling with the nitrogen atom of the NH_2 group attached to the β-carbon. The coupling with the β-protons is almost the same as that observed for the corresponding HO-adduct radicals.

133

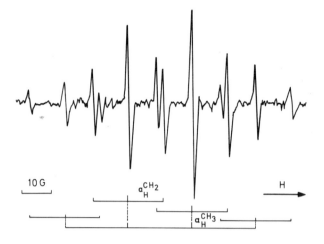

Fig. 4.89. ESR spectrum from isopropenyl acetate in the system $TiCl_3-H_2O_2$. Isopropenyl acetate (0.046 mole l^{-1}), $TiCl_3$ (0.007 mole l^{-1}), H_2O_2 (0.15 mole l^{-1}) and H_2SO_4 (0.022 mole l^{-1})[2137]

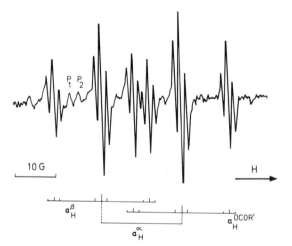

Fig. 4.90. ESR spectrum from vinyl butyrate in the system $TiCl_3-H_2O_2$. Vinyl butyrate (0.035 mole l^{-1}), $TiCl_3$ (0.007 mole l^{-1}), H_2O_2 (0.15 mole l^{-1}) and H_2SO_4 (0.022 mole l^{-1})[2137]

When methyl radicals are used as initiator for vinyl acetate, no detectable amounts of radicals (8.93) are obtained under any experimental conditions. Methyl radicals ($\cdot CH_3$) are less reactive than $HO\cdot$ and $\cdot NH_2$. The reactivity of vinyl acetate is apparently very low compared with that of acrylic monomers[662]. The coupling constants obtained for $HO\cdot$ and $\cdot NH_2$ adduct radicals of various vinyl esters are listed in Table 4.7. As seen from Table 4.7 all vinyl ester monomer radicals show almost the same β-coupling constant independent of the ester group.

There is no significant variation in the a_{H_β} values for vinyl acetate monomer radicals with temperature, i.e. 12 G (281 K), 12.2 G (293 K), and 12.8 G (323 K). These results indicate that the vinyl ester monomer radicals have the $\beta-CH_2$ group linked in respect to the α-carbon group in such a way that the two hydrogens remain

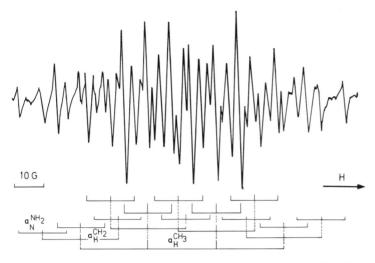

Fig. 4.91. ESR spectrum from isopropenyl acetate in the system $TiCl_3-NH_2OH$. Isopropenyl acetate (0.069 mole l^{-1}), $TiCl_3$ (0.007 mole l^{-1}), NH_2OH (0.25 mole l^{-1}) and H_2SO_4 (0.02 mole l^{-1})[2137]

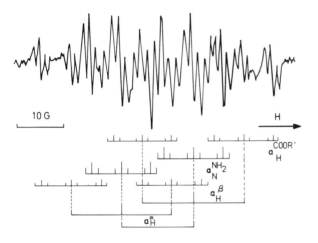

Fig. 4.92. ESR spectrum from vinyl proprionate in the system $TiCl_3-NH_2OH$. Vinyl proprionate (0.056 mole l^{-1}), $TiCl_3$ (0.007 mole l^{-1}), NH_2OH (0.25 mole l^{-1}) and H_2SO_4 (0.02 mole l^{-1})[2137]

equivalent[2137]. The free rotation angle Φ calculated from the a_{H_β} values using equation (4.107) are as follows:

$$
\begin{array}{ccc}
\text{H} & & \\
| & & \\
\text{HO}-\text{CH}_2-\text{C}\cdot & 0° & (4.93) \\
| & & \\
\text{OCOCH}_3 & &
\end{array}
$$

$$
\begin{array}{ccc}
\text{CH}_3 & & \\
| & & \\
\text{HO}-\text{CH}_2-\text{C}\cdot & 14° & (4.94) \\
| & & \\
\text{OCOCH}_3 & &
\end{array}
$$

135

Table 4.7. Coupling constants for various vinyl ester monomer radicals[2136]

Monomer	Initiator	Radicals	Coupling Constants (G)				
			a_{H_α}	$a_{H_{CH_3}}$	a_{H_β}	$a_{H_{OCOR}}$	$a_{N_{NH_2}}$
Vinyl acetate	HO·	$HO-CH_2-\dot{C}H(OCOCH_3)$	20.3 ± 0.1	—	12.2 ± 0.1	1.30 ± 0.04	—
Isopropenyl acetate	HO·	$HO-CH_2-\dot{C}(CH_3)(OCOCH_3)$	—	22.5 ± 0.1	12.5 ± 0.1	0.40 ± 0.03	—
	$H_2N·$	$H_2N-CH_2-\dot{C}(CH_3)(OCOCH_3)$	—	23.0 ± 0.1	13.1 ± 0.1	unresolved	8.2 ± 0.1
Vinyl proprionate	HO·	$HO-CH_2-\dot{C}H(OCOCH_2CH_3)$	20.2 ± 0.1	—	12.3 ± 0.1	1.61 ± 0.05	—
	$H_2N·$	$H_2N-CH_2-\dot{C}H(OCOCH_2CH_3)$	20.5 ± 0.1	—	14.2 ± 0.1	1.62 ± 0.05	8.3 ± 0.1
Vinyl butyrate	HO·	$HO-CH_2-\dot{C}H(OCOCH_2C_2H_5)$	20.1 ± 0.1	—	12.1 ± 0.2	1.34 ± 0.04	—
Vinyl crotonate	HO·	$HO-CH_2-\dot{C}H(OCOCH=CHCH_3)$	20.4 ± 0.3	—	12.3 ± 0.3	unresolved	—
Ethyl acetate	HO·	$CH_3-\dot{C}H(OCOCH_3)$	18.8 ± 0.7	24.0 ± 0.9	—	1.44 ± 0.8	—

H
|
HO–CH$_2$–C·   0°   (4.95)
|
OCOCH$_2$CH$_3$

H
|
H$_2$N–CH$_2$–C·   15°   (4.96)
|
OCOCH$_2$CH$_3$

And for radical where the methyl group is directly attached to a tervalent carbon[2041]:

H
|
CH$_3$–C·   45°   (4.97)
|
OCOCH$_3$

This means that the angle Θ is nearly 60° for both methylene hydrogens (see Fig. 4.86) and that the R group (R = HO, NH$_2$) in the radicals (4.93–4.96) is above or below the radical plane. This phenomenon cannot be understood as being caused by the enhanced size of substituent R, attached to the β-carbon. It seems more likely that some type of intramolecular interaction leading to a locked conformation operates in this case. Two explanations may be suggested[2137]:

1. The formation of an intramolecular hydrogen bond between the hydrogen of the β-hydroxyl group or β-amino group and the carbonyl oxygen of the ester group giving a seven-membered ring structure. This conformation seems to be feasible for a molecular model, although the ring structure is not planar.

2. A titanium chelate of two vinyl ester monomer radicals in which titanium ions may be coordinated between the two polar side groups.

The formation of HO· adduct radicals of vinyl acetate is mainly related to the concentration of initiator, monomer, and sulphuric acid. The intensity of the signal is strongly affected by the molar ratio of H$_2$O$_2$ to TiCl$_3$. The maximum intensity was obtained with a ratio close to r = 15[2137]. When the ratio decreases below r < 10, either by increasing the TiCl$_3$ concentration or by decreasing the H$_2$O$_2$ concentration, the intensity of the ESR signal was considerably reduced. According to simple kinetics, however, the highest intensity is expected at a molar ratio of unity. When the H$_2$O$_2$/TiCl$_3$ ratio was kept constant at the optimal value r = 16 with increasing concentration of the initiator, the intensity at first increased, passed through a maximum, and then decreased. After adding a small amount of ferric chloride, which is known to be a radical scavenger in the reaction system, the signal intensity decreased markedly down to the ESR noise level. These results suggest that species like Ti(IV) ions, derived from the redox reaction, may take part in radical termination, e.g. through a process of electron transfer.

As seen from the spectrum in Figure 4.88b weak signals caused by vinyl acetate polymer radicals were observed at saturated monomer concentration. The coupling

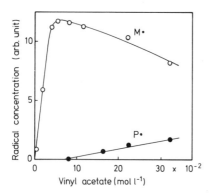

Fig. 4.93. Effect of vinyl acetate monomer concentration on signal intensity of vinyl acetate radicals (M·) and polymer radicals (P·): TiCl$_3$ (0.06 mole l^{-1}), H$_2$O$_2$ (0.10 mole l^{-1}) and H$_2$SO$_4$ (0.02 mole l^{-1})[2137]

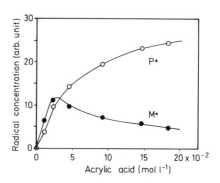

Fig. 4.94. Effect of acrylic acid monomer concentration on signal intensity of acrylic acid radicals (M·) and polymer radicals (P·): TiCl$_3$ (0.06 mole l^{-1}), H$_2$O$_2$ (0.10 mole l^{-1}) and H$_2$SO$_4$ (0.02 mole l^{-1})[2136]

constants obtained from this spectrum are a_{H_α} = 20.3 G and a_{H_β} = 17.5 G and a_{HOCOCH_3} = 1.3 G. The formation of vinyl acetate and polymer radicals at increasing monomer is shown in Figure 4.93.

In most cases no polymer-like substance could be detected in the reaction mixture immediately after mixing both solutions. The very low concentration of vinyl acetate polymer radicals is in marked contrast to the results for acrylic monomers[662], for which intense, well-resolved spectra arising from polymer radicals (Fig. 4.87a) are easily obtained even at monomer concentration below 5 x 10^{-2} mole l^{-1}. For comparison, data obtained for acrylic acid are given in Figure 4.94.

These results suggest that rather than undergo propagation, the highly reactive vinyl acetate radicals could, in the flow system terminate preferentially with species such as HO· radicals, vinyl acetate monomer radicals, or titanium ions. This is probably an effect of the low reactivity of vinyl acetate monomer.

As reported by Takakura and Rånby[2133] the introduction of small amounts of more reactive second monomers, such as acrylonitrile, into the vinyl acetate system gives a well-resolved spectrum containing two new components, one assigned to new monomer radicals (e.g. acrylonitrile radicals) and one attributed to initial copolymer radicals resulting from addition of the second monomer to the vinyl acetate radicals (see next Chapter).

4.6.3.1.5. Copolymerization of Vinyl Acetate and Acrylonitrile with Various Co-monomers.
Copolymerization of vinyl acetate (monomer M$_1$) with acrylonitrile, acrylic acid, acrylamide, maleic acid, and fumaric acid (monomer M$_2$) has been investigated in detail by Takakura and Rånby[2133, 2135]. The ESR spectra obtained from the binary monomer systems containing the vinyl ester and small amounts of monomer M$_2$ show a predominant new signal assigned to the radical species (*4.98*) (Fig. 4.95) and (Fig. 4.96).

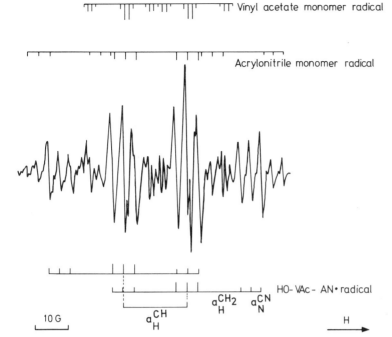

Fig. 4.95. ESR spectrum from a polymerizing system of vinyl acetate and acrylonitrile in aqueous solution at 293 K. The predominant spectral component is assigned to the radical

$$
\begin{array}{ccc}
& \text{H} & & \text{H} \\
& | & & | \\
\text{HO--CH}_2\text{--C} & \text{---} & \text{CH}_2\text{--C}\cdot \\
& | & & | \\
& \text{OCOCH}_3 & & \text{CN}
\end{array}
$$

The concentration of: vinyl acetate (0.055 mole l^{-1}), acrylonitrile (0.019 mole l^{-1}), molar ratio of vinyl acetate: acrylonitrile = 75 : 25[2133])

HO--(vinyl acetate)--M_2. $\qquad$ (4.98)

These spectra are distinguishable from the corresponding monomer radical spectra (4.99) by the difference in the a_{H_β} values.

HO--$M_1\cdot$ and HO--$M_2\cdot$ $\qquad$ (4.99)

The initial copolymer radicals are characterized by their low a_{H_β} values. No evidence of the reverse type of copolymer radicals, i.e. HO--M_2-(vinyl acetate)· was observed. This is a clear demonstration that vinyl acetate monomer has low reactivity, while vinyl acetate monomer radicals are highly reactive.

The coupling constants for the initial copolymer radicals are given in Table 4.8, together with those of the corresponding monomer radicals:

139

Table 4.8. Coupling constants for free radicals obtained from single and binary monomer systems[2136]

Monomer substrate	Radical	Coupling constant (G)		Others
		a_{H_α}	a_{H_β}	
Acrylonitrile	$HOCH_2\dot{C}H(CN)$	20.1 ± 0.1	28.2 ± 0.1	$a_N C = 3.45 \pm 0.07$
Vinyl acetate + Acrylonitrile	$HOCH_2CHCH_2\dot{C}H(CN)$ $\;$ OCOCH$_3$	20.3 ± 0.2	20.6 ± 0.2	$a_N C = 3.42 \pm 0.07$
Isopropenyl acetate + Acrylonitrile	$HOCH_2CCH_3CH_2\dot{C}H(CN)$ $\;$ OCOCH$_3$	20.0 ± 0.2	19.3 ± 0.2	$a_N CN = 3.33 \pm 0.07$
Acrylic acid	$HOCH_2\dot{C}H(COOH)$	20.6 ± 0.1	27.8 ± 0.2	
Acrylic acid (Dimer radical)	$HOCH_2CHCH_2\dot{C}H(COOH)$ $\;$ COOH	21.1 ± 0.2	22.6 ± 0.2	
Vinyl acetate + Acrylic acid	$HOCH_2CHCH_2\dot{C}H(COOH)$ $\;$ OCOCH$_3$	20.4 ± 0.2	21.2 ± 0.2	
Maleic acid	$HOCH(COOH)CH(COOH)$	20.80 ± 0.05	12.7 ± 0.1	
Vinyl acetate + Maleic acid	$HOCH_2CHCH(COOH)\dot{C}H$ $\;$ OCOCH$_3$ $\;$ COOH	20.75 ± 0.05	11.1 ± 0.1	
Methacrylic acid	$HOCH_2\dot{C}CH_3(COOH)$	—	19.9 ± 0.1	$a_H CH_3 = 23.0 \pm 0.1$
Methacrylic acid (Dimer radical)	$HOCH_2CCH_3CH_2\dot{C}CH_3$ $\;$ COOH $\;$ COOH	—	13.8 ± 0.2 ; 11.0 ± 0.2	$a_H CH_3 = 22.4 \pm 0.2$
Vinyl acetate + Methacrylic acid	$HOCH_2CHCH_2\dot{C}CH_3(COOH)$ $\;$ OCOCH$_3$	—	15.5 ± 0.2 ; 9.5 ± 0.2	$a_H CH_3 = 22.4 \pm 0.2$; $a_H COCH_3 = 1.8 \pm 0.3$
Acrylamide	$HOCH_2\dot{C}H(CONH_2)$	20.1 ± 0.3	26.6 ± 0.3	$a_N CONH_2 = 2.0 \pm 0.3$
Vinyl acetate + acrylamide	$HOCH_2CHCH_2CH(CONH_2)$ $\;$ OCOCH$_3$	20.0 ± 0.3	21.0 ± 0.3	$a_H CONH_2 = 1.8 \pm 0.3$; $a_N CONH_2 = 2.0 \pm 0.3$

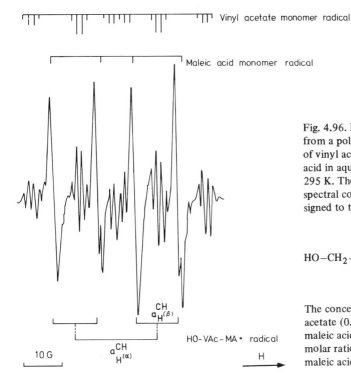

Vinyl acetate monomer radical

Maleic acid monomer radical

Fig. 4.96. ESR spectrum from a polymerizing system of vinyl acetate and maleic acid in aqueous solution at 295 K. The predominant spectral component is assigned to the radical:

$$\underset{\underset{OCOCH_3}{|}}{HO-CH_2-\overset{\overset{H}{|}}{C}}\underset{\underset{COOH}{|}}{\overset{\overset{H}{|}}{C}}\underset{\underset{COOH}{|}}{\overset{\overset{H}{|}}{C}}\cdot$$

The concentration of: vinyl acetate (0.055 mole l^{-1}), maleic acid (0.021 mole l^{-1}), molar ratio of vinyl acetate: maleic acid = 72 : 28[2133])

HO-VAc-MA • radical

$a_{H}^{CH_{(\beta)}}$

$a_{H}^{CH_{(\alpha)}}$

10 G

H

The small a_{H_β} values for copolymer radicals indicate that the rotation of the RCH_2 group around the C–C bond is restricted to a greater extent than in the corresponding monomer radicals. This is probably due to steric hindrance of the large vinyl acetate units R, attached to the β-carbon atom. In comparison the HO groups on the corresponding monomer radicals are small.

The relative concentration of the various radical species during copolymerization was estimated from the intensities of the ESR spectral components. Typical results obtained from vinyl acetate-maleic acid and vinyl acetate- fumaric acid systems are shown in Figure 4.97, giving the relative radical concentration as a function of co-monomer concentration.

The addition of small amounts of comonomer M_2 (molar ratio $M_2/M_1 \leqslant 0.05$) causes in each case a sharp decrease in the concentration of vinyl acetate monomer radicals, while the concentration of the corresponding copolymer radical (4.98) increases. This indicates a rapid reaction of vinyl acetate monomer radicals with comonomer (M_2), resulting in an almost instantaneous transformation to radicals (4.98). As expected, maleic acid and fumaric acid give the same ESR spectrum. Fumaric acid (trans isomer) reacts much faster than maleic acid (cis isomer) with vinyl acetate monomer radicals. Plots of the free radical concentration vs. concentration of co-monomer (M_2) are shown in Figure 4.98.

The data give straight lines with a characteristic slope, which is a measure of the relative rate of conversion of HO$-M_1\cdot$ to HO$-M_1-M_2\cdot$ for different M_2 comonomers,

141

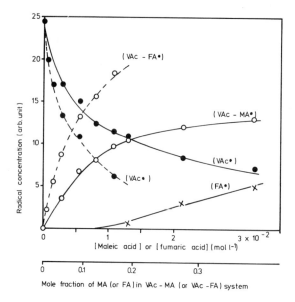

Fig. 4.97. Concentration of different radicals measured from ESR spectra during copolymerization of vinyl acetate (VAc) with maleic acid (MA) and vinyl acetate (VAc) with fumaric acid (FA) at different molar concentrations of MA in the VAc-MA system and FA in the VAc-FA system, respectively. Concentration of vinyl acetate (0.055 mole l^{-1}) whereas concentration of maleic acid and fumaric acid variable[2133]

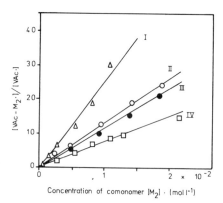

Fig. 4.98. Plots of concentration ratio VAcM$_2$·/VAc· vs concentration of comonomer M$_2$ for the systems: I-vinyl acetate (VAc)-fumaric acid, II-vinyl acetate-acrylonitrile, III-vinyl acetate-acrylic acid, IV-vinyl acetate maleic acid. Vinyl acetate concentration (0.055 mole l^{-1})[2136]

i.e. the slope measures the relative reactivity of M_2 comonomers to vinyl acetate radicals.

Izumi and Rånby[1040, 1809] reported that copolymerization studies of acrylonitrile (M_1) with various comonomers (M_2) gave interesting results for comparison. The reactivity of the acrylonitrile monomer is high and that of the acrylonitrile radical ($M_1 \cdot$) is low. Therefore copolymer radicals HO$-M_2-M_1 \cdot$ are formed and their relative concentration can be used to measure the reactivity of different monomer radicals ($M_2 \cdot$).

4.6.3.1.6. Allyl Monomers. ESR measurements using the flow technique are particularly useful for polymerization studies of allyl and methallyl monomers[481, 564, 1039, 1041, 1042, 2040, 2043]

Smith et al.[2043] studied allyl alcohol using the $TiCl_3-H_2O_2$ system and observed from ESR spectra at low allyl alcohol concentration the following free radicals (4.100) and (4.101):

$$\begin{matrix} CH_2-\overset{.}{C}H-CH_2 \quad (4.100) \\ | \qquad\quad | \\ OH \qquad OH \end{matrix} \qquad \begin{matrix} \cdot CH_2-CH-CH_2 \quad (4.101) \\ | \quad\; | \\ OH\;\; OH \end{matrix}$$

and a relatively weak signal from the hydrogen-atom abstraction product (4.102) or (4.103):

$$CH_2{=}\overset{.}{C}H{-}CH \rightleftharpoons \cdot CH_2{-}CH{=}CH \qquad (4.103) \qquad\qquad\qquad\qquad (4.108)$$
$$\quad\;\; | \qquad\qquad\qquad\qquad |$$
$$\quad\;\; OH \qquad\qquad\qquad\quad\;\; OH$$

(4.102)

As the concentration of allyl alcohol was raised, an additional, less intense, spectrum became increasingly apparent and was attributed to allyl alcohol polymeric radicals.

Several authors[481, 1039, 1041, 2040] have presented extensive studies of the effects of three types of free radicals $HO\cdot$, $\cdot NH_2$, and $\cdot CH_3$ from redox systems on the polymerization of different allyl and methallyl monomers as alcohols, carboxylic esters, ethers, sulphonated and nitrogen-containing monomers. The main result of these studies is the observation that all allyl compounds (except methallyl alcohol) add $HO\cdot$ radicals both at the tail and the head end of the monomer, the tail-addition (giving head radicals) always being the most frequent. With methallyl alcohol, however, only head radicals were observed. These authors have presented valuable data on the coupling constants and g-values for the observed free radicals. The coupling constants of $\beta-CH_2$ protons vary considerably with the substituents. The $a_{H\beta_1}$ coupling constants (CH_2 to which the initiator is added) decrease in the order $HO > CH_3 > NH_2$ and the $a_{H\beta_2}$ coupling constants (allyl hydrogen, CH_2 with substituent) decrease in the order $CH_2OH > CH_2NH_2 > CH_2OCOCH_3 > CH_2SO_3Na$, i.e. decrease with increased size of the substituent groups[1041].

Izumi and Rånby[1041, 1042] have made a detailed ESR study of the mechanism of radical polymerization of methallyl monomers with redox systems. In the case of $\cdot NH_2$ and $HO\cdot$ addition to methallyl alcohol, methallyl amine, and sodium methallyl sulphonate, the ESR spectra of the reacting species are interpreted as being due to monomer head radicals only ($\cdot NH_2$ and $HO\cdot$ are added to the monomer tail). Methallyl acetate with $HO\cdot$ is an exception, giving hydrogen abstraction to form an allyl type radical. This reaction may influence the polymerization behavior of methallyl alcohol. This monomer behaves differently from the other allyl monomers, in which appreciable amounts of monomer tail radicals were found, in addition to the head radicals which are the main species. For methallyl monomers, this may be due to steric hindrance caused by the two substituents on the α-carbon. Methyl radicals add only to positively polarized reactive double bonds. The coupling constant of $\beta-CH_2$ protons vary considerably with the substituents. For H_{β_1}, the coupling constant ($a_{H_{\beta_1}}$) decreases in the order $HO > CH_3 > NH_2$. For H_{β_2} (allyl hydrogen) the

143

coupling constants ($a_{H_{\beta_2}}$) decrease in the order $CH_2OH > CH_2NH_2 > CH_2OCOCH_3 > > CH_2SO_3Na$, i.e. the constants decrease in the order of increased size of the groups.

Using the ESR technique Izumi and Rånby[1040] have also investigated the co-polymerization of several allyl and methallyl monomers (M_1) with acrylonitrile. All results show the very low reactivity of the monomers M_1 and their radicals towards copolymerization with acrylonitrile. The predominant copolymer radical formed is M_1(acrylnitrile)·, which is characterized by lower a_{H_β}-values than the acrylonitrile monomer radical. This is interpreted as being due to restricted rotation caused by steric hindrance from the M_1 units attached to the β-carbon of the acrylonitrile radicals.

Interesting results have been obtained in the ESR study of trimethylolpropane monoallyl ether[564]. The observed ESR spectra (Fig. 4.99) were assigned to the following free radicals (*4.104*)(*4.105*) and/or (*4.106*)(*4.107*):

$$HO-CH_2-\overset{\cdot}{C}H \qquad (4.104) \qquad \cdot CH_2-CH-OH \qquad (4.105)$$
$$| \qquad\qquad\qquad\qquad\qquad\quad |$$
$$CH_2 \qquad\qquad\qquad\qquad\qquad CH_2$$
$$| \qquad\qquad\qquad\qquad\qquad\quad |$$
$$OR \qquad\qquad\qquad\qquad\qquad\quad OR$$

$$CH_2{=}CH \qquad \cdot CH_2-CH \qquad\qquad\qquad\qquad\qquad\qquad (4.109)$$
$$| \qquad\qquad\qquad \|$$
$$\cdot CH \rightleftharpoons CH$$
$$| \qquad\qquad\qquad |$$
$$OR \qquad\qquad\quad OR$$

(*4.106*)   (*4.107*) CH_2OH
$$|$$
where R is the group $-CH_2-C-CH_3$
$$|$$
$$CH_2OH$$

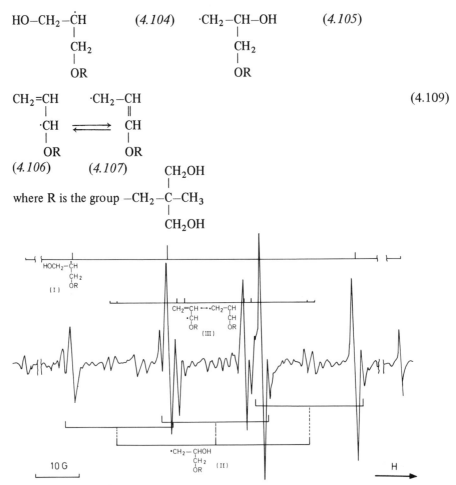

Fig. 4.99. ESR spectrum of trimethylolpropane monoallyl ether radicals. Concentration of monomer (5.5 ml l^{-1}), TiCl$_3$ (0.009 mole l^{-1}), H$_2$O$_2$ (0.15 mole l^{-1}) and H$_2$SO$_4$ (0.05 mole l^{-1})[564]

144

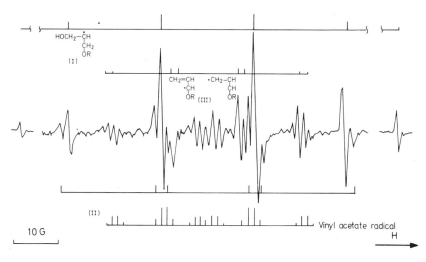

Fig. 4.100. ESR spectrum from a copolymerizing system of vinyl acetate and trimethylolpropane monoallyl ether. Concentrations of: vinyl acetate (0.041 mole l^{-1}), trimethylolpropane mono-allyl ether (7.5 ml l^{-1}), $TiCl_3$ (0.009 mole l^{-1}), H_2O_2 (0.041 mole l^{-1}) and H_2SO_4 (0.05 mole l^{-1})[564]

ESR spectra from copolymerizing system of vinyl acetate (Fig. 4.100) and methyl acrylate with trimethylol-propane monoallyl ether were also studied.

The ESR spectrum shown in Figure 4.100 shows the presence of three radicals (*4.104*), (*4.105*), and/or (*4.106*) (*4.107*). In the system methyl acrylate/trimethylol-propane monoallyl ether, the ESR spectrum was mainly due to the radicals (*4.104–4.107*) in which the allylic radicals were predominant. This result is consistent with the observation that polymerization of methyl acrylate is strongly retarded by the

Table 4.9. Coupling constants for HO-adduct radicals of butadiene, vinyl fluoride and vinyl chloride[2136]

Radicals	Coupling Constants (G)			
	$^aH_\alpha$ (doublet)	$^aH_\beta$ (triplet)	a_x	References
$HO-CH_2-\overset{\cdot}{C}H-CH=CH_2$ ⬇⬆	14.9	13.8, 12.6	–	2470)
$HO-CH_2-CH=CH-\overset{\cdot}{C}H_2$	14.4, 4.0	13.4, 12.4	–	
$HO-CH_2-\overset{\cdot}{C}H$ | F	19.1	11.2	57.6 (X=F)	821)
$HO-CH_2-\overset{\cdot}{C}H$ | Cl	17.7	14.4	3.0 (X=Cl)	821)

addition of trimethylol-propane monoallyl ether, which is interpreted as being due to the high reactivity of methyl acrylate radicals towards the trimethylol-propane monoallyl ether monomer.

4.6.3.1.7. Miscellaneous Monomers. ESR measurements using the flow technique have also been applied to polymerization studies of ethylene, vinyl chloride, and vinyl fluoride[550, 821]. The spectra obtained are assigned to monomer radicals formed by hydroxyl radical addition to the methylene group of the monomers. However, no spectrum attributed to propagating radicals has been observed, probably because of the very limited solubility, with a resulting low concentration of these monomers in aqueous media. The coupling constants for some of the monomer radicals are summarized in Table 4.9.

The small a_{H_β} values recorded for HO-adduct radicals of vinyl fluoride and vinyl chloride may be interpreted as being due to steric effects.

Yoshida and Rånby[2470] studied polymerization of butadiene using HO· radicals as initiator. The observed spectrum is assigned to the radical (*4.108*) or (*4.109*) formed by hydroxyl addition to the monomer:

$$HO-CH_2-\dot{C}H-CH=CH_2 \rightleftharpoons HO-CH_2-CH=CH-\dot{C}H_2 \quad (4.109) \qquad (4.110)$$
$$(4.108)$$

The main product of the reaction of hydroxyl radicals generated from the system of ferrous ions and H_2O_2 with butadiene is known to be 1,8-dihydroxy-2,6-octadiene[821]. This product is formed by dimerization of radical (*4.109*). Griffiths et al.[821] also studied butadiene initiated with HO· and obtained almost the same ESR spectrum. They recorded, moreover, the spectrum obtained by addition of ·NH$_2$ to butadiene. This spectrum was, however, highly complex, presumably because isomeric radicals were formed, and an unequivocal assignment was not possible.

ESR measurements using flow technique have proved to be very useful for studies of initiation reactions of 1,2-substituted unsaturated monomers such as: maleic and fumaric acid[557, 662, 678], crotonyl alcohol, crotonic acid, crotonitrile and croton-aldehyde[557, 662, 821, 1043, 2039], and copolymerization of these monomers with acrylonitrile[392].

Smith et al.[2036] investigated the addition of hydroxyl radicals to several oximes: acetaldoxime, acetoxime, and 2,3-butadienone monoxime in flow system. The basic reaction in two steps may be represented as follows:

$$(CH_3)_2C=N-OH + \cdot OH \longrightarrow (CH_3)_2\underset{\underset{OH}{|}}{\overset{\cdot}{C}-N}-OH \qquad (4.110) \qquad (4.111)$$

$$(CH_3)_2\underset{\underset{OH}{|}}{C-\dot{N}}-OH \longrightarrow (CH_3)_2\underset{\underset{OH}{|}}{-C}-NH-O\cdot \qquad (4.111) \qquad (4.112)$$

In each case a β-hydroxy nitroxide radical (*4.111*) is formed.

146

In conclusion the flow technique combined with ESR spectroscopy is an extremely powerful method, because it allows direct observations and measurements of the radical formation in the initiation stage of redox reactions in polymerization. The method also allows the characterization of several free radicals simultaneously present, which is important, e.g. in copolymerization studies.

4.7. Anionic Polymerization

4.7.1. Introduction

Two types of anionic polymerization are of special interest for ESR study:
  1. Alkali metal initiated anionic polymerization:

$$
\begin{array}{c}
R \\
| \\
CH_2=C^+ \\
| \\
R
\end{array}
+
\begin{array}{c}
\text{alkali metal (M)} \\
\text{propagating cation} \\
\text{(Li, Na, K)}
\end{array}
\longrightarrow
\begin{array}{c}
R \\
| \\
-CH_2-C^- \\
| \\
R
\end{array}
+ M^+ \qquad (4.113)
$$

where M^+ represents here the positive counterion ("Gegenion").
  2. Anion radial initiated anionic polymerization: The transfer of an electron from an "electron donor" (D or D^-) to monomer (M) ("acceptor") (anion radical initiation):

$$D + M \longrightarrow D^+ + M^{\overline{\cdot}} \qquad (4.114)$$
$$D^- + M \longrightarrow D + M^{\overline{\cdot}} \qquad (4.115)$$

where $M^{\overline{\cdot}}$ represents the "primary radical ion" formed in the "electron transfer process":

$$(4.116)$$

$$(4.117)$$

$$(4.112)$$

The primary radical-ion ($M^{\overline{\cdot}}$) is neither a conventional radical nor a carbanion and its structure may be represented by two resonating forms (*4.112*), which implies that one end of $M^{\overline{\cdot}}$ acts as a radical while the other acts as a carbanion (or vice versa) (4.117). By addition of one monomer molecule to $M^{\overline{\cdot}}$ a dimer is formed, possessing one "true" radical end and one "true" carbanion end:

147

$$\overset{\cdot}{C}H-CH_2^- \; + \; CH=CH_2 \longrightarrow \overset{\cdot}{C}H-CH_2-CH_2-CH^-$$

(4.118)

(4.112)   (4.113)

There is no interaction between these two centers. Such a dimer can result in further polymerization of a dual character. The radical end might grow according to a radical mechanism, while the carbanion end as an anionic polymerization. Monomeric radical ions ($M^{\bar{}}$) may also dimerize and form dicarbanions (4.114):

$$2 \; \overset{\cdot}{C}H-CH_2^- \longrightarrow {}^-CH-CH_2-CH_2-CH^-$$

(4.119)

(4.112)   (4.114)

which may continue to grow by the anionic mechanism only.

A large number of radical anions have been prepared by reducing the parent hydrocarbon with alkali metals[749, 1052, 1350, 1687, 1827, 2119]. ESR studies of radical ions e.g. of benzene and toluene formed in reactions with Na or K metal mirrors in the presence of crown ethers, have been reported[1180]. Panayotov et al.[1695] found experimental evidence for the formation of the anion radical of benzene as a product of the interaction between benzene and potassium in the presence of poly(ethylene oxide). The ESR spectrum of a fresh "red" colored solution is shown in Figure 4.101.

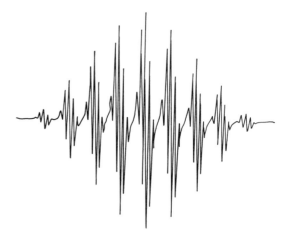

Fig. 4.101. ESR spectrum of the "red" solution of benzene and potassium in the presence of poly(ethylene oxide)[1695]

Fig. 4.102. ESR spectrum of the "brown-green" solution obtained after UV irradiation of the "red" solution (see Fig. 4.101)[1695]

Under UV-light the red solution changes to brown-green giving a new ESR spectrum (Fig. 4.102).

148

The interpretation is that the benzene anion radical (*4.115*) passes into the more stable biphenyl anion radical (*4.116*):

$$\left[\bigcirc\right]^{\cdot-} \longrightarrow \left[\bigcirc\!\!-\!\!\bigcirc\right]^{\cdot-} \tag{4.120}$$

(*4.115*)   (*4.116*)

ESR spectroscopy of radical ions is given in several monographs[20, 109, 259, 1072] and will not be discussed in detail here. Only a few examples directly attributed to ionic polymerization will be given.

4.7.2. ESR study of Ion Pairs, Free Ions and Electron-transfer Reactions

The structure of ion pairs and free ions in solution and the mechanism of electron-transfer reactions between a radical ion and the corresponding molecule, pose some of the most interesting problems concerning the relationship between reactivities and structures of radical ions in solutions.

Weissman et al.[17, 93] have shown that the original hyperfine lines of radical anions split into multiples on their association with cations possessing nuclear spin. The magnitude of the new hyperfine splitting constants may yield valuable information about the location of the cation within the pair, and the width and shape of the lines may reveal details of the dynamic processes involving these species[2136].

Weissman and his coworkers[1064, 2385, 2482] have been among the first to establish methods for measuring the rapid rates of electron-transfer reactions by ESR. As an example they have shown that the rate of electron exchange between the naphthalenide radical ions and naphthalene depends on the state of ionic association of the radical ions. The rate of electron transfer can be determined from the broadening of ESR lines at slow and at fast exchange rate limits. Figures 4.103 and 4.104

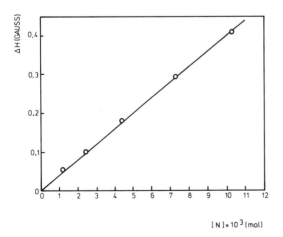

Fig. 4.103. Dependence of ESR line width ΔH on naphthalene concentration [N] in the slow exchange $N^{\cdot-} + N \rightleftarrows N + N^{\cdot-}$ in hexamethylphosphoroamide[952]

149

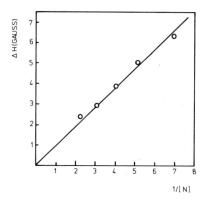

Fig. 4.104. Dependence of ESR line width ΔH on $1/[N]$ (Compare Fig. 4.103) in the fast exchange $N^{\cdot-} + N \rightleftharpoons N + N^{\cdot-}$ in hexamethylphosphoro-amide[952]

show typical plots of $\Delta H(G)$ vs $[N]$ derived from experiments performed at a slow (4.103) and a fast (4.104) exchange limit, respectively, for the electron-transfer reaction between naphthalenide radical anions and naphthalene molecules in hexamethyl-phosphoramide[952].

The surrounding of an ion pair by solvent molecules is dependent on the nature of both the pair and the solvent and also on the temperature. The hyperfine splitting constant (a_M), calculated from the ESR spectra of paramagnetic ion pairs, usually increases at higher temperatures. The structure of an ion pair may be nearly fixed at a certain temperature ("static model"). This means that the position of a cation with respect to the anion in the pair and the locations of the surrounding solvent molecules do not deviate substantially from their average configuration. The fluctuations are rapid with a frequency greater than 10^{10} sec^{-1} [952, 2120].

In some systems ion pairs may have two or more distinct structures, i.e. two or more distinct patterns produced by the pair and the neighboring solvent molecules may coexist, each pattern lasting for a relatively long time ($\sim 10^{-10}$ s) ("dynamic model"). The observed properties of the system again depend on temperature, as well as on the lifetime of each type of ion pair. For example, two sets of lines may be seen in the ESR spectrum of paramagnetic ion pairs, if they form two distinct and slowly interconverted types. The usual broadening, coalescence, and subsequent sharpening of the lines should be observed as the temperatures rises and the rate of interconversion increases. The distinction between the two models (static and dynamic) of ion pairs is not easy when only one splitting constant (a_M) is deduced from the ESR spectrum.

Hirota et al.[947, 948] have investigated ESR spectra of alkali salts of naphthalenide and anthracenide ions and reported that the dynamic model of ion pairs shows a characteristic dependence of a_M on temperature. Thus, for equilibrium of: ion pair (I) $\rightleftharpoons$ ion pair (II), the related equilibrium constants can be calculated from the experimental data if a_{M_I} and $a_{M_{II}}$, pertaining to the individual ion pairs, are temperature-independent. The dynamic model predicts also dependence of the width of the ESR lines on M_z (atomic number) of the cation[948].

The character of ion pairs may be modified either by varying the solvent or the temperature of the solution. Szwarc et al.[952] have shown that it is also possible to modify the character of an ion at constant temperature in a virtually unchanged

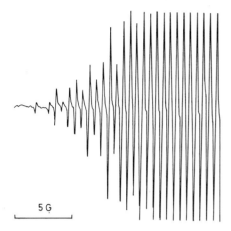

Fig. 4.105. ESR spectrum of sodium naphthalenide ($N^{\cdot -} Na^{+}$) in tetrahydropyran at 300 K[952)]

5 G

medium, by the addition of small amounts of agents, e.g. polyadenate ethers (glymes) which strongly interact with the ion pairs.

The ESR spectrum of sodium naphthalenide ($N^{-} Na^{+}$) in tetrahydropyran is shown in Figure 4.105.

The hyperfine lines of naphthalenide are split into quadruplets by the presence of ^{23}Na. The magnitude of the respective splitting varies from about 1.4 G at 320 K to 1.0 G at 236 K (Table 4.10).

Table 4.10. Temperature dependence of hyperfine splitting constant of Na in tight ion pair and in the glymated pair in tetrahydropyran[952)]

Temp. K K	a_{Na} (tight pair), G	a_{Na} (glymated pair) G
316	1.37	0.39
300	1.23	0.39
297	1.19	0.39
279	1.16	0.33
275.5	–	0.23
270.5	–	0.17
258	1.08	–
241	1.01	0.0
235.5	1.01	0.0

On the addition of tetraglyme (~ 0.24 M) the a_{Na} is greatly reduced. The ESR spectrum at room temperature is shown in Figure 4.106.

The relevant splitting constant (a_{Na}) is about 0.4 G. Lowering the temperature of the solution reduces a_{Na} still further, and at 236 K the four lines merge into one (Table 4.10). The variation of line width permits calculation of the rate constant of interconversion.

151

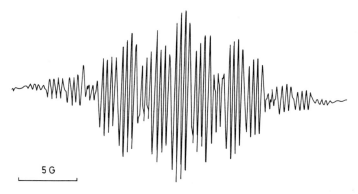

5 G

Fig. 4.106. ESR spectrum of $N^{\cdot-}$, G, Na^+ in tctrahydropyran in the presence of 0.24 mole l^{-1} tetraglyme at 300 $K^{952)}$ (see text)

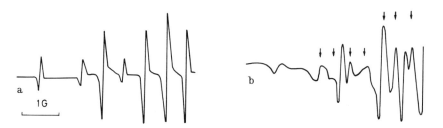

a

1 G

b

Fig. 4.107. Low-field wing of ESR spectrum of $N^{\cdot-}$, Na^+ in tetrahydrofuran at 300 K: a. in the absence of tetragylme, b. in the presence of 0.006 mole l^{-1} tetraglyme[952)]

At room temperature and for intermediate concentrations of glyme, two sets of lines were recognized in the spectrum; those characteristic of the "tight" pairs present in the pure tetrahydropyran, and those identified as the lines of the glymated pairs which were observed in the presence of an excess of glyme (Fig. 4.107).

The lines corresponding to the glymated pairs are marked in Figure 4.107b by arrows. In this system the following equilibrium occurs:

$$N^{\cdot-} Na^+ + G \underset{k_2}{\overset{k_1}{\rightleftharpoons}} N^{\cdot-}(G)Na^+ \tag{4.121}$$

The width of the lines seen in Figure 4.107b gives the values of k_2 and k_1, namely $\sim 10^6 \ M^{-1} \ s^{-1}$ and $10^8 \ M^{-1} \ s^{-1}$, respectively.

Szwarc et al.[21)] have applied the ESR method with great sucess to the study of electron-transfer exchange between durosemiquinone triplet-ions ($DQ^{\cdot-} 2Na^+$) and the parent quinone in tetrahydrofurane. Earlier ESR studies of semiquinone radical ions[799, 800)] provided direct evidence for the existence of symmetric triple-ions ($Na^+DQ^{\cdot-}Na^+$) where two sodium cations interact with the unpaired electron. The ESR spectrum of durosemiquinone triple-ion is shown in Figure 4.108. Both sodium ions are transferred in the process of exchange between durosemi-quinone triple-ions ($DQ^{\cdot-} 2Na^+$) and the parent quinone. The ESR spectrum, in the

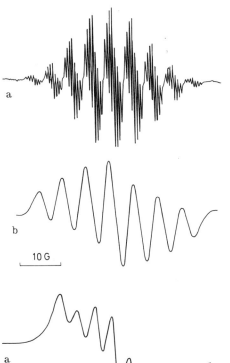

Fig. 4.108. ESR spectrum of durosemiquinone triple-ion in tetrahydrofurane at 303 K: a. Complete spectrum, b. Expanded spectrum showing the splitting from two equivalent sodium cations

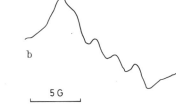

Fig. 4.109. Fast electron transfer ESR spectra at 310 K in the presence of 2 M duroquinone: a. Triple-ion, b. Ion-pair[21]) (see text)

presence of 2 M duroquinone, collapses into 7 lines. It was suggested[21]) that the complexes (DQ$^{\cdot-}$ Na$^+$DQ) and (DQ$^{\cdot-}$ 2 Na$^+$DQ) are formed under these conditions. Fast electron-transfer spectra have been observed in solution saturated with duroquinone (2 M) as shown in Figure 4.109.

It is probable that the exchange rates for the ion pairs and triple-ions are about the same ($\sim 5 \times 10^9$ M^{-1} s^{-1}).

The rate of an anionic polymerization in liquid phase depends on intermolecular collisions between radical-ion and monomer molecules. A radical ion continually changes its shape, its conformation, because thermal motion induces rotation around its various C–C bonds. These processes depend on the temperature, the viscosity of the solvent, the structure of the radical ion, and the end-groups. If the radical ion has a sufficiently long chain, intramolecular collisions between its end-groups may occur. Such intramolecular collisions have been investigated for the first time, by applying ESR spectroscopy, by Szwarc and coworkers[470, 1978, 2121, 2122]. A series of hydrocarbons radical ions, having the structure (α-naphtyl$^{\cdot-}$)-(CH$_2$)$_j$-(α-naphtyl) with j varying from 3 to 12, have been synthesized. The resulting ESR spectra are shown in Figure 4.110–4.115.

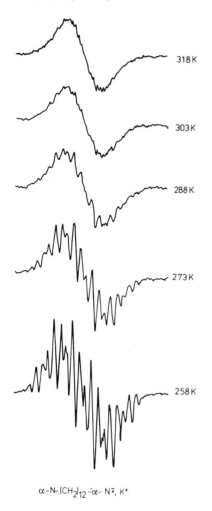

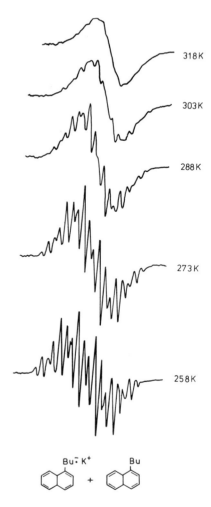

α-N-(CH₂)₁₂-·α- N·, K⁺

Fig. 4.110. ESR spectrum of (α-naphthyl·⁻)-(CH₂)₁₂-(α-naphthyl) in hexamethylphosphoroamide. Concentration of: radical-ions = 0.0002 mole l^{-1}, and hydrocarbon = 0.001 mole l^{-1} 1978)

Fig. 4.111. ESR spectrum of n-butyl-α-naphthalenide in 0.015 M l^{-1} solution of n-butyl-α-naphthalene in hexamethylphosphoroamide. Concentration of radical-ions = 0.0001 mole l^{-1} 1978)

Figure 4.110 shows the spectra of (α-naphtyl⁻)-(CH₂)₁₂-(α-naphtyl) recorded at temperatures varying from 258 K to 321 K. The shape of the recorded spectra reveals the effect of the intramolecular transfer only. At low rates of exchange the spectra of (α-naphtyl·⁻)-(CH₂)ⱼ-(α-naphtyl) are almost identical with the ESR spectra of α-n-butylnaphthalenide·⁻ dissolved in hexamethylphosphortriamide of an appropriate concentration (0.015 mole l^{-1} solution), where intermolecular transfer only occurs (4.111). Hence, for j = 12 and at T = 258 K, the rates of the intramolecular collisions of (α-naphtyl·⁻)-(CH₂)₁₂-(α-naphtyl) are the same as the rates of the bimolecular collisions at about 0.015 mole l^{-1} concentration of α-n-butylnaphthalene. At higher temperatures, as the rate of collisions increases, the spectra of these two systems

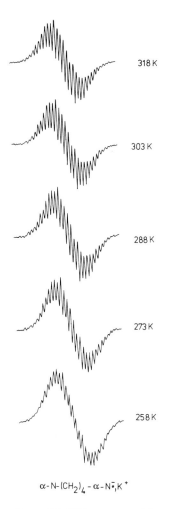

α-N-$(CH_2)_4$-α-N$\dot{\bar{}}$,K$^+$

Fig. 4.112. ESR spectrum of (α-naphthyl$\cdot^-$)-$(CH_2)_4$-(α-naphthyl) in hexamethylphosphoroamide. Concentration of radical-ions = 0.0002 mole l^{-1}, and hydrocarbon = 0.002 mole l^{-1} 1978)

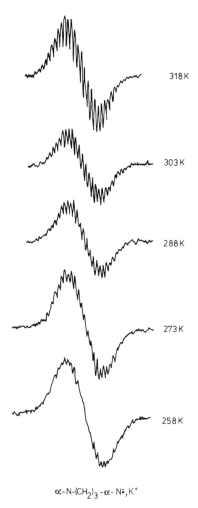

α-N-$(CH_2)_3$-α- N$\dot{\bar{}}$,K$^+$

Fig. 4.113. ESR spectrum of (α-naphthyl$\cdot^-$)-$(CH_2)_3$-(α-naphthyl) in hexamethylphosphoroamide. Concentration of radical-ions = 0.0003 mole l^{-1}, and hydrocarbon = 0.002 mole l^{-1} 1978)

diverge. A new pattern of lines develops at 318 K in the spectrum of the intramolecular transfer system (Fig. 4.110), whereas the hyperfine structure is lost in the spectrum of the intermolecular system at the same temperature (Fig. 4.111). For j = 4 (Fig. 4.112) the rate of the intramolecular collisions at 258 K seems to be comparable to that observed for j = 12 at 318 K (Fig. 4.110). As the temperature of the j = 4 solution rises, the increase in the rate leads to sharpening of the lines. At 318 K, the spectrum of a hypothetical "dimer" appears clearly.

For j = 3 the rate of collisions is very fast and all the recorded spectra (Fig. 4.113) are interpreted as those of a hypothetical "dimer" at all temperatures. The spectra of radical ions derived from hydrocarbons with j = 5 (Fig. 4.114) and j = 6

155

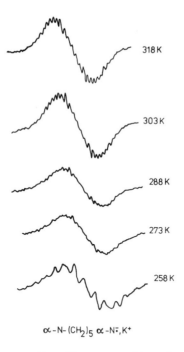

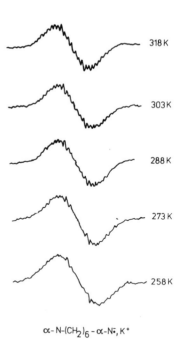

α-N-$(CH_2)_5$ α-N$\dot{\bar{}}$, K^+

α-N-$(CH_2)_6$-α-N$\dot{\bar{}}$, K^+

Fig. 4.114. ESR spectrum of (α-naph-thyl$\dot{\bar{}}$)-$(CH_2)_5$-(α-naphthyl) in hexa-methylphosphoroamide. Concentration of radical-ions = 0.00015 mole l^{-1} and hydrocarbon = 0.001 mole l^{-1} 1978)

Fig. 4.115. ESR spectrum of (α-naph-thyl$\dot{\bar{}}$)-$(CH_2)_6$-(α-naphthyl) in hexa-methylphosphoroamide. Concentration of radical-ions = 0.00015 mole l^{-1} and hydrocarbon = 0.0017 mole l^{-1} 1978)

(Fig. 4.115) clearly reveal the transition from a slow to a fast exchange region. The ESR spectra presented here demonstrate quantitatively how the rates of intra-molecular collisions increase with decreasing length of the chain (j) and increasing temperature.

For other problems related to the ESR study of structure of ion pairs and electron-transfer reactions, the readers are referred to[90, 471, 598, 751, 952, 1629, 1631, 1742, 1841, 1976, 1977, 1990].

4.7.3. Initiation of Polymerization by Electron Transfer

The ESR spectrum of the radical anion of 1,3-butadiene has been observed by con-tinous electrolysis in liquid ammonia[1358, 1359]. Russian scientists have reported a special technique in which simultaneous condensation of alkali metal vapors and monomers, like acrylic and methacrylic compounds[1066, 1694] and dienes[754], give adducts which have characteristic ESR spectra (Fig. 4.116)

These spectra have been attributed to radical anions. In the case of butadiene the ESR spectrum consists of five lines with the intensities 1:4:6:4:1 and $\Delta H = 7.617\ G$[754]. A much better-resolved hyperfine structure is found in the ESR spectra

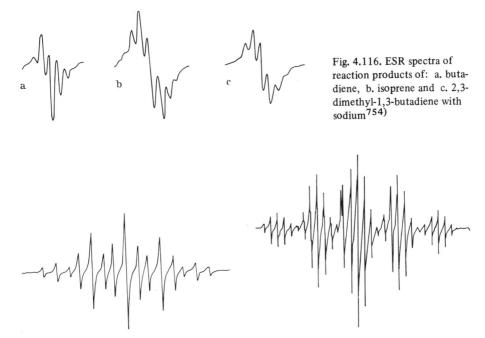

Fig. 4.116. ESR spectra of reaction products of: a. butadiene, b. isoprene and c. 2,3-dimethyl-1,3-butadiene with sodium[754]

Fig. 4.117. ESR spectrum of butadiene radical anion in liquid ammonia at 195 K[1358]

Fig. 4.118. ESR spectrum of 2,3-dimethyl-1,3-butadiene radical anion[1359]

of radical anions of 1,3-butadiene (Fig. 4.117) and 2,3-dimethyl-1,3-butadiene (Fig. 4.118) obtained in liquid ammonia.

Several ESR studies have been reported of different radical anions which can be applied in vinyl polymerization, e.g. radical ions from reactions of alkali metals (Li, Na and K) in tetrahydrofuran with hexamethyl phosphoric triamide[1702], alkali metals with triphenylamine, triphenylphosphine and triphenylphosphine oxide[483, 1697, 2438], potassium in 1,2-dimethoxy-ethylene[380, 2319], potassium in tetrahydrofuran in the presence of polyethylene oxide[1700], and rubidium in bis-2-(2-methoxyethoxy)ethyl ether[380].

ESR spectroscopy has also been applied to studies of mono- and disodium adducts of several compounds with C=C, C=O, C=N, N=N, N=NO and N=O double bonds which form radical anions and dianions, which may initiate anionic polymerization of acrylonitrile, methyl methacrylate, styrene, and vinyl acetate in tetrahydrofuran[274, 1693].

ESR study of ion-radical complexes between carbon black with alkali metals (Li, Na, K) in tetrahydrofuran have also been reported[1502]. Free radicals with long lifetime present in heated carbon black react with sodium in tetrahydrofurane giving an ion-radical complex. This reaction can be registered by changes in shape and line width of the ESR spectra (Fig. 4.119).

A carbon black-alkali metal complex was allowed to initiate the polymerization of styrene in tetrahydrofuran at room temperature in a nitrogen atmosphere. The grafted polymer was 35% of the product. The authors[1502] suggest that "living" polystyrene

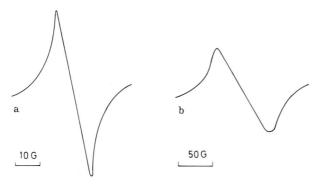

Fig. 4.119. ESR spectra of: a. carbon black and b. ion-radical complex[1502]

may be added to carbonyl or quinone groups, which are functional groups on the surface of a carbon black particle:

$$\Large >C{=}O \;+\; Na^{+}\;\; {}^{-}\underset{\substack{|\\[-2pt]\bigcirc}}{C}H{-}CH_2{-} \;\longrightarrow\; >\underset{\substack{|\\[-2pt]\bigcirc}}{C}\overset{ONa}{{-}}CH{-}CH_2{-}$$

(4.122)

4.7.4. Initiation of Polymerization with Stable Alkali Metal Complexes

4.7.4.1. Polycyclic Aromatic Complexes

Application of ESR spectroscopy to studies of "living" anionic polymerization of styrene initiated by naphthalene radical ion has been reported by Levy and Szwarc[1360]. In polynuclear systems a more extensive delocalization of the unpaired electron results in a greater stability of aromatic anions. ESR studies of anionic polymerization of styrene by radical anions of acenapthylene[922, 1513] and 9,10-dimethyl-anthracene[63] and of polymerization of 1-vinylpyrene adducts with alkali metals (K, Na, Li)[1661] have also been reported. Russian scientist published ESR studies of the anionic polymerization of bis(triethyleneglycol)phthalate dimethylacrylate[1377] and of acrolein and its derivatives[1203], initiated by Na-naphthalene.

4.7.4.2. Ketyl Complexes

Special attention has been given to aromatic ketyls which are formed in the reduction of aromatic ketones by alkali metals[1693, 2308]. The great reactivity of these highly colored solutions has been attributed to the presence of paramagnetic mono-anions, diamagnetic di-anions and various free radicals.

Mono-alkali (Li, Na, K) metal complexes of thiobenzophenone (thioketyls) have been studied in detail by ESR spectroscopy[1047, 1504]. A typical ESR spectrum of monosodium thiobenzophenone in THF is shown in Figure 4.120.

158

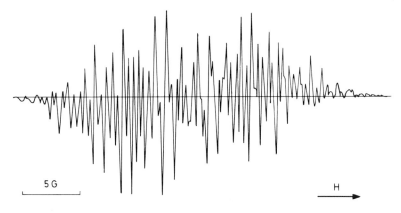

Fig. 4.120. ESR spectrum of monosodium thiobenzophenone in tetrahydrofurane[1504]

This ESR spectrum consisting of about 70 lines, is very complex and its total spectral width is about 25 G[1504]. The reactions of thiobenzophenone with equimolar amounts of alkali metals (Li, Na and K) give deep red complexes which were shown to be thioketyls (4.117). It is difficult to attribute the ESR spectrum in Figure 4.120 to the radical anion. The reaction of thiobenzophenone, with more than twice the equimolar amount of alkali metals (Me), showed the formation of dark red dialkali metal adducts (4.118):

$$\text{(}\overset{}{C}=S\text{)} + Me \longrightarrow \text{(}\dot{C}-S^- Me^+\text{)} \xrightarrow{+Me} \text{(}\overset{+}{Me}\ \overset{+}{Me}\text{)}\ C^- - S^- \qquad (4.123)$$

$$(4.117) \qquad\qquad (4.118)$$

These dimetallic adducts (4.118) have a greater reactivity than the thioketyls (4.117) and induce polymerization of acrylonitrile, methyl methacrylate, styrene, isoprene, and butadiene[1503]. In conclusion, the polymerization with thioketyl catalysts is concluded to proceed and terminate as shown by the equation:

$$\dot{C}-S^-Na^+ + M \longrightarrow \dot{C}-SMM---M^-Na^+ \xrightarrow[HCl, O_2]{CH_3OH} \overset{OH}{C}-SMM---MH \qquad (4.124)$$

$$(4.117) \qquad\qquad\qquad\qquad\qquad\qquad\qquad (4.124)$$

where M denotes the monomer (methyl methacrylate or acrylonitrile). Styrene does not polymerize with a thioketyl radical anion (4.117), but polymerization occurs by initiation of the dialkali metal adducts (dicarbanion) (4.118):

159

4. ESR Study of Polymerization

$$\text{(diphenyl)C=S}^-\ \overset{Na^+\ Na^+}{} + M \longrightarrow Na^+S^-\text{—}C\text{(phenyl)}\text{—MM---M}^-Na^+ \overset{HCl}{\longrightarrow}$$

(4.118)

$$HS\text{—}\overset{|}{C}\text{(phenyl)}\text{—MM---MH} \overset{I_2}{\underset{O_2}{\longrightarrow}} HM\text{—MM---}\overset{|}{C}\text{(phenyl)}\text{—S—S—}\overset{|}{C}\text{(phenyl)}\text{—MM---MH} \qquad (4.125)$$

where M denotes styrene.

4.7.4.3. Aromatic Nitrile Complexes

ESR studies of vinyl polymerization, initiated by radical anions and di-anions of aromatic nitriles[1699, 1701, 1703, 1819] and radical ions from electrochemical reduction[1423, 1696], have also been published.

4.7.4.4. Pyridyl Complexes

The sodium adduct of pyridine N-oxide in tetrahydrofurane yields a violet-colored solution and exists as anion radical of which the ESR spectrum is shown in Figure 4.121[2437].

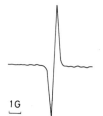

1 G

Fig. 4.121. ESR spectrum of sodium adduct of pyridine N-oxide in tetrahydrofurane[2437]

The structure of the anion radical could not be elucidated from the ESR signal, as it is only a singlet-line spectrum. The anionic polymerization mechanism of vinyl monomers (styrene, methyl methacrylate, and acrylonitrile) with the sodium adduct of pyridine N-oxide can be explained as a charge-transfer reaction to the monomer from pyridyl, 2,2'-bipyridyl and 4,4'-bipyridyl anion radicals. The formation of "living" α-methylstyrene tetramer has also been observed[2435].

160

4.7.4.5. Organic Sulphur Complexes

An ESR study of the reactions of carbon disulphide with one or two equivalents of alkali metals (Na, K) was published[2436]. The structures of the colored reaction products were considered to be mixtures of anion radical (*4.119*), di-anion (*4.120*), di-anion dimer (*4.121*), tri-anion radical (*4.122*) and tetra-anion (*4.123*):

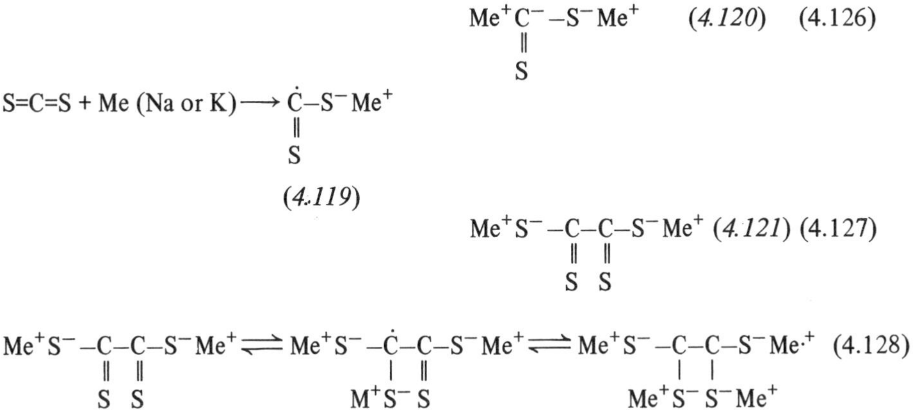

$$Me^+C^- - S^- Me^+ \qquad (4.120) \quad (4.126)$$
$$\overset{\|}{S}$$

$$S=C=S + Me \text{ (Na or K)} \longrightarrow \overset{\cdot}{C}-S^- Me^+$$
$$\overset{\|}{S}$$

(*4.119*)

$$Me^+S^- - C - C - S^- Me^+ \quad (4.121) \ (4.127)$$
$$\overset{\|}{S} \ \overset{\|}{S}$$

$$Me^+S^- - C - C - S^- Me^+ \rightleftharpoons Me^+S^- - \overset{\cdot}{C} - C - S^- Me^+ \rightleftharpoons Me^+S^- - C - C - S^{\cdot +} Me \quad (4.128)$$
$$\overset{\|}{S} \ \overset{\|}{S} \qquad\qquad \overset{|}{M^+S^-} \ \overset{\|}{S} \qquad\qquad \overset{|}{Me^+S^- S^- Me^+}$$

(*4.121*) \qquad\qquad (*4.122*) \qquad\qquad (*4.123*)

These carbon disulphide-alkali metal adducts induce polymerization of N-phenyl-maleimide, methyl vinyl ketone, and acrylonitrile, but not polymerization of methyl methacrylate and styrene.

Diphenyl sulphoxide reacts with an equimolar amount of potassium in tetra-hydrofurane at 195 K to form a reddish-black solution, giving an ESR signal only below 200 K[933]. This singlet-line spectrum (similar to that shown in Fig. 4.122 for the diphenylsulfone anion radical) has been attributed to a very labile diphenyl-sulphoxide anion radical (*4.124*), which initiates the anionic polymerization of acrylonitrile, but not polymerization of methyl methacrylate, styrene, and isoprene.

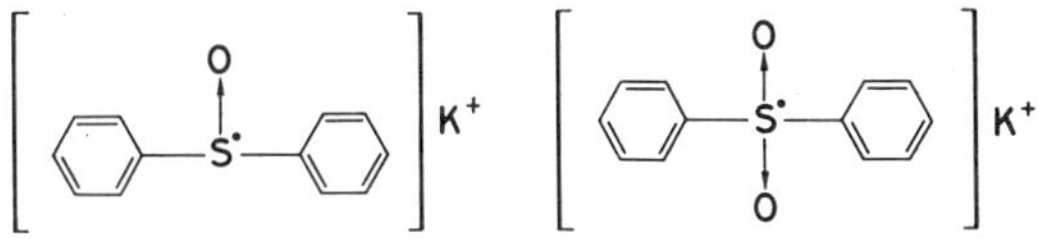

(*4.124*) \qquad\qquad (*4.125*)

Diphenylsulfone was found to react with an equimolar amount of potassium in tetrahydrofurane and other solvents (e.g. diglyme), yielding reddish-black solution and giving an ESR single-line spectrum (Fig. 4.122)[934].

This signal is attributed to the formation of a relatively labile diphenylsulfone anion radical (*4.125*), which can initiate polymerization of acrylonitrile but not polymerize methyl methacrylate, styrene, or isoprene.

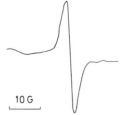

$\underbrace{\qquad}_{10\,G}$ Fig. 4.122. ESR spectrum of diphenylsulfone-monopotassium complex in diglyme solvent[934]

4.8. Cationic Polymerization

The transfer of an electron from a nucleophile monomer (an electron donor (D)) to an electron acceptor (A), converts the monomer to a cation radical. In this initiation step of cationic polymerization the formation of a charge-transfer complex (CT) is an important intermediate step:

$$D + A \rightleftharpoons \underset{(CT)}{(D\text{–}A)} \rightleftharpoons D^{.+}\, A^{-} \rightleftharpoons D^{+}\,(\text{solv}) + A^{-}\,(\text{solv}) \qquad (4.129)$$

The electron-donating character of the monomer results from the presence of nucleophilic substituents or heteroatoms containing unshared electron pairs, e.g. nitrogen or oxygen. The electron acceptors (A) which participate in charge-transfer interactions include organic molecules containing electrophilic substituents and inorganic molecules, including halogens, e.g. I_2 and metal salts in their higher valence states, carbonium ions, radicals, and cation radicals. Solvation of the charged species can accelerate the electron transfer and ionization becomes easier in polar solvents.

For example, it was suggested that the tetracyanoethylene (A)-initiated polymerization of N-vinylcarbazole (D) in methylene chloride results from the radical cation of N-vinylcarbazole (4.126), whereas in benzene, the intermediate radical anion and radical cation coupled to a "zwitterion" (4.127) are assumed to polymerize monomers[172, 2166]:

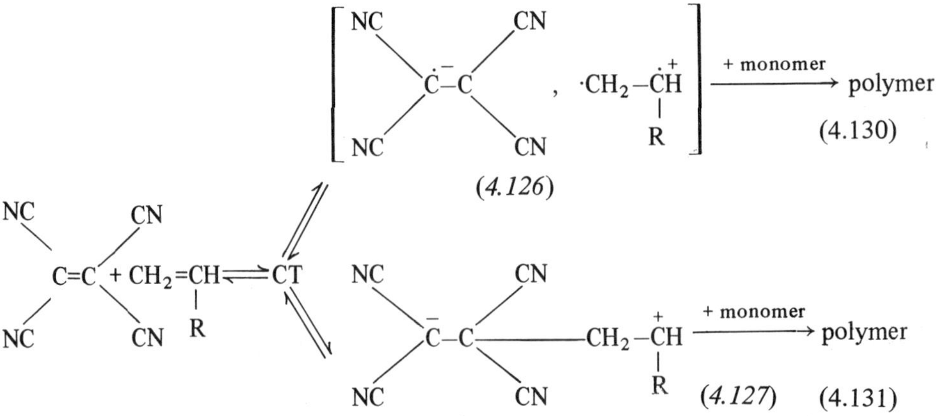

$$\left[\begin{array}{c} \underset{NC}{\overset{NC}{\diagdown}}\dot{C}^{-}\text{–}\underset{CN}{\overset{CN}{\diagup}}C \quad , \quad \cdot CH_2\text{–}\overset{\cdot\,+}{CH}\underset{R}{|} \end{array} \right] \xrightarrow{+\ monomer} \text{polymer} \qquad (4.130)$$

$$(4.126)$$

$$\underset{NC}{\overset{NC}{\diagdown}}C{=}C\underset{CN}{\overset{CN}{\diagup}} + CH_2{=}CH\underset{R}{|} \rightleftharpoons CT$$

$$\underset{NC}{\overset{NC}{\diagdown}}\overset{-}{C}\text{–}C\underset{CN}{\overset{CN}{\diagup}}{-\!-\!-}CH_2\text{–}\overset{+}{CH}\underset{R}{|} \xrightarrow{+\ monomer} \text{polymer} \quad (4.127)\quad(4.131)$$

It has also been proposed that the polymerization of N-vinylcarbazole (D) initiated by chloranil (A) proceeds by a mechanism shown in reaction (4.132):

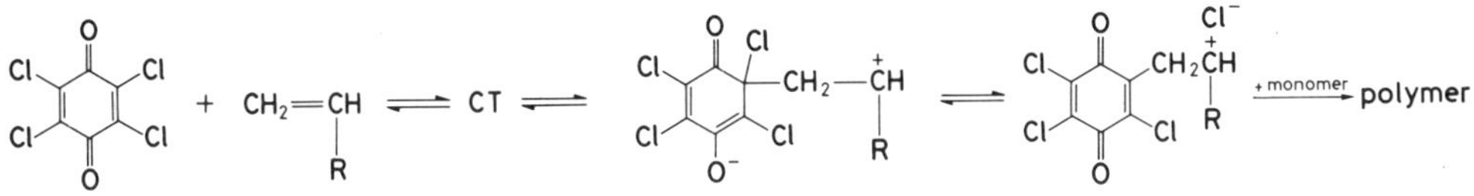

(4.132)

Neither the tetracyanoethylene radical anion nor the alkyl vinyl ether radical cation was observed with flow ESR experiments in methylene chloride or acetonitrile[2166]. This seems reasonable for a reaction in which the intermediate radicals rapidly couple before they are solvent-separated. Similar results have been obtained during polymerization of isopropyl vinyl ether (D) initiated by tetracyanoquinonodimethane (A)[2166]. No free radicals were observed by ESR in this system. The above presented results, however, do not eliminate the transient existence of tetracyanoquinonodimethyl radical anion. If the rate-controlling step involves an electron transfer, and the subsequent reactions are relatively fast and occur before the radical anion and radical cation are solvent-separated, the tetracyanoquinonodimethane radical anion would not be observed.

Successful ESR results have been obtained in the case of 2,3-dichloro-5,6-dicyano-p-benzoquinone (A), which reacts with isobutyl vinyl ether (D) much faster than the analogous reaction with tetracyanoquinonodimethane. The ESR spectrum of a radical ion was recorded by flow technique and is shown in Figure 4.123.

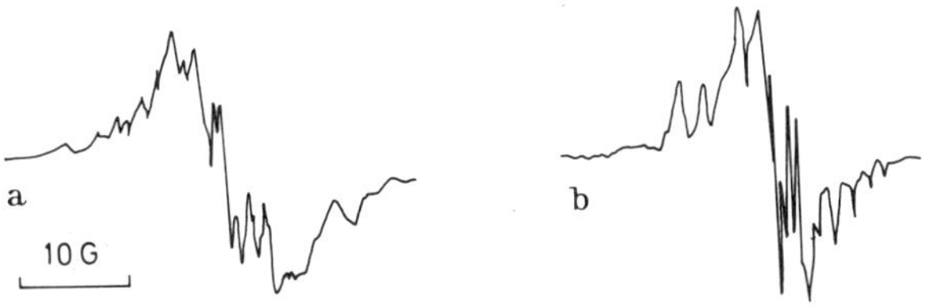

Fig. 4.123. ESR spectra of mixtures of 2,3-dichloro-5,6-dicyano-p-benzoquinone and isobutyl vinyl ether in acetonitrile: a. 1:400 and b. 1:1[2166]

In acetonitrile, this acceptor and alkyl vinyl ethers undergo a rapid one-electron transfer through a charge-transfer complex, to yield a radical anion of 2,3-dichloro-5,6-dicyano-p-benzoquinone (4.128) and a radical cation of the alkyl vinyl ether (4.129), and a subsequent cationic polymerization (4.133):

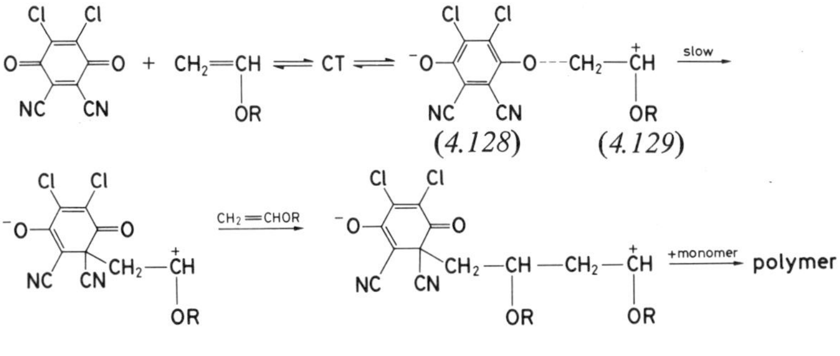

(4.128)　　(4.129)　　　　　(4.133)

163

Table 4.11. ESR spectra of charge-transfer complexes in solid state[2282]

Donor	Acceptor	Colour of complex	ESR spectra	a(G)
N-vinylcarbazole	Cl_2	dark green	singlet	7.6
	Br_2	black	singlet	9.5
	I_2	brown	singlet	11.4
	BF_3	dark blue	singlet	7.6
	$SnCl_4$	white green	singlet	10.0
	SO_2	yellow	asymmetric doublet	–
Carbazole	Cl_2	yellow	singlet	12.4
	SO_2	yellow	not clear	–
Poly-N-vinylcarbazole	Cl_2	gray	singlet	9.4

Initiation with tetracyanoquinonodimethane is assumed to proceed by an analogous mechanism to that in acetonitrile.

ESR studies have been made showing that charge-transfer complexes (CT) may play an important role in initiating the polymerization of N-vinylcarbazole (D) with gaseous catalysts such as halogen or sulphur dioxide (A)[1604, 2282, 2283]. The results are summarized in Table 4.11.

The ESR spectra obtained for almost all complexes were singlets, without hyperfine splitting structure, and were attributed to radical cations (4.130) or (4.131):

(4.130)   (4.131)

(4.134)

The solid-state polymerization of trioxane and tetraoxane[1604], allyl alcohol[1016, 1017], and bicyclo[2.2.1]-hept-2-ene[2519] initiated by SO_2 has also been examined by ESR spectroscopy.

An electron-transfer reaction between iodine and an alkyl vinyl ether has also been proposed from ESR studies[2257]:

$$2\,I_2 \rightleftharpoons I^+ \cdot I_3^-$$ (4.135)

$$I^+ \cdot I_3^- + CH_2{=}CH \rightleftharpoons (Complex) \longrightarrow I\cdot + \cdot CH_2{-}\overset{+}{CH}\cdots I_3^-$$ (4.136)
$$\quad\quad\quad\;\; |\quad\quad\quad\quad\quad\quad\quad\quad\quad\quad\quad\quad\quad\quad |$$
$$\quad\quad\quad OR \quad\quad\quad\quad\quad\quad\quad\quad\quad\quad\quad\quad OR$$

(4.132)

It is feasable that the radical cation (4.132) is stabilized by the alkoxy group. Its ESR spectrum is shown in Figure 4.124.

Fig. 4.124. ESR spectrum of isobutyl vinyl ether-iodine system measured at 77 K[2257)]

10 G

The observed g-value 2.0057, somewhat larger than that for a free spin 2.0023, indicates a considerable spin density on oxygen due to conjugation, which makes the radical cation reluctant to recombine.

ESR signals have also been observed in systems of styrene-I_2, alkyl vinyl ether-$BF_3O(C_2H_5)_2$, and styrene-$BF_3O(C_2H_5)_2$, but the radical concentration was small in comparison with the alkyl vinyl ether-I_2 system[2257)].

4.8.1. Photo-Induced Cationic Polymerization

Photo-induced ionic polymerization of α-methylstyrene in polar solvents, in the presence of tetracyanobenzene[1011, 1013)] or pyromellitic dianhydride[1014)] and of cyclohexane oxide, occurs via cationic propagation mechanism. Tetracyanobenzene and pyromellitic dianhydride are known as strong electron acceptors, while α-methylstyrene and cyclohexane oxide are weak electron donors. A pair of these together forms an electron donor-acceptor (EDA) complex.

The ESR spectrum observed during photo-irradiation of tetracyanobenzene-α-methylstyrene in n-butanol at room temperature is composed of nine hyperfine lines (Fig. 4.125).

Fig. 4.125. ESR spectrum of α-methylstyrene (1.5 mole l^{-1}) and tetracyanobenzene (0.0006 mole l^{-1}) UV-irradiated at 298 K in n-butanol[1013)]

1 G

This ESR spectrum has been attributed to spin coupling with the four equivalent nitrogen nuclei in tetracyanobenzene anion. This result shows that the electron-donor complex dissociates from its excited state to the ions: tetracyanobenzene anion and α-methylstyrene radical cation[1011, 2462)]. The initiation process occurs via the following mechanism:

165

$$D + A \rightleftharpoons (DA) \tag{4.137}$$

$$A \xrightarrow{+h\nu} {}^1A \longrightarrow {}^3A \tag{4.138}$$

$${}^1A + D \longrightarrow {}^1(DA) \longrightarrow {}^3(DA) \tag{4.139}$$

$$(DA) \xrightarrow{+h\nu} {}^1(DA) \longrightarrow {}^3(DA) \tag{4.140}$$

$${}^1(DA) \text{ or } {}^3(DA) \longrightarrow (D^+_{solv} \cdots A^-_{solv}) = D^+_{solv} + A^-_{solv} \tag{4.141}$$

$$D^+_{solv} + D \longrightarrow \cdot D - D^+_{solv} \xrightarrow{+D} \text{polymer} \tag{4.142}$$

$$(D^+_{solv} \cdots A^-_{solv}) + D \longrightarrow (\cdot D - D^+_{solv} \cdots A^-_{solv}) \xrightarrow{+D} \text{polymer} \tag{4.143}$$

where [1] denotes excited singlet and [3] excited triplet states, D denotes α-methylstyrene or cyclohexane oxide, and A denotes tetracyanobenzene or pyromellitic dianhydride.

4.8.2. Cationic Polymerization Initiated with Lewis Acids

Lewis acids are used to generate carbonium ions which are applied to initiate cationic polymerization. The Lewis acids used as Friedel-Crafts catalysts are metal halides of Group III metals — most frequently the following: $ZnCl_2$, $CuCl_2$, $AlCl_3$, BF_3, $GaCl_3$, $SnCl_4$ and $SbCl_5$. ESR studies of Lewis acids have been presented in several papers [753, 787, 788, 939, 940, 945, 1229, 1354, 2213, 2517]. It is well known that metal halides are inactive catalysts, and that cocatalysts such as water, protonic acids, and alkyl halides are required to activate the catalyst.

It has been found that some of the pure Lewis acids have a catalytic activity in free radical polymerization. For example, during γ-irradiation of methyl methacrylate-$ZnCl_2$ complex ($1:1$) unsymmetrical ESR spectrum have been observed as consisting of nine lines (Fig. 4.126a), whereas γ-irradiation of pure methyl methacrylate gives seven-line spectra (Fig. 4.126b)[940].

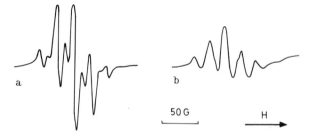

Fig. 4.126. ESR spectrum of γ-irradiated: a. methyl methacrylate-$ZnCl_2$ complex at 183 K and b. methyl methacrylate at 93 K[940]

50 G   H

The spectra of methyl methacrylate-$ZnCl_2$ complex become sharper and more symmetrical with increasing temperature. The nine-line spectrum has been assigned to radicals of type (4.67), containing two or three monomer units.

4.9. Heterogenous Chain-Growth Polymerization

4.9.1. Ziegler-Natta Catalysts

Ziegler-Natta type catalyst systems are extensively used as catalysts for stereospecific polymerization of α-olefins and dienes. Although many studies have been carried out, the nature of the active polymerization centers is not completely established.

Aluminium alkyls are the most commonly used organometallic compounds in Ziegler-Natta catalytic systems. ESR studies have been reported for the following aluminium alkyls: $AlRCl_2$[1251], $AlEt_2Cl$, $AlEtCl_2$, $AlEt_3$[2261], $AlEt_3-ZnCl_2$[2453], and $AlEt_2Cl-ZnCl_2$[752, 788] for initiating polymerization and copolymerization. Especially interesting is the bimetallic complex system $AlEt_3-ZnCl_2$ which is the basis for alternating complex copolymerization[1870, 1874, 2453]. The behavior of the radicals derived from this catalytic system was examined by ESR spectra using spin trapping technique[2453] (see Chapter 11.3).

Several different homogeneous and heterogeneous Ziegler catalysts have been examined by ESR spectroscopy[12, 22-24, 36, 48, 49, 64, 156, 305, 310, 335, 491, 558, 633, 912-915, 936, 937, 942-944, 971, 1172, 1221, 1297, 1410, 1436, 1619, 1622, 1659, 1663, 1740, 2141, 2142, 2195, 2197, 2198, 2394, 2395, 2485].

ESR studies are very useful for catalyst systems which involve transition metals such as Ti and V, but the interpretation of the ESR spectra is difficult. It has been found that only "Ziegler complexes" which contain Ti(III) with an unpaired electron ($3 d^1$ configuration) form ESR spectra. Most Ti(III) compounds have g values in the range of $1.86-2.01$[1302]. $TiCl_3$ itself shows two absorptions[778, 1170, 1658, 1659] (Fig. 4.127).

One line is sharp (Fig. 4.127b) with a g-value of 1.94 and the other is broad (Fig. 4.127c)[1659]. When triethylaluminium ($AlEt_3$) solution was added to $TiCl_3$, an ESR spectrum (Fig. 4.128) with two new absorptions was obtained[1659, 1663]. The solid phase from the mixture of $TiCl_3$ and diethylaluminium chloride ($AlEt_2Cl$) gives the same signal as in Figure 4.128b, which suggests the identity of the action of di- and tri-alkyl aluminium on the titanium trichloride surface. The ESR signals (g = 1.97) obtained from the liquid phase of both systems were assigned to reaction products of alkyl aluminium with titanium tetrachloride occluded in the solid. The intensities of the singlet-line absorption with g = 1.97 increased considerably in the presence of olefins, in the same order as the polymerization activity of the catalyst systems, i.e. $AlEt_3 > AlEt_2Cl > AlEtCl_2 \approx 0$[1659]. This observation suggests that the ESR spectrum with g = 1.97 is due to the polymerization centers for the initial polymerization of propylene in presence of $TiCl_3-AlEt_3$.

ESR spectra of heterogeneous Ziegler-Natta catalysts of the types $TiCl_4-AlR_3$. $TiCl_4-AlRCl_2$, $TiCl_4-AlR_2Cl$, or $Ti(OR')_4-AlRCl_2$ are dependent on the structure of the titanate and the type of solvent. The reduction of tetravalent titanium to trivalent in a Ziegler-Natta catalyst system has been examined in detail by IR spectroscopy[2196].

Trivalent titanate from α- and β-$TiCl_3$ type complexes. Primary titanates in aliphatic and aromatic solvents give α-$TiCl_3$ type, while secondary titanates from the β-$TiCl_3$ type only in aliphatic solvents. Tertiary titanates give the α-$TiCl_3$ type in

167

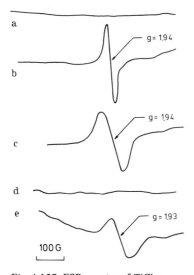

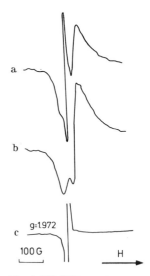

Fig. 4.127. ESR spectra of TiCl₃:
a. A-TiCl₃, b. ground A-TiCl₃,
c. AA-TiCl₃, d. H-TiCl₃ and
e. ground H-TiCl₃ (where: A, AA,
and H are different types of
TiCl₃)[1659]

Fig. 4.128. ESR spectra
from $TiCl_3-Al(C_2H_5)_3$
system: a. $TiCl_3-Al(C_2H_5)_3$
mixture, b. solid phase,
c. liquid phase[1663]

both aliphatic and aromatic solvents, but secondary titanates give the α-TiCl₃ type only in aromatic solvents[971].

The α-TiCl₃ type catalyst species has an ESR signal with a g-value of 1.940 ± 0.003. The 3 d^1 electron in β-TiCl₃ complex is spin-paired with other β-TiCl₃ molecules and does not give an ESR signal[22-24, 491, 971, 2195]. α-TiCl₃ type catalysts produce trans-1,4-polymerization of dienes, while β-TiCl₃ type produces cis-1,4-polymerization of dienes[971].

The homogeneous Ziegler catalysts of the type n-butyl titanate-AlEt₃ at room temperature in solution show ESR spectra attributed also to Ti(III) with a g-value of 1.951 (Fig. 4.129)[936].

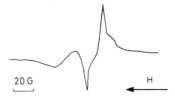

Fig. 4.129. ESR spectrum of n-butyl titanate-triethyl-
aluminum catalyst system at 195 K[936]

A series of seven kinds of ESR signals was detected at Al/Ti ratios less than two, where no polymerization of the conjugated dienes took place. Another series of four signals with g-values of 1.934, 1.952, 1.966, and 1.979 was detected at Al/Ti ratios larger than 2.9, where the conjugated dienes were polymerized. The signals of the latter series were ascribed to the active species for polymerization of the conjugated dienes, and these structures were polymerized. The signals of the latter series were attributed to the active species for polymerization of the conjugated dienes, and these structures were proposed to have at least two active titanium-ethyl bonds and one alkoxy group as shown in structure (4.133). Structure (4.134) was proposed for species responsible for the ESR signal with g-value of 1.952 and a hyperfine structure of 11 components[636]:

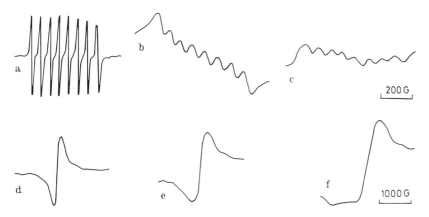

(4.133)   (4.134)

A detailed discussion of the ESR study of these systems is presented in a later paper of this series[937].

Another important group of Ziegler-Natta catalysts consists of the vanadium type catalysts VCl_4-AlR_3, $VCl_4-AlRCl_2$, and VCl_4-AlR_2Cl. Their ESR spectra in solid and liquid phases are also dependent on the structure of the vanadium halide and the type of solvent used[633, 1334, 1580]. Some of these ESR spectra are shown in Figure 4.130.

Fig. 4.130. ESR spectra of the liquid (a, b, c) and solid (d, e, f) phases from a. and d. VCl_4-AlEt_3, b. and e. VCl_4-AlEt_2Cl, c. and f. $VCl_4-AlEtCl_2$[1122]

The liquid phase of VCl_4-AlEt_3 shows an eight-component spectrum (Fig. 4.130a) due to the 7/2 nuclear spin of V, and the solid phase gives a singlet-spectrum (Fig. 4.130a). From the VCl_4-AlEt_2Cl and $VCl_4-AlEtCl_2$ similar spectra were obtained (Fig. 4.130b and 4.130e and Figs. 4.130c and 4.130f, respectively). Since the width of the spectral lines corresponds to the electron density around the vanadium atom, these results suggest that the reducing power of the alkylaluminium compounds are in the order $AlEt_3 < AlEt_2Cl \approx AlEtCl_2$.

169

ESR studies of another vanadium type catalyst $V(acac)_3-AlEt_3$ for polymeriza-tion[944] and $VO(acac)_2-AlEt_3(AlEt_2Cl)$ for inducing alternating copolymerization[738] have also been reported.

4.9.2. Phillips Catalysts

The Phillips catalyst (chromium oxide supported on silica-alumina) is used as a com-mercial catalyst for polymerization of ethylene. The catalyst is activated by calcina-tion at 773—873 K in a dry air stream.

Several authors have investigated the narrow resonance line of the ESR spectra of chromium-alumina or chromium-silica-alumina catalysts[112, 337, 533, 619, 620, 1080, 1104, 1431, 1434, 1667, 1722, 1764, 1845, 2348]. The activated catalyst has an ESR spec-trum containing a sharp and symmetrical component (Fig. 4.131) near 3500 G which can be assigned to Cr^{5+}, and a very broad component between 1500 G and 3500 G assigned to Cr^{3+} [1431].

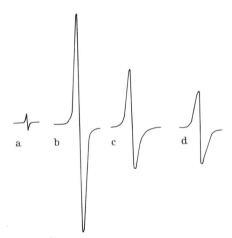

a   b   c   d

Fig. 4.131. ESR spectra of various Phillips catalysts: a. signal of the original catalyst before heat treatment, b. signal of the activated catalyst before heat treatment, c. signal of the activated and heat-treated catalyst, d. signal of the used catalyst for ethylene polymerization for 40 min at 313 K under 300 mm Hg pressure[1431]

Heat treatment (573 K in vacuum) changes the ESR spectrum. The symmetrical signal with large amplitude from the activated catalyst changes to a small, asymmetri-cal, broad one with less area. When ethylene is added, the amplitude of the signal distinctly decreases following the polymerization. The role of heat treatment is con-sidered to be the removal of traces of impurities from the surface of the catalyst, exposing the Cr^{5+} ions on the surface. During the polymerization of ethylene at 418 K the ESR signal intensity reaches a maximum value and then decreases continu-ously[39, 1103, 2063, 2965]. The maximum is observed during the induction period which precedes the maximum rate. The decrease of relative intensity of the ESR spectra during polymerization is considered to be the result of a reduction of Cr^{5+} [1431]. ESR studies indicate that there exists some correlation between the intensity of the sharp ESR signal and the catalytic activity[2063, 2064].

4.9.3. Other Metal Catalysts for Polymerization

Czechoslovak scientists[160–163, 969, 970] have devoted several papers to the ESR study of copper chelates, e.g. bis-(−)ephedrine copper(II)chelate. These Cu(II)complexes catalyze the decomposition of different organic peroxides e.g. benzoyl peroxide, cumene peroxide[160, 163], and azo-bis(isobutyronitrile)[969]. The reactivity of these complexes with α,α′-diphenyl-β-picrylhydrazyle (DPPH) has also been investigated[161, 970]. Ephedrine copper(II)chelate initiate polymerization of di(methacrylate)ethylene glycol[1462].

An ESR study has been reported on the polymerization of 2,6-dimethyl-phenol by the complex of copper(I) salt with pyridine[1836, 1837] and 2,6-dichlorophenol by the copper(II) complex with hexamethylphosphoramide[2303].

Formation of complexes of copper ions in the reaction of copper oleate with a polyketone[974] and poly-4-vinylpyridine[1132] has been proved by ESR spectroscopy. Japanese scientists[999] have reported an ESR study of vinyl polymerization initiated by amine-copper(II) complexes in the presence of tetrahydrofurane.

Tkač et al.[1767, 2198, 2206, 2207] made ESR studies on the heterogeneous catalytic system Ni(0)colloid-Ni(I)-Ni(II) for low-pressure polymerization of butadiene. Co-ordinated radical polymerization of butadiene by rhodium nitrate in ethanol has also been investigated with the application of ESR spectroscopy[1626].

ESR spectra of complex-bonded radicals formed by reaction of cobalt(II) acetyl-acetonate with t-butyl hydroperoxide[2208] and ESR study of the ageing effect of tris(acetonylacetonate)chromium(II)-triethylaluminium catalysts have been reported[941]. ESR study of vanadyl acetate and triethylamine complex as catalyst for acetylene polymerization has been presented by Russian scientists[1577].

An ESR study of the polymerization of di(methacrylate)ethylene glycol in the presence of t-butyl magnesium chloride was also published[842].

Bulgarian scientists studied the vinyl polymerization initiated by metal complexes of dicarbonyl compounds[1698] and aromatic nitryles[1703].

4.9.4. Polymerization Initiated by Clays

It has been stated that neutral clays such as montmorillonite, kaolinite, and illite have the ability to promote the polymerization of vinyl-pyridine and butadiene[722]. The ESR study shows that these clay minerals have unpaired electrons which are responsible for inducing a "spontaneous" free radical polymerization.

4.9.5. Polymerization Initiated by Silica Gel

Mechanical grinding of silica gel leads to the formation of free radicals as evidenced by ESR signals. These radicals can initiate the polymerization of styrene, vinyl acetate, ethyleneimine, and even thiophene which is difficult to polymerize[2079–2081]. Also unground silica gel heated to 453–473 K exhibits high chemical activity, believed to be due to surface radicals formed during the elimination of water, as shown by ESR signals.

4.9.6. Polymerization Initiated by Molecular Sieves

During γ-irradiation at 77 K of molecular sieves free electrons are formed, which can further produce α-vinyl-substituted radicals from α-chlorovinyl (4.144) and acetylenic (4.145) compounds absorbed on molecular sieves[192, 245, 475, 1427, 1428].

$$\underset{\underset{R_2}{|}}{\overset{\overset{R_1}{|}}{CH_2=C}} + e \longrightarrow \underset{\underset{R_2}{|}}{CH_2=\dot{C}} + R_1^- \qquad (4.144)$$

$$CH{\equiv}C-R_1 + \dot{e} \longrightarrow CH{\equiv}C-\bar{R}_2 + H\cdot \qquad (4.145)$$

$$CH{\equiv}C-R_1 + H\cdot \longrightarrow CH_2=\dot{C}-R_1 \qquad (4.146)$$

The main problem in these studies is that the free radicals are very reactive. For that reason they must be kept at low concentration for separating the molecules in the matrix and at low temperature to prevent radical migration. As a result the ESR spectra are poorly resolved. The vinyl acetyl radical gives three different spectra, depending on temperature[1553, 1554]. The ESR spectrum of the α-styryl radical (4.135) obtained by γ-irradiation at 77 K of phenylacetylene is shown in Figure 4.132[245].

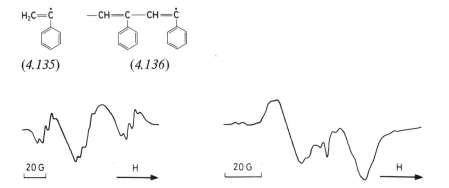

(4.135)   (4.136)

Fig. 4.132. ESR spectra of γ-irradiated phenylacetylene: a. at 77 K and b. at 178 K[245]

When the sample is heated to 178 K a broad doublet, attributed to a polymer radical (4.136), appears.

Chapter 5

ESR Study of Degradation Processes in Polymers

5.1. Radiation and Photo-Degradation of Polymers

The formation of free radicals in polymers exposed to the simultaneous action of γ-radiation and light has been studied by Russian scientists[1134, 1486, 1488, 1489, 1771, 2487]. The interpretation of ESR spectra to determine the radicals formed in these conditions is complicated, on account of the simultaneous reactions of the formation and transformation of radicals. For that reason, we prefer to give in this chapter separate descriptions of ESR studies of free radicals formed in vacuum by ionizing and light irradiation respectively. The oxidation processes of free radicals formed during the degradation in the presence of air (oxygen) will be discussed separately in Chapter 7.

5.1.1. Polyolefines

5.1.1.1. n-Alkane Single Crystals

γ-Irradiation of single crystal of n-paraffins containing 4–20 carbon atoms, e.g. hexane, decane, hexadecane, and eicosane give rise to ESR spectra from secondary alkyl radicals $(5.1$ and $5.2)$[731, 773, 775, 776, 1393, 1394].

$$CH_3-\overset{\cdot}{C}H-CH_2-R \quad (5.1) \qquad R'-CH_2-\overset{\cdot}{C}H-CH_2-R'' \quad (5.2)$$

The relative concentration of radicals (5.2) increases as the number of carbon atoms becomes larger. The quantity of radicals (5.1) is in agreement with the mechanism where all C–H bonds in an n-alkane molecule are ruptured with the same probability, and in which an isomerization of primary to secondary alkyl radicals follows. For n-decane this mechanism predicts 45.5% of radical, (5.1) as compared to the experimental value of 42.5%[775].

ESR spectra of trapped electrons in γ-irradiated polycrystalline hydrocarbons as n-nonane, n-decane, and n-hexadecane[1037], and radical pairs in single crystal of n-eicosane[856], were also reported.

5.1.1.2. Polyethylene

5.1.1.2.1. Radical Formation Under Ionizing Radiation. Free radical formation in polyethylene by high energy radiation (mainly γ-rays)[14, 35, 101, 104, 179, 373, 384,

436, 450, 592, 826, 927, 1061, 1087, 1117, 1137, 1275, 1362, 1569, 1573, 1574, 1589, 1602, 1640, 1642, 1669, 1812, 1891, 1982, 1985–1988, 2116, 2151, 2154, 2156, 2242, 2304, 2402, 2402) and electron beam radiation[99, 102, 103, 439, 730, 1363, 2026, 2154, 2248, 2392, 2459) has been the subject of a large number of ESR studies.

Ionizing radiation of polyolefines induces excitation and ionization of the molecules. Irradiation of polyethylene at 77 K gives the formation of an alkyl radical (5.3), which is observed as a six-line ESR spectrum[14, 101, 439, 1082, 1083, 1087, 1088, 1137, 1191, 1309, 1310, 1363, 1388, 1638, 1640, 1641, 1650, 1812, 2027, 2154, 2248, 2392), and shown in Figure 5.1.

$$-CH_2-CH_2-\overset{\cdot}{C}H-CH_2-CH_2- \qquad (5.3)$$
$$\gamma \qquad \beta \qquad \alpha \qquad \beta \qquad \gamma$$

In this type of radical the unpaired electron occupies a carbon p-orbital and only the protons in the α- and β-positions are expected to have measurable coupling constants. The hyperfine coupling constant (a_{H_β}) can be described by the following relationship[904, 2123):

$$a_{H_\beta} = B_{H_\beta}\rho \cos^2\Theta \qquad\qquad\qquad (5.1)$$

where B_{H_β} is a constant, ρ is the spin density on the α-carbon (C_α), and Θ is the angle between the H–C_β plane and the axis of the p-orbital at the C_α carbon containing the unpaired electron (see Fig. 4.86). The constant B_{H_β} is small (a few gauss or less) and independent of Θ.

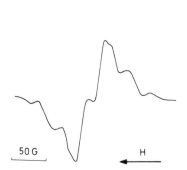

Fig. 5.1. ESR spectrum of electron-irradiated polyethylene[2248).

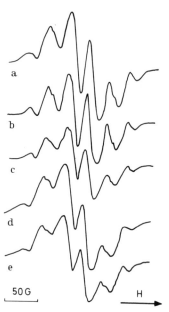

Fig. 5.2. ESR spectra of polyethylene containing different crystallinities: a. 96%, b. 91%, c. 83%, d. 63% and e. 59%[2154)

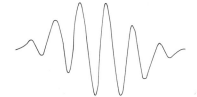

Fig. 5.3. Calculated ESR spectrum of alkyl radical (5.3) [2154]

The spectral shape of alkyl radical (5.3) differs a little for samples of different crystallinities (Fig. 5.2)[2154].

The hyperfine splitting constant of the H_β is about 30 G (H_α- and H_β-protons are equal), while the line width varies between 15 and 25 G, because of the different orientation of the H_β-protons with respect to the magnetic field. Using the coupling constants of alkyl radicals of polyethylene obtained by Kiselev et al.[1137] $a_\alpha = 22.4$ G and $a_\beta = 33.1$ G and assuming Gaussian shape and the line width 15.5 G, Tamura and Shinohara[2154] calculated the spectrum of alkyl radical (5.3) which is shown in Figure 5.3.

Sohma et al.[1572, 1574] examined in detail the steric configuration of alkyl radicals in connection with molecular motion of the matrix polymer. Variation in the steric configuration of the free radicals are due to the variations in the steric conformation of the molecular chain sites at which unpaired electrons are formed. This occurrence influences the interpretation of observed ESR spectra.

Alkyl radicals (5.3) may be formed by dissociation of excited states (5.2)[1713–1715] or by ion-molecular reactions $(5.3–5.4)$[1030, 2185]:

$$RH \xrightarrow{\gamma} RH^{\cdot+} + e^- \longrightarrow (RH)^* \longrightarrow R\cdot + H\cdot \tag{5.2}$$

$$RH^{\cdot+} + RH \longrightarrow R\cdot + RH_2^+ \tag{5.3}$$

$$RH_2^+ + e^- \longrightarrow R\cdot + H_2 \tag{5.4}$$

$$RH_2^+ + e^- \longrightarrow RH + H\cdot \tag{5.5}$$

$$RH + e^- \longrightarrow RH^{\cdot-} \tag{5.6}$$

$$RH^{\cdot+} + RH^{\cdot-} \longrightarrow R-R + H_2 \tag{5.7}$$

$$H\cdot + H\cdot \longrightarrow H_2 \tag{5.8}$$

$$H\cdot + RH \longrightarrow H_2 + R\cdot \tag{5.9}$$

where RH is a polyethylene macromolecule.

Polyethylene samples irradiated at 291 K give a septet-line ESR spectrum, which is generally attributed to the allyl radical (5.4) with the hyperfine splitting constant of 21 G[101, 439, 1082, 1191, 1309, 1310, 1638, 1641, 1649, 1650, 1812, 2027].

$-CH=CH-\dot{C}H-CH_2-$ (5.4)

This spectrum may be partially overlapped by the spectrum of the alkyl radical (5.3) and is dependent on the steric configuration of the radical (5.4) (Fig. 5.4b)[1649].

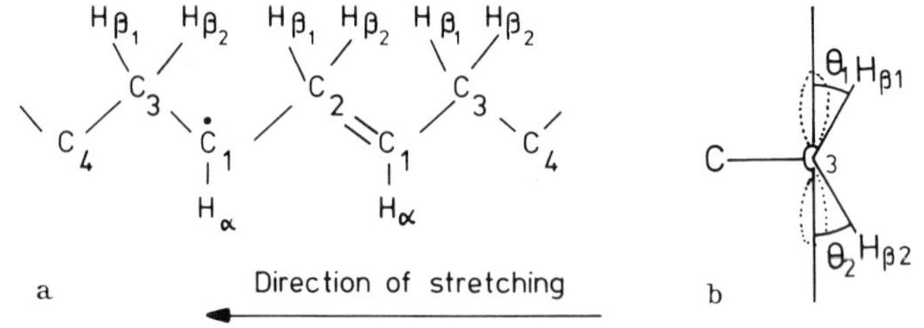

Fig. 5.4. a. Configuration of allyl radical (5.4), b. View of the allyl radical along the C_1-C_3 bond direction[1649]

Auerbach[102, 103] has proposed an ionic mechanism for the formation of the allyl radical (5.4). This mechanism seems less probable, because the positive radical ions largely disappear, especially after heating to room temperature[2392].

When polyethylene is irradiated by doses of several thousand megarads, the observed ESR spectrum becomes a singlet line which has been assigned to a conjugated polyene radical (5.5)[1310, 1641, 1650]:

$-CH_2-(CH=CH)_n-\dot{C}H-CH_2-$ (5.5)

The line width of this singlet, with increasing doses, converges towards 17 G. The delocalization of the electron associated with a long chain of conjugated unsaturation is expected to produce a singlet-line ESR spectrum[861]. Alternatively, the ESR spectra due to various polyene radicals may overlap and yield a broad single peak[239]. The polyenyl radicals are very stable under vacuum. In the presence of oxygen, the concentration of polyenyl radicals decreases.

5.1.1.2.2. Radical Formation under UV-Irradiation. ESR of free radicals produced by UV-irradiation of polyethylene has been investigated by many workers[322, 436, 440, 850, 1053, 1652, 1817, 1981, 1983, 1984, 2144, 2145, 2156, 2241, 2242, 2244, 2247, 2248, 2250, 2262, 2263, 2265, 2267, 2269–2274, 2277, 2284, 2285]. The polymer irradiated with UV-light at 77 K in nitrogen atmosphere has an ESR spectrum as presented in Figure 5.5. a.

Browning et al.[322] and Rånby and Yoshida[1812, 1817] obtained a six-line spectrum with a hyperfine splitting constant of 33 G, and traces of a quintet with a hyperfine splitting constant of 30 G. This spectrum was interpreted by these authors, and later also by Tsuji and Seiki[2271], as being due to the superposition of two kinds of free radicals: alkyl radical (5.3) (a six-line spectrum) and end radical (5.6) (quintet-line spectrum):

$-CH_2-CH_2-\overset{.}{C}H_2$ (5.6)

Tsuji[2247, 2265] suggested that this spectrum should be interpreted as an eight-line spectrum, because two more lines appear on both sides of the six-line spectrum[322, 1812, 1817], and can be attributed to the free radical (5.7) which could be formed from radical (5.6) by internal hydrogen migration.

Fig. 5.5. ESR spectra of polyethylene UV-irradiated at 77 K and measured at 77 K. a. Sample tube filled with nitrogen gas at 760 mm Hg, b. Pressure in the sample tube 10^{-6} mm Hg. The separation between the to Mn^{2+} is 86.7 G[2267]

$-CH_2-\overset{.}{C}H-CH_3$ (5.7)

Another possible free radical which can give an eight-line spectrum is the free radical (5.8)

$-CH_2-\overset{.}{C}-CH_2-$ (5.8)
$\quad\quad\;\; |$
$\quad\quad\; CH_2$
$\quad\quad\;\; |$

A small broad singlet ESR spectrum has been observed during vacuum irradiation of polyethylene at 77 K[1652, 2267] (Fig. 5.5b).

5.1.1.2.2.1. ESR Studies on Photosensitized Degradation of Polyethylene.

The photosensitized decomposition of polymers is well known and was extensively reviewed by Rabek[1775, 1779, 1780, 1782–1784, 1814]. The mechanism of photosensitized generation of free radicals in polyethylene was investigated by Tsuji et al.[2144, 2145, 2244, 2246, 2281]. In the case of aromatic sensitizers, e.g. naphthalene, anthracene, phenanthrene, and pyrene, it has been found that energy transfer from higher triplet states of excited sensitizer molecules to unsaturated bonds in polyethylene may occur[2145, 2281]. An excited unsaturated group may release its allylic hydrogen atom, giving an allylic radical (5.4), for which ESR spectra have been recorded.

In the case of polyethylene-containing ferric stearate, observable ESR signals of free radicals (5.7) were obtained under UV and visible light irradiation[2250]. These results show that ferric stearate can sensitize polyethylene decomposition in that range of visible light which is not absorbed by the polymer molecule alone.

177

Cernia et al.[411] published an ESR study of photosensitized degradation of polyethylene in the presence of 1,1,4,4-tetraphenylbutadiene.

5.1.1.2.3. Conversion of Free Radicals Under Light Irradiation and Warming.

An interesting feature is the changes of radical structures, obtained by gamma-, electron- and light-irradiation, on further irradiation with light and on warming[384, 520, 638, 1086, 1151, 1467, 1479, 1483, 1519, 1602, 1613, 1652, 1981, 1983–1987, 2156, 2246, 2247, 2391, 2393].

Allylic radicals (5.4) can be converted by two mechanisms dependent on the experimental conditions[1983, 1985]:

1. When the allylic radicals (5.4) are irradiated with UV-light, they are converted to alkyl radicals (5.3). Further heating reconverts the alkyl radicals to stable allylic radicals (5.4):

$$-CH_2-CH=CH-\dot{C}H-CH_2-CH_2- \underset{heat}{\overset{UV}{\rightleftharpoons}} -CH_2-CH=CH-CH_2-\dot{C}H-CH_2- \quad (5.10)$$

(5.4) (5.3)

(septet-line spectrum) (sextet-line spectrum)

2. When the allylic radicals (5.4) are irradiated with visible light ($\lambda > 3900$ Å), main chain scission and the presence of alkyl radicals (5.7) are observed (5.11). This alkyl radical (5.7) can then be converted to another alkyl radical (5.3) by warming for a short time to 273 K (5.12). Long warming of the alkyl radicals (5.7) to room temperature (273 K) converts them back to allylic radicals (5.4) (5.13). The transformation of free radicals can be described by the following reactions:

$$-CH_2-CH=CH-\dot{C}H-CH_2-CH_2-CH_2-CH_2 \xrightarrow[(\lambda > 3900\text{ Å})]{light}$$

(5.4)

(septet-line spectrum)

$$-CH_2-CH=CH-CH=CH_2 + CH_3-\dot{C}H-CH_2- \quad (5.11)$$

(5.7)

(octet-line spectrum)

$$CH_3-\dot{C}H-CH_2-CH_2 \xrightarrow[\text{warming to 273 K}]{short} CH_3-CH_2-\dot{C}H-CH_2- \quad (5.12)$$

(5.7) (5.3)

(octet-line spectrum) (sextet-line spectrum)

$$CH_3-\dot{C}H-CH_2-CH_2-CH_2-CH=CH-CH_2 \xrightarrow[\text{warming to 273 K}]{long}$$

(5.7)

(octet-line spectrum)

$$CH_3-CH_2-CH_2-CH_2-\overset{\cdot}{C}H-CH=CH-CH_2- \qquad\qquad (5.13)$$

$$(5.4)$$

(septet-line spectrum)

The activation energy of these processes has been found to be 18 kcal/mol[986].

The second poblem concerns the conversion of alkyl radicals (5.7). Two radical conversions were elucidated by Tsuji et al.[2242, 2246–2248, 2269, 2271]:

1. Conversion of alkyl radicals (5.7) in vacuum:

$$-CH_2-\overset{\cdot}{C}H-CH_3 \xrightarrow{\text{warming to 133 K}} -CH_2-\overset{\cdot}{C}H-CH_2- \xrightarrow{\text{warming to 273 K}} \cdots$$

$$(5.7) \qquad\qquad\qquad\qquad (5.3)$$

(octet-line spectrum)   (sextet-line spectrum)

$$\cdots \longrightarrow -CH=CH-\overset{\cdot}{C}H-CH_2- \xrightarrow[\text{(}\lambda > 3900\text{ A)}]{\text{light}} -CH_2-\overset{\cdot}{C}H-CH_3 \qquad (5.14)$$

$$(5.4) \qquad\qquad\qquad\qquad\qquad (5.7)$$

(septet-line spectrum)   (octet-line spectrum)

2. Conversion of alkyl radicals (5.7) in the presence of carbon monoxide[2284, 2285]:

$$-CH_2-\overset{\cdot}{C}H-CH_3 \xrightarrow{\text{warming to 133 K}} -CH_2-\overset{\cdot}{C}H-CH_2- \xrightarrow{+\text{CO at 195 K}} \cdots$$

$$(5.7) \qquad\qquad\qquad\qquad (5.3)$$

(octet-line spectrum)   (sextet-line spectrum)

$$\overset{\displaystyle\overset{\cdot}{C}O}{\underset{\displaystyle |}{}}$$

$$\cdots \longrightarrow -CH_2-CH-CH_2- \xrightarrow{\text{warming to 273 K}} -CH=CH-\overset{\cdot}{C}H-CH_2- \xrightarrow[\text{(}\lambda > 3900\text{ A)}]{\text{light}} \cdots$$

$$(5.9) \qquad\qquad\qquad\qquad\qquad (5.4)$$

(single-line spectrum)   (septet-line spectrum)

$$\cdots \longrightarrow -CH_2-\overset{\cdot}{C}H-CH_3 \qquad (5.7) \qquad\qquad\qquad\qquad\qquad (5.15)$$

(octet-line spectrum)

The acyl radical (5.9) gives a sharp singlet spectrum (g = 2.001) which is formed after warming the sample to 195 K. The singlet decayed at about 233 K, and at 263 K a well-resolved septet-line spectrum (g = 2.003) is formed, which has been attributed to allyl radical (5.4) and after visible light irradiation ($\lambda > 3900$ Å) converts into alkyl radical (5.7) (5.15).

Commercial polyethylene contains carbonyl groups in the main chain which absorb UV-light at about 2800 Å. It has been shown that an ESR spectrum of polyethylene is formed when a polymer sample is irradiated with UV-light of wavelength longer than 2800 Å[2271, 2272]. After absorption of photons at 77 K carbonyl groups are excited to the triplet state and then a Norrish type I reaction may occur:

$$-CH_2-CH_2-\overset{\overset{\textstyle O}{\|}}{C}-CH_2-CH_2 \xrightarrow{h\nu} -CH_2-CH_2-\overset{\overset{\textstyle O}{\|}}{C}-CH_2-CH_2-\longrightarrow$$

$$2 -CH_2-\dot{C}H_2 + CO \quad (5.6) \tag{5.16}$$

End alkyl radicals (5.6) are easily converted at 77 K to alkyl radicals (5.7):

$$-CH_2-CH_2-\dot{C}H_2 \longrightarrow -CH_2-\dot{C}H-CH_3 \tag{5.17}$$
$$(5.6) \qquad\qquad (5.7)$$

During warming of the sample from 77 K to 273 K, changes of the ESR spectra occur (Fig. 5.6a and b).

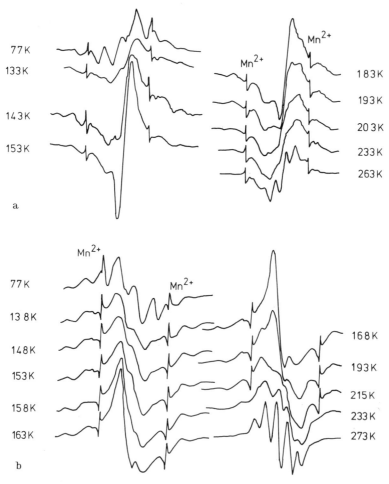

Fig. 5.6. Change of ESR spectra of: a. Low-density and b. High-density polyethylene (UV-irradiated 2 hr in nitrogen atmosphere at 77 K) with increasing measuring temperature. The separation between the Mn^{2+} peaks is 86.7 G[2271]

180

At about 133 K the free radicals (5.7) change into alkyl radicals (5.3). Carbon oxide formed in reaction 5.16 reacts with radical (5.3) at about 153 K and the rate is controlled by diffusion of CO molecules into the polymer phase at this temperature:

$$-CH_2-\overset{.}{C}H-CH_2- + CO \longrightarrow -CH_2-\overset{\overset{\displaystyle \overset{.}{C}O}{|}}{C}H-CH_2- \qquad (5.18)$$

(5.3)   (5.9)

The acyl radical (5.9) has been observed as a sharp singlet at 150 K in the ESR spectra shown in Figure 5.6. A similar singlet spectrum (g = 2.001) due to acyl radicals was observed immediately after γ-irradiation of an ethylene-acrolein copolymer at 77 K[2249].

Hama et al.[849, 850] suggested that the excitation of carbonyl groups in polyethylene is followed by chain-bond dissociation and formation of acyl radical of type (5.10). These radicals are converted by the absorption of visible light to free radicals (5.11):

$$-CH_2-CH_2-\overset{.}{C}O \rightleftharpoons -CH_2-\overset{.}{C}H-CHO \qquad (5.19)$$

(5.10)   (5.11)

5.1.1.2.4. Morphological Effects on the Formation and Behavior of Radicals. The identity and behavior of radical species formed by UV-irradiation of crystalline polymers have been extensively investigated by the ESR method[1275, 1388, 2262, 2263].

Loy[1388] and Tsuji[2242, 2263, 2265, 2271] suggested that preferential radical formation occurs in the amorphous regions of the irradiated polyethylene. On the other hand, in a study of solution-grown polyethylene single crystals, Salovey et al.[1889, 1891] and Takayanagi et al.[1275] have suggested that the free radicals are preferentially trapped in the intermediately disordered regions of the crystalline texture.

Samples in which the polymer chains are oriented by stretching give two different ESR spectra after irradiation[926, 1088, 1137, 1268]. One is attributed to radicals on polymer chains oriented perpendicularly to the magnetic field, showing a 10-line spectrum with a hyperfine splitting constant of 15 G and a line width of 11.5 G. The other spectrum, obtained with chains parallel to the magnetic field, shows a sharp sextet with a hyperfine splitting constant of 32.5 G. An interesting correlation between the degree of orientation and the hyperfine splitting constant of the α-proton has been found[1088]. Salovey and Yager[1891] observed the overlapping six-line and ten-line patterns of the ESR spectra from irradiated solution-grown crystals of polyethylene. The ESR spectra were clearly resolved, because of the high crystallinity and excellent orientation of polyethylene chains at right angles to the lamellar crystals.

Sohma et al.[1982, 1988] have recorded the variation of the alkyl radical (5.3) spectra at 77 K as a function of the angle (ψ) between the direction of magnetic field and the c-axis of polyethylene crystals (Fig. 5.7).

From the observed ESR spectra, the principal values of the hyperfine coupling tensor were estimated to be $|A_z| = 13.6$ G, $|A_x| = 26.0$ G and $|A_y| = 40.6$ G, where

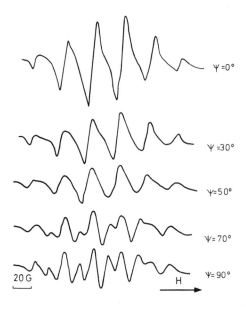

$\Psi = 0°$

$\Psi = 30°$

$\Psi = 50°$

$\Psi = 70°$

20 G   H $\Psi = 90°$

Fig. 5.7. ESR spectra of solution grown crystals of polyethylene γ-irradiated at 77 K as a function of the angle (ϕ) between applied magnetic field and direction of c-axis of the crystal[1988)

x is the direction of the π-orbital of the free electron, y is the c-axis (chain direction), and z the direction of the C—H bond.

5.1.1.2.5. Decay of Free Radicals.

When polyethylene containing free radicals was heated to room temperature, a substantial decrease in radical concentration was observed by several authors[100, 104, 486, 568, 569, 1084, 1309–1311, 1191, 1388, 1574, 1612, 1638, 1688, 1936, 1937, 2045, 2049, 2391, 2392).

Free radicals present in polyethylene have different lifetimes. The fast-decaying free radicals (about 1 day) are alkyl radicals (5.3). The allyl radicals (5.4) have a lifetime of several months at 77 K. The polyene radicals (5.5) are very stable even at room temperature[1650).

Russian scientists[1191) suggested that, in the presence of double bonds, alkyl radicals (5.3) change into allyl radicals (5.4) by hydrogen migration. A previous suggestion is that alkyl radicals can migrate in polyethylene by a random-walk process[568, 569).

Work made by Ormerod[1668) on the decay of alkyl radicals at room temperature indicates that the decay is a composite effect caused by the reactions:

$$R_1{\cdot} + R_2{\cdot} \longrightarrow R_1-R_2 \tag{5.20}$$

$$R{\cdot} + -CH_2-CH{=}CH- \longrightarrow RH + -\dot{C}H-CH{=}CH- \tag{5.21}$$

$$R{\cdot} + -\dot{C}H-CH{=}CH- \longrightarrow -CH(R)-CH{=}CH- \tag{5.22}$$

Waterman and Dole[2392) presented evidence that alkyl radicals (5.3) which persist at room temperature after an electron beam irradiation at 77 K are quantitatively converted to allyl radicals (5.4) by the reaction with trans-vinylene or vinyl double

182

bonds. Dienyl and trienyl radicals are formed by a similar process involving an alkyl radical and diene and triene structures, respectively. About 70% of the alkyl radicals (5.3) decayed by reacting with vinyl type unsaturation and the remainder with vinylene groups.

During the irradiation of polyethylene the alkyl radicals are probably not produced in a uniform distribution but in tracks, spurs, etc. The initial distribution of the allyl radicals, on the other hand, depends on the localization of unsaturation, chiefly at vinyl end groups in polyethylene. Allyl radicals (5.4) are formed subsequently to irradiation at 77 K by migration of alkyl radicals (5.3) to groups adjacent to a double bond[2402].

A generally accepted opinion is that radical decay occurs by recombination, although there are some differences in opinion concerning the decay mechanism. Thus, there is a question whether the kinetics is first- or second-order[104, 366, 383, 486, 566, 1309].

Decay curves of alkyl radicals (5.3) can be interpreted by second-order kinetics[104, 439, 1574, 1668, 1985], although the initial and the last parts of decay curve do not conform to second-order kinetics (Fig. 5.8).

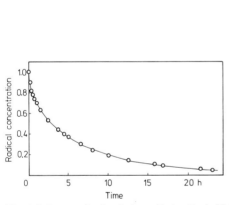

Fig. 5.8. Decay of polyethylene alkyl radicals (5.3) in air at 293 K[1936]

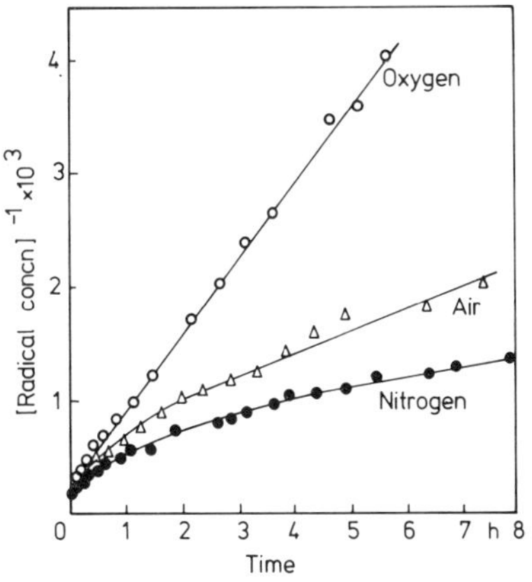

Fig. 5.9. Comparative effects of nitrogen, air and oxygen on radical decay in polyethylene[104]

When interpreted as first-order kinetics, the decay is presumed to involve more than two stages with different rate constants[486, 568, 1309, 1310, 2392]. The decay constant of alkyl radicals (5.3) increases with the amorphous fraction of polyethylene. The decay is markedly catalyzed by hydrogen gas which is soluble only in the amorphous fraction[517, 1060, 1061, 1235, 1236, 1668, 2392, 2401]. The most reasonable suggestion is that alkyl radicals (5.3) under vacuum are converted to allyl radicals (5.4) with first-order kinetics, and a residue of the radicals decay with second-order kinetics[1936]. Allyl radicals decay according to diffusion-controlled second-order kinetics[2402].

Oxygen provides an alternative decay mechanism of alkyl radicals to peroxy radicals. Auerbach and Saunders[104] have shown the effect of oxygen on the second-order decay of alkyl radicals (5.3) in polyethylene (Fig. 5.9).

The amorphous and crystalline phases provide different radical recombination rates in a diffusion-controlled reaction[366]. For better understanding of this mechanism there exists a useful theory of Lebedev[1321], which is based on a model of two kinetic processes:

1. A slow motion of the radicals in the solid phase occurs until they reach a favorable position for the reaction followed by:
2. A fast recombination of the radicals in the activated volume unit where they are located.

Alkyl radicals trapped in crystalline regions can migrate to the surfaces of crystallites by hydrogen abstraction and then decay by reaction with oxygen at the surface[1936]. The kinetics of the radical decay can well be explained by a diffusion theory, assuming that the decay rate is controlled by the rate of radical migration in the crystallites.

5.1.1.2.6. Ethylene Copolymers. ESR studies of ethylene-acrolein[2245, 2249] and ethylene-styrene[2253] copolymers UV-irradiated at 77 K have given similar results as those for polyethylene.

5.1.1.3. Polypropylene

5.1.1.3.1. Radical Formation Under Ionizing Radiation. In the reported studies of radicals formed in polypropylene by ionizing[115, 179, 679, 708, 838, 851, 855, 1031, 1149, 1265, 1326, 1363, 1392, 1481, 1483, 1487, 1569, 1570, 1640, 1652, 1664, 1665, 1812, 1994, 2307, 2467, 2488] and neutron[837] radiation, many radical structures have been proposed.

Irradiation of polypropylene at 77 K gives formation of radicals (5.12) which are observed as eight- or nine-line ESR spectra (Fig. 5.10)[115, 679, 708, 1149, 1483, 1569, 1571, 2302].

$$-CH_2-\overset{\overset{\textstyle CH_3}{|}}{\underset{\cdot}{C}}-CH_2- \qquad (5.12)$$

Fig. 5.10. ESR spectrum of isotactic polypropylene γ-irradiated at 77 K and masured at 77 K[1031]

50 G   H

There is no good agreement between the results. Charlesby et al.[1363] have reported formation of a six-line spectrum of alkyl radicals (5.3) during γ-irradiation of polypropylene at 77 K. Ohnishi et al.[1640] found that the spectrum of polypropylene irradiated and measured at 195 K consists of four different components, which could be characterized as eight-, seven-, six- and one-line spectra. No attempt was made to assign these spectra to specific radicals. Forrestal and Hodgson[708] interpreted their low temperature spectra as being composed of equal amounts of eight-line and broad four-line spectra, which were assigned to radical (5.12) and radical (5.13) or (5.14), respectively:

$$-CH_2-CH-CH_2- \qquad (5.13) \qquad -CH-\overset{\cdot}{C}H-CH- \qquad (5.14)$$
$$\qquad\quad | \qquad\qquad\qquad\qquad\qquad\qquad | \qquad\quad |$$
$$\qquad\quad CH_2 \qquad\qquad\qquad\qquad\qquad\quad CH_3 \qquad CH_3$$
$$\qquad\quad \overset{\cdot}{}$$

Loy[1392] using a deuterium-substituted polymer, concluded that the spectrum consists of a quartet due to radical (5.13) and a trace of nonet-line spectrum due to radical (5.12).

Isotactic polypropylene which is γ-irradiated at room temperature, gives a well-resolved ESR spectrum of 17 lines at the same temperature. From experiments with oriented samples, Fischer and Hellwege[679] reported that this 17-line spectrum is due to an allyl radical with two resonance forms (5.15) and (5.16):

$$-CH-\overset{\cdot}{C}H-C=CH-CH- \rightleftarrows -CH-CH=C-\overset{\cdot}{C}H-CH- \qquad (5.23)$$
$$\quad | \qquad | \qquad | \qquad\qquad\qquad | \qquad | \qquad |$$
$$\quad CH_3 \quad CH_3 \quad CH_3 \qquad\quad CH_3 \quad CH_3 \quad CH_3$$
$$\qquad (5.15) \qquad\qquad\qquad\qquad (5.16)$$

The formation of allyl radical is strongly supported by ESR measurements made by Milinchuk et al.[1150, 1157, 1161, 1480, 1481, 1483, 1485, 1486]. ESR studies of γ-irradiated isotactic polypropylene by Iwasaki et al.[1031] and Milinchuk et al.[1151, 1159, 1480] have shown that the 17-line spectrum appears as a 9-line spectrum when measured at 77 K. When the irradiated sample is exposed to UV-light for a short time, this 9-line spectrum is converted to a four-line spectrum due to radical (5.13). Yoshida and Rånby[2467] have also analyzed the low temperature spectrum of oriented polypropylene and interpreted it as a sum of a quartet due to radicals (5.13) or (5.14) and an anisotropic spectrum due to radical (5.15). Ayscough and Munari[117] re-examined the ESR spectra obtained at room temperature with stretched and unstretched polypropylene. They compared these spectra with those of irradiated model hydrocarbons at 77 K, which gave an allylic radical (5.4) with a spectrum differing from that of the proposed radical (5.15–5.16). Ohnishi et al.[1640] also proposed the formation of radical (5.7). The lack of agreement reported for ESR spectra and their interpretation may partially be caused by differences in samples used by the various investigators. It is expected that the difference in stereoregularity of isotactic polypropylene and atactic polypropylene affects the ESR spectra[1265, 1665]. The difference in hyperfine structure of the spectra of isotactic polypropylene and of atactic polypropylene can be explained by different radical conformation related to the stereo-regularity. The hyperfine coupling and conformations of radical (5.12) in isotactic and atactic polypropylene are shown in Tables 5.1 and 5.2, respectively.

185

Table 5.1. Hyperfine coupling constants and radical conformation for isotactic polypropylene[1665]

$$H_{\beta1} \quad CH_3 \quad H_{\beta3}$$
$$-C_{\beta1}-\overset{\bullet}{C_\alpha}-C_{\beta2}-$$
$$H_{\beta2} \qquad H_{\beta4}$$

H.f. coupling constant (G)		Conformational angle (degrees)		Radical conformation
$\Delta H(CH_3)$	21.4			
$\Delta H(C-H_{\beta_1})$	39.6	θ_1	15.5	
$\Delta H(C-H_{\beta_2})$	15.1	θ_2	53.5	
$\Delta H(C-H_{\beta_3})$	42.1	θ_3	6.6	
$\Delta H(C-H_{\beta_4})$	9.1	θ_4	62.5	

Table 5.2. Hyperfine coupling constants and radical conformation for atactic polypropylene[1665]

H.f. coupling constant (G)		Conformational angle (degrees)		Radical conformation
$\Delta H(CH_3)$	21.4			
$\Delta H(C-H_{\beta_1})$	1.8	θ_1	78.2	
$\Delta H(C-H_{\beta_2})$	30.0	θ_2	33.0	
$\Delta H(C-H_{\beta_3})$	30.0	θ_3	33.0	
$\Delta H(C-H_{\beta_4})$	1.8	θ_4	78.2	

The spectral change from a 9-line to a 17-line spectrum with elevation of temperature for isotactic polypropylene can be explained by hindered oscillation of the β-methylene protons.

Japanese scientists[1270] have made ESR studies giving information on the sites of trapped radicals in the different phases of isotactic polypropylene samples. They treated the polypropylene samples with fuming nitric acid which removed the disordered regions stepwise. The free radical sites of etched samples were largely concentrated in the defects of crystal regions.

5.1.1.3.2. Radical Formation under UV-Irradiation.
Irradiation of polypropylene with UV-light[322, 632, 854, 855, 1812, 1817, 2243, 2268, 2275, 2278, 2466] gives different ESR spectra from those previously described for γ-irradiation.

One spectral component is a quartet with sharp lines (Fig. 5.11) which has been interpreted as being due to methyl radical (5.17)[322, 854, 855, 2268, 2466]:

·CH$_3$ (5.17)

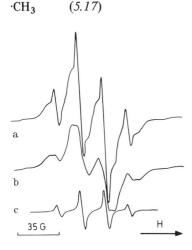

a

b

c

35 G

H

Fig. 5.11. ESR spectra of polypropylene UV-irradiated at 77 K and measured at 77 K: a. Immediately after irradiation, b. Two days after irradiation, c. The component which decayed during the two first hours after irradiation[2466]

Tsuji and Seiki[2268] have reported that the formation and decay of methyl radical (5.17) are dependent on pressure. The other components of spectra, containing broad lines, were assigned to radicals of the type (5.3), which are the counterparts of the methyl radicals and to radicals (5.13) and (5.14)[322, 855, 1812, 1817, 2278]. In addition, Rånby and Yoshida[1817] have suggested that the broad-line quartet could also be caused by radicals (5.18) formed by main chain scission:

$$-CH_2-CH-\overset{\cdot}{C}H_2 \qquad (5.18)$$
$$\underset{CH_3}{|}$$

A detailed mechanism of radical formation during UV-irradiation of polypropylene is described by Hama et al.[854, 855] and Tsuji and Seiki[2246, 2275, 2278]. In their interpretation, the UV-light is mainly absorbed by carbonyl groups in polyproylene produced by oxidation[1784, 1814]. The wavelength dependence of radical formation[2275] indicates that methyl radicals and other free radicals (5.13) and (5.14), responsible for the broad four-line spectrum, are produced by secondary reactions of free radicals, which show a singlet spectrum. Photolysis with light of longer wavelength changed the singlet spectrum to the four-line spectrum. The singlet spectrum is supposed to be due to free acyl radicals (5.19) which are formed when a carbonyl group absorbs UV-light at 77 K. Scission of the main chain occurs and acyl radicals (5.19) are formed:

$$-\overset{\overset{O}{\|}}{C}-CH_2-\overset{\overset{O}{\|}}{C}-CH_2-CH- \xrightarrow{+h\nu} -CH-CH_2-\overset{O}{\overset{\|}{C}}\cdot + \cdot CH_2-C \qquad (5.24)$$

$$\underset{CH_3}{|} \qquad \underset{CH_3}{|} \qquad \underset{CH_3}{|} \qquad \underset{CH_3}{|}$$

$$\qquad\qquad (5.19) \qquad (5.18)$$

187

Radicals (5.19) may absorb UV-light during the irradiation and change to radicals (5.13):

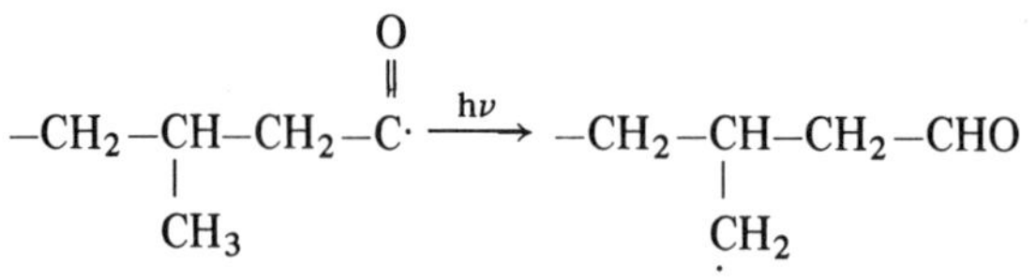

$$-CH_2-CH-CH_2-C \cdot \xrightarrow{\;h\nu\;} -CH_2-CH-CH_2-CHO \qquad (5.25)$$

with C bearing $=O$ above, CH_3 below the left CH, and CH_2 (with radical dot) below the right CH.

$$(5.19) \qquad\qquad\qquad (5.13)$$

By internal hydrogen transfer radical (5.18) may change to radical (5.13):

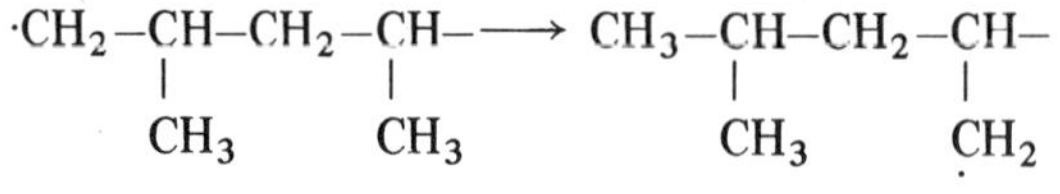

$$\cdot CH_2-CH-CH_2-CH- \longrightarrow CH_3-CH-CH_2-CH- \qquad (5.26)$$

with CH_3 and CH_3 below the left structure, and CH_3 and CH_2 (radical dot) below the right structure.

$$(5.18) \qquad\qquad\qquad (5.13)$$

but it may also undergo hydrogen abstraction, forming a double bond.

An ESR study of free radicals formed during the degradation of polypropylene photosensitized by hydropyrene, was also reported[123].

5.1.1.4. Molecular Motions in Solid Polyolefines

Relations between the decay of trapped radicals and the molecular motion of the polymer matrix have been discussed by many authors[101, 104, 1082, 1084, 1086, 1274, 1269, 1309, 1379, 1571, 1572, 1574, 1645, 1798, 2049, 2147, 2156].

The kinetic curves of radical amount *vs* temperature have typical fast and slow portions. At certain temperatures, considerable amounts of radicals become trapped in the solid polymer and the decay is practically stopped (Fig. 5.12).

The decay characteristic of free radicals in irradiated polyethylene indicate that the decay varies with the motion of polymer molecules of the matrix, trapping the free radicals[1084]. Similar phenomena have been discussed for hydrocarbon molecules of low molecular weight[195]. An increase of pressure hinders molecular motion by reducing the voids. The stability of free radicals in polyethylene[2116] and polypropylene[2115] increases, therefore, proportionally with pressure. At about 8000 atm, most of the molecular motion is eliminated, so that further increase does not significantly affect the decay rate constant.

Sohma et al.[1086, 1572, 1574] stated that there are three temperature regions, in which the free radicals decay very rapidly. These regions are at about 120, 200, and 250 K and were designated as T_A, T_L, and T_B, respectively (Fig. 5.12). The decay of the free radicals in high-density polyethylene at these temperatures has activation energies of 0.4 kcal/mole for T_A, 9.4 kcal/mole for T_L, and 18.4 kcal/mole for T_B. For low-density polyethylene these values are 0.7 kcal/mol for T_A, 23.1 kcal/mole for T_L, and 24.8 kcal/mole for T_B. Comparison of the time constants for the decay reactions, and for the molecular motions of the polymer matrix, indicates that the decay at T_A and T_B is closely related to molecular motion in the amorphous regions

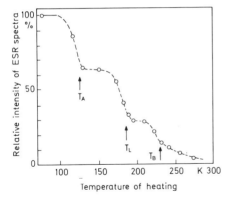

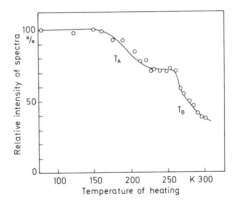

Fig. 5.12. Decay curve for free radicals produced in polyethylene γ-irradiated at 77 K[1574]

Fig. 5.13. Decay curve for free radicals produced in isotactic polypropylene γ-irradiated at 77 K[1571]

of the polymer. The decay of the free radicals at T_L in high-density polyethylene is due to molecular motion associated with local mode relaxation at lamellar surfaces, while that of low-density polyethylene is due to local mode relaxation in the completely amorphous regions. This study shows that the free radicals which decayed in the lower temperature region have a less uniform steric configuration than that of free radicals which decayed in higher temperature regions.

Steric configuration of alkyl radical (5.12) in irradiated polypropylene, in relation to molecular motion of the polymer matrix, has also been proposed by Sohma et al.[1078, 1570, 1571]. The free radicals decay rapidly in two temperature regions; 170 K (T_A) and 260 K (T_B) (Fig. 5.13) called α-, and β-dispersion regions of molecular motion.

The activation energies of decay were found to be 11 kcal/mole at the lower temperature and 48 kcal/mole at the higher temperature. The activation energy of the radical decay if of the same order as that of molecular motion in the same temperature region[1254, 1645]. The tacticity of the polymer affects the nature and the behavior of free radicals trapped in the polymer[679, 723, 1392, 1664]. Takayanagi et al.[1270] suggested that free radicals formed by γ-irradiation of isotactic polypropylene are preferentially trapped in the intermediate disordered region of the crystalline texture.

5.1.1.5. Radical Pairs in γ-Irradiated Polyolefines

The formation of radical pairs in γ-irradiated polymers has been reported in several papers[728–730, 732, 823, 851, 923, 1028–1030, 2421]. The radical pairs are detected as the ESR signal due to $\Delta M_S = 2$ transitions and the spectra are shown in Figures 5.14 and 5.15.

The radical pairs can be classified into two groups:
1. Those formed between two adjacent molecules.
2. Those formed in a single molecule.

189

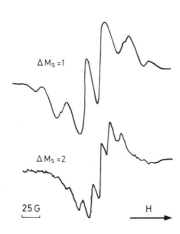

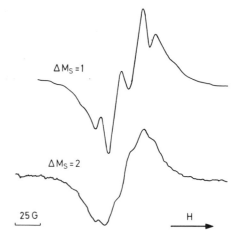

Fig. 5.14. $\Delta M_S = 1$ and $\Delta M_S = 2$ spectra of γ-irradiated polyethylene at 77 K[1029]

Fig. 5.15. $\Delta M_S = 1$ and $\Delta M_S = 2$ spectra of γ-irradiated polypropylene at 77 K[1029]

The two radicals, e.g. in polyethylene (5.20)

$$-CH_2-\overset{\cdot}{C}H-CH_2-CH_2-CH_2-\overset{\cdot}{C}H-CH_2- \qquad (5.20)$$

are produced with specific distance and orientation in the crystalline units. They are reported to be separated by a distance of 5–10 Å (e.g. for polyethylene −5.3 Å)[923]. On the other hand, Grinberg et al.[823] have suggested that radical pairs in polyethylene are formed from randomly distributed radicals.

The two unpaired electrons of a radical pair are coupled through exchange inter- action to form a triplet state. The ESR spectrum of a radical pair has the characteristic of a triplet. For the $\Delta M_S = 1$ transition it shows the zero-field splitting due to mag- netic interaction of the two unpaired spins. It also gives the weak signal at $g = 4$ due to the forbidden transition $\Delta M_S = 2$, which is allowed by the mixing of the state $M_S = 0$ and $M_S = \pm 1$. For both transitions $\Delta M_S = 1$ and $\Delta M_S = 2$, the hyperfine separations of the ESR spectra of the radical pairs are approximately half of those observed for isolated radicals. The spectrum due to the $\Delta M_S = 2$ transition does not show the zero-field splitting. The $\Delta M_S = 2$ transition is commonly measured with a $10^3 - 10^4$ times higher gain than in the case of the $\Delta M_S = 1$ transition.

The $\Delta M_S = 2$ spectra of polyethylene are well-resolved (Fig. 5.14), while $\Delta M_S = 2$ spectra for polypropylene have similar structure but are poorly resolved (Fig. 5.15). The average hyperfine separations of the $\Delta M_S = 2$ spectra of polyethyl- ene and polypropylene are 16 and 15 G, respectively[1029].

The $\Delta M_S = 1$ spectra of radical pairs in polymers have not been clearly observed. The failure to observe the $\Delta M_S = 1$ spectra in polymers has been interpreted as being due to the lack of a specific distance between the two radicals, resulting from the complicated structure of the macromolecules. It was possible to measure the $\Delta M_S = 1$ spectra (Fig. 5.16) of a vacuum-irradiated polyethylene sample which was drawn to 1000% of its initial length[728, 730]. When the magnetic field was parallel to the draw

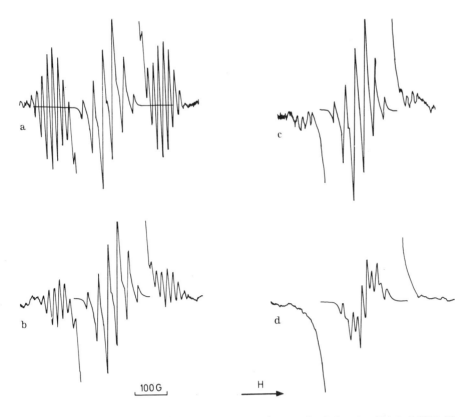

Fig. 5.16. Angular dependence of $\Delta M_S = 1$ spectra in drawn polyethylene irradiated at 77 K. The angle between magnetic field and draw direction is: a. 0°, b. 10°, c. 15°, d. 90° [732]

direction or to the chain axis of polyethylene, well-resolved multiples consisting of eleven lines were observed on both sides of the main sextet spectrum (5.16b).

The sextet is assigned to the isolated alkyl radicals (5.3). The separation between the two outer multiplets is 371 G and the hyperfine splittings of each multiplet are 16 G, amounting to just half of the value (32 G) of the main sextet. When the magnetic field is perpendicular to the draw direction, the shape and the position of the outer spectra change considerably, and the hyperfine structure disappears (5.16d). The change of the main spectrum from six lines to ten lines has been well explained as being due to the change of H_α coupling of the alkyl radicals [1137]. The $\Delta M_S = 2$ spectrum of drawn polyethylene is shown in Figure 5.17.

When the magnetic field is parallel to the chain axis, the hyperfine splitting is in this case 16 G. Both $\Delta M_S = 2$ and $\Delta M_S = 1$ spectra are thermally unstable and when the sample is annealed at elevated temperature they decay simultaneously (Fig. 5.18) [729, 732, 1030, 1338].

For small molecules, radical pairs are only formed in single crystals. This indicates that there might be a preference for crystalline structure. For polymers, however, the formation of radical pairs is not a particular phenomenon occurring in the

191

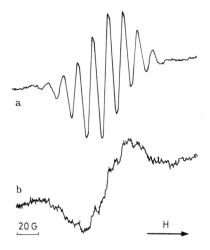

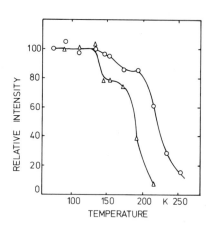

Fig. 5.17. ΔM_S = 2 spectra in drawn polyethylene irradiated at 77 K. The angle between magnetic field and draw direction is: a. 0°, b. 90°[732]

Fig. 5.18. Decay curves of ESR spectra due to radical pairs: (o) ΔM_S = 2 and (△) ΔM_S = 1[732]

crystalline regions[1029]. However, the mechanism of the pairwise formation of radicals is not well elucidated. There are two hypotheses:

1. Hydrogen-free radical produced in the dissociation of C—H bond may react with the adjacent molecule producing a second radical. This may result in the pairwise formation of radicals between the two adjacent molecules[1030]:

$$RH \xrightarrow{\gamma} \boxed{R\cdot} + H\cdot$$
$$H\cdot + RH \longrightarrow \boxed{R\cdot} + H_2$$

(5.27)

This mechanism refers to the stereospecific formation of paired radicals found in single crystals, and it is rather difficult to understand that the hydrogen radical may be formed at a random orientation.

2. The radical pairs may originate from the recombination of a cation and an electron[923, 1030]:

$$RH \xrightarrow{\gamma} RH^{\cdot+} + e^-$$
$$RH^{\cdot+} + RH \longrightarrow \boxed{R\cdot} + RH_2^+$$
$$RH_2^+ + e^- \longrightarrow \boxed{R\cdot} + H_2$$

(5.28)

A similar mechanism, where the charge neutralization reaction of an anion and a cation is assumed to produce a pair of radicals, may also be possible:

192

$$RH \xrightarrow{\gamma} RH^{\cdot+} + e^-$$

$$RH + e^- \longrightarrow RH^{\cdot-} \tag{5.29}$$

$$RH^{\cdot+} + RH^{\cdot-} \longrightarrow \boxed{R\cdot \ \cdot R} + H_2$$

On the other hand, radical pairs are not believed to be ion-electron pairs, since the $\Delta M_S = 2$ signal cannot be bleached out by illumination with visible light, as observed for trapped electrons. Therefore, the mechanism for the pairwise radical trapping is still open to discussion. No reasonable explanation of the stereospecific formation of the paired radicals is as yet proposed.

5.1.1.6. Poly-1-butene

Irradiation of poly-1-butene with ionizing radiation at 77 K gives an ESR spectrum, containing a six-line structure with a hyperfine splitting constant of 21 G (Fig. 5.19)[737, 978, 979, 1269, 1392, 1863].

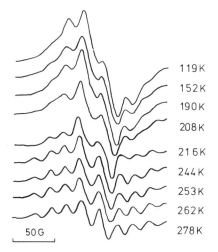

119 K
152 K
190 K
208 K
216 K
244 K
253 K
262 K
278 K

50 G

Fig. 5.19. Change of ESR spectra of poly-1-butene γ-irradiated at 77 K with increasing measuring temperature[979]

This spectrum indicates that the radical (5.21) is formed:

$$-CH_2-CH-CH_2- \tag{5.21}$$
$$\qquad\ \ |$$
$$\qquad \cdot CH$$
$$\qquad\ \ |$$
$$\qquad CH_3$$

Above 208 K an octet and a quartet structure are dominant[978, 979, 1268]. Loy[1392] reported that, after warming to 245 K for 3 minutes only, a doublet spectrum was observed, which was not assigned to a specific radical structure. The octet was interpreted as arising from an allylic radical (5.22):

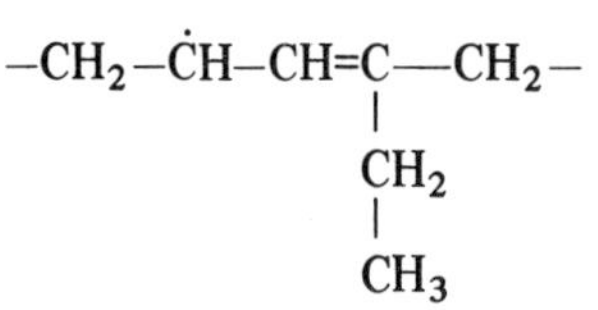

$$-CH_2-\overset{\cdot}{C}H-CH=C-CH_2- \qquad (5.22)$$
$$\overset{|}{C}H_2$$
$$\overset{|}{C}H_3$$

The hyperfine splitting constants (20 G and 7 G) found for radical (5.22) agree well with the splitting constants found for the allylic radical in polyethylene (5.4). The sextet spectrum observed at 77 K has been attributed to radicals (5.3) and/or (5.13) which might be formed by abstraction of ethyl side groups or by C–C bond scission in the side chain, respectively.

The quartet spectrum observed for amorphous samples[979] at temperatures higher than 203 K may be due to the rearrangement of radical (5.13) to radical (5.23):

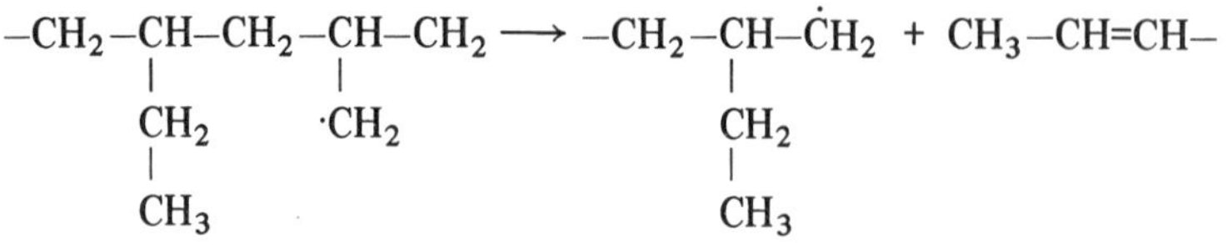

$$-CH_2-CH-CH_2-CH-CH_2 \longrightarrow -CH_2-CH-\overset{\cdot}{C}H_2 \; + \; CH_3-CH=CH- \qquad (5.30)$$
$$\overset{|}{C}H_2 \qquad \overset{|}{\cdot}CH_2 \qquad\qquad\qquad \overset{|}{C}H_2$$
$$\overset{|}{C}H_3 \qquad\qquad\qquad\qquad\quad \overset{|}{C}H_3$$
$$(5.13) \qquad\qquad\qquad\qquad\quad (5.23)$$

Rubin and Huber[1863] obtained an octet spectrum which was attributed to alkyl radical (5.24), proposed previously by Takayanagi et al.[1269]:

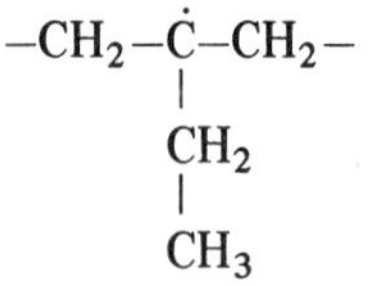

$$-CH_2-\overset{\cdot}{C}-CH_2- \qquad (5.24)$$
$$\overset{|}{C}H_2$$
$$\overset{|}{C}H_3$$

Irradiation of poly-1-butene with UV-light[322] gives spectra containing twelve lines with a hyperfine splitting constant of 27 G (g = 2.0026), which have been attributed to ethyl radicals (5.25):

$$\cdot CH_2-CH_3 \qquad (5.25)$$

The spectrum of polymer alkyl radicals (5.3) is overlapping the spectrum of ethyl radicals (5.25).

5.1.1.7. Polyisobutylene

ESR study of polyisobutylene has been carried out by several authors[373, 1392, 1798, 2169, 2170, 2304, 2307, 2491]. Irradiation of polyisobutylene with high energy radiation at 77 K gives a spectrum showing a broad doublet with a hyperfine splitting constant of 20 G. This spectrum is attributed to free radicals of the type (5.26):

$$
\begin{array}{ccc}
CH_3 & & CH_3 \\
| & & | \\
-C-CH-C- & & (5.26) \\
| \quad \cdot & & | \\
CH_3 & & CH_3
\end{array}
$$

On warming the irradiated sample to 213 K and exposing it to UV-light, a complex spectrum was recorded[2169, 2170] containing seven basic lines with some additional weak lines. This spectrum is reversibly converted to the initial doublet on standing and is assigned to the radical (5.27):

$$
\begin{array}{c}
CH_3 \\
| \\
-CH_2-C\cdot \qquad (5.27) \\
| \\
CH_3
\end{array}
$$

As shown by Carstensen and Rånby[397, 398, 403] irradiation of polyisobutylene with UV-light gives completely different spectra (Fig. 5.20).

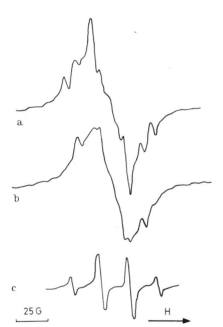

Fig. 5.20. ESR spectra of polyisobutylene UV-irradiated at 77 K, measured at 77 K. a. immediately after irradiation, b. the same sample after being kept at 77 K for 5 days showing the broad component of the spectrum (a), c. the spectral component which decayed during the first hourst at 77 K, i.e. the sharp component of spectrum (a)[403]

25 G   H

The spectrum consists of two components: one sharp quartet, which has a lifetime of 1.5 hours at 77 K and is attributed to methyl radicals (5.17), and one broad spectral component constituted of many superimposed lines and interpreted as being due to three different radicals, all of which are stable at 77 K. One of these radicals (5.12) is the counterpart of the methyl radical (5.17), and the other two are the radicals (5.26) and (5.28):

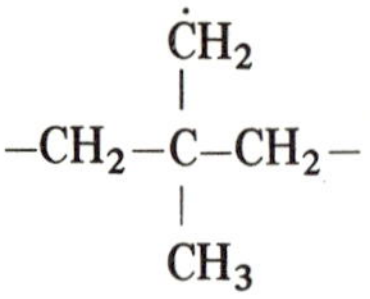

$$\begin{array}{c} \overset{\displaystyle \cdot}{C}H_2 \\ | \\ -CH_2-C-CH_2- \\ | \\ CH_3 \end{array} \qquad (5.28)$$

5.1.1.8. Poly-3-methyl-1-butene

γ-Irradiation of poly-3-methyl-1-butene at 77 K gives a seven-line spectrum which is attributed to radical $(5.29)^{1266, \, 1274)}$:

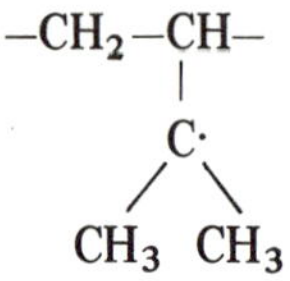

$$\begin{array}{c} -CH_2-CH- \qquad (5.29) \\ | \\ \overset{\displaystyle \cdot}{C} \\ \diagup \quad \diagdown \\ CH_3 \quad CH_3 \end{array}$$

A narrow four-line spectrum can additionally be observed and is assigned to methyl radicals (5.17), whereas a broad quartet is attributed to radical (5.30):

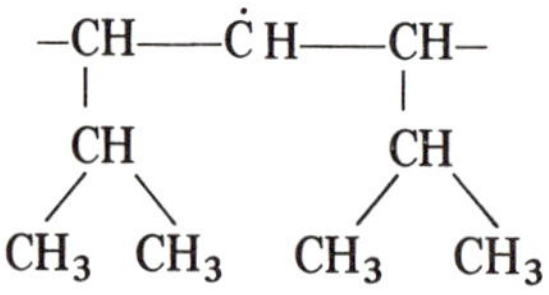

$$\begin{array}{c} -CH\!-\!\!-\!\!\overset{\displaystyle \cdot}{C}H\!-\!\!-\!\!CH- \qquad (4.30) \\ | \qquad\qquad | \\ CH \qquad\quad CH \\ \diagup \,\diagdown \qquad \diagup\,\diagdown \\ CH_3 \ CH_3 \quad CH_3 \ CH_3 \end{array}$$

The latter radical has also been found by ESR study of γ-irradiated poly-1-pentene[1271].

5.1.1.9. Poly-4-methyl-1-pentene

Irradiation of poly-4-methyl-1-petene with γ- and electron-radiation give one kind of ESR spectra which is significantly different from ESR spectra obtained in the UV-degradation[322, 434, 792, 1267, 1272, 1273, 1748, 2409).

After electron irradiation of a polymer sample at 77 K an octet-line spectrum is formed (Fig. 5.21).

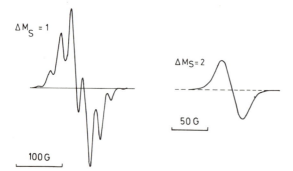

Fig. 5.21. $\Delta M_S = 1$ and $\Delta M_S = 2$ spectra of poly-4-methyl-1-pentane γ-irradiated at 77 K[434)

It has been attributed to radicals (4.31)[792]:

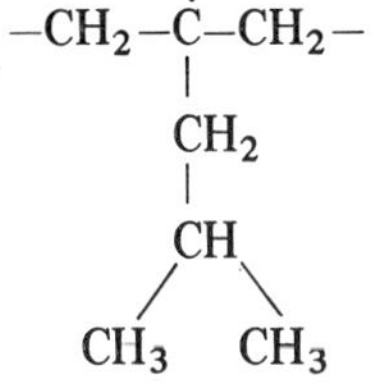

$$-CH_2-\overset{\cdot}{C}-CH_2- \qquad (4.31)$$

with side chain:
$$\begin{array}{c} | \\ CH_2 \\ | \\ CH \\ \diagup\,\diagdown \\ CH_3 \quad CH_3 \end{array}$$

Figure 5.21 presents two spectra, one sharp for $\Delta M_S = 1$ and one weak and broad for $\Delta M_S = 2$ transition characteristic for radical pairs (see Chapter 5.1.1.5). The spectrum for $\Delta M_S = 2$ disappears rapidly when the polymer is warmed to room temperature and recooled to 77 K. This fact indicates a recombination of free radical pairs.

Takayanagi et al.[1272] reported that, in addition to the octet-line spectrum, odd-number hyperfine structures (5, 7 and 9 lines) are observed at liquid nitrogen temperature. By increasing the temperature to about 253 K, the octet lines split clearly into a doublet substructure, which, by cooling to liquid nitrogen temperature, returns to the original octet. The octet-line spectrum was suggested by these authors to be due to allylic radicals (5.32) or to the alkyl radicals in the side chain (5.33):

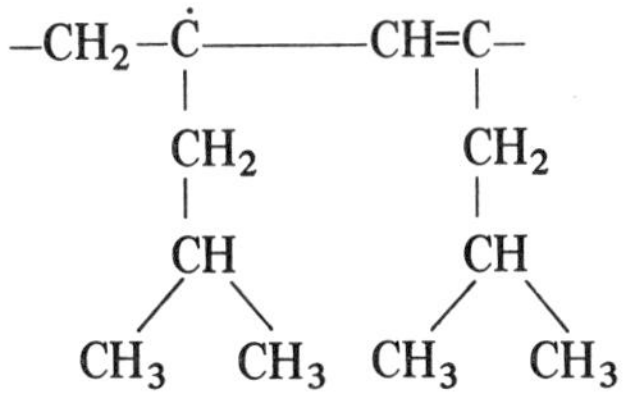

$$-CH_2-\overset{\cdot}{C}\text{———}CH=C- \qquad (5.32) \qquad\qquad -CH-CH_2- \qquad (5.33)$$

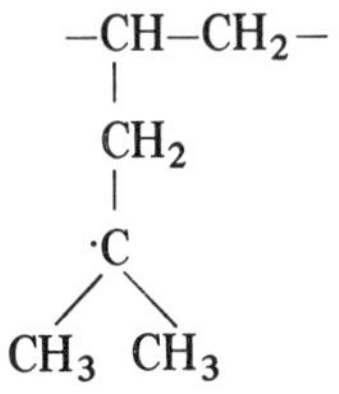

Under UV-irradiation at 77 K, a sharp four-line spectrum with the hyperfine splitting constant of 22.5 G and a broad four-line spectrum are formed[322, 792, 1273]. The former spectrum is attributed to methyl radicals (5.17) which decay by second order reaction[1273]. The latter spectrum is due to the following radicals (5.34) and/or (5.35) and (5.36)

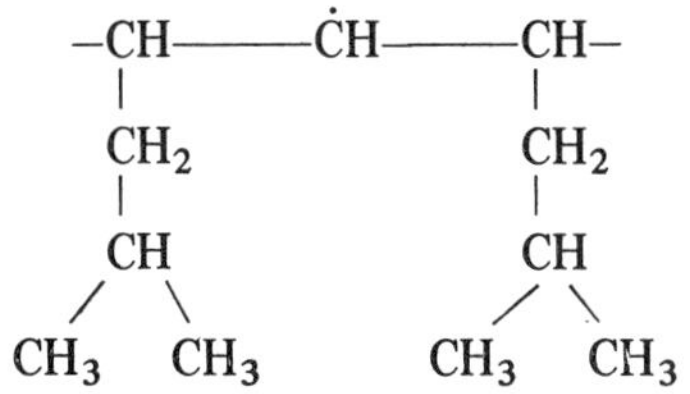

$$-CH\text{———}\overset{\cdot}{C}H\text{———}CH- \qquad (5.34) \qquad\qquad -CH_2-CH- \qquad (5.35)$$

$$-CH-CH_2- \qquad (5.36)$$

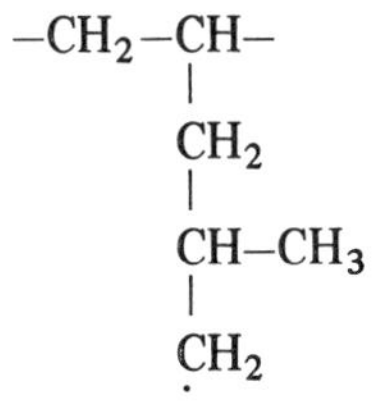

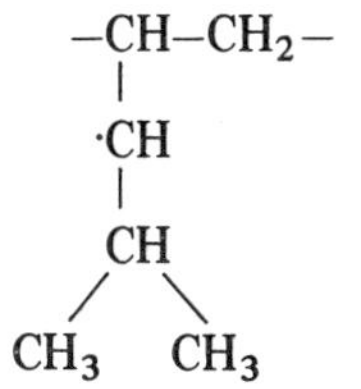

Goodhead[792] has reported conversion of radical (*5.33*) to radical (*5.34*) and/or (*5.36*) under UV-irradiation.

5.1.1.10. Trapped Electrons in Polyolefines

Irradiation of such polyolefines as polyethylene, polypropylene, poly-4-methyl-1-pentene, polyisobutylene with low energy γ-irradiation at 77 K produces an additional sharp singlet in the central region of the ESR spectrum, which was attributed to trapped electrons[851, 1117–1119, 2409].

Figure 5.22 shows such type of spectrum for polyethylene (g = 2.002 ± 0.001, $\Delta H \approx 4$ G).

A similar spectrum, but much broader ($\Delta H \approx 20$ G), was obtained by Russian scientists[35, 1602]. The formation of this singlet spectrum may be explained in that the electron is trapped by a defect in the crystalline regions of the polymer.

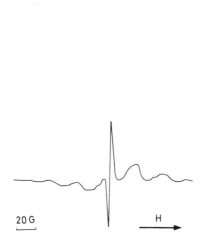

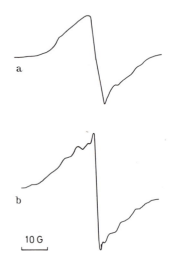

Fig. 5.22. ESR spectrum of trapped electron in polyethylene γ-irradiated at 77 K[1117]

Fig. 5.23. ESR spectra of cis-1,4-polybutadiene (95.8% cis) γ-irradiated at 77 K: a. amorphous, b. crystalline sample[2338]

5.1.2. Polydienes

5.1.2.1. Polybutadiene

ESR studies of free radicals formed in γ-irradiated cis-1,4-polybutadiene have been published by many authors[644, 1218, 1283, 1291, 1363, 2338]. ESR spectra of amorphous cis-1,4-polybutadiene obtained at 77 K show a singlet-line (Fig. 5.23 a).

γ-Irradiation of crystalline cis-1,4-polybutadiene under the same conditions gives a spectrum (Fig. 5.23 b) in which a septet structure appears more clearly than in amorphous samples (Fig. 5.23 a). The singlet is superimposed by the seven-line spectrum which is attributed to the allyl radical (5.37)[2338].

$$-CH_2-\overset{\cdot}{C}H-CH=CH-CH_2- \qquad (5.37)$$

The decay rates of the radicals were found to differ depending on the aggregation state of the trapping matrix. For amorphous samples ESR spectra disappeared when the samples were warmed up to 200 K, which is close to the glass-transition temperature ($T_g \sim 198$ K).

Cis-1,4-polybutadiene, trapped in a molecular sieve and irradiated with γ-irradiation at room temperature, gives a nearly structure-less broad singlet spectrum (Fig. 5.24) which is quite stable[2338].

The different decay behavior of otherwise identical radicals from different samples suggests that the decay reaction of the radicals is mainly controlled by the aggregation state and the molecular motion of the trapping matrix[2049, 2240].

During γ-irradiation of 1,2-polybutadiene at 77 K a broad spectrum is formed (Fig. 5.25 a) which is identical with the spectrum of poly-1,4-butadiene presented in Figure 5.24.

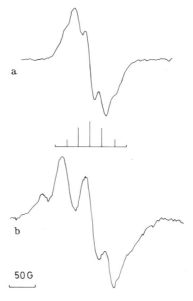

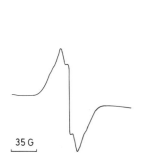

Fig. 5.24. ESR spectrum of free radicals from γ-irradiated cis-1,4-polybutadiene at 77 K, trapped on molecular sieve[2338]

Fig. 5.25. ESR spectra of 1,2-polybutadiene γ-irradiated at 77 K: a. Measured at 273 K, b. Difference spectrum of two 1,2-polybutadiene samples tempered at 133 K and 203 K[2515]

This spectrum consists of a septet with a hyperfine splitting constant of 16 G and is assigned to the radical (5.38)[2515]:

199

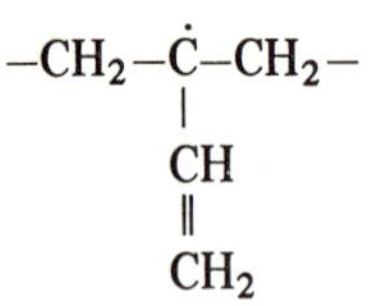

$$-CH_2-\overset{\displaystyle\cdot}{C}-CH_2- \qquad (5.38)$$
$$\overset{\displaystyle |}{\underset{\displaystyle\parallel}{CH}}$$
$$CH_2$$

On warming to 200 K the sample exhibits a quintet of lines separated by about 27 G (Fig. 5.25b). This spectrum is attributed to alkyl radicals (5.39) trapped in the polymer matrix at low temperatures.

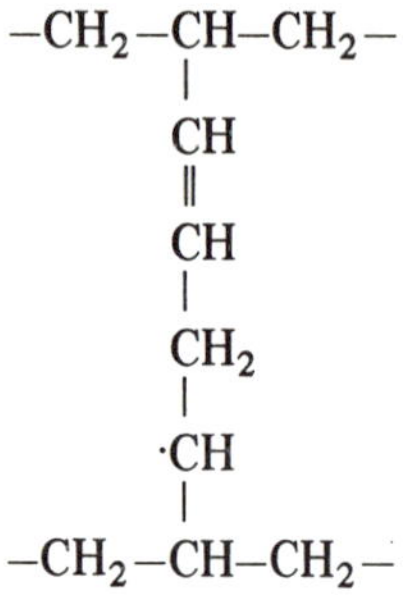

$$-CH_2-CH-CH_2- \qquad (5.39)$$
$$|$$
$$CH$$
$$\parallel$$
$$CH$$
$$|$$
$$CH_2$$
$$|$$
$$\cdot CH$$
$$|$$
$$-CH_2-CH-CH_2-$$

The radical (5.39) can only be formed by crosslinking.

Irradiation of cis-1,4-polybutadiene with UV-light at 77 K gives ESR spectra with some resolved hyperfine structure (Fig. 5.26)[400, 402].

The spectra consist of a broad main signal with a g-value approximately equal to that of a free spin, and two narrow signals with a separation of 510 G. The narrow lines are due to hydrogen atoms formed on the surface of the quartz[403]. The radical concentration increases rapidly at the beginning of the irradiation and then levels off. By increasing doses the spectra become asymmetric. In Carstensen's interpretation[398, 400] the broad main signal consists of an even number of lines, of which four lines in the centre are observed. A six-line spectrum with about 12 G hyperfine splitting constant is suggested from the allyl radical (5.37). The radical concentration decreases by heating the sample to temperatures just below the glass-transition temperature (T_g). Above T_g the ESR signal changes to a weak singlet. This singlet has been attributed to polyenyl radicals (5.40):

$$-CH_2-(CH{=}CH)_n-\overset{\displaystyle\cdot}{C}H_2 \qquad (5.40)$$

There is also a report on the ESR study of free radicals formed during the thermal degradation of polydienes[2199].

5.1.2.2. Polyisoprene

Irradiation of cis-1,4-polyisoprene (natural rubber) with γ-rays at 77 K gives a singlet spectrum with no resolved hyperfine structure[321, 1290, 1368, 1640, 1641, 2304, 2307] (Fig. 5.27).

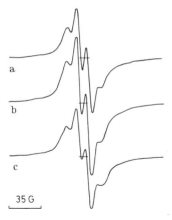

Fig. 5.26. ESR spectra of
cis-1,4-polybutadiene UV-irradiated
at 77 K: a. 10 min, b. 20 min and
c. 50 min[400)]

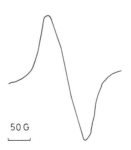

Fig. 5.27. ESR spectrum of
cis-polyisoprene γ-irradiated
at 77 K[1368)]

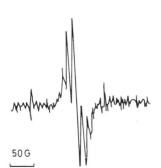

Fig. 5.28. The hidden
multiplet in ESR spectrum
of cis-polyisoprene shown
in Fig. 5.27[1368)]

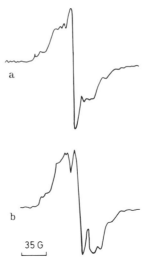

Fig. 5.29. ESR spectra of
cis-1,4-polyisoprene γ-irradiated at 77 K with dose:
a. 0.6 Mrad and b. 3.15
Mrad[1092)]

When the resolution enhancement technique is applied to this broad spectrum, a hidden multiplet is revealed (Fig. 5.28)[1368)].

The analysis of the spectrum shows that its main component can be a septet, in which the outermost and weakest peaks are masked with noise. The separation of the septet lines is 13 G and this value is nearly equal to the separation of multiplets due to an allyl radical (5.41):

$$-CH_2-\overset{\cdot}{C}H-C=CH-CH_2- \qquad (5.41)$$
$$\overset{|}{C}H_3$$

The next component is a sextet which can be assigned to one of the following radicals (5.42) and/or (5.43):

$$-CH_2-CH=CH-\overset{\cdot}{C}H_2 \qquad (5.42) \qquad\qquad -CH_2-CH=C-CH_2- \qquad (5.43)$$
$$\overset{|}{C}H_3 \qquad\qquad\qquad\qquad\qquad\qquad \overset{|}{\cdot CH_2}$$

Katzer und Heusinger[1092] obtained ESR spectra of γ-irradiated polyisoprene which had a different hyperfine structure depending on the radiation dose (Fig. 5.29).

At lower doses they obtained a quintet-line spectrum with a hyperfine splitting constant of 23.5 G which was assigned to the alkyl radical (5.44)[1092, 2515]:

$$-CH_2-CH-CH_2- \qquad (5.44)$$
$$\quad|$$
$$\quad C-CH_3$$
$$\quad\|$$
$$\quad CH$$
$$\quad|$$
$$\quad CH_2$$
$$\quad|$$
$$\quad \cdot C-CH_3$$
$$-CH_2-CH-CH_2-$$

At higher doses a septet-line spectrum with a hyperfine splitting constant of 15 G is formed and assigned to radicals (5.41).

When trans-1,4-polyisoprene is irradiated in vacuum at ambient temperatures ESR spectra with some resolution are observed[1283].

Only a few ESR studies have been reported on the formation of free radicals in polyisoprene irradiated with UV-light[398, 399, 402, 1126, 1473]. The observed spectra have similar hyperfine structures for cis- and trans-1,4-polyisoprene[399]. The signals are also similar in shape to those observed for cis-1,4-polybutadiene. Some differences were however observed: 1) The intensity ratio between the central lines of the spectrum is higher for polybutadiene than for polyisoprene, 2) The total width of the signal is larger for polyisoprene than for polybutadiene. The spectrum of UV-irradiated polyisoprene consists of eight lines with a hyperfine separation of 12 G and is attributed to the radicals (5.42) and (5.43). By heating to temperatures above T_g the spectra change to a singlet, which was assigned to polyenyl radicals (5.45):

$$-CH_2-(C=CH)_n-\overset{\cdot\cdot}{C}H_2 \qquad (5.45)$$
$$\quad\quad \overset{|}{C}H_3$$

Kosek and Zielinski[1199, 1200] have reported ESR studies of free radicals formed during the radiolysis of natural latex in the presence of methane halides such as CCl_4, $CHCl_3$, CH_2Cl_2.

5.1.2.3. Polypiperylene

ESR spectra for cis-1,4-polypiperylene are reported only by Carstensen[398, 401, 402], who used UV-light at 77 K for the irradiation. The observed spectrum was a singlet without any trace of hyperfine structure. It is, therefore, concluded that no methyl radicals are formed. The second derivative ESR spectrum is an octet-line spectrum attributed to the main chain scission radicals (5.46) and/or (5.47):

$$-CH_2-CH=CH-\overset{\cdot}{C}H \rightleftharpoons -CH_2-\overset{\cdot}{C}H-CH=CH \qquad (5.31)$$
$$\underset{CH_3}{|} \qquad\qquad \underset{CH_3}{|}$$
$$(5.46) \qquad\qquad\qquad (5.47)$$

The intensity distribution, however, implies the presence of a quintet spectrum, which is assigned to other main chain scission free radicals (5.48) and/or (5.49)

$$\cdot CH_2-CH=CH-CH- \rightleftharpoons CH_2=CH-\overset{\cdot}{C}H-CH- \qquad (5.32)$$
$$\underset{CH_3}{|} \qquad\qquad\qquad \underset{CH_3}{|}$$
$$(5.48) \qquad\qquad\qquad (5.49)$$

5.1.2.4. Polychloroprene

ESR spectra of polychloroprene exposed to γ-radiation at 77 K[1214, 2163] and UV-light[402] show a broad singlet. The second derivative spectra have only traces of hyperfine structure. Despite the poorly resolved hyperfine structure in the ESR spectra, Carstensen[402] suggested that the primary radicals are produced by a chain scission (5.50) and/or (5.51) and (5.52) and/or (5.53):

$$\underset{\underset{-CH_2-C=CH-\overset{\cdot}{C}H_2}{}}{\overset{\overset{Cl}{|}}{}} \rightleftharpoons \underset{\underset{-CH_2-\overset{\cdot}{C}-CH=CH_2}{}}{\overset{\overset{Cl}{|}}{}} \qquad (5.33)$$
$$(5.50) \qquad\qquad\qquad (5.51)$$

$$\underset{\underset{\cdot CH_2-C=CH-CH_2-}{}}{\overset{\overset{Cl}{|}}{}} \rightleftharpoons \underset{\underset{CH_2=C-CH-CH_2-}{\underset{\cdot}{}}}{\overset{\overset{Cl}{|}}{}} \qquad (5.34)$$
$$(5.52) \qquad\qquad\qquad (5.53)$$

5.1.3. Poly(methyl acrylate) and Poly(methyl methacrylate)

The ESR spectra of radicals trapped in γ-irradiated poly(acrylic acid) have been examined by a few authors[14, 464, 1672, 1993].

During the past two decades, the ESR spectra of irradiated poly(methyl acrylate) and poly(methyl methacrylate) have been an interesting subject for a number of workers[13, 54, 97, 98, 134, 137, 145, 352–355, 373, 387, 441, 507, 508, 696, 736, 740, 763, 764, 845, 882, 1006, 1033, 1035, 1131, 1206, 1207, 1313, 1344, 1425, 1466, 1471, 1484, 1640, 1642, 1645, 1672, 1681, 1682, 1916, 2050, 2054, 2118, 2123, 2125, 2191, 2215, 2304, 2339, 2363, 2428, 2459, 2464].

Photo-irradiated poly(methyl methacrylate) samples have only been studied by a few authors using ESR[354, 441, 1091, 1465, 1466, 2428]. Russian scientists[83] have also studied the formation of paramagnetic centers in poly(methyl methacrylate) exposed to the action of laser radiation.

ESR spectra of poly(methyl acrylate) and poly(methyl methacrylate) irradiated with ionizing radiation or with ultraviolet-radiation are similar, and for that reason they are discussed together. Both polymers give a nine-line spectrum which consists of an intense five-line and a weak four-line component[137, 354, 387, 422, 441, 845, 1035, 1131, 1206, 1207, 1466, 1471, 1480, 1672, 1685, 1682, 2050, 2054, 2123, 2125, 2171, 2339, 2428].

This nine-line spectrum (Fig. 5.30) is attributed to radicals (5.54) and (5.55):

$$
\begin{array}{cc}
\begin{array}{c}
CH_3 \\
| \\
-CH_2-C\cdot \\
| \\
COOH
\end{array}
\quad (5.54)
&
\begin{array}{c}
CH_3 \\
| \\
-CH_2-C\cdot \\
| \\
COOCH_3
\end{array}
\quad (5.55)
\end{array}
$$

The origin and assignment of the nine-line ESR spectrum is still under discussion[383, 1754]. Many investigators have tried to identify the radicals from which the

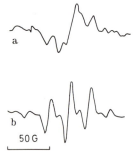

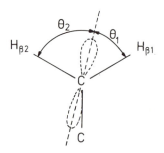

Fig. 5.30. ESR spectra of samples UV-irradiated (2527 Å) at 77 K:
a. Monomerfree poly(methyl methacrylate),
b. Poly(methyl methacrylate) with methylmethacrylate[1465]

Fig. 5.31. Conformation of the propagating radical (5.55)[1035]

204

nine-line spectrum originates. The line width of the hyperfine components in this spectrum is well interpreted as being due to line-broadening caused by hindered oscillation of the methyl group around the C_α–C_β bond[1035, 2050, 2054, 2125] (Fig. 5.31).

Several authors[1465, 1672, 2054, 2339] observed that the nine-line spectrum is produced only when excess monomer is present (5.30b).

Hajimoto et al.[845] reported that a commercial poly(methyl methacrylate) sample, irradiated with electrons at 77 K, shows, at the same temperature, a spectrum consisting of seven lines. Poly(methyl methacrylate) free from monomer gives spectra with the same overall splitting as commercial samples, but the spectra are more complicated in shape and vary with the irradiation dose. At about 313 K (below the glass transition $T_g \sim 363$ K) radical (5.55) decays according to second-order kinetics with an activation energy of about 30 kcal/mol[845, 1645, 2464]. Tino and Szöcs[2191] investigated the effect of pressure on the ESR spectra of radical (5.55). They concluded that the formation of secondary chain radicals of the type (5.56) was:

$$-CH_2-\overset{\overset{\displaystyle CH_3}{|}}{\underset{\underset{\displaystyle COOCH_3}{|}}{C}}-\hspace{-0.3em}-\hspace{-0.3em}-\hspace{-0.3em}-\overset{\cdot}{C}H-\overset{\overset{\displaystyle CH_3}{|}}{\underset{\underset{\displaystyle COOCH_3}{|}}{C}}-\hspace{2em}(5.56)$$

and also of allyl and polyene radicals. Davis[508] and Szöcs[2104–2106, 2108, 2111] established that the decay rate of free radicals is sensitive to pressure.

The interpretation of ESR spectra of irradiated poly(methyl methacrylate) at 4–6 K is not completely solved[1033, 1681, 2054]. Such a spectrum is shown in Figure 5.32.

35 G

Fig. 5.32. ESR spectrum of poly(methyl methacrylate) γ-irradiated at 6 K[1033]

Many possible interpretations may be applied to this well-resolved multiplet at 6 K, considering several steric conformations of the protons in both the methylene and the methyl groups (Fig. 5.33).

Iwasaki et al.[1033] concluded that this type of spectrum cannot be interpreted by a static mixture of a number of conformations. There is probably a non-classical quantum mechanical "tunneling" of the methyl groups in a poly(methyl methacrylate) radical.

Other radicals trapped at 77 K after irradiation are probably $CH_3OOC\cdot$ or $HOOC\cdot$ which give singlet spectra, $OHC\cdot$ (doublet spectrum) and $\cdot CH_3$ (quartet spectrum), with splitting of the latter two of 130 G and 23 G respectively[763, 845, 1091].

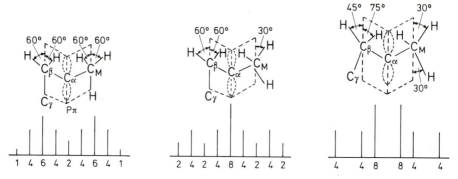

Fig. 5.33. Assumed conformation of free radical formed during γ-irradiation at 6 K of poly(methyl methacrylate) and the expected stick spectra[2054]

ESR spectroscopy has also been applied to studies of photodegradation of poly(methyl methacrylate) sensitized by $FeCl_3$[1469, 1705], studies of trapped ions from additives in poly(methyl methacrylate) and the role of energy migration in this polymer[248, 249, 507, 2419], and to studies of the following polymers: poly(methyl acrylate)[924], poly(glycol methacrylate)[2118, 2110, 2492] and copolymers of methyl methacrylate and styrene[1314, 2419] and methyl methacrylate and maleic anhydride[2188, 2189].

5.1.4. Polystyrenes

When polystyrene is exposed to ionizing radiation at 77 K, free radicals are formed which can be recorded by ESR spectroscopy[14, 373, 682, 698, 702−704, 869, 1001, 1914, 2096, 2304, 2307, 2358, 2420]. The observed triplet-line spectrum (Fig. 5.34) has been assigned to free radicals (5.57):

$$-H_2C-\overset{\bullet}{C}-CH_2- \qquad (5.57)$$

Verma and Peterlin[2358] concluded that the β-methylene protons in radical (5.57) are nonequivalent. They report that, in the spectrum shown in Figure 5.34, each component of the main triplet with hyperfine splitting constant 20 G is further split into another triplet with hyperfine splitting constant 5 G. The main triplet is due to one set of two β-protons which interact more strongly with the unpaired electron than the other two β-protons. The observed nonequivalence of β-protons in radical (5.57) is attributed to a restricted rotation at the $C_\alpha-C_\beta$ bond. As it was shown earlier, the hyperfine coupling constant (ΔH_β) (5.1) varies with a $\cos^2\Theta$. Calculation made by Verma and Peterlin[2358] show that the angle Θ_1 (for the set of two protons giving the larger splitting of 20 G) is about 48°39′, and the angle Θ_2 for the other set of protons is 70°37′. The assumption that ortho-hydrogen atoms

206

HIGH FIELD SIDE

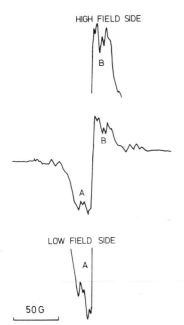

LOW FIELD SIDE

50 G

Fig. 5.34. ESR spectra of polystyrene γ-irradiated at 298 K

in the phenyl ring can influence the hyperfine splitting constant and increase it to the observed 35 G[702)] is questioned by several authors[1472, 1653, 2096)]. On the other hand, it has been shown that radicals formed by γ-irradiation of benzene and substituted benzenes[651, 657, 704, 1653, 2210, 2382, 2369)] have a large hyperfine splitting of the order 50 G, which is due to a cyclohexadienyl radical (5.58)[658)] formed by hydrogen addition to the benzene ring:

$$-CH_2-CH-CH_2- \quad + \; H\cdot \quad \longrightarrow \quad -CH_2-CH-CH_2- \tag{5.35}$$

$$(5.58)$$

Szöcs and Placek[2112)] have made an ESR study of the influence of pressure (1–8000 atm) on the free-radical decay in γ-irradiated polystyrene. They have pointed out the correlation between molecular motion in the relaxation region and the radical decay. The pressure increases the glass-transition temperature, increases density, and decreases the free volume of the polymer, which causes a temperature shift in the relaxation maxima. These parameters are related to the molecular motion which affects the stability of free radicals in a polystyrene matrix.

ESR spectra of different substituted deuterated polystyrenes have been discussed in a few other papers[702, 704, 1001, 2382)].

Russian scientists[35, 1112, 2209)] reported that, after irradiation of polystyrene at 77 K with fast electrons, the ESR spectrum consists of a singlet superimposed on the normal spectrum. The formation of this singlet-line spectrum may be attributed to trapped electrons in the polystyrene matrix.

Iwasaki et al.[1029] have reported formation of radical pairs in γ-irradiated polystyrene for which ESR spectra are shown in Figure 5.35.

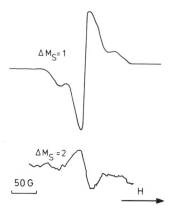

Fig. 5.35. ΔM_S = 1 and ΔM_S = 2 spectra of polystyrene γ-irradiated at 77 K[1029]

Fig. 5.36. ESR spectrum of UV-irradiated benzene[2161]

A broad singlet spectrum with the width of about 20 G has been found during UV-irradiation of polystyrene at 77 K[322, 1472, 1941]. This spectrum has been attributed to radical (5.57). When the sample was warmed to room temperature in the presence of air, the spectrum changed into an asymmetric singlet spectrum with a width of about 16 G and was assigned to peroxy radicals[1941].

Fox et al.[484] obtained an ESR spectrum of polystyrene in tetrahydrofurane glass at 77 K which had a seven-line structure and was assigned to radical (5.57) with two different chain conformations. Some polymer segments have a conformation containing three protons with hyperfine coupling constant of 18 G and one proton with 54 G, while other polymer segments have a conformation with all four hyperfine coupling constants equal to 18 G.

UV-irradiation of polystyrene may also cause formation of phenyl radicals[1814]. The ESR spectrum of phenyl radical obtained during UV-irradiation of benzene at 77 K consists of six lines with a splitting constant of 6 G and with an intensity ratio 1 : 1.8 : 2.6 : 2.6 : 1.8 : 1 (Fig. 5.36)[2161].

Phenyl radicals have also been observed during electron irradiation of benzene at 149 K[1000].

ESR spectroscopy has also been applied to studies of the mechanism of degradation of polystyrene in the presence of peroxides[1193, 1194, 2373], of photodegradation of brominated polystyrene[1070], and of trapped ions from additives in polystyrene and the role of energy migration in this polymer[949, 1712].

Irradiation of poly(α-methylstyrene) with ionizing radiation at 77 K gives the ESR spectrum shown in Figure 5.37[682].

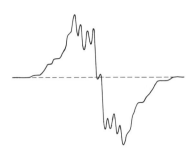

Fig. 5.37. ESR spectrum of poly-α-methylstyrene γ-irradiated at 77 K[682]

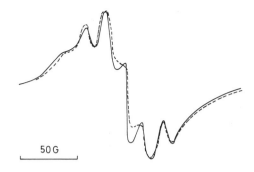

Fig. 5.38. ESR spectra of poly(vinyl chloride) γ-irradiated (dotted line) and electron-irradiated (solid line) at 77 K[1463]

This spectrum has not been analysed in detail, but it was assigned to radical (5.57). ESR studies have also been made of γ-irradiated poly(α-methyl-styrene) in tetrahydrofurane glass[52, 484].

5.1.5. Poly(vinyl chloride) and Poly(vinylidene chloride)

The ESR study of free radicals formed in electron- and γ-irradiated poly(vinyl chloride) have been discussed by several authors[736, 882, 1057, 1204, 1308, 2389, 1463, 1491, 1640, 1644, 2183, 2304]. The typical spectrum of γ-irradiated poly(vinyl chloride) at 77 K is shown in Figure 5.38.

This septet-line spectrum with a hyperfine splitting constant of 18 G is attributed to allyl radical (5.59):

$$-CH-CH_2-\dot{C}H-CH=CH-CH_2-CH- \qquad (5.59)$$
$$\quad | \qquad\qquad\qquad\qquad\qquad\qquad | $$
$$\quad Cl \qquad\qquad\qquad\qquad\qquad\qquad Cl$$

After warming to room temperature, the septet-line spectrum disappears and a new singlet-line spectrum is formed, which is assigned to a conjugated polyene structure of the macroradical (5.60)[1644]:

$$-CH-CH_2-\dot{C}H-(CH=CH)_n-CH-CH_2-CH- \qquad (5.60)$$
$$\quad | \qquad\qquad\qquad\qquad\qquad | \qquad\quad | $$
$$\quad Cl \qquad\qquad\qquad\qquad\qquad Cl \qquad\quad Cl$$

Lawton and Balwit[1308] suggested that the predominant radical from poly(vinyl chloride) is of the type (5.61). In this case a six-line spectrum should be formed.

$$-CH-CH_2-\dot{C}H-CH_2-CH- \qquad (5.61)$$
$$\quad | \qquad\qquad\qquad\quad | $$
$$\quad Cl \qquad\qquad\qquad\quad Cl$$

209

Instead of a six-line spectrum (which was not observed) a poorly resolved eight-line spectrum was obtained after γ-irradiation of poly(vinyl chloride) in the dark at 77 K[462]. This spectrum was assigned to radical (5.61). The same authors[1308] have suggested that γ-irradiated poly(vinyl chloride) in 2-methyltetrahydrofuran glass solution acts as an electron scavenger to produce the radicals (5.61). They proposed two probable mechanisms for the formation of the primary radical (5.61):

1. Homolysis followed by geminate recombination:

$$-CH_2-\underset{\underset{Cl}{|}}{CH}- \xrightarrow{\gamma} \left[-CH_2-\underset{\underset{Cl}{|}}{CH}-\right]^{+\cdot} + e^- \longrightarrow -CH_2-\dot{CH}- + \dot{Cl} \qquad (5.36)$$

2. Dissociative electron structure:

$$-CH_2-\underset{\underset{Cl}{|}}{CH}- \xrightarrow{+e^-} -CH_2-\dot{CH}- + Cl^- \qquad (5.37)$$

Several authors[901, 1560, 1641, 1888] reported that γ-irradiated poly(vinyl chloride) in vacuum at room temperature gives a singlet-line spectrum of 25 G peak-to-peak width (Fig. 5.40).

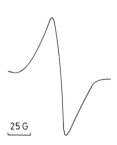

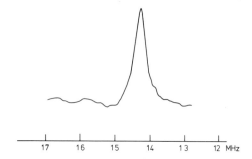

Fig. 5.39. ESR spectra of poly(vinyl chloride) γ-irradiated at room temperature in vacuum[1888]

Fig. 5.40. Matrix ENDOR spectrum of γ-irradiated poly(vinyl chloride)[901]

This spectrum was attributed to polyene radicals (5.60). A similar singlet-line spectrum has been obtained during γ-irradiation of polyene acids[861].

The ENDOR spectrum of heavily irradiated (>200 Mrad) poly(vinyl chloride) is a singlet centered at 14.4 MHz (Fig. 5.40).

This ENDOR spectrum provides substantial proof for the formation of polyenyl radical (5.60) and extensive delocalization of unpaired electrons in the radical (5.60)[901].

The formation of a singlet-line ESR spectrum was also observed during thermal[227, 882, 1364, 1365, 1666, 1680, 1989] and photo[1788, 2239, 2246, 2441] degradation, and chemical dehydrochlorination[562] of poly(vinyl chloride).

In Figure 5.41 single-line spectra are shown which have been obtained during UV-irradiation of poly(vinyl chloride) in vacuum and in presence of air.

It has been found by Rabek and Rånby[1788, 1816] that conjugated polyene sequences, when present in poly(vinyl chloride), can "photosensitize" the formation of radicals during UV-irradiation. ESR spectra of UV "pre-irradiated" poly(vinyl chloride) samples containing double bonds show a much higher intensity after continued UV-irradiation (Fig. 5.41). This means that the concentration of free radicals grows more rapidly for pre-irradiated samples (Fig. 5.42).

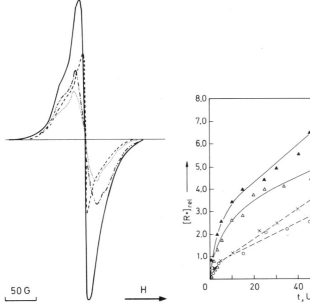
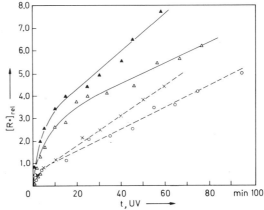

Fig. 5.41. ESR spectra of poly(vinyl chloride) UV-irradiated (2534 A) at 77 K in vacuum (‒ ·· ‒ ··) and in air (‒ ·· ‒ ··). Two other samples were previously UV-irradiated 90 min at 77 K, then heat-treated at 368 K, and further once again UV-irradiated at 77 K (‒ ‒ ‒) for measuring ESR spectra[1788]

Fig. 5.42. Kinetics of free radical formation during UV-irradiation at 77 K of poly(vinyl chloride). Notation the same as in Fig. 5.41[1798]

The decay of free radicals formed during irradiation of poly(vinyl chloride) has been investigated in detail by Atchinson[85, 86] and Rabek and Rånby[1788]. Figure 5.43 shows the decay curves of free radicals during warming of a UV-irradiated poly(vinyl chloride) sample.

The results obtained indicate the formation of two types of free radicals during UV-irradiation of poly(vinyl chloride); namely alkyl (5.61) and polyenyl (5.60) radicals, which decay at different rates. Alkyl radicals disappear at a much faster rate than polyenyl radicals, which exist even above 323 K.

Rabek and Rånby[1786, 1787] used ESR spectroscopy for the study the role of commercial thermostabilizers in photodegradation of poly(vinyl chloride).

211

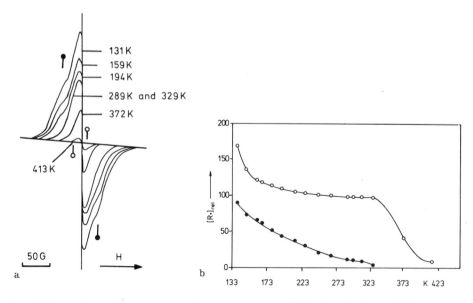

Fig. 5.43. a. Change with temperature of ESR spectrum of UV-irradiated poly(vinyl chloride) sample. b. Free radical decay calculated from ESR spectra: (•) attributed to alkyl radical and (○) attributed to polyenyl radical[1798]

ESR spectroscopy has also been applied to studies of the mechanism of thermal degradation of poly(vinyl chloride) in the presence of a second polymer[843] and to the study of electron capture by acceptors such as anthracene, p-terphenyl, p-benzo-quinone, and chloranil during the radiolysis of poly(vinyl chloride)[1205, 2332, 2449].

Hay[882] and Fukuda[736] presented ESR spectra of poly(vinylidene chloride) irradiated with γ-rays at 77 K, and spectra obtained during thermal degradation of polymer samples. In the latter case a broad singlet spectrum was obtained and attributed to polyene radical. A similar ESR spectrum was observed during UV-irradiation of poly(vinylidene chloride)[1677].

5.1.6. Fluorinated polymers

ESR spectra of irradiated poly(tetrafluoroethylene) were studied by several workers[14, 38, 58, 62, 236, 450, 699, 736, 889, 1025, 1026, 1148, 1317, 1327–1329, 1353, 1432, 1433, 1478, 1639–1640, 1642, 1685, 1840, 1915, 2006, 2147–2150, 2152, 2153, 2160, 2216, 2304–2307, 2329, 2375]. Ionizing radiation of poly(tetrafluoroethylene) produces free radicals which are extremely resistant to heat (up to 523 K). Figure 5.44 shows ESR spectra obtained during electron irradiation of poly(tetrafluoroethylene) in vacuum at 77 K and 300 K.

This spectrum contains two components. One is a doublet of quintets with the hyperfine splitting constant 87 G and 32 G and is attributed to radicals (5.62)[2147–2149, 2304]. The other is a three-line spectrum assigned to radical (5.63)[2216].

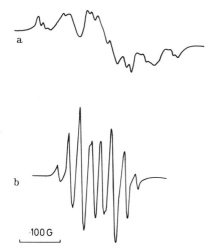

Fig. 5.44. ESR spectra of electron-irradiated poly(tetrafluoroethylene) in vacuum: a. at 77 K and b. at 300 K[2148]

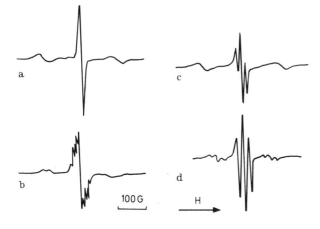

Fig. 5.45. ESR spectra of the end-radical -CF$_2$ĊF$_2$ trapped in oriented films of poly(tetrafluoroethylene): a. and b. measured at 77 K, c. and d. measured at room temperature a. and c. magnetic field parallel and b. and d. perpendicular to the stretching axis[2216]

$$-CF_2-\overset{\cdot}{C}F-CF_2- \qquad (5.62) \qquad\qquad -CF_2-\overset{\cdot}{C}F_2 \qquad (5.63)$$

The triplet decays on heating to 403 K. Milinchuk et al.[1148] and Siegel and Hedgpeth[2006] suggested that the triplet in the centre is only a part of the main spectrum and that its wing peaks are due to the hyperfine anisotropy of α-fluorine coupling. These wing peaks are characteristic for radicals having α-fluorine atoms[1024].

In the case of free radical (5.63), a change of spectrum interpreted by the molecular motion around chain axis (Fig. 5.45) was observed[1026, 2216].

The geometrical structure of free radical (5.63) is determined from the hyperfine structure measured at 77 K and at room temperature. The principal values of the hyperfine tensor, $a_\parallel$ and $a_\perp$, are 225 G and 17 G at 77 K and 110 G and 74 G at room temperature. The direction of $a_\parallel$ is about 45° from the molecular chain axis at 77 K and parallel to the chain axis at room temperature (Fig. 5.46).

213

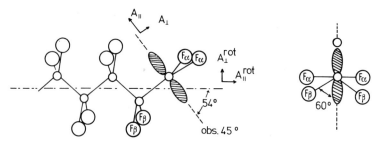

Fig. 4.46. Conformational structure of the radical -CF$_2$ĊF$_2$ and the principal directions of the hyperfine tensor[2216]

A few ESR studies are also reported for other fluorine-containing polymers, such as: poly(vinylfluoride) and poly(vinylidene fluoride)[1932, 1933, 2155], copolymer of tetrafluoroethylene with ethylene[273], copolymer of tetrafluoroethylene with hexa-fluoropropylene[699, 1036, 1038], poly(chlorotrifluoroethylene)[14, 699], poly(trifluoro-ethylene)[699], copolymer of chlorotrifluoroethylene and vinylidene fluoride[699], copolymer of hexafluoropropylene and vinylidene fluoride[699], poly(α,β,β-trifluoro-styrene)[699], and poly(2,3,4,5,6-pentafluorostyrene)[699].

ESR study of the conversion of free radicals in different fluorocarbon polymers under light irradiation has also been made[1214, 1217].

Polyenyl radicals, stable at room temperature, were trapped in irradiated poly-(vinyl fluoride) and poly(vinylidene fluoride) and detected by ESR spectroscopy[901, 2155] (Fig. 5.47).

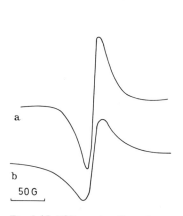

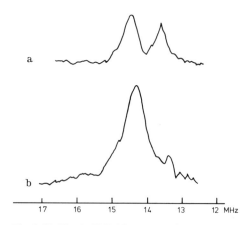

Fig. 5.47. ESR spectra after γ-irra-diation at 298 K: a poly(vinyl chloride) and b. poly(vinylidene fluoride)[901]

Fig. 5.48. Matrix ENDOR spectra of γ-irradiated: a. poly(vinylidene fluoride) and b. poly(vinyl fluoride)[901]

The ENDOR spectra of heavily γ-irradiated (>200 Mrad) poly(vinyl fluoride) and poly(vinylidene fluoride) consist of two lines; one at 13.6 MHz (free fluoro-radical), and another at 14.4 MHz (free proton) (Fig. 5.48)[901].

The narrow matrix width of the ENDOR lines strongly supports the polyenyl character of radicals obtained during γ-irradiation of these polymers.

Free radicals formed during the fluorination of polymers were also investigated by ESR spectroscopy[701].

5.1.7. Poly(vinyl acetate)

ESR spectra of free radicals generated in poly(vinyl acetate) by γ-radiation have been described by a few authors[14, 1640, 2114]. The triplet-line spectrum (Fig. 5.49) was assigned to the free radical (5.64) or (5.65)[2114]:

$$-CH_2-\overset{\centerdot}{C}-CH_2- \quad (5.64) \qquad -CH_2-\overset{\centerdot}{C}H \quad (5.65)$$
$$\quad\quad\ | \qquad\qquad\qquad\qquad\qquad\quad |$$
$$\quad\quad OCOCH_3 \qquad\qquad\qquad\qquad OCOCH_3$$

After thermostating the sample exposed to high pressure (7000 atm) a relative intensification of a quartet spectrum (Fig. 5.50) corresponding to a new radical (5.66) appears:

$$-CH\!-\!\!-\!\!-\!\overset{\centerdot}{C}H\!-\!\!-\!\!-\!CH- \quad (5.66)$$
$$\ \ |\qquad\qquad\qquad\ |$$
$$OCOCH_3 \qquad\quad OCOCH_3$$

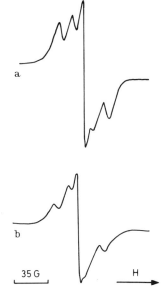

a

b

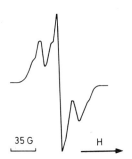

35 G   H

Fig. 5.49. ESR spectrum of γ-irradiated poly(vinyl acetate)[2114]

Fig. 5.50. ESR spectrum of γ-irradiated poly(vinyl acetate): a. after 90 min heating at 343 K and 7000 atm, and b. after 20 min heating at 363 K at 7000 atm[2114]

215

This type of radical (5.66) is more sensitive to the stabilizing effect of pressure than radical (5.64).

When poly(vinyl acetate) is irradiated with UV-light at 77 K (Fig. 5.51), a sharp quartet with a hyperfine splitting constant of about 23 G due to methyl radical (5.67) is observed, and a triplet-line spectrum tentatively assigned to free radicals (5.65)[1473, 2251].

$$\cdot CH_3 \qquad (5.67)$$

After subsequent treatment at 195 K the triplet-line spectrum becomes predominant. Tsuji[2251] suggested that this spectrum was composed of two kinds of spectra; a triplet and a singlet. These spectra were attributed to free radicals (5.65) and (5.68),

$$\begin{array}{c} -CH_2-CH- \qquad (5.68) \\ | \\ O \\ | \\ C=O \\ \cdot \end{array}$$

5.1.8. Poly(vinyl alcohol)

After γ-irradiation of poly(vinyl alcohol) at 77 K, the sample becomes purple-colored. The ESR spectrum obtained is shown in Figure 5.52a[874].

Illumination with visible light bleaches the color, leaving a spectrum as shown in Figure 5.52b. After photobleaching the spectrum is a quintet with a hyperfine splitting constant of 10 ± 2 G, and its intensity increases. Subsequent warming to room temperature produces a new spectrum, which is a triplet with a hyperfine splitting constant of 32 ± 2 G (Fig. 5.52c). The same spectrum was obtained when poly(vinyl alcohol) films were irradiated at room temperature[874, 1632]. The radical produced at 77 K is probably the precursor of the stable radical produced at room temperature. The observed quintet-line spectrum has been attributed to radicals (5.69):

$$\begin{array}{c} -CH_2-CH-\overset{\cdot}{C}H-CH-CH_2- \qquad (5.69) \\ \quad\;\; | \qquad\quad | \\ \quad\;\; OH \qquad\; OH \end{array}$$

The triplet-line spectrum has been assigned to free radicals (5.70)[824, 874, 1632, 1640, 1642].

$$\begin{array}{c} -CH_2-\overset{\cdot}{C}-CH_2-CH- \qquad (5.70) \\ \quad\;\; | \qquad\qquad | \\ \quad\;\; OH \qquad\;\; OH \end{array}$$

After abstraction of the dotted line spectrum from the solid line spectrum in Figure 5.52, a single-line spectrum can be obtained (Fig. 5.52c). This singlet spec-

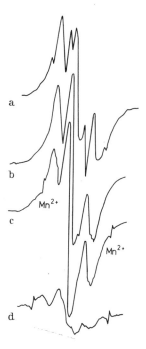

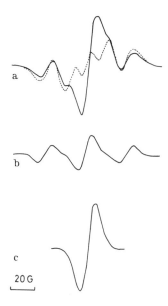

Fig. 5.51. ESR spectra of poly(vinyl acetate) UV-irradiated at 77 K: a. immediately after 46 min UV-irradiation, b. after 12 min heat treatment at 195 K, c. after 30 min heat treatment at 195 K, d. measured at room temperature[2251]

Fig. 5.52. ESR spectra of poly(vinyl alcohol) γ-irradiated at 77 K: a. immediately after irradiation, b. after subsequent illumination with visible light, and c. after warming to room temperature and cooling again to 77 K – this spectrum is the difference between spectra a and b, its intensity was normalized to that of (a) at the outermost peak[874]

trum, which is associated with the purple color, has a line width of $15 \pm 1\,G$ and g-factor of about 2.00. These results show that the singlet-line spectrum can be attributed to trapped electrons in irradiated poly(vinyl alcohol)[2140]. The disappearance of the color after illumination with visible light has been interpreted in terms of photodetachment of electrons from traps.

During photo-irradiation of poly(vinyl alcohol) three radicals, a singlet, a triplet, and a quartet have been observed by ESR spectroscopy[1636].

5.1.9. Polynitroethylene

An ESR spectrum of polynitroethylene γ-irradiated at 77 K is shown in Figure 5.53a[2255].

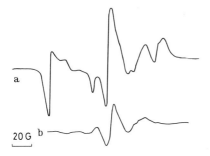

Fig. 5.53. ESR spectra of polynitroethylene of γ-irradiated at 77 K: a. immediately after irradiation, b. after 30 min heat treatment at 273 K[2255]

This spectrum is attributed to the radical (5.71):

$$\cdot NO_2 \qquad (5.71)$$

It is relatively stable at 195 K but decays after warming to 273 K for 30 minutes. The remaining ESR spectrum, (Fig. 5.53b) is tentatively assigned to radical (5.72):

$$-CH_2-\overset{\displaystyle \cdot}{C}-CH_2- \qquad (5.72)$$
$$\underset{\displaystyle NO_2}{|}$$

No definite indication for the formation of alkyl radicals was found.

5.1.10. Polyvinylpyridines

Interpretation of ESR spectra of γ-irradiated polyvinylpyridines is difficult. For this reason it is necessary to present in few words the result obtained for γ-irradiated pyridine in crystalline state or in organic matrices[257, 413, 414, 506, 1955, 2288, 2294, 2296, 2297]. Three kinds of spectra are observed in irradiated polycrystalline pyridine at 77 K[2288]:

1. A triplet spectrum with a coupling constant of about 30 G, observed at 77 K, which decays at about 173 K (I)
2. An eight-line spectrum with a coupling constant of about 12 G which is observed at 183 K decays at 208 K (II)
3. An apparent singlet which decays at about 223 K (III)

When a mixture of pyridine and alcohol is γ-irradiated, a ten-line spectrum with coupling constant of 6 G (IV) is observed at about 130 K, after the alcohol free radicals disappear[414, 506]. Assignments of the above mentioned spectra are presented in Table 5.3. (Free radicals I–IV.)

γ-irradiation at 77 K of poly-2-vinylpyridine and poly-4-vinylpyridine gives triplet-line spectra with a predominant central line (Fig. 5.54)[866, 1850].

These spectra appear to be formed by the superposition of the signals of two radicals. The side peaks disappear either on UV-irradiation or on warming to 180 K with a decrease in the overall concentration of radicals. The spectra of the remaining

Table 5.3. Assignment of radicals produced in irradiated pyridine[2288)]

Free radicals				References
I	II	III	IV	

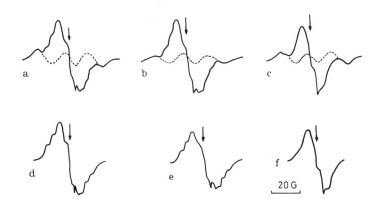

Fig. 5.54. ESR spectra of polyvinylpyridines γ-irradiated at 77 K: a. and d. spectra of atactic poly-2-vinylpyridine, b. and e. spectra of isotactic poly-2-vinylpyridine, c. and f. spectra of poly-4-vinyl-pyridines. The spectra a., b. and c. are recorded at 123 K, the spectra d. e. and f. are recorded at 243 K. The dotted lines correspond to the difference between spectra initially recorded at 123 K (upper line) and the spectra obtained at this temperature after warming at 243 K. Vertical arrows: g = 2.0032[866)]

radicals can be observed up to 350 K. From the structure of these spectra, it may be concluded that they do not correspond to alkyl radicals (5.73):

$$-CH-\overset{\bullet}{C}H-CH- \qquad (5.73)$$

Some authors[866, 1850] consider the possible formation, under γ-irradiation, of azabenzyl, pyridyl, pyridinyl or azacyclohexadienyl radicals.

Pyridinyl-free radicals (5.74) are derived from polyvinylpyridines by addition of hydrogen atoms to the nitrogen of the pyridine ring. These radicals can only be obtained by irradiation of the polymer in alcoholic glassy matrices at 77 K (Fig. 5.55).

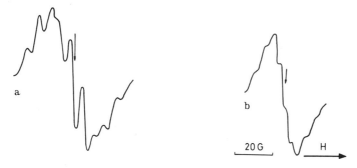

Fig. 5.55. ESR spectra of polypyridinyl radicals trapped in alcohol matrices at 150 K: a. from poly-2-vinylpyridine and b. from poly-4-vinylpyridine. Vertical arrow g = 2.0032[866]

Pyridinyl radical (5.74) results from an electron capture followed by protonation of the radical anion, as shown by the disappearance of the signal of the solvated electron normally observed in irradiated glassy alcohol:

$$-CH_2-CH- \xrightarrow{+e^-} -CH_2-CH- \qquad (5.38)$$

$$-CH_2-CH- + ROH \longrightarrow -CH_2-CH- + RO^- \qquad (5.39)$$

$$(5.74)$$

The hyperfine coupling constant of the hydrogen bonded to nitrogen is given by the difference in width of the spectra of the radicals formed in protonated and deuterated matrices.

Azacyclohexadienyl radicals (5. 75) and (5. 76) are formed by γ-irradiated of polyvinylpyridines at 77 K in acid medium necessary to protonate the pyridine ring. The observed ESR spectrum is a triplet-line if the matrix is protonated, and a doublet-line in the case of deuterated matrix (Fig. 5.56).

The structure of 2-azacyclohexadienyl (5. 75) and 4-azacyclohexadienyl (5. 76) radicals obtained from polyvinylpyridines is presented below:

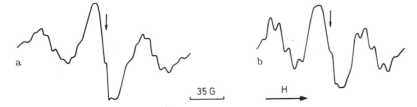

Fig. 5.56. ESR spectra of azacyclohexadienyl radicals trapped in HCl-H$_2$O matrix at 105 K: a. from poly-2-vinylpyridine and b. from poly-4-vinylpyridine. Vertical arrow g = 2.0032[866)]

$$-CH_2-CH- \qquad -CH_2-CH-$$

(5. 75)   (5. 76)

Free radicals (5. 75) and (5. 76) are not selectively formed by γ-irradiation of poly-2-vinylpyridine and poly-4-vinylpyridine. It cannot be excluded, however, that they are formed together with other radicals, since the recorded spectra seem to contain several superimposed signals.

5.1.11. Poly(vinyl pyrrolidone)

During γ-irradiation of poly(vinyl pyrrolidone) at 77 K a five-line spectrum is obtained (Fig. 5.57)[1523)].

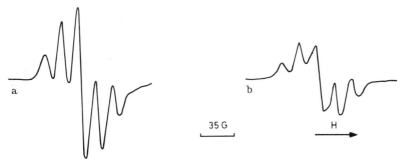

Fig. 5.57. ESR spectra of poly(vinyl pyrrolidone) γ-irradiated at 77 K: a. immediately after irradiation, b. after 5 hr warming to 293 K and then measured at 77 K[1523)]

221

A similar spectrum has also been reported when poly(vinyl pyrrolidone) is UV-irradiated at 77 K[725]. This spectrum is attributed to the free radical (5.77):

$$-CH_2-\overset{\cdot}{\underset{|}{C}}-$$

(5.77)

After warming the previously UV-irradiated sample to room temperature, a single-line spectrum is observed and assigned to unpaired electron in conjugated double bonds[725]. Mönig et al.[1523] reported that warming of a γ-irradiated sample to room temperature leads to decrease of radical concentration (5.57b).

5.1.12. Polyethers

5.1.12.1. Polyoxymethylene

ESR spectra of free radicals produced in polyoxymethylene by ionizing radiation were reported by several authors[696, 1345, 1426, 1585, 1897, 2088, 2266, 2307, 2314, 2439, 2444, 2469]. During the electron-irradiation of polyoxymethylene at 77 K a broad singlet is formed (Fig. 5.58a) which is attributed to radicals (5.78):

$$-CH_2-O\cdot \qquad (5.78)$$

At 273 K the singlet spectrum is converted into a triplet (Fig. 5.58b) which is considered to be due to radical (5.79):

$$-O-\overset{\cdot}{C}H_2 \qquad (5.79)$$

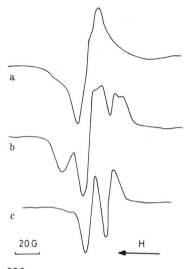

Fig. 5.58. ESR spectra of polyoxymethylene (Cellol pellet) electron-irradiated at 77 K: a. immediately after irradiation, b. after 70 min heat treatment at 195 K, c. after 100 min heat treatment at 273 K[2266]

20 G

H

Sohma et al.[1533, 2049] studied the molecular motion of the end radical (5.79) during the warming of the sample.

When the radiation dose is very small, an asymmetric doublet spectrum was observed at room temperature (Fig. 5.58c) and was found to be caused by superposition of the doublet and the singlet spectrum. The singlet has been assigned to the free radical (5.80)[2266] and the doublet spectrum to free radical (5.81)[2469]:

$$-O-\overset{\cdot}{C}=O \quad (5.80) \qquad -O-\overset{\cdot}{C}H-O- \quad (5.81)$$

Tsuji et al.[2266] reported a doublet spectrum with a separation of about 86 G and assigned to free radicals (5.82), and another doublet spectrum with separation of 125 G due to formyl radicals (5.83):

$$HO-\overset{\cdot}{C}=O \quad (5.82) \qquad H-\overset{\cdot}{C}=O \quad (5.83)$$

Yoshida and Rånby[2469] have found that the doublet and triplet spectra of polyoxymethylene show anisotropy, while the singlet spectrum does not. The doublet asymmetric spectrum is considered to be related to an anisotropic hyperfine coupling tensor and an anisotropic g-value. The chain conformation corresponds to that of crystalline polyoxymethylene.

ESR studies of polyoxymethylene photolyzed by UV-light show that the same types of free radicals are formed as those with γ-irradiation[741, 742, 976, 2091].

5.1.12.2. Polyoxyethylene

γ-irradiated[53, 373, 723, 1605, 1642] and photolyzed with UV-light in the presence of sensitizers[681, 725], polyoxyethylene gives singlet spectra assigned to free radicals (5.78). Iwasaki et al.[1029] have reported formation of radical pairs in γ-irradiated polyoxyethylene, of which ESR spectra are shown in Figure 5.59.

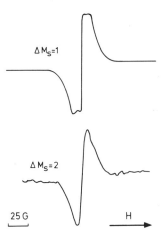

Fig. 5.59. $\Delta M_S = 1$ and $\Delta M_S = 2$ spectra of polyoxymethylene γ-irradiated at 77 K[1029]

223

5.1.12.3. Poly-3,3-bis(chloromethyl)oxethane

A few studies have been made on ESR spectra of poly-3,3-bis(chloromethyl)oxethane irradiated with ionizing radiation[460, 2250, 2258, 2259] and UV-light[2256, 2276, 2277].

A three-line spectrum with a hyperfine splitting constant of about 21 G, and a two-line spectrum with a hyperfine splitting constant of about 18 G, were observed after irradiation with an electron beam in vacuum (Fig. 5.60)[2256].

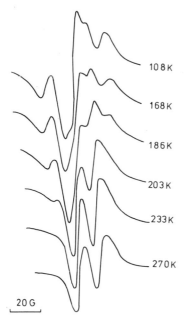

Fig. 5.60. Change in ESR spectra with increasing temperature of poly-3,3-bis(chloromethyl)ox-ethane electron-irradiated at 77 K[2256]

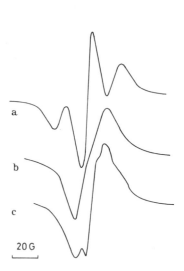

Fig. 5.61. Change in ESR spectra with increasing temperature of poly-3,3-bis(chloromethyl)oxetane UV-irradiated at 77 K: a. immediately after UV-irradiation, b. after 132 min heat treatment at 195 K, c. after 280 min subsequent treatment at 273 K. All measurements made at 77 K[2256]

These spectra were attributed to free radicals (5.84) and (5.85), respectively:

$$\begin{array}{c} CH_2Cl \\ | \\ -CH_2-C-\dot{C}H_2 \\ | \\ CH_2Cl \end{array} \qquad (5.84) \qquad\qquad \begin{array}{c} CH_2Cl \\ | \\ -CH_2-C-\dot{C}H-O- \\ | \\ CH_2Cl \end{array} \qquad (5.85)$$

On the other hand, a three-line spectrum with a hyperfine splitting constant of about 20 G and an asymmetric singlet spectrum were observed after UV-irradiation in vacuum (Fig. 5.61). These spectra were assigned to free radicals (5.84) and (5.78).

224

5.1.12.4. Poly(2,6-dimethyl-1,4-phenyleneoxide)

The ESR spectra of γ-irradiated[2128] and UV-irradiated[1113, 2128, 2279, 2280, 2287] poly(2,6-dimethyl-1,4-phenyleneoxide) have been interpreted as being due to the formation of a dimethylphenoxy radical (5.86):

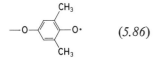 (5.86)

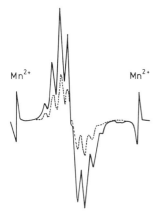

Fig. 5.62. ESR spectra of γ-irradiated poly(2,6-dimethyl-1,4-phenyleneoxide) powder sample observed at room temperature in the presence of air before (dotted line) and after UV-irradiation for 35 min (solid line). The separation between the two Mn^{2+} peaks is 86.7 G[2280]

Fig. 5.63. ESR spectrum of UV-irradiated poly(2,6-dimethyl-1,4-phenyleneoxide) in benzene solution at room temperature[2280]

The ESR spectrum of the radical (5.86) is shown in Figure 5.62.

Irradiation of the sample with UV-light at room temperature causes an increase of the ESR spectral intensity. This eight-line spectrum shows an effective isotropic hyperfine coupling of 5.7 G to the two nearly equivalent methyl groups and anisotropic g-values ($g_\perp = 2.002$ and $g_\parallel = 2.006$)[2128].

Poly(2,6-dimethyl-1,4-phenyleneoxide) in benzene solution, UV-irradiated at room temperature under nitrogen atmosphere, gives a well-resolved spectrum (Fig. 5.63)[2128, 2279, 2280].

The hyperfine splitting constants are $a_{CH_3(I)} = 5.13$ H, $a_{CH_3(II)} = 6.20$ G and $a_{H(meta)} = 1.14$ G[983, 2128, 2280] and the spectrum (Fig. 5.63) is attributed to radical (5.86). The different values obtained for the hyperfine splitting constant indicate that the methyl groups are not magnetically equivalent at room temperature and this

225

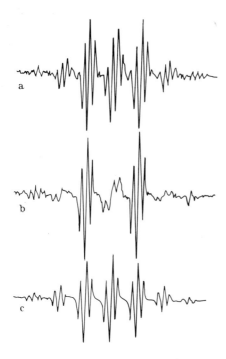

Fig. 5.64 Temperature dependence of ESR spectrum of poly(2,6-dimethyl-1,4-phenyleneoxide) in xylene: a. at 333 K, b. at 373 K and c. at 423 K[2280])

fact is due to hindered rotation around the C–O bond[2280, 2286]). When the temperature for the measurements was higher, the ESR spectra changed (Fig. 5.64).

The results derived from the spectrum obtained at 423 K show that both methyl groups become equivalent at this temperature. The hyperfine splitting constants are a_{CH_3} = 5.63 G and $a_{H(meta)}$ = 1.23 G. These results indicate that free rotation about the C–O bond occur at temperatures $\geqslant$423 K. A detailed mechanism of radical (5.86) formation and possible reactions of the phenoxy radicals (5.86) have been proposed by Tsuji et al.[2279, 2299].

5.1.12.5. Polyglycols

Poly(propylene glycols) give, under ionizing irradiation, mainly seven-line ESR spectra well-resolved at 140 to 180 K with a 15.5 G hyperfine splitting constant[2181]. These spectra have been attributed to radical (5.87):

$$-O-CH_2-\overset{\cdot}{C}H \qquad (5.87)$$
$$\hspace{1.4cm}|$$
$$\hspace{1.4cm}CH_3$$

Poly(ethylene glycols) irradiated with UV-light in aqueous solution at 77 K and at room temperature give ESR spectra due to formyl radicals (5.83) and radicals (5.88)[1605].

$$-CH_2-O-\overset{\cdot}{C}H-CH_2- \qquad (5.88)$$

It has been shown that formyl radicals have an important role in the photo-degradation of poly(ethylene glycol).

5.1.13. Polycarbonates

A few ESR studies of organic carbonates and polycarbonates have been reported[572, 828, 852, 853, 1400, 1451].

Poly(bisphenol-A carbonate) has been investigated in detail by ESR spectroscopy by Hama et al.[852, 853] and by Lyons et al.[1400]. ESR spectra of γ-irradiated polycarbonates are composed of a sharp singlet, some broad singlets, and a weak signal with hyperfine structure. These spectra were assigned to trapped electrons (the sharp singlet), positive radical ions, phenoxy radicals, phenyl radicals (the broad singlets), and to radical (5.89)[853].

$$-O-C_6H_4-\dot{C}(CH_3)_2 \qquad (5.89)$$

ESR spectra obtained from diaryl carbonates (diphenyl, di-p-tolyl, phenyl-p-tolyl and bis-diphenyl-propyl carbonates) γ-irradiated at 77 K show that one of the major results of irradiation is the pairwise trapping of phenoxy radicals[1451].

The ESR spectrum of polycarbonates irradiated with UV-light at 77 K was a singlet with the width 16 G and g-factor 2.0045. This spectrum was attributed to three kinds of free radicals: phenoxy, phenyl, and polyene radicals[853]. Lyons et al.[1400] observed, besides the singlet spectrum assigned mainly to phenoxy radicals, two weak peaks separated by about 150 G, which were attributed to radical pairs.

ESR studies of phototransformation of poly(bisphenol-A carbonate) at 77 K sensitized by FeCl$_3$[512, 2214] or benzoyl peroxide[1708] have also been published.

5.1.14. Polyesters

5.1.14.1. Poly(β-propiolactone)

Poly(β-propiolactone) electron-irradiated in vacuum at 77 K has an ESR spectrum (Fig. 5.65) assigned to radical (5.90)[1614, 1647]:

$$-CH_2-\dot{C}H-CO-O- \qquad (5.90)$$

The spectrum shows an anisotropy in the yz plane (z is the chain direction), whereas in the xy plane it is nearly the same at all angular positions. The main four-line spectrum at the angle $\Phi = 0°$ changes to sextet at $\Phi = 90°$, where Φ is the rotation angle measured at the z axis. The feature of the spectral change suggests that this is a double-triplet spectrum and that the doublet splitting changes with the angle, while the triplet remains almost unchanged.

227

5.1.14.2. Linear Aliphatic Polyesters

The ESR spectra of γ-irradiated at 77 K linear aliphatic polyesters of the structure (5.91):

$$-(CH_2)_m-O-\underset{\underset{O}{\|}}{C}-(CH_2)_n-\underset{\underset{O}{\|}}{C}-O- \qquad (5.91)$$

(m, n = 2.2, 2.4, 2.6, 2.7, 2.8 and 6.8)

show the formation of two types of spectra[1620]. The first, a six-line spectrum obtained at low temperatures, is relatively short-lived and increases in signal intensity with increasing number (m, n) of methylene groups in the polymer chains. A second well-resolved five-line spectrum is stable even at room temperature. Figure 5.66 shows the ESR spectra of γ-irradiated 2.8 polymer.

The change of ESR spectra with increasing temperature is interpreted as a result of changing conformation of the polymer radicals.

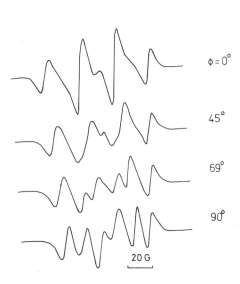

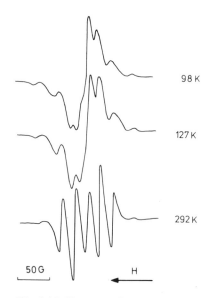

Fig. 5.65. ESR spectra of poly(β-propiolactone) γ-irradiated at 77 K showing angular variation in the zx plane of the ESR spectrum of the irradiated polymer[1647]

Fig. 5.66. Change in ESR spectra with increasing temperature of linear aliphatic polyester (2.8) γ-irradiated at 77 K[1620]

5.1.14.3. Poly(ethylene terephthalate)

Several ESR studies of electron and γ-irradiated poly(ethylene terephthalate) have been published[57, 179, 385, 389, 852, 909, 1252, 1353, 1379, 1549, 1551, 1640, 1642, 1847, 1848, 2047]. ESR spectra of amorphous poly(ethylene terephthalate) or polymer

228

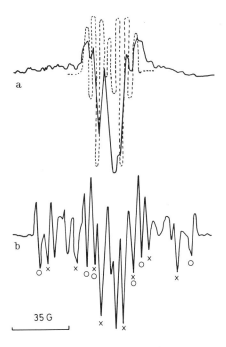

Fig. 5.67. ESR spectra of γ-irradiated at 77 K poly(ethylene terephthalate): a. amorphous sample (dashed line shows spectrum of radical *5.93*), b. highly crystalline sample, peaks corresponding to (x) 8 line spectra and (o) peaks corresponding to 6-line spectra of radical (*5.92*)[389]

with varying levels of crystallinity (5—50%) present an insufficient image of fine structure (Fig. 5.67a) to allow for the identification of free radicals[852, 1252, 1640, 2047]. Only the ESR spectrum from γ-irradiated (50%) crystalline poly(ethylene terephthalate) (®Mylar film) gives well-resolved six-line and eight-line structures (Fig. 5.67b)[57, 385].

This spectrum was assigned to two major radicals (*5.92*) and (*5.93*):

$$-\langle\bigcirc\rangle-CO-O-CH_2-\overset{\cdot}{C}H-O-CO- \qquad -O-CO-\langle\bigcirc\rangle-CO-O-$$

$$(5.92) \qquad\qquad\qquad (5.93)$$

Radical (*5.93*) can be separated from radical (*5.92*) by heat treatment at 423 K. At that temperature radical (*5.92*) decays[1847].

Poly(ethylene terephthalate) irradiated with UV-light at room temperature gives a poorly resolved asymmetric five-line spectrum[1419], which is attributed to radical (*5.93*) and free radical (*5.94*) (a weak central line):

$$-CH_2-CH_2-O\cdot \qquad (5.94)$$

The radical (*5.92*), which is produced by ionizing radiation, decayed when exposed to UV radiation.

Free radical conversion in poly(ethylene terephthalate) by photoinduced post-irradiation has been studied by Campbell and Turner[390].

5.1.14.4. Poly(ethylene-2,6-naphthalene dicarboxylate)

Two types of radicals in γ-irradiated poly(ethylene-2,6-naphthalene dicarboxylate) have been indentified by ESR as radical (5.95) and a radical located on the naphthalene ring (5.96)[1848]:

$$-\text{O}-\overset{\cdot}{\text{C}}\text{H}-\text{CH}_2-\text{O}- \qquad\qquad -\text{O}-\text{OC}\underset{}{\bigcirc\!\!\bigcirc}\overset{\cdot}{}\text{-CO}-\text{O}-$$

$$(5.95) \qquad\qquad\qquad (5.96)$$

The relative concentrations of the two radicals in bulk polymer powder or pressed film are 10–20% of radical (5.95) and 80–90% of radical (5.96).

5.1.15. Polyamides

ESR spectra of γ-irradiated polyamides have been studied by several authors[179, 312, 313, 357, 373, 504, 680, 803, 811, 1089, 1135, 1379, 1474, 1475, 1477, 1480, 1640, 1868, 1952, 1954, 1992, 2117, 2162, 2192, 2336, 2372]. The ESR spectra observed for different polyamides (Fig. 5.68) were attributed to free radicals (5.97):

$$-\text{CO}-\text{NH}-\overset{\cdot}{\text{C}}\text{H}-\text{CH}_2- \qquad (5.97)$$

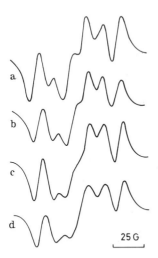

Fig. 5.68. ESR spectra of γ-irradiated at 77 K uniaxially oriented polyamides: a. nylon 610, b. nylon 66, c. nylon 57 and d. nylon 6[1089]

25 G

It has been shown[2192] that a change in orientation of the polyamide sample affects the ESR spectrum by the anisotropy of the α-proton of radical (5.97). Szöcz et al.[2117] have reported that ESR spectra of polyamides can be assigned to three radicals, which can be determined by measurements over a wide range of pressures (1–16000 atm) and temperatures (353–393 K). While the spectrum of radical (5.97) is predominant in the overall spectrum at room temperature, other components distinctly appear at higher temperatures. At 393 K and atmospheric pressure a triplet spectrum was obtained and assigned to free radicals (5.98). The third component was the spectrum of allyl radicals (5.99) which appeared especially after treating the samples at elevated temperatures and high pressures:

$$-CH_2-\overset{|}{\underset{OH}{C}}-NH-CH_2- \quad (5.98) \qquad -CH_2-\overset{\cdot}{C}H-CH=CH-CO-NH- \quad (5.99)$$

It has been shown that polypropiolamide has an ESR spectrum associated with a conjugated bond system[1464]. Russian scientists[224] have reported ESR studies of polyamides irradiated with fast neutrons. Four-line spectra have been obtained for polycaprolactam and six-line spectra for nylon 6 and nylon 68.

UV-irradiation of nylon 6 at 77 K gives a two-component spectrum (Fig. 5.69) 925, 1090, 1829).

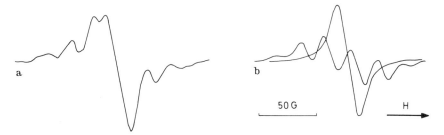

Fig. 5.69. ESR spectra of UV-irradiated at 77 K nylon 6: a experimental spectrum, b. The two component calculated ESR spectrum[925]

One component is a six-line spectrum attributed to free radicals (5.100) and the other component is a singlet spectrum assigned to radical (5.101):

$$-CH_2-\overset{\cdot}{C}H-CH_2- \quad (5.100) \qquad\qquad -CH_2-\overset{\cdot}{C}=O \quad (5.101)$$

Rafikov et al.[1803] observed a singlet-line spectrum and a triplet-line spectrum after UV-irradiation of polycaproamide at 77 K. When the sample was warmed, the triplet spectrum changed into a quintet-line spectrum which was attributed to the free radical (5.102):

$$-CH_2-\overset{\cdot}{C}H_2 \quad (5.102)$$

On further warming the quintet-line spectrum changed to a singlet-line spectrum, which was assigned to free radicals produced by elimination of a hydrogen atom in α-position to the NH group.

Pariskii et al.[1706, 1707] also studied the formation and the phototransformation of macroradicals in polycaprolactam exposed to the action of light in the presence of FeCl$_3$.

An ESR study of the fading of dyed nylon has recently been published[466].

5.1.16. Polyurethanes

ESR studies of polyurethanes were mainly made by Russian scientists[184, 199, 1415, 2371]. Beachel and Chang[176] studied the photodegradation of urethane model systems. They obtained ESR spectra for the UV-irradiated ethyl-N-ethylcarbamate, but not for ethylphenylcarbamate. The radical from ethyl-N-ethylcarbamate was identified as radical (5.103).

$$CH_3-\overset{\cdot}{C}H-NH-COOC_2H_5 \qquad (5.103)$$

It was concluded that the degradation may be followed by radical attack on the α-hydrogen and by the formation of the hydroperoxide.

The ESR study of polyurethanes, based on polyester diol-p,p'-diphenylmethane-diisocyanate and ethylphenylcarbamate irradiated with UV-light under vacuum, shows the formation of singlet spectra at 77 K (Fig. 5.70) which change to four-line spectra and finally to singlet again on warming[1673].

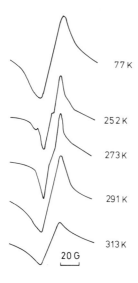

77 K

252 K

273 K

291 K

313 K

20 G

Fig. 5.70. Change in ESR spectra of UV-irradiated at 77 K of polyurethane with increasing of temperature[1673]

Berlin et al.[184, 199] reported that on UV-irradiation of aromatic polyurethanes a singlet-line spectrum is obtained, which is attributed to delocalized electrons in the conjugated double bonds. They also reported that ESR spectra result from secondary reactions of free radicals.

5.1.17. Sulphur-Containing Polymers

During the γ-irradiation of poly(ethylene sulphide)[2478] and poly(propylene sulphide)[1821] with thiol and thioether end groups the following free radicals (5.104–5.106) have been identified by ESR spectroscopy:

$$-CH_2-S-\overset{\cdot}{C}H \quad (5.104) \qquad -S-\overset{\cdot}{C}-CH_2-S- \quad (5.105) \qquad -CH_2-CH-S\cdot \quad (5.106)$$
$$\underset{CH_3}{\mid} \qquad\qquad\qquad \underset{CH_3}{\mid} \qquad\qquad\qquad\qquad \underset{CH_3}{\mid}$$

The ESR spectra of poly(butene-1-sulphone)[115, 116, 319], poly(hexene-1-sulphone)[319], and aromatic polysulphone (prepared from 2,2'-bis(4-hydroxyphenyl)-propane and 4,4'-dichlorophenyl sulphone)[1399] have been reported. Poly(vinyl mercaptan) and its copolymer with vinylpyrrolidone were extensively studied by ESR spectroscopy by Mönig et al.[1522, 1524]. γ-Irradiation of these polymers at room temperature give spectra which are characteristic for randomly oriented neutral sulphur radicals of the type (5.106).

γ-Irradiation of copolymers of ethoxyvinylthiomethane and vinylpyrrolidone at 77 K gives a five-line spectrum with an asymmetric signal[1523]. This spectrum was attributed to a free radical identical with the pyrrolidone radical (five lines) and to a radical cation localized at the sulphur atom (asymmetric line).(5.107):

$$R-\overset{.+}{S}-R \qquad (5.107)$$

This kind of free radical consists of a positive hole trapped in the sulphur non-bonding orbital. The existence of sulphur radical cations (5.107) was established by measuring ESR spectra of γ-irradiated sulphur-containing compounds with thioether groups[261, 379, 1183, 1184, 2400].

After γ-irradiation at 77 K of poly[1-(tert-butoxy carbonylthio)ethylene] and its copolymer with vinylpyrrolidone, tert-butyl radicals and other species were detected by ESR spectroscopy[1525]. One of them is the sulphur radical cation (5.107). Another suggestion is that electrons formed during γ-irradiation can be captured by the carboxyl group, giving a molecular anion which could be responsible for the central singlet line.

5.1.18. Polysiloxanes

Few papers have been published on the ESR study of free radicals in polysiloxanes produced by ionizing radiation[47, 321, 1216, 1283, 1285, 1286, 1671, 2304, 2307, 2512].

Electron irradiation gives a well-resolved ESR spectrum (Fig. 5.71), which consists of two groups of lines: a quartet with the hyperfine splitting constant 23 G and a triplet with the hyperfine splitting constant 20 G[307].

The four-line spectrum was assigned to methyl radicals (5.108) and the three-line spectrum to free radicals (5.109):

$$\cdot CH_3 \qquad (5.108)$$

$$\overset{\displaystyle \cdot CH_2}{\underset{\displaystyle CH_3}{\overset{\mid}{-O-Si-O-}}} \qquad (5.109)$$

γ-Irradiation of poly(dimethylsiloxane) at room temperature gives a weak narrow singlet (Fig. 5.72) which is attributed to radicals (5.110):

$$-O-\overset{\cdot}{Si}-O- \qquad (5.110)$$
$$\underset{CH_3}{|}$$

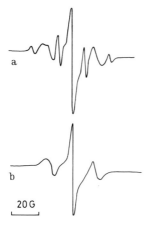

Fig. 5.71. ESR spectra of electron-irradiated at 77 K polymethylsiloxanes: a. immediately after irradiation, b. 24 hours later[1671]

Fig. 5.72. ESR spectrum of γ-irradiated at room temperature polymethylsiloxane[1671]

Poly(dimethylsiloxane) polymer is very stable towards UV-light irradiation. Free radicals can be observed only in sensitized photodegradation, e.g. by naphthalene[2005] and triphenylamine[2511]. Poly(phenylmethylsiloxane) is less stable towards UV-irradiation than poly(dimethylsiloxane). ESR studies of free radicals formed during photodegradation of poly(phenylmethylsiloxane) have been published by Russian scientists[1215, 2510, 2513].

ESR was also used for detecting the presence of anion radicals of poly(phenyl-methylsiloxane) (5.111) formed by metal reduction and cleaving of the chain[597]:

$$\underset{CH_3}{\overset{CH_3}{|}}$$
$$\cdot Si-O^-(K^+) \qquad (5.111)$$
$$\underset{C_6H_5}{|}$$

The interpretation of ESR spectra of anion radicals (5.111) encountered substantial difficulties due to several factors: temperature and concentration dependencies, inability to obtain good spectra caused by overlapping components, and other gross effects that might be expected from multiple reduction sites within a single polymer molecule.

5.1.19. Biopolymers

The application of ESR spectroscopy for the study of free radicals in biological systems has been reviewed in several papers[139, 181, 712, 804, 906, 1396, 1417, 1772, 1966, 2034, 2387, 2447)] and monographs[181, 231, 244, 246, 477, 609, 887, 1595, 1818, 1920, 2034, 2046, 2103)].

5.1.19.1. Cellulose

The presence of free radicals in irradiated cellulose was shown in ESR studies by many workers[14, 28, 34, 77–79, 169, 215, 216, 382, 551, 628, 630, 700, 779, 840, 1120, 1121, 1391, 1406, 1634, 1709, 1869, 1979, 1980, 2321, 2333, 2334, 2341, 2342, 2431)]. Reviews of the use of ESR in studying the structure of free radicals in cellulose and its derivatives initiated by ionizing radiation, light, heat, oxidation-reduction systems, and mechanical action were published by Arthur[68, 69, 75)]. The line shape and width, and the hyperfine splitting constants of the ESR spectra of γ- and UV-irradiated cellulose depend on type of cellulose, moisture content, temperature, frequency distribution and intensity of the irradiation, and the resulting spin concentration[28, 77, 79, 700, 1741)].

On γ-irradiation at 77 K in vacuum a pure cotton sample yields a poorly resolved triplet of approximately 24–26 G hyperfine splitting constant (Fig. 5.73).

Irradiation at 300 K in vacuum or in air causes mainly a broadening of this triplet. Two more satellite lines appeared and the intensity of the ESR spectrum decreased with a loss of resolution.

The presence of moisture reduces the number of free radicals. Crystalline regions of the cellulose contained larger number of more stable (long-lived) free radicals than amorphous regions[28)].

It has been suggested[79, 169)] that the triplet ESR spectrum of cellulose (pure cotton) is due to the equal interaction of the two hydrogen atoms at C_6 with the unpaired electron formed by the removal of the hydrogen atom at C_5.

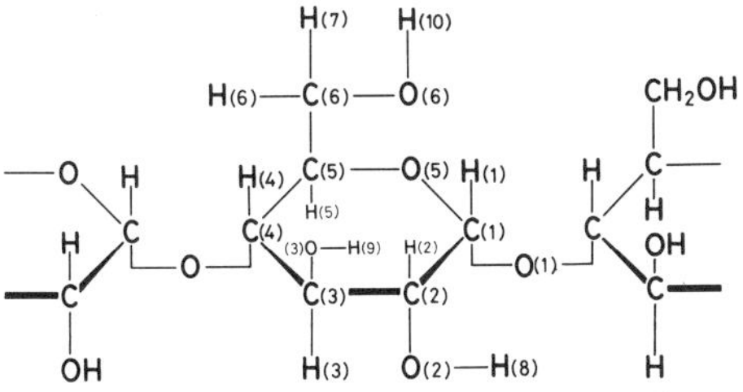

$$(5.112)$$

Mercerized cellulose (obtained by soaking in aqueous 20% NaOH solution for 30 min at room temperature) gives, after γ-irradiation at 77 K, an ESR spectrum of two triplets with hyperfine splitting constants of 18 G and 9 G, respectively (Fig. 5.74)[454)].

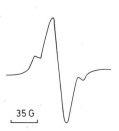

Fig. 5.73. ESR spectrum of
pure cotton γ-irradiated at
77 K in vacuum[454]

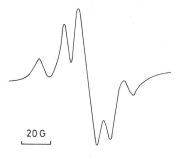

Fig. 5.74. ESR spectrum of mercerized
cellulose γ-irradiated at 77 K in vacuum[454]

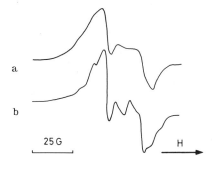

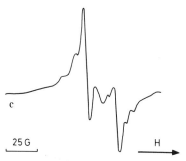

Fig. 5.75. ESR spectra of acetylated
cotton cellulose with different degree of
substitution γ-irradiated at 77 K: a. 0.74,
b. 1.77 and c. 2.34[455]

Fig. 5.76. ESR spectrum
of bezoylated cotton cellulose
γ-irradiated at 77 k in
vacuum[454]

It was suggested that the ESR spectrum of γ-irradiated mercerized cotton arises
from a superposition of two triplets due to unpaired electrons at C_5 and C_6 in different anhydroglucose units. In addition, radicals at C_1 and C_4 have also been postu-
lated[79, 169, 454, 1979]. In view of these conclusions, as well as of the changes in the
ESR spectra of mercerized sample irradiated at 77 K and warmed to 300 K, it may
be inferred that one of the primary species decays faster than the other.

Detailed ESR studies of cellulose acetate[373, 391, 455, 521, 700, 2341, 2450, 2451], furoates[2017], thenoates[2018], benzoylated and allyated cotton cellulose[454] after γ-irradiation were made.

The ESR spectrum of irradiated cellulose acetate, obtained by acetylating cellulose with acetic anhydride in the presence of glacial acetic acid, amyl acetate as solvent, and perchloric acid as catalyst, depends on the degree of substitution (Fig. 5.75)[455].

At low doses of γ-irradiation, a sample with degree of substitution = 0.74 gives a poorly resolved doublet with the hyperfine splitting constant 24 G. The resolution and intensity of spectra increases with the increasing of the degree of substitition. The observed doublet spectrum is attributed to radical (5.113)

$$R_{cell}-\dot{C}H-O-CO-CH_3 \qquad (5.113)$$

The triplet is assigned to free radicals formed by dehydrogenation at the C_5 carbon of the anhydroglucose units. The quartet spectrum with a hyperfine splitting constant of 13 ± 2 G and a g-value of 2.0044 ± 0.004 might eventually be due to radicals (5.114) and (5.115):

$$CH_3\dot{C}O \qquad (5.114) \qquad\qquad CH_3COO \cdot \qquad (5.115)$$

After warming the γ-irradiated sample to 178 K another spectrum was obtained and assigned to peroxy radicals resulting from the decomposition of acetyl radicals.

ESR spectra of benzoylated cottons (obtained by reaction of pure cellulose with benzyl chloride in pyridine for 30 min at 338 K) are similar to those of unsubstituted cellulose at all conditions of irradiation (Fig. 5.76).

The nature and intensity of ESR spectra of allylated samples (obtained by reaction of pure cellulose, presoaked in NaOH, with allyl bromide in xylene) recorded after irradiation at 77 K in vacuum is found to depend on the degree of substitution. The allylated sample with the lowest degree of substitution (= 0.11) gave an almost equally split quintet with the hyperfine splitting constant 15 ± 1 G (Fig. 5.77a) at all doses of irradiation.

Samples with higher degree of substitition gave additional lines which showed a better resolution when higher doses of radiation were employed[454]. Warming an irradiated sample with the highest degree of substitution (= 1.25) produced changes which differed considerably from those found for samples with low degree of substitution (Fig. 5.77b). The results obtained for allylated cotton suggest the presence of certain new radical species which are comparatively unstable. The allyl cellulose unit may be represented as (5.116):

$$R_{cell}-O-CH_2-CH=CH_2 \qquad (5.116)$$

During γ-irradiation, hydrogen atoms liberated from cellulose may saturate the double bond in the allyl group, yielding free radicals (5.117) and (5.118):

$$R_{cell}-O-CH_2-CH_2-\dot{C}H_2 \quad (5.117) \qquad R_{cell}-O-CH_2-\dot{C}H-CH_3 \quad (5.118)$$

237

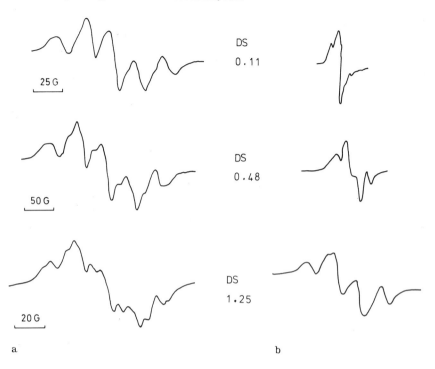

DS 0.11

DS 0.48

DS 1.25

25 G

50 G

20 G

a

b

Fig. 5.77. a. ESR spectra of allylated cotton cellulose with different degrees of substitution γ-irradiated at 77 K in vacuum. b. after warming subsequently to 248 K and recording at 77 K[454]

ESR spectroscopy has been sucessfully applied in studying the mechanism of radiative protection of chemically modified cellulose[80, 454, 2017, 2018]. Thus, the protective action of benzoyl groups is parallel to the yields of radicals at all doses used. Substitution with allyl groups, which also gives radiative protection to cotton cellulose at low doses of irradiation, enhances the overall radical yield. This increase is due to the localization of radiation damage by the substituent groups[454]. This observation indicates that the nature of primary radicals and their intermediary reactions play a significant role in the radiation degradation and the radiative protection of cotton cellulose.

ESR spectra of irradiated cotton cellulose, which is interacting with anhydrous ammonia, have also been reported[931]. Most of the trapped free radicals in irradiated cellulose are scavenged when cellulose is immersed in anhydrous liquid ammonia. This indicates that liquid ammonia penetrates the highly ordered regions of cellulose. A small fraction of trapped free radicals in irradiated cellulose is scavenged when cellulose is treated with gaseous anhydrous ammonia.

ESR spectra of traces of copper ions in purified cotton cellulose after the treatment of cotton with solutions of sodium hydroxide were recorded[129, 130]. Complexes of cupriethylenediamine with cellobiose and cellulose generate similar ESR spectra[71, 170]. The ESR spectrum of a well-defined copper-fructose complex that contained sodium nitrate was recorded[1]. Furthermore, ESR studies of interaction of ammonia, copper, and cuprammonium with cellulose are reported[170, 929]. Cupric ions dissolved in solutions of strong bases, such as concentrated ammonium hydroxide, form complexes with fibrous cotton cellulose. These complexes have paramagnetic resonance properties and generate characteristic ESR spectra.

The characteristics of cellulose radicals produced by photo-irradiation are also studied using ESR spectra[68–70, 171, 964, 1142–1145, 1566, 1633, 1634, 1741]. During UV-irradiation of cellulose two kinds of photo-induced reactions occur[1633, 1634]:

1. The direct scission of glucosidic bonds and oxidation of the cellulose due to the degradation.
2. Photosensitizing reactions depending upon the contaminants, moisture content, and various oxidized groups present in the sample.

All spectra of UV-irradiated cellulose samples contain a single line with a line width of about 28 G[1634]. In the initial stage of irradiation this singlet is the main spectrum. With increasing time of irradiation, two absorption lines with a line width of 8 G were added to the original singlet spectrum, leading to a symmetrical three-line spectrum with line width of 8 G, 18 G and 8 G, respectively. The resulting composite ESR spectrum has a middle line with a line width of 28 G showing the maximum intensity (Fig. 5.78).

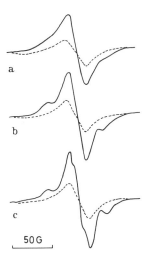

Fig. 5.78. ESR spectra of cellulose UV-irradiated at 77 K: a. 30 min, b. 120 min, c. 180 min. Dotted lines represent the spectra after warming to room temperature for 2 min[1634]

According to Arthur et al.[79] free radical formation in cellulose may be expected at C_1, C_4 or C_5 through the abstraction of H or OH. In case of reaction of C_1 or C_4 a single-line spectrum results, whereas the cleavage of H at C_5 or of the OH group at C_6 or H from the OH group at C_6 results in three-line spectra[77]. At higher temperature, only singlet and triplet spectra are detectable, whereas at low temperatures, such as at 77 K, two doublet spectra of formyl radicals and hydrogen atoms are also observed[966].

Hon[964] investigated in detail the effect of wavelength for the formation of free radicals in photo-irradiated cellulose. No radicals were detected by ESR recordings when cellulose was irradiated with light of wavelength longer than 3300–3400 Å.

Kleinert et al.[1143, 1566] found that oxygen slightly reduces the ESR signal intensity of UV-irradiated cellulose, while atomic hydrogen reduces it to zero.

Several contaminants, moisture content, and various oxidized groups influence shape and intensity of the ESR spectra observed[700, 963, 1633, 1634, 1828, 1979, 1980]. Moisture content in the range of 5–7% significantly inhibits the radical formation. When the moisture content is lower or higher than in that range, the formation of free radicals is increased[963].

The effect of several added inorganic and organic photosensitizers on the formation of free radicals in cellulose was investigated by ESR spectroscopy[646, 965, 967, 1633, 1634, 1741]. Hon[965] reported that cellulose during irradiation with light of $\lambda > 3400$ Å in the presence of benzoyl peroxide, hydrogen peroxide, benzophenone, riboflavin, azobisisobutyronitrile, and metal ions such as Cr^{2+} and Ni^{2+} exhibits single-line spectra with a line width of 16 G to 26 G. On irradiation with light of $\lambda > 2537$ Å, most of the sensitized samples exhibit five-line spectra with different relative signal intensities. Among these, samples treated with Fe^{3+} formed spectra with strongest relative signal intensities[965–967]. A similar ESR study was made for cellobiose[513].

Nanassy et al.[1566] describe ESR studies of the effect of lignin and various gases on UV-photodegradation of cellulose (Fig. 5.79).

The ESR spectrum of lignin is a single narrow line. When lignin is present in a pulp, a narrowing of the ESR line for the pulp occurs and the effect increases with the lignin content. Hence, it can be concluded that the strongly UV-absorbing lignin exhibits a screening effect, which results in a reduction of free spin concentration of a UV-irradiated pulp. Rånby et al.[1813] studied the influence of daylight on the free radical concentration of lignin and wood. Virgin lignin, e.g. in conifer roots unexposed to light, contains practically no radicals. Most radicals are formed during UV-irradiation and therefore are photochemically induced.

Virnik et al.[2364–2366] observed that during UV-irradiation of graft-copolymers of cellulose with poly(vinyl chloride), polyacrylonitrile, and poly(methylvinyl pyridine), a singlet-line spectrum is formed. This spectrum is attributed to free radicals formed from photodegraded cellulose. No free radicals from the polymers grafted to the cellulose substrate were observed.

5.1.19.2. Starch

ESR spectra of γ-irradiated starch samples were studied by several authors[9, 14, 18, 19, 228, 1621, 1662, 1894]. After γ-irradiation of starch at 128 K a triplet-line spectrum with the hyperfine splitting constant of about 18 G and line width of 12–14 G was observed (Fig. 5.80)[18].

After warming the sample to room temperature a doublet with hyperfine splitting constant of about 15 G and line width of 10–12 G is the most intense signal. The triplet line spectrum is interpreted similarly, as in the case of cellulose and attributed to free radicals at the C_6 atom. During warming of the sample radical transformation with the migration of radical from C_6 to C_5 may occur. The unpaired electron in C_5 produces the doublet-line spectrum by the hyperfine interaction with H_5 (5.112).

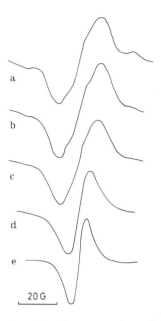

Fig. 5.79. ESR spectra of samples
UV-irradiated at 77 K: a. pulp
without lignin, b. pulp with 4%
lignin, c. pulp with 10% lignin,
d. ground wood, and e. lignin[1566]

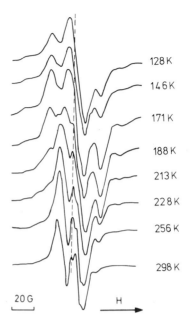

Fig. 5.80. Change in ESR spectra
of potato amylose γ-irradiated at
77 K with increasing temperature[18]

5.1.19.3. Other Polysaccharides

ESR studies of other γ-irradiated polysaccharides such as glycogen, dextran, maltose, cellobiose, lactose, sucrose, glucose, fructose, galactose, arabinose, ribose, and rhamnose were also reported[467, 622, 627, 705, 785, 786, 808, 1238, 1969, 2090, 2238, 2322, 2326].

Photo-induced radicals in glucose and cellobiose, model compounds of the cellulose molecule, were studied by ESR spectrometry[1238]. Very poor formations of radicals in glucose, as compared to that in cellobiose, were observed. A single line was easily produced by the use of short wavelength light. The spectrum was attributed to a radical formed at the reducing C_1 position of the glucose molecule. Photo-irradiated cellobiose shows an ESR spectrum similar to that of cellulose.

5.1.19.4. Lignin

Free radical content of native lignin and of humic acids isolated from the soil and lignite was examined by ESR[877, 1563, 2071]. The spin concentration measured in spin g^{-1} was as follows:
native lignin < degraded lignin < fulvic acid < humic acid.

The spin concentration for lignin indicates a free radical content of 5×10^{19} spins mole^{-1}, whereas the spin concentration for humic acid was 4×10^{22} spins mole^{-1}. ESR study suggests that the radical species are ortho- and para-semiquinone, which are coexistent with quinohydrone and quinone moieties. A significant increase in free radical concentration was observed for all samples when they were converted to the Na salts[2071], or hydrolyzed[2237]. Rånby et al.[1813] reported that most lignin radicals are photochemically induced, e.g. by daylight.

5.1.19.5. Wool

The ESR spectra of γ-irradiated wool are dependent on the temperature at which the irradiation is performed[226, 601, 1111]. Irradiation at room temperature gives glycine, alanine, and cystine radicals. At 77 K a broad unresolved singlet has been reported[1717].

Most ESR spectra obtained during UV-irradiation of wool have been attributed to cystyl (thiyl) radicals (RS·) formed by breaking of S—S bonds[1315, 1316, 1349, 1493, 1956–1960, 2423]. ESR studies of solvent- and heat-treated wool were also reported[600].

5.1.19.6. Polynucleotides

Nucleic acids or polynucleotides are polymers containing the repeating unit (5.119):

$$(5.119)$$

where: X = OH in ribonucleic acid (RNA) and X = H in deoxyribonucleic acid(DNA). A polymer of a sugar phosphate forms the backbone: ribose phosphate in RNA and deoxyribose phosphate in DNA. The side chains consist of heterocyclic bases such as adenine, guanine, thymine, and cytosine in DNA and adenine, guanine, uracil, and cytosine in RNA[443, 1201, 2383].

DNA, the well-known biochemical substance isolated from cell nuclei which is the genuine source of genetic information in living cells, is at present intensively studied. In 1953 Crick and Watson[487] elaborated the famous model of the "double helix" for DNA structure. The DNA double helix is composed of two intertwined polymer strands, each containing alternating sugar and phosphate groups. A planar side group called a base is attached to each sugar ring. A pair-wise coupling through hydrogen bonds between the bases belonging to the two strands stabilizes the structure of the macromolecule to the double helix. Four different bases are present in DNA: adenine, guanine, thymine, and cytosine. Base pairing through hydrogen bonds takes place between adenine and thymine as one pair, and guanine and cytosine as the other pair. The specific sequence of the bases in a DNA chain provides the chemical basis for the genetic code, which governs the protein synthesis in the cell and thus is the origin of all cellular properties and activities. The destruction of a single strand of these macromolecules through radiation may be biologically important. Irradiation of the DNA macromolecules in a solvent, such as water, produces direct and indirect effects. The direct effects are caused by absorption of γ-irradiation by macromolecules, whereas the indirect are caused by reactive radiation-induced species from the solvent. Both effects are probably responsible for the radiation damage of chromosomes carrying hereditary units in the form of DNA complexes with nucleoproteins.

Primary γ-induced free radicals in DNA were attributed to the thymine base (5.120), formed by hydrogen-addition to carbon atom C_6, thus breaking up the C_5-C_6 double bond to the thymine radical (5.121)[1967, 1968]:

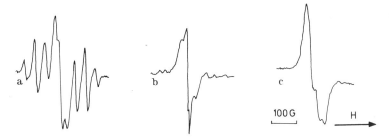

$$(5.40)$$

(5.120)   *(5.121)*

The ESR spectra of γ-irradiated DNA and their single strand components are generally well resolved[8, 608, 615, 617, 806, 958, 815, 816, 817, 1340, 1541, 1542, 1768, 1890, 1943, 1944, 2019–2021, 2349]. An ESR spectrum obtained by γ-irradiation of a sample of salmon sperm DNA at room temperature under vacuum is shown in Figure 5.81 a[474, 617].

Fig. 5.81. ESR spectra of salmon sperm DNA γ-irradiated: a. at room temperature, b. sample a, measured at 300 K after annealing for 5 min at 353 K, c. γ-irradiated and measured at 77 K[617]

The outer narrow lines were attributed to thymine radical (5.121). At least two radical species are present in the γ-irradiated DNA, which differ in line-width and have different decay rate on thermal annealing.

After annealing at 353 K the thymine radical spectrum (Fig. 5.81 a) disappeared, leaving the central line as a residual dominnat component (Fig. 5.81 b). When samples of DNA are irradiated at 77 K the thymine radical ESR spectrum is absent (Fig. 5.81 c). After annealing the sample to 210 K the eight-line spectrum identical to Figure 5.81 a, appears. These results show that the formation of thymine radicals (5.121) is not a primary result of γ-radiation, but depends upon a chemical change which is only appreciable above about 200 K. It has been suggested that the primary

ESR spectrum originates from an anionic base radical[167, 806, 902, 918, 960, 1339, 1340, 1670, 1942].

The ESR spectra recorded with the magnetic field of the spectrometer parallel and perpendicular to the direction of helical axis of the sample were found to be composed of two major components with different line-widths[610, 805, 807]. The difference in hyperfine structure was interpreted in terms of a mixture of an anionic radical of thymine (or cytosine) and cationic base radical of guanine (or cytosine).

Computer analysis and reconstruction of ESR spectra of γ-irradiated DNA were described in papers[815, 816].

Radical formation in γ-irradiated DNA or their single-strand components in frozen aqueous matrices has been a subject of many investigations[2, 8, 315, 631, 919, 1116, 1339, 1341, 1922, 1953]. ESR studies of DNA in frozen aqueous solutions show that radicals are produced by protonation of intermediate anionic species, thymine anions in isomeric forms. The anions are formed by the addition of radiation-produced electrons[1339, 1341]:

Thymidine γ-irradiated in sulphuric acid:

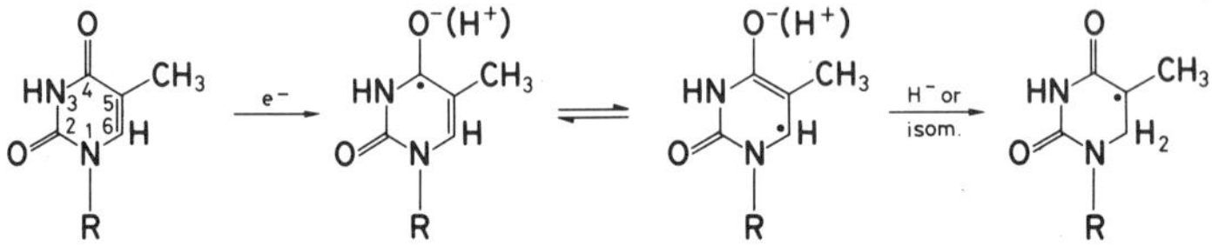

$$(5.41)$$

(5.122)   (5.123)

Thymidine γ-irradiated in sodium hydroxide:

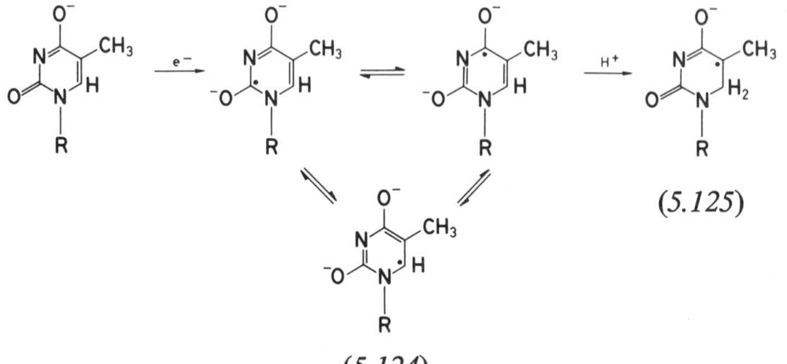

$$(5.42)$$

(5.125)

(5.124)

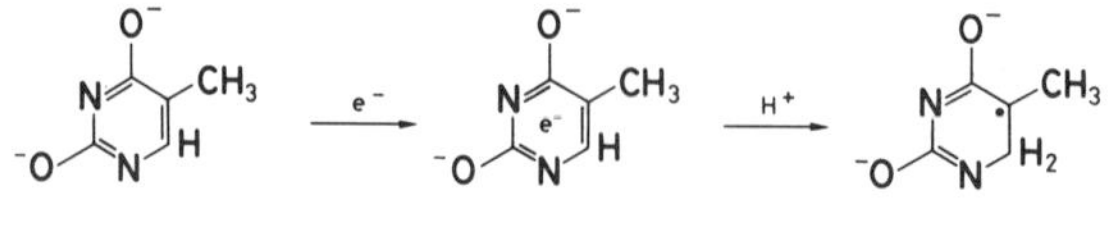

$$(5.43)$$

(5.126)   (5.127)

where: anion radicals (5.122) (5.124) and (5.126), and 5-thymyl radicals (5.123) (5.125) and (5.127).

The structure of the radical anions is dependent on pH of the irradiated DNA solutions. In thymidine at low pH the spin is best located as being shared mainly between C_4 and C_6, giving the maximum interaction with the C_6 proton (5.41). At high pH the spin is further shared with C_2, thus reducing the spin density at C_6 and consequently the ESR doublet splitting[1341] (5.42). Finally, in thymine at high pH, where the molecules are doubly ionized before irradiation (5.43), the free spin density is delocalized and the interaction with the C_6 proton is at minimum, giving an ESR singlet or unresolved doublet of small splitting. The presence and nature of thymine anion radicals in DNA seems to be fully established.

An interesting problem in the study of polynucleotides is the application of ESR spectroscopy for detecting light-induced free radicals in DNA-acridine complexes. It is generally assumed that the strong internal binding of acridines to DNA occurs via intercalation[1352], i.e. insertion of the planar dye molecules between successive base pairs in the double helix. In addition, further dye molecules may be weakly associated on the outside of the DNA helix. Even without light, the strongly bound acridines have a profound effect on the properties of DNA, and are well known as mutagenic agents. When light is introduced into the system, its energy can be transferred from the dye to the DNA. The long-lived triplet states of the dye are believed to be important for their stability to energy transfer, causing, in this way, the photosensitization of the DNA macromolecule. The formation of free radicals in a DNA-acridine complex illuminated at 77 K was detected[531, 809, 810, 1239]. The detailed ESR study shows that the strongly bound intercalated dye molecules involved in the free radical formation, in contrast to the inactive dye molecules weakly bound on the outside of the DNA helix[809].

It can be concluded from the ESR study of DNA systems that the formation of ionic base radicals seems to be the fundamental process by which DNA responds to radiation. Further reactions of these radicals, either with the DNA or with external agents, may be important for the damage of DNA system produced by radiation.

5.2. Free Radicals Formed in Glow Discharge of Polymers

Glow discharge on the surface of polyethylene, poly(vinyl chloride), polystyrene, and poly-p-xylene samples causes the formation of macroradicals. These macroradicals react further with oxygen and give peroxy radicals, which have been examined by ESR spectroscopy[647, 1532, 2455]. During glow discharge of polymer samples crosslinking and degradation processes were observed.

5.3. Anionic Degradation of Vinylaromatic Polymers

Electron transfer from sodium to polymers containing aromatic centers with high electron-affinity, e.g. polystyrene, polyphenylene, poly(benzo-p-xylene)[594] poly-

Table 5.4. Types of mechanisms of chain scissions during anionic degradation of vinylaromatic polymers[1834]

Polymer	Polyradicalanion	Anions
Poly-N-vinyl carbazole (PVCa)	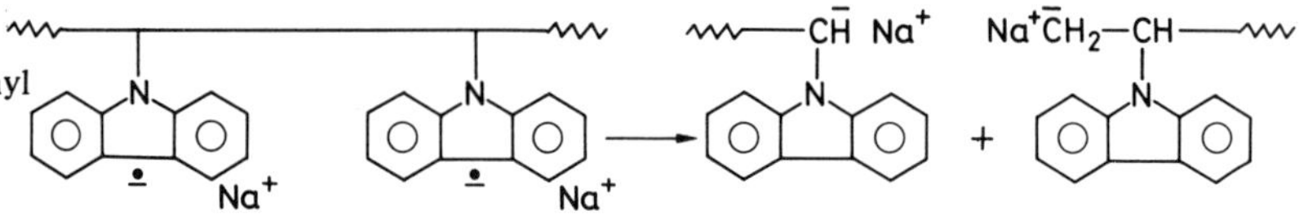	$\longrightarrow$ ─CH Na⁺ Na⁺CH₂─CH─
Poly-β-vinyl naphthalene (PβVN)	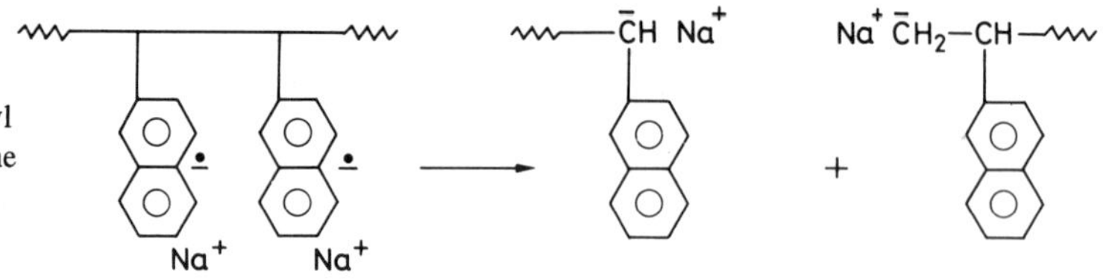	$\longrightarrow$ ─CH Na⁺ + Na⁺ CH₂─CH─
Poly-α-vinyl naphthalene (PαVN)	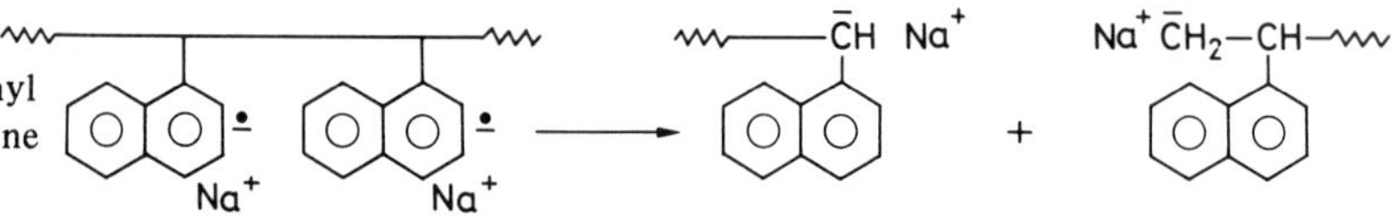	$\longrightarrow$ ─CH Na⁺ + Na⁺ CH₂─CH─
Poly-4-vinyl biphenyl (PVB)	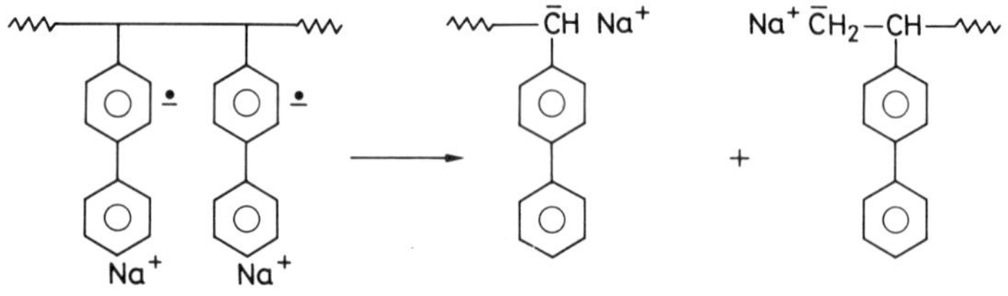	$\longrightarrow$ ─CH Na⁺ + Na⁺ CH₂─CH─
Polyace-naphthylene (PAcN)	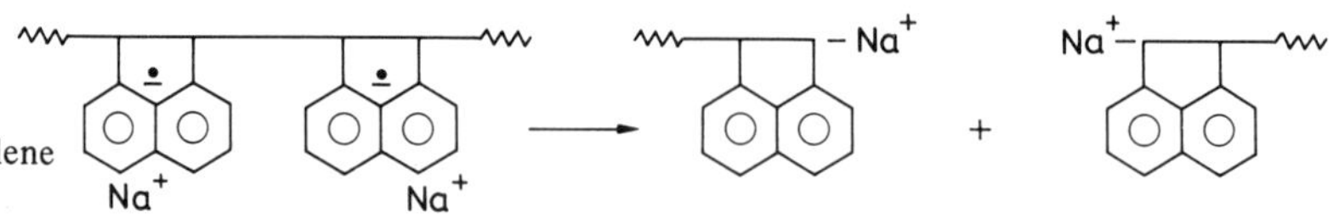	$\longrightarrow$ ─ Na⁺ + Na⁺─

(vinylbiphenyl), and poly(vinylnaphthalene)[1830-1834] results in polymer degradation. ESR study shows that, during the chain-breaking process, polymeric fragments are formed, which have the properties of polyradical anions equivalent to "living polymers". They are capable of initiating polymerization of a number of vinyl monomers. The following mechanism of chain scission was proposed and presented in Table 5.4[1834].

5.4. Reactions of Polymers with Hydroxy Radicals

Using flow techniques (described in detail in Chapter 3.6.2) hydroxy radicals can be generated in an ESR cavity. In the presence of water-soluble polymers, e.g. polyethylene oxide, poly(vinyl alcohol), poly(acrylic acid), poly(methacrylic acid), poly(ethylene imine), poly(propylene oxide), dextran, dextrin, soluble starch, and maltose, hydroxyl radicals may abstract hydrogen and form macroradicals[491, 696, 2003]. In Figures 5.82–5.85 ESR spectra of polymer radicals obtained by this method are present.

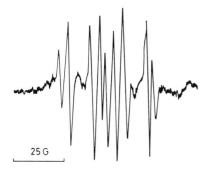

Fig. 5.82. ESR spectrum of free radicals from poly(vinyl alcohol) (0.25 mole 1^{-1}) produced by TiCl$_3$-H$_2$O$_2$ system[696]

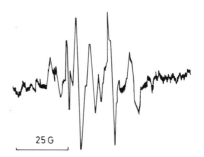

Fig. 5.83. ESR spectrum of free radicals from poly(acrylic acid) (0.23 mole 1^{-1}) produced by TiCl$_3$-H$_2$O$_2$ system[696]

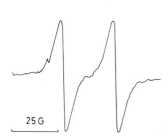

Fig. 5.84. ESR spectrum of free radicals from poly(methacrylic acid) (0.23 mole 1^{-1}) produced by TiCl$_3$-H$_2$O$_2$ system[696]

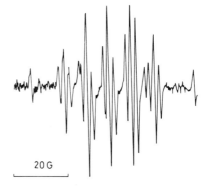

Fig. 5.85. ESR spectrum of free radicals from poly(ethylene oxide) (0.5 mole 1^{-1}) produced by TiCl$_3$-H$_2$O$_2$ system[696]

The detailed analysis of hyperfine structure of the radicals formed, may be found in the original papers.

ESR studies of the reaction of cellulose with hydroxy radicals generated by Fe^{2+}–H$_2$O$_2$ in a flow system, have been described by Arthur et al.[76]. It has been suggested, considering ESR spectrum shown in Figure 5.86, that hydrogen atoms

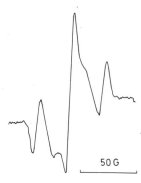

50 G

Fig. 5.86. ESR spectrum of cotton cellulose radicals produced by Fe^{2+}-H_2O_2 system[76]

have been abstracted from hydroxyl groups on carbon C_6, or possibly hydrogen from carbon C_5 (*5.112*).

The ESR spectrum generated on microcrystalline cellulose was less intense than that generated on cotton cellulose.

Chapter 6

ESR Study of Polymers in Reactive Gases

6.1. Molecular Oxygen

Using quantum mechanics, the electronic configuration of molecular oxygen (O_2) in the ground state $^3\Sigma_g^-$ may be written as follows[770]:

$$O_2[KK(\sigma_g 2s)^2(\sigma_u 2s)^2(\sigma_g 2p_x)^2(\pi_u 2p_y)^2(\pi_u 2p_z)^2(\pi_g 2p_y)(\pi_g 2p_z)] \qquad (6.1)$$

Two unpaired electrons are in orbitals, each one having the orbital angular momentum on the molecular axis equal to unity, but resolving in different directions. This gives a total orbital angular momentum of zero. This state contains a molecular electron cloud with rotational symmetry, and it is called a sigma (Σ) state. The last two electrons have parallel spins, leading to a triplet state with paramagnetic properties. Molecular oxygen has a biradical nature and it reacts easily with other organic free radicals giving peroxyradicals.

The ESR spectrum observed in gaseous oxygen at a pressure of a few mm Hg is very complicated. A total of 140 lines have been observed (Fig. 6.1).

This spectrum has been fully analyzed and compared with a theoretical spectrum[197, 258, 946, 2186, 2194]. The relative intensity of particular lines changes with temperature, following the temperature response of the molecular rotational states. The observed absorption lines broaden rapidly with increasing pressure (Fig. 6.2) and are not resolved at atmospheric pressure.

In ESR resonance cavities oxygen gives rise only to a very broad, and generally undetectable, absorption signal.

Povich[1766] reported applications of ESR spectroscopy for the measurement of concentration and diffusion coefficient of oxygen dissolved in organic solvents.

6.2. Singlet Oxygen

Singlet oxygen has attracted increased attention in photooxidation processes occurring in polymers[1780, 1781, 1784, 1790, 1791, 1814, 1815].

When sufficient energy is absorbed by molecular oxygen it may change its electronic configuration. Two states of excited oxygen are formed and are known as:[1107]:

1. Singlet oxygen $^1O_2(^1\Delta_g)$ which has the electronic configuration:

$$O_2[KK(\sigma_g 2s)^2(\sigma_u 2s)^2(\sigma_g 2p_x)^2(\pi_u 2p_y)^2(\pi_u 2p_z)^2(\pi_g 2p_y)^2] \qquad (6.2)$$

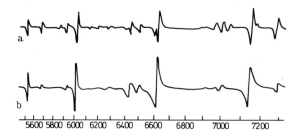

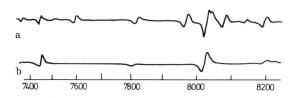

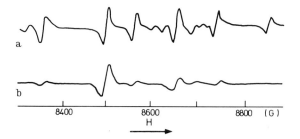

Fig. 6.1. Effect of temperature on the ESR spectrum of molecular oxygen under 4 mm Hg pressure: a. at 300 K and b. 85 K[197)]

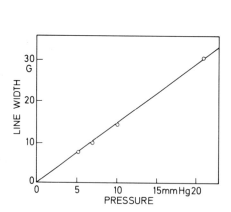

Fig. 6.2. Variation of line width at 6000 G (Fig. 6.1) with O_2 pressure at 300 K[197)]

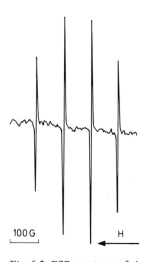

Fig. 6.3. ESR spectrum of singlet oxygen ($^1\Delta_g$) at 0.4 mm Hg[284)]

The electrons in the highest orbitals are paired and their spins are antiparallel. The first excited state-singlet oxygen $^1O_2(^1\Delta_g)$-exceeds the ground state of molecular oxygen $(^3\Sigma_g^-)$ by 22.5 kcal/mol (0.98 eV). Since the electrons in this state are paired (have antiparallel spins) they show a characteristic ESR spectrum[67, 284, 590, 639, 640, 1108, 1109, 2389, 2390] (Fig. 6.3).

2. Singlet oxygen $^1O_2(^1\Sigma_g^+)$ has an identical electron configuration with molecular oxygen $(^3\Sigma_g^-)$. The difference is that in the $^1O_2(^1\Sigma_g^+)$ state the electrons in the antibonding (π^*) orbitals have antiparallel spins, whereas in the $^3\Sigma_g^-$ state these electrons have parallel spins. The second excited state-singlet oxygen $^1O_2(^1\Sigma_g^+)$-exceeds the ground state of molecular oxygen $(^3\Sigma_g^-)$ by 37.5 kcal/mol (1.63 eV).

6.3. Atomic Oxygen

Atomic oxygen is formed during electric discharges in molecular oxygen[956, 1452], and in flames. Most recently, increased attention has been devoted to the ageing of polymers in natural environments where atomic oxygen may be present[847, 862, 863, 1312, 1814, 1851]. Atomic oxygen has an even number of electrons, is highly paramagnetic, and can easily be detected by ESR spectroscopy. Its spectrum (Fig. 6.4) contains six fine-structure lines[284, 1226].

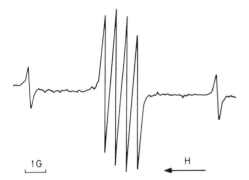

Fig. 6.4. ESR spectrum of atomic oxygen at 0.1 mm Hg[284]

Measurements at several pressures from 1.0 to 2.0 mm Hg indicate an exponential decay which increases with pressure.

ESR studies of the interaction of atomic oxygen with polystyrene "fluffs"[817] and polyethylene, polypropylene, poly(methyl methacrylate), poly(acrylonitrile)[136, 144], and cellulose[867] were reported.

6.4. Atomic Hydrogen

During the reaction of atomic hydrogen with different organic compounds[421, 465, 1961, 1962], synthetic polymers[144, 574, 575, 579, 582, 583, 1001, 2182, 2376, 2377, 2381,

[2382] and biopolymers[930, 1367] at low temperatures, free radicals are formed which may be detected by the ESR method. Because the depth of the penetration of hydrogen atoms into polymers is limited, polymers have to be exposed in "fluffs" obtained by freeze-drying of benzene solutions[1361, 2362]. It has been observed that polystyrene, poly(α-methylstyrene), poly(ethylene terephthalate), and polyacrylonitrile react with atomic hydrogen by addition[575, 582] and large concentrations of polymer radicals can be detected by ESR. In other polymers such as polyethylene, polypropylene, polyisobutylene, poly(vinyl alcohol), poly(vinyl acetate), poly(methyl methacrylate), and polyamides, atomic hydrogen abstracts hydrogen from the polymer molecules. The reactions of atomic hydrogen with polymers take place in a thin surface layer of about 10^{-6} cm[574]. Crosslinking reactions were also observed.

ESR studies of deuterium atom[136, 2376, 2377] and tritium reactions[2448] with polystyrene "fluffs" were also reported.

6.5. Atomic Nitrogen

Solid nitrogen at 4.2 K under γ-irradiation dissociates into atoms, which give well-developed ESR spectra (Fig. 6.5)[316, 2375, 2378, 2379].

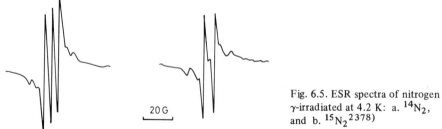

20 G

Fig. 6.5. ESR spectra of nitrogen γ-irradiated at 4.2 K: a. $^{14}N_2$, and b. $^{15}N_2$[2378]

ESR study of the reaction between atomic nitrogen obtained from microwave discharge and polyethylene, polypropylene, poly(methyl methacrylate), polyacrylonitrile[236], and polystyrene[2376] were reported.

6.6. Noble Gases

It was found that active species of He and Xe from microwave discharge induce crosslinking of polystyrene "fluffs"[817, 1001, 1002, 2376, 2381]. Based on ESR spectra it was reported that the formation of free radicals in this case is not different from what is observed in reaction of hydrogen atoms with polystyrene or during the radiolysis of that polymer.

6.7. Other Gases

ESR studies of reactions of free radicals formed during γ-irradiation of such polymers as poly(vinyl chloride), polypropylene, polytetrafluoroethylene, and cellulose with different gases, e.g, O_3, SO_2, NO, NO_2, Cl_2, Br_2, I_2, H_2S, NH_3, CO, CO_2, CH_4, C_2H_6 were also recorded[701, 1252–1254, 1433, 1823, 2324, 2327, 2329–2331].

Using ESR spectroscopy, it has been found that stable radicals present in γ-irradiated polyethylene initiate the chlorination of that polymer via the chain mechanism[1356].

ESR studies of plasma synthesis of fluorocarbon films have also been reported[1490].

Chapter 7

ESR Studies of the Oxidation of Polymers

7.1. Radiation-Induced Oxidation of Polymers

Several papers have been published on ESR studies of oxygen radicals formed during the irradiation of the following polymers:

polyethylene[14, 456, 1137, 1388, 1473, 1476, 1589, 1642, 1648, 1650, 1936, 2267, 2272)

polypropylene[332, 456—458, 603, 623, 683, 684, 838, 1034, 1154, 1158, 1159, 1330, 1473, 1476, 1481, 1591, 1639, 1648, 1650, 1651, 2243)

,

polyisobutylene[1473)],

polytetrafluoroethylene[62, 888—891, 893, 1025, 1034, 1137, 1148, 1152, 1158, 1160, 1330, 1432, 1478, 1480, 1520, 1534, 1639, 1686, 1840, 2006, 2152, 2160, 2216, 2304, 2306, 2490)

,

poly(methyl methacrylate)[293, 375, 395, 1137, 1473, 1645)],

polystyrene[293, 375, 836, 1941)],

poly(vinyl chloride)[1254, 1390, 1642, 1648, 1651, 1788)],

poly(vinyl acetate)[293, 375, 1476, 1483)],

poly(oxymethylene)[2091, 2252, 2266)],

poly(oxyethylene)[1590)],

poly(2,6-dimethyl-1,4-phenyleneoxide)[2128, 2280)],

poly-3,3-bis(chloromethyl)oxethane[460, 2256, 2276, 2277)],

polyquinones[2023)],

polyamides[451, 803, 1476)], and polyurethanes[1593)].

Some problems in the ESR study of photochemical oxidation of polymers were reviewed by Rabek and Rånby[1789)].

Most polymers show asymmetric singlet-line spectra attributed to ROO· (RO·) radicals, which are formed after introduction of air to samples containing free radicals (R·). There is some difficulty in distinguishing between alkylperoxy (ROO·) and alkoxy (RO·) radicals, since both have the unpaired electron mainly concentrated on oxygen atoms and neither show a hyperfine interaction with alkyl protons[1166)]. During study of photodissociation of some alkyl hydroperoxides, Piette and Landgraf[1745)] obtained alkoxy (RO·) radicals showing a hyperfine splitting constant of 3 G. This hyperfine splitting value is too small for observation even in well-resolved peaks of ESR spectra of solid polymers. These results have later been disclaimed[1003, 2124)]. Oxygenated radicals have smaller spin-lattice relaxation times than the majority of other free radicals trapped in polymers. ESR spectra of alkyl radicals (R·) can be recorded without saturation broadening at a microwave power of about 1—3 mW, whereas oxygenated free radicals need a very high power, e.g. 100 mW[1650)]. Figure 7.1 shows a plot of the detected signal voltage (V) as a function of the square root of the microwave power (P) for alkyl and oxygenated radicals in polypropylene, demonstrating in this way the differences in saturation behavior.

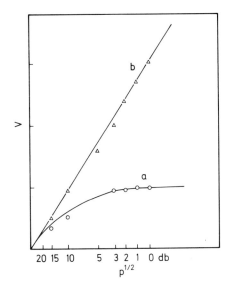

Fig. 7.1. Saturation effect of free radicals trapped in γ-irradiated polypropylene: a. in vacuum and b. in the presence of oxygen[1650])

The characteristic ESR signal of the oxygenated radicals (ROO·) varies with the angle between the external field and the stretching direction of the polymer sample (i.e. the chain axis). This effect is interpreted as arising from the density of the free spin in the p-orbital of the oxygen atom of peroxy radicals. Also the spin density distribution in the p-orbital possibly causes the g-factor anisotropy. The apparent principal values of the g-tensor, $g_\parallel$ and $g_\perp$, can be expressed in terms of the principal values g_1, g_2 and g_3 for the rigid state (see Chapter 2.5). The g-tensor of peroxy radicals in the rigid state is approximately axially symmetric, and the value of $g_\parallel$ is much larger than $g_\perp$, because of the large spin-orbit coupling constant of an oxygen atom and the small energy separation of the two $2p(\pi)$ orbitals, one of which contains the unpaired electron. In many cases, the value of $g_\perp$ is close to the free spin value (~ 2.0023)[2160], while $g_\parallel$ is around $2.035-2.045$[684, 1034, 1686, 1927]. From theoretical considerations[995, 1683] the symmetry axis of the tensors, that is the direction of $g_\parallel$, is expected to be parallel to the O–O-bond. In order to elucidate the axis of rotational motion, it is desirable to determine the principal values and directions of the g-tensor, both at room temperature and 77 K (see Chapter 7.4).

The peroxy radicals formed in various polymers decayed after different lifetimes (Fig. 7.2).

The decay of alkyl radicals (R·) is not observed, as they easily react with molecular oxygen at diffusion controlled rates, and are not present in detectable amounts.

7.2. Polyethylene

ESR studies of peroxy radicals formed during the reaction of oxygen with free radicals formed under ionizing and UV-irradiation (see Chapter 5.1.1.2) were published by many authors[14, 456, 1137, 1388, 1473, 1476, 1589, 1642, 1648, 1650, 1936, 2267, 2272].

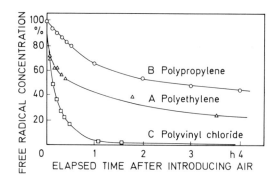

Fig. 7.2. Decay of polymer free radicals after breaking of the vacuum[1651]

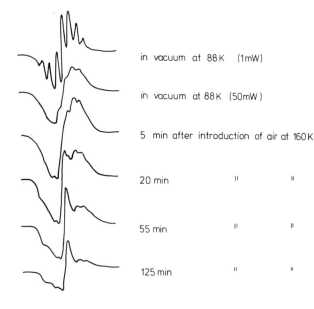

in vacuum at 88 K   (1mW)

in vacuum at 88 K   (50mW)

5 min after introduction of air at 160 K

20 min   "   "

55 min   "   "

125 min   "   "

Fig. 7.3. Change of ESR spectra during reaction of allyl radicals (7.2) present in polyethylene with oxygen at 160 K[1650]

Alkyl (5.3) and allyl (5.4) radicals react with oxygen immediately and give peroxy radicals which decay with half-lives of about 10 and 20 min, respectively[1650]. Free radicals decay quickly to a level of about 80%, and thereafter decay is rather slow. It is difficult to prepare a sample in which only alkyl or allyl radicals are present. It is also difficult to recognize from the type of free radicals, what peroxy radicals are formed. In Figure 7.3 the change in ESR spectra is shown for allyl radicals (5.4) directly after introducing air.

Introduction of air to a polyethylene sample containing polyenyl radicals (5.5) did not give the asymmetric peroxide singlet spectrum (Fig. 7.4), as in the case of allyl radicals (Fig. 7.4).

Polyenyl peroxy radicals decay with a half-life of 20 min[1650]. It has been found that the asymmetric spectrum of alkyl peroxy radicals in a polyethylene sample shows an anisotropy (Fig. 7.5).

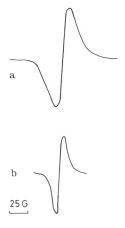

Fig. 7.4. Change of ESR
spectra during reaction of
polyenyl radicals (7.3) present in polyethylene with
oxygen: a. in vacuum (gain
1) and b. after 5 days after
introduction of air (gain
100)[1650]

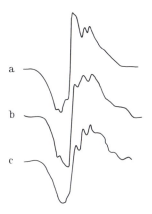

Fig. 7.5. Anisotropy of ESR
spectrum of peroxide radicals in polyethylene at various angles between the external field and the direction
of stretching: a. 0°, b. 45°
and c. 90°[1650]

A detailed analysis of the anisotropy of peroxy radicals is given in the description
of peroxy radicals of poly(tetrafluoroethylene) in Chapter 7.4.

The predominating reaction mechanism of peroxy radicals is the intermolecular
hydrogen abstraction:

$$ROO\cdot + RH \longrightarrow ROOH + R\cdot \tag{7.1}$$

According to this mechanism a new alkyl free radical can be formed. In spite of
great efforts to study the oxidation mechanism of polyolefines, this problem and its
interpretation has not yet been fully resolved[518, 519, 1784].

7.3. Polypropylene

A number of workers recorded the ESR studies of radiation-produced peroxy radicals of polypropylene[332, 456–458, 603, 623, 683, 684, 838, 1034, 1154, 1158, 1159, 1330,
1476, 1481, 1591, 1639, 1648, 1560, 1651, 2243].

ESR spectra of peroxy radicals produced by γ-irradiation of polypropylene in
the presence of oxygen at room temperature and at 77 K are asymmetric and do not
show any distinct dependence upon temperature (Fig. 7.6).

Figure 7.7 shows ESR spectra of polypropylene peroxy radicals obtained during
thermal oxidation in the range of 383–413 K.

257

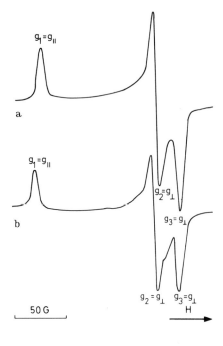

$g_1 = g_{\parallel}$

a

$g_1 = g_{\parallel}$

$g_2 = g_{\perp}$

$g_3 = g_{\perp}$

b

$g_2 = g_{\perp}$ $g_3 = g_{\perp}$
H

50 G

Fig. 7.6. ESR spectra of peroxy radicals in γ-irradiated polypropylene: a. at room temperature and b. at 77 K[1034)

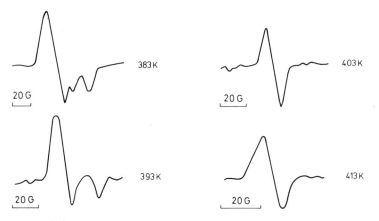

383 K

20 G

393 K

20 G

403 K

20 G

413 K

20 G

Fig. 7.7. ESR spectra of peroxy radicals in thermally oxidized polypropylene[456)

The spectra at 383 K show marked anisotropy independent of molecular weight or crystallinity of the polypropylene used. The g-values are: g_1 = 2.0135, g_2 = 2.0245 and g_3 = 2.0356[456)]. At 393 K the spectra showed either partial averaging of g anisotropy as shown in Figure 7.7 for crystalline polypropylene or nearly complete averaging to resemble at 403 K (amorphous samples). For the former, $g_{\perp}$ = 2.0061 and $g_{\parallel}$ = 2.0187. At 403 K, most spectra are symmetric and they have the following parameters g = 2.0067 and $\Delta H_{1/2}$ = 8.6 G. At 413 K spectra are symmetric and have g = 2.0067 and $\Delta H_{1/2}$ = 7.4 G. At 423 K the line-width is about 8 G.

258

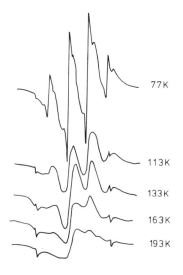

77 K

113 K

133 K

163 K

193 K

Fig. 7.8. ESR spectra of free radicals formed during UV-irradiation of polypropylene in air at 77 K and their transformation into peroxyradicals during warming of the sample to 193 K[2243]

Neudörfel[1591] desribes the oxygen consumption of irradiated samples as a function of the decrease of the radical concentration. The rate constants for the oxidation of polypropylene were also determined by ESR method[458].

The formation of peroxy radicals during UV-irradiation of polypropylene was investigated in detail by Tsuji[2243]. Typical ESR spectrum obtained during UV-irradiation of polypropylene at 77 K in the presence of air is shown in Figure 7.8.

This spectrum was attributed to the free radicals (5.13) and methyl (5.17) radicals. During warming of the sample to room temperature spectral changes were observed. The four-line spectrum due to methyl radicals disappeared at 113 K. The broad four-line spectrum began to change at about 133 K and a broad singlet spectrum was observed at 193 K. When the spectrum was recorded at 113 K an asymmetric component, characteristic of peroxy radicals in a rigid matrix, was observed ($g_\perp = 2.004$ and $g_\parallel = 2.032$). This result indicates that the peroxy radicals were produced below 133 K, but are shifted somewhat at this temperature. The peroxy radicals are expected to move relatively freely as they are located on a side chain (7.1):

$$
\begin{array}{ccc}
\overset{\bullet}{C}H_2 & & CH_2 \\
| & & | \\
-CH_2-CH-CH_2- \ + \ O_2 & \longrightarrow & -CH_2-CH-CH_2-
\end{array}
\tag{7.2}
$$

(7.1)

When polypropylene was UV-irradiated through a UV-33 filter a broad singlet spectrum with a g-value of about 2.005 and with some shoulders was observed (Fig. 7.9).

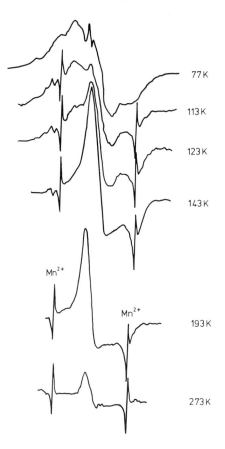

77 K

113 K

123 K

143 K

Mn²⁺

Mn²⁺

193 K

273 K

Fig. 7.9. ESR spectra of free radicals formed during UV-irradiation of polypropylene using UV-33 filter, in air at 77 K and their change during warming of the sample[2243]

A similar asymmetric ESR spectrum due to peroxy radicals was observed at 123 K and remained up to 273 K. The temperature dependence shows that the peroxy radicals in this case do not move so freely. This type of peroxy radicals (7.2) may be produced from the alkyl radicals (5.12) by the following reaction:

$$-CH_2-\overset{\cdot}{C}-CH_2- \ + \ O_2 \ \longrightarrow \ -CH_2-\overset{O-O}{\underset{|}{C}}-CH_2- \qquad (7.3)$$
$$\underset{CH_3}{|} \qquad\qquad\qquad \underset{CH_3}{|}$$

(5.12)   (7.2)

Milinchuk and Pshezhetskii[1481] have found that peroxy radicals in polypropylene at 77 K are converted under UV-irradiation (<3000 Å) into alkyl radicals.

Tkač et al.[1470] have recently studied the decomposition of polymeric hydroperoxides of polypropylene with amines in a styrene matrix and obtained an asymmetric spectrum (Fig. 7.10).

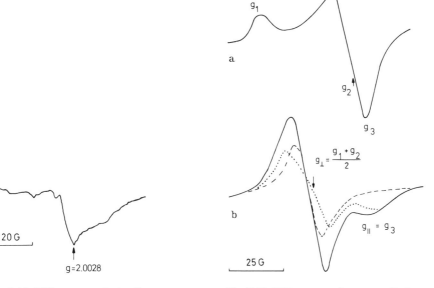

Fig. 7.10. ESR spectrum during the decomposition of polypropylene hydroperoxide with triethylene-tetramine in styrene at 296 K[1470]

Fig. 7.11. ESR spectra of peroxy radicals in γ-irradiated polytetrafluoroethylene: a. at 77 K and b. at room temperature[1534]

This spectrum was attributed to alkoxy radicals resulting from homolytic scission of the O–O bonds in polypropylene hydroperoxide or to nitroxy radicals formed during the decomposition of the system hydroperoxide-amine.

A similar ESR study was reported on the reactions of aromatic amines with peroxide groups in solid polymers in the temperature range from 77 K to 293 K[581].

Thermal decomposition of benzoyl peroxide[1796] and di-tert-butyl peroxalate[1597] in polypropylene and also in other polymers produce peroxy radicals which were examined by ESR spectroscopy.

7.4. Poly(tetrafluoroethylene)

ESR spectra of peroxy radicals in poly(tetrafluoroethylene) were investigated by several authors[62, 888–891, 893, 1025, 1034, 1137, 1148, 1152, 1158, 1160, 1330, 1432, 1478, 1480, 1520, 1534, 1639, 1686, 1840, 2006, 2152, 2160, 2216, 2304, 2306, 2490].

The line shape and the characteristics of ESR spectra due to peroxy radicals in poly(tetrafluoroethylene) are difficult to interpret and are not yet well understood. Temperature dependence and orientation effects on the spectral line shapes were studied and interpreted on the basis of general theory for randomly oriented samples and for molecular motion[1034, 1534]. Figure 7.11 shows the variation with temperature of the ESR spectra of peroxy radicals trapped in poly(tetrafluoroethylene) powder.

As it has been reported by several authors[1034, 1433, 2160], the spectrum, at room temperature, of the peroxy radicals trapped in poly(tetrafluoroethylene) consists of two spectral components, one asymmetric and the other symmetric[1034]. The symmetric component indicated by the broken line in Figure 7.11b is attributed to peroxy radicals which have enough motional freedom to average out the entire anisotropy in g-value. The asymmetric component indicated by the dotted line has a characteristic line shape arising from the axially symmetric g-tensor. From the line shape the principal values were obtained: $g_{\parallel}^r = 2.0061$ and $g_{\perp}^r = 2.0221$, where $g_{\parallel}^r$ and $g_{\perp}^r$ represent the apparent principal values at room temperature in directions parallel and perpendicular, respectively, to the symmetry axis of the g-tensor. In the spectrum observed at 77 K (Fig. 7.11a) only one spectral component is observed. The following principal values of the g-tensor were obtained at 77 K from the line shape of the spectrum: $g_1 = 2.0384$, $g_2 = 2.0071$, $g_3 = 2.0026$[1034] and $g_1 = 2.0385$, $g_2 = 2.0079$ and $g_3 = 2.0023$[1534]. Since g_2 is nearly equal to g_3 at 77 K and g_1 is assigned approximately to $g_{\parallel}$, it is concluded that g_2 and g_3 represent $g_{\perp}$. The symmetric component of the ESR spectrum at room temperature was attributed to free radical (7.3) and the asymmetric component to free radical (7.4)[889, 1034, 1148, 1433, 2006, 2152]:

$$-CF_2-CF_2-O-O\cdot \quad (7.3) \qquad \qquad \overset{\displaystyle \overset{\textstyle \dot{O}}{|}}{\underset{\displaystyle -CF_2-\overset{\textstyle |}{\underset{\textstyle O}{C}F}-CF_2-}{}} \quad (7.4)$$

The temperature variations of the ESR line shape of the peroxy radicals (7.3) and (7.4) are shown in Figures 7.12 and 7.13, respectively.

It has been found that the peaks corresponding to g_1 and g_2 change into a single line at 223 K. This coalescence indicates that the g_1 and g_2 of the g-tensor are averaged out by rapid motion.

ESR spectra of peroxy radicals of poly(tetrafluoroethylene) obtained at room temperature at various orientations of the stretch axis to the magnetic field are shown in Figure 7.14.

Angular dependence of the observed g-value for the anisotropic component of the room temperature spectra of an oriented poly(tetrafluoroethylene) sample is given in Figure 7.15.

In general, the direction of $g_{\parallel}$ does not always agree with the molecular chain axis. The g-value g = 2.0059, which has been found with the field parallel to the stretch axis, is very close to $g_{\parallel}^r = 2.0061$ obtained from the line shape in the powdered spectrum, and g = 2.0213 measured in the perpendicular direction is near to $g_{\perp}^r = 2.0221$. This means that the symmetry axis of the g-tensor is parallel to the molecular chain axis at room temperature.

The angular dependence of the spectra of oriented films was also observed at 77 K (Fig. 7.16).

The ESR spectrum observed with the field direction parallel to the stretch axis has a range of g-values from 2.0019 to 2.0376 (Fig. 7.16a). The spectrum of the radical giving a symmetric component at room temperature becomes anisotropic at 77 K, because of decreasing molecular motion. The peroxy radicals (7.3) and (7.4)

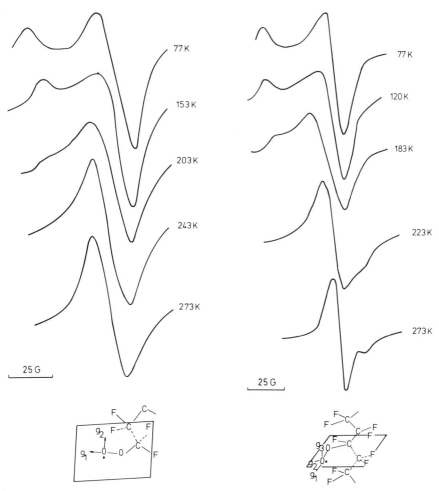

Fig. 7.12. Change of the ESR spectra from the peroxy end radical in polytetrafluoroethylene during warming of the sample[1534]

Fig. 7.13. Change of the ESR spectra from the peroxy chain radical of polytetrafluoroethylene during warming of the sample[1534]

trapped in amorphous regions are probably not oriented in respect to the chain axis, and thus produce no orientation in the spectrum. From detailed analysis of the changes in g-values it has been assumed that the $-COO\cdot$ groups lie in a plane perpendicular to the molecular chain axis, as shown in Figure 7.13. The principal values and directions of the g-tensor are collected in Table 7.1 and the g-values in oriented film in Table 7.2.

The apparent g-tensor of a freely rotating chain radical should have axial symmetry in respect to the axis of rotation, because the magnitude of g-value with the field direction along the axis of rotation is invariant to rotation. The g-value for any field direction perpendicular to the axis of rotation is the same when the correlation frequency of the motion is sufficiently larger than the difference between the maxi-

263

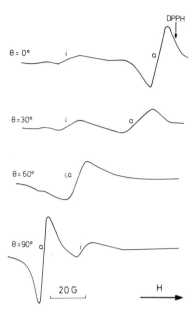

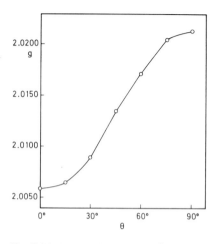

Fig. 7.14. Angular dependence of the room temperature ESR spectra of peroxy radicals in γ-irradiated films of polytetrafluoroethylene. The angles indicated are the field directions measured from the stretch axis. The symbols i and a indicate the isotropic and anisotropic components respectively[1034]

Fig. 7.15. Angular dependence of the room temperature g-value of the asymmetric component of peroxy radicals in γ-irradiated oriented films of polytetrafluoroethylene[1034]

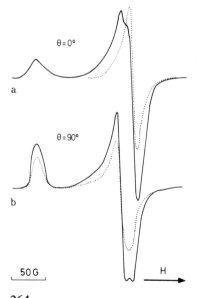

Fig. 7.16. ESR spectra at 77 K of peroxy radicals in γ-irradiated oriented films of polytetrafluoroethylene observed: a. with the field parallel to the stretch axis and b. with the field perpendicular to the stretch axis. Dotted line curves were obtained by subtracting from the solid curves the contribution in the powder spectrum due to randomly oriented radicals[1034]

Table 7.1. Principal values and directions of the g-tensor of peroxy radicals in γ-irradiated poly-tetrafluoroethylene at 77 K and room temperature[1034])

Room temperature		77 K	
Principal values	Principal directions	Principal values	Principal directions
$g_\parallel^r$ 2.006$_1$	$\parallel$ chain axis	$g_1 (\approx g_\parallel^1)^{a)}$ 2.038$_4$	$\perp$ chain axis
$g_\perp^r$ 2.022$_1$	$\perp$ chain axis	$g_2 (\approx g_\parallel^1)^{a)}$ 2.007$_1$	($\parallel$ chain axis)[b)]
		$g_3 (\approx g_\perp^1)$ 2.002$_6$	($\perp$ chain axis)[b)]
g_0^r 2.016$_8$		g_0 2.016$_0$	

[a)] $g_\parallel^1$ and $g_\perp^1$ denote approximate principal values of the g tensor at 77 K.
[b)] the directions of g_2 and g_3 are not conclusive because of the nature of the nearly axially symmetric tensor.

Table 7.2. Observed g values in oriented samples of γ-irradiated polytetra-fluoroethylene at 77 K and room temperature[1034])

Field direction	g	
	Room temperature	77 K
$\parallel$ stretch axis	2.005$_0$	2.003$_1$
$\perp$ stretch axis	2.021$_3$	2.005$_1$–2.038$_3$[a)]

[a)] The spectrum extends between these two limiting values.

mum and minimum g-values during one revolution. Supposing that the very rapid motion around C–O bond occurs at room temperature, the g-tensor should then have axial symmetry about the C–O bond, and consequently the direction of $g_\parallel^r$ should become perpendicular to the chain axis. The conclusion contradicts the results obtained. On the other hand, if rapid motion around the chain axis takes place, the g-tensor should be axially symmetric around the molecular chain axis, and this explanation is in agreement with experimental data. Consequently, g_3 is invariant to rotation around the chain axis and then is equivalent to $g_\parallel^r$. For the direction perpendicular to the chain axis, the average of g_1 and g_3 can be observed as $g_\perp^r$, if the frequency of the rotational motion is sufficiently larger than $g_1 - g_2$. Thus,

$$g_\parallel^r = g_3$$

$$g_\perp^r = \frac{g_1 + g_2}{2}$$

(7.4)

As it is shown above, the principal values and directions of g-tensor and the temperature change of the spectra of peroxy radicals have very valuable information related to the rotational motion around the molecular chain axis.

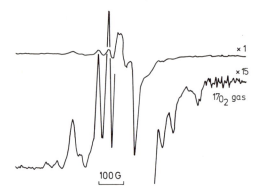

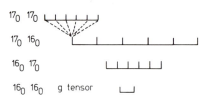

Fig. 7.17. ESR spectrum of peroxy radicals in polytetrafluoroethylene at 77 K. The central line arises from $R\text{-}^{16}OO\cdot$ and the stick diagram shows the lines arising from interaction with a nucleus with $I = 5/2$ and assigned to $R\,^{17}O\,^{16}O\cdot$ and $R\,^{16}O\,^{17}O\cdot$ radicals[447]

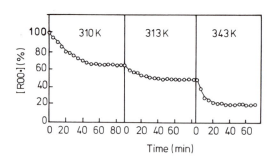

Fig. 7.18. Decay of peroxy radicals in poly(chlorotrifluoroethylene)[893]

Che and Tench[447] investigated ESR spectra of ^{17}O-labeled peroxy radicals in poly(tetrafluoroethylene). Admission of oxygen of 58% ^{17}O at room temperature to γ-irradiated poly(tetrafluoroethylene) leads to a large number of new lines (recorded at 77 K), resulting from the coupling of the unpaired electron with the ^{17}O nucleus with $I = 5/2$ (Fig. 7.17).

The hyperfine spectrum is centered on the $g_\perp$ component and two sets of six equally spaced lines can be resolved with hyperfine splittings of $a_{1\perp} = 46 \pm 1$ G and $a_{2\perp} = 107 \pm 1$ G.

In several papers[1148, 1152, 1160, 1478, 1480, 2006, 2216] the photoconversion of peroxide radicals in poly(tetrafluoroethylene) has been reported. Under UV-light ($\lambda < 3000$ Å) the peroxy radicals (7.3) are dissociated to fluoroalkyl radicals (5.63) and CF_2O:

$$-CF_2-CF_2-CF_2-CF_2-OO\cdot \xrightarrow{+h\nu} -CF_2-\dot{C}F_2 + 2\,CF_2O \quad (7.3) \qquad (7.5)$$

266

At every stage of the decay of the end peroxide radical (7.3) the free valency migrates along the polymer chain to a distance equal to two lengths of the C–C bond (ca. 3 Å)[1478].

A stepwise decay curve for peroxy radicals in poly(chlorotrifluoroethylene) was observed[893] (Fig. 7.18).

By comparing the results with molecular mobility measurements performed by dielectric and mechanical relaxation methods it was concluded that the stepwise character of the radical decay cannot be interpreted by a bimolecular radical-radical recombination process. In the case of poly(chlorotrifluoroethylene) a reaction of trapped radicals with thermally mobilized electrons has been suggested[893]. This means that a polymer may behave as an electret.

7.5. ESR Studies of Antioxidants

A free radical chain mechanism for the oxidation of polymers is generally accepted as follows:

$$\text{Initiator} \longrightarrow R\cdot \tag{7.6}$$

$$R\cdot + PH \longrightarrow P\cdot + RH \tag{7.7}$$

$$P\cdot + O_2 \longrightarrow POO\cdot \tag{7.8}$$

$$POO\cdot + PH \longrightarrow POOH + P\cdot \tag{7.9}$$

$$POOH \longrightarrow PO\cdot + \cdot OH \tag{7.10}$$

where PH is polymer.

The main role of antioxidants is to transform highly reactive polymer alkyl ($P\cdot$), alkoxy ($PO\cdot$), and peroxy ($POO\cdot$) radicals into less reactive aromatic radicals ($Ar\cdot$) or phenoxyl radicals ($ArO\cdot$).

Phenolic antioxidants terminate the chain reaction by a hydrogen abstraction mechanism:

$$P\cdot + ArOH \longrightarrow PH + ArO\cdot \tag{7.11}$$

$$POO\cdot + ArOH \longrightarrow POOH + ArO\cdot \tag{7.12}$$

The phenoxyl radicals formed can either react with another alkylperoxy radical or dimerize. The reason why certain phenols are better antioxidants than others was examined by many investigators. In addition to the ease of oxidation of a certain phenol, the stability of its phenoxyl radical is also important for the antioxidant activity. Reasonably long-lived phenoxyl radicals are required for reactions to occur between $ArO\cdot$ and alkyl or peroxy radicals. Radical reactivity and lifetime are dependent on the degree to which the unpaired electron can be delocalized. In general, the aromatic free radicals are more stable than aliphatic radicals because the unpaired electron can interact with the π-electrons in a phenyl ring. The second important factor influencing radical stability is the steric hindrance. In a series of phenoxyl radicals, the lifetime increases as the steric hindrance in the ortho and para positions increases[330]:

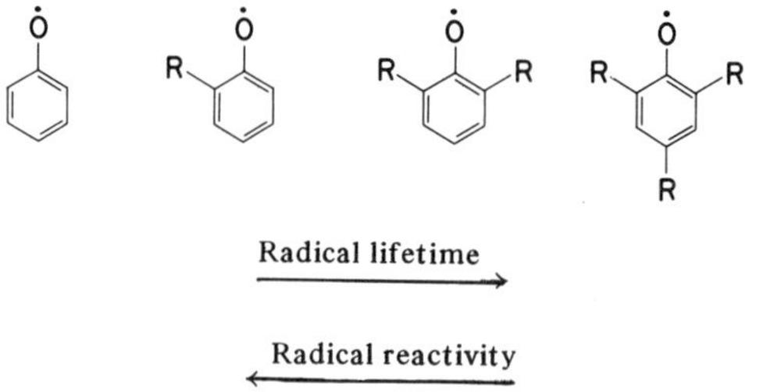

(7.13)

Radical lifetime

Radical reactivity

ESR spectroscopy has widely been applied to determine molecular structure of phenoxyl radicals and also to measure relative concentrations of free radicals by phenolic compounds[133, 330, 982–984, 1765, 2203, 2204, 2407, 2408].

Tkač and Omelka[2203, 2204] investigated the reactivity of antioxidants, e.g. phenols, biphenols, thiophenols, naphthols, and benzophenones, towards peroxy and alkoxy radicals in the presence of transition metal ions. Traces of cobalt induce the decomposition of hydroperoxides generated in the thermal oxidation of a polymer. The principle of this method is shown in reactions:

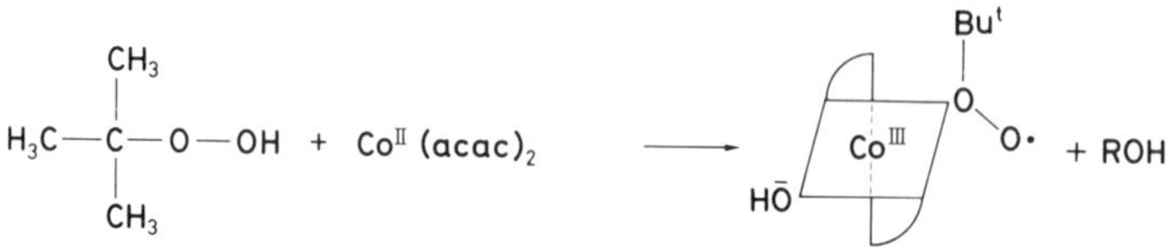

(7.14)

(7.5)

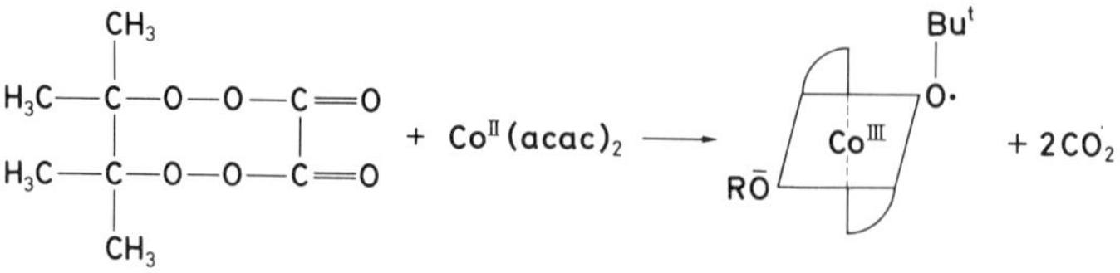

(7.15)

(7.6)

In the reaction (7.14) cobalt(II) acetyloacetonate ($Co^{II}(acac)_2$) reacts with tert-butyl-hydroperoxide (Bu^t–OOH). The $Bu^t OO\cdot$ radicals (g = 2.0147) are fixed in the ligand field of cobalt (7.5) and have relatively long lifetimes at temperatures 278–303 K. Their ESR spectrum is shown in Figure 7.19a.

268

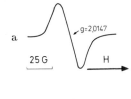

a

25 G

g=2.0147

H

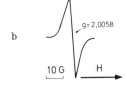

b

g= 2.0058

10 G   H

Fig. 7.19. ESR spectra of
free radicals: a. Bu^{t}OO·
(7.5) and b. (Bu^{t}O· 7.6)
fixed in ligand field of
cobalt[2204]

In the reaction (7.15) cobalt(II) acetyloacetonate reacts with di-tert-butyl peroxalate. The Bu^{t}O· radicals (g = 2.0046) are also fixed in the ligand field of cobalt
(7.6) and are more stable than (7.5). Their ESR spectrum is shown in Figure 7.19b.
Both types of free radicals (7.5) and (7.6) react rapidly with antioxidants and new
ESR spectra characteristic for the antioxidant radicals are formed (Fig. 7.20).

$$
\begin{array}{ccc}
\text{(7.5)} & + \text{ArH} \longrightarrow & \text{(7.7)} \quad + \text{Ar·}
\end{array}
\qquad (7.16)
$$

$$
\begin{array}{ccc}
\text{(7.6)} & + \text{ArH} \longrightarrow & \text{(7.8)} \quad + \text{Bu}^t \text{OH}
\end{array}
\qquad (7.17)
$$

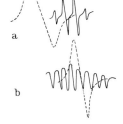

a

b

Fig. 7.20. ESR spectra of
free antioxidant radicals
formed in reactions:
a. (7.7) and b. (7.8)[2204]

269

Interaction of the free electron with the magnetic field of the cobalt nucleus $I = \dfrac{7}{2}$ gives an octet spectrum (Figure 7.20b) with equal intensity of lines and a hyperfine splitting constant of 10.11 G. In both spectra of Figure 7.20 the protons of the aromatic ring cause further hyperfine splitting of the spectra ($a_{H_{para}}$ = 5.1 G, $a_{H_{ortho}}$ = 3.4 G, and $a_{H_{metha}}$ = 1.7–1.8 G)[2204]. In Tables 7.3 and 7.4 are collected ESR spectra of complex-bonded phenoxyl radicals in the ligand field of cobalt generated in reactions (7.16) and (7.17).

The radical concentration and the ratio of alkoxy and peroxy radicals in a chain oxidation process in the presence of β-phenyl or α-naphthylamine (Ar_1–NH–Ar_2) have also been determined from the different ESR spectra of fixed (Co^{III})-(Ar_1–$\dot{N}$–Ar_2) and free stable nitroxy (Ar_1–N–Ar_2) radicals[2203].

$$Ar_1 - \underset{\underset{\dot{O}}{|}}{N} - Ar_2$$

According to ESR spectra of the generated phenoxy radicals, the differently substituted phenols may be divided into five main groups[2204]:

1. Phenols which form stable primary free radicals, e.g. phenols with two tert-butyl groups in ortho-position and para-position substituted with alkyl groups such as $-CH_3$, $-CH_2CH_3$, or $-C(CH_3)_3$ (Table 7.3).

2. Phenols which form stable secondary free radicals but not in para-position (Table 7.3 A). Primarily formed unstable radicals change rapidly into secondary more stable radicals, as registered by ESR spectroscopy. When the two methyls are in ortho-position, the oxidative coupling leads to the formation of polyphenyl ether radicals (Table 7.3 B). In the case of 2-tert-butyl-4-methylphenol in contrast to other phenols nonsubstituted in ortho-position, the delocalization of the free electrons in this position leads to a rapid dimerization (Table 7.3 C).

3. Phenols which give stable phenoxy radicals after coordination in the ligand field of the transition metal. Phenols, biphenols, naphthols, and dihydroxybenzo-phenons belong to this group (Table 7.4).

4. Phenols whose radicals undergo disproportionation, e.g. dihydroxythiophenols.

5. Phenols substituted with NO_2 groups, e.g. nitrophenols which do not form radicals in nonpolar solvents.

Westfahl et al.[2408] measured relative radical concentrations, i.e. radical stability of phenols as a function of temperature using ESR, and found that the radical concentrations are a good measure of the activity as antioxidants of twelve different compounds studied. Westfahl[2407] also applied another reaction for studying ESR spectra of phenols using ferricyanide ions as oxidative agent:

$$HO-\langle\bigcirc\rangle-CH_2OH \ + \ Fe(CN)_6^{-3} \ + \ OH^- \ \longrightarrow \ \cdot O-\langle\bigcirc\rangle-CH_2OH \ + \ Fe(CN)_6^{-4} \ + \ H_2O \qquad (7.18)$$

$$(7.9)$$

Considering the ESR spectra Westfahl proposed a division of the different phenols studied into two classes: one which gives stable free radicals (7.9) and another which forms relatively unstable free radicals.

The ESR spectroscopy was also used for the study of the inhibition effect of some antioxidants of the polymerization of methyl methacrylate which proceeds

Table 7.3. ESR spectra of stable free phenoxy radicals generated with Bu^{t}OO· and Bu^{t}O· radicals from ortho-substituted phenols[2204]

Table 7.4. ESR spectra of complex-bonded phenoxy radicals of Co(III) (acac)$_2$OH (7.5) and (7.6) generated with Bu^{t}OO· and Bu^{t}O· from phenols with a free ortho position[2204]

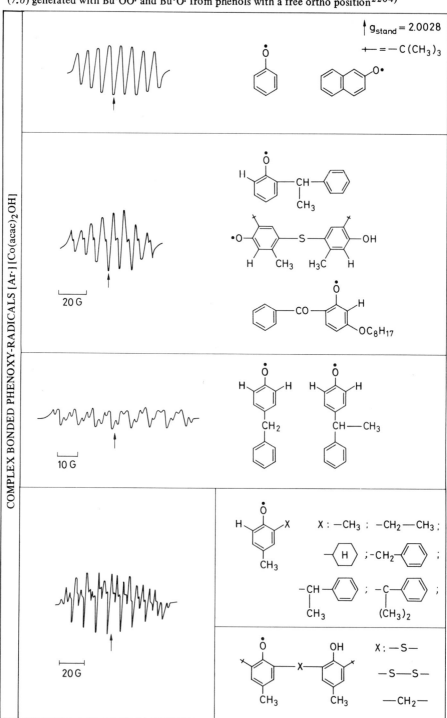

COMPLEX BONDED PHENOXY-RADICALS [Ar·][Co(acac)$_2$OH]

on the surface of synthetic zeolites[1865]. From the equilibrium levels of trapped radicals the inhibitor efficiency was deduced to be the following:
hydroquinone > diphenylamine > 2,6-di-tert-butyl-4-methylphenol > pyrogallol > N-phenyl-2-naphthylamine > 1-naphthol.

In the presence of oxygen, ESR spectra show the formation of peroxyradicals.

Russian scientists[1948] applied ESR spectroscopy for the study of antioxidant and light-protective properties of piperidinoxy stable radicals and sterically hindered piperidines. An ESR study of antioxidants in synthetic rubbers was also published[1802].

Chapter 8

ESR Studies of Molecular Fracture in Polymers

8.1. Introduction

Molecular fracture in polymers is a very complicated phenomenon. The basic mechanism of fracture may involve chain scission, slippage, unfolding, chain pulling, phase segregation, dewetting, and failure of fillers and reinforcements. The theory, mechanism, and kinetics of molecular fracture are discussed in many publications[43, 44, 146, 158, 220, 336, 365, 367, 377, 488, 538, 542, 549, 1093, 1094, 1097, 1098, 1732, 1733, 1734, 1736, 1737, 1852, 1862, 2012, 2025, 2168, 2359–2362, 2416, 2472, 2496].

The formation of free radicals, sometimes called "mechanoradicals" in polymers under mechanical action, was first discovered in 1959 by Bresler et al.[306] using ESR spectroscopy. It was further shown that the ESR method may be sucessfully applied for a direct study of bond rupture in highly oriented fibers and films of the following polymers:

polyethylene[306, 510, 536, 540, 1099, 1100, 1557, 1800, 1878, 2474–2476, 2479, 2504],

polypropylene[1100, 1378, 1800, 1881, 1877, 1878, 2504],

polyisobutylene[306],

polytetrafluoroethylene[306, 1099, 1100, 1878, 1880, 2481],

polydienes (see references in Chapter 8.4),

poly(methyl methacrylate)[306, 363, 540, 1100, 1826, 1876, 1878],

polystyrene[306, 1468, 1596, 1826, 2187],

poly(vinyl alcohol)[360, 1799],

poly(vinyl chloride)[1378],

polyoxymethylene[1100, 1799],

poly(2,6-dimethyl-p-phenylene oxide)[1559],

polyamides[178, 306, 388, 452, 488, 536, 538, 540, 542, 848, 1058, 1095, 1378, 1385, 1437, 1774, 1905, 2107, 2190, 2193, 2212, 2359, 2360, 2472, 2474, 2477, 2499, 2504],

polyesters[452, 540, 1378, 1556, 1558, 2013, 2480],

polysulphides[2506],

polysiloxanes[29],

wool and silk[2425]

and lignin[797],

polynucleotides (DNA)[6, 7].

8.2. Generation of Submicrocracks

The state of the amorphous phase, the structure of the crystalline phase, defects in the crystals, and external impurities have influence on the formation and behavior of radicals in solid polymers.

274

The extent of bond breakage along the surface of fracture is smaller in cast or moulded materials than in fibers, because a fracture involves a higher degree of unraveling of chains than of chain scission. Chain rupture is more localized in the crack surface of the cast or moulded materials than in drawn fibers, where chain breaks are distributed more uniformly within the stressed material[540]. When significant amounts of bond ruptures occur along certain fracture surfaces, microcracks are formed.

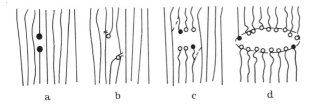

Fig. 8.1. Zhurkov model of submicrocrack generation: a. Primary thermal fluctuation break of a stressed macromolecule with formation of two end radicals (•-denotes end radical), b. interaction of end radicals with adjacent macromolecules resulting in formation of internal free radicals(-x-) and stable endgroups(○-), c. Scission of internal free radicals yielding stable atomic groups and end radicals, d. formation of submicrocracks as a result of submicrocracks a result of chain degradation of macromolecules[2504]

The strength and fracture properties of semicrystalline polymers are dependent on the character of the regions between the crystallites or well-ordered micelles, i.e. the number, orientation, and length distribution of the "tie chains" connecting crystalline regions[122, 540]. Peterlin et al.[1738, 1739] have shown that these regions change significantly during drawing and processing. The most strained tie molecules, bridging through the amorphous layers which are sandwiched between the crystal cores of consecutive lamellae, are broken[1730].

There are three models of generation of submicrocracks:

1. Zhurkov model[2494–2509]
2. Peterlin model[388, 488, 1730, 1731, 1737, 2359, 2360]
3. Sohma model[1100, 2057]

1. In solid polymers in glassy state, where mobility of the macromolecules is limited, scission of chemical bonds is localized near the site of the formation of primary radicals. For the same reason chain end free radicals may only abstract hydrogen atoms from adjacent molecules. Following the chain reaction may neighboring macromolecules be degraded in fast reactions, and local sites of disintegration are formed in a stressed polymer. Such local degraded sites may be considered as submicrocracks. Figure 8.1 shows the Zhurkov model of submicrocrack formation in an oriented polymer.

The formation of submicrocracks was detected by small angle X-ray scattering technique[2497, 2498, 2504]. This method allows the measuring of their dimension and concentration. The observed crack dimensions correspond to the measured microfibril diameters in deformed fibres of polyethylene, polypropylene, and polyamides[1243]. The Zhurkov mechanism presented above does not consider cage recom-

275

bination and cage escape of chain end free radicals. This problem was discussed in detail by Butyagin[368] and Davis et al.[509]. The Zhurkov model has recently been developed by Kausch et al.[1093, 1095-1098].

2. The Peterlin model is based on the observation that the number of radicals formed is dependent on the strain, but not on the stress. The radical concentration remains unchanged when the applied stress is removed. New radicals are only formed when additional stretching in the second loading cycle reaches the maximum strain of the previous loading. Considering these results, Peterlin[1731, 1737] proposed the model presented in Figure 8.2.

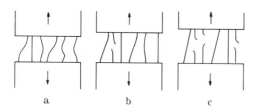

Fig. 8.2. Peterlin model of submicrocracks generation. In this figure the squares represent crystallites and the strings are the tie molecules connecting with crystallites[1737]

From experimental results, Peterlin suggested that the applied stress is mostly concentrated in the tie molecules, which stretch to their maximum length at a given strain. Increased tensile force leads to chain rupture producing free radicals. This model of fracture assumes that the crystallites are sufficiently strong to remain unaffected in large deformation of the sample, and that the polymer chains are not pulled out from the crystallites by the applied stress. This model was supported by Sohma et al.[1100, 2057], who suggested that free radicals may only be formed by scission of tie molecules in the case of crystalline polymers like polyethylene, polypropylene, and polytetrafluoroethylene. Polyethylene which has been treated with nitric acid removing tie molecules does not give ESR signals under stressing the sample, although nontreated samples give easily detected ESR spectra. The above-described Peterlin model cannot be applied to the fracture of amorphous polymers, e.g. poly(methyl methacrylate) and polybutadiene, because these polymers do not contain tie macromolecules.

3. In order to explain the mechanism of the formation of free radicals in amorphous solid polymers, Sohma[1100, 2057] proposed that chain scission in amorphous solid polymers results from large scale shearing motions of the macromolecules.

During tensile stressing of polymers strong ESR signals are detected (Fig. 8.3–Fig. 8.6).

While the tensile strength of polyethylene and nylon is increased roughly ten times by the drawing operation, the radical concentration at fracture increases from $10^{11}-10^{13}$ free radicals/cm^3 to approximately 10^{17} free radicals/cm^3 [540]. The intensity of an ESR spectrum recorded is proportional to the number of broken bonds in a degraded sample. The maximum number of radicals observed is particularly large in nylons[388, 540, 2502] and may reach $1-5 \times 10^{17}$ free radicals/cm^3, and is substantially smaller in polyethylene, i.e. 10^{16} free radicals/cm^3, or even less[2502]. Figure

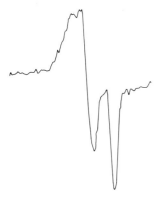

Fig. 8.3. ESR spectrum obtained during tensile loading to fracture of polyethylene fibers[540]

Fig. 8.4. ESR spectrum obtained during tensile loading to fracture of nylon 6 fibers[540]

Fig. 8.5. ESR spectrum (very weak) from poly(methyl methacrylate) after fracture[540]

Fig. 8.6. ESR spectrum obtained during tensile loading to fracture of polyester fibers[540]

8.7a shows the increase in stress and free-radical concentration for a constant load rate test.

For constant stress (the creep test) the free-radical concentration increases linearly with time (Fig. 8.7b).

It was also reported that heat treatment of solid polymers, e.g. polypropylene, produces an anomalous increase in radical concentration formed by mechanical fracture[1877].

When impurities are present in polymer materials, they can react with primary radicals formed by fracture and give more stable free radicals, which may be detected after several days at room temperature[388]. It was reported that chloranil and other quinones were used for trapping radicals produced during the stretching of polyamides[848]. This method provides a quantitatives measurement of the integrated number of free radicals produced during a stressing process.

277

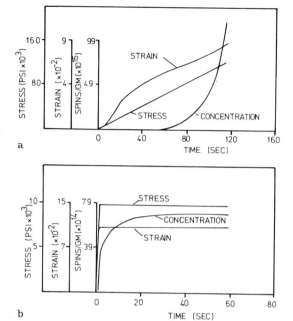

Fig. 8.7. Computer-reduced data from: a. Constant load rate and b. Creep test of nylon 6 fibers[540]

8.3. Generation of Free Radicals in Mechanically Deformed Polyethylene

Mechanochemical destruction of polyethylene under applied stress in vacuum may be described by the following mechanism[2473, 2475, 2479, 2504]:

$$-CH_2-CH_2-CH_2-CH_2- \longrightarrow -CH_2-\dot{C}H_2 + \dot{C}H_2-CH_2- \tag{8.1}$$

$$-CH_2-\dot{C}H_2 + -CH_2-CH_2-CH_2- \longrightarrow -CH_2-CH_3 + -CH_2-\dot{C}H-CH_2- \tag{8.2}$$

$$-CH_2-\dot{C}H-CH_2-CH_2-CH_2- \longrightarrow -CH_2-CH{=}CH_2 + \cdot CH_2-CH_2- \tag{8.3}$$

Under a uniaxial tension the mechanical force deforms bonds in the polymer chain[2500, 2507–2509]. In this process the energy required for breaking a C–C bond decreases (ca. 72 kcal/mol at zero stress for a perfect molecule or slightly less if rupture occurs at preexisting defects, e.g. internal stress[327]) and the bonds are scissioned by thermal fluctuation. More detailed examination of the observed activation energies of deformation are given in later papers[41, 509, 1692]. The homolytic breakage of bonds in the main chain of a polymer leads to formation of end-macroradicals. These radicals are very reactive and abstract hydrogen atoms from neighboring macromolecules; this is shown to occur in the solid state even at temperatures as low as 130 K[1800, 2474]. The abstraction of hydrogen weakens the interatomic bonds of adjacent atoms considerably. The activation energy for C–C bond scission in a polymer chain in β-position to the unpaired electron is by about one third higher than the activation

278

energy needed for the rupture of a common bond[1403]. For that reason, the C–C bonds near the abstracted hydrogen may be considered as weaker bonds in the macromolecule. The free radical formed after the abstracting of hydrogen atom initiates a new reaction of scission. In this way mechanochemical decomposition in a stressed polymer may be considered as a rapid chain reaction[1244, 2497]. One radical may initiate the scission of about 1000 other bonds in polymer molecules[2504]. Consequently a small number of primary radicals formed by mechanical forces and thermal fluctuations results in the destruction of many adjacent polymer chains and in the decrease of mechanical strength.

As the supporting evidence for this deformation model, Zhurkov et al.[2504] report that the concentration of submicroscopic cracks in a polyethylene sample at fracture is about the same (ca. 10^{16} cracks/cm^3) as the concentration of radicals accumulated. This fact is consistent with the chain reaction model where permanent rupture produces no further increase in radical concentration. From infrared absorption studies, the concentration of methyl groups and groups with unsaturated bonds was found to be ca. 10^{19}/cm^3. Hence, each crack should intersect about 1000 broken molecules, corresponding to a crack diameter of roughly 150 Å, i.e. about the value of the microfibril diameter in polyethylene.

Sohma[1978] investigated the formation of free radicals by ESR during the mechanical desintegration of paraffins and low molecular weight polyethylenes by milling. Mechanical fracture of n-paraffin of 32 carbons did not produce radicals. For polyethylene with about 100 monomer units per chain an ESR signal was observed which was too weak to be analysed (Fig. 8.8).

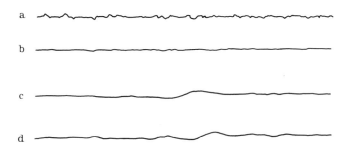

Fig. 8.8. ESR spectra at 77 K from: a. n-paraffin with a chain of 32 carbon atoms, b. Polyethylene with a chain of 71 carbon atoms, c. Polyethylene with a chain of 100 carbon atoms, d. Polyethylene with a chain of 136 carbon atoms. All samples were milled in vacuum at 77 K[1878]

For polyethylene with 136 monomer units an ESR spectrum was recorded. The experimental results show that critical size for mechanical chain break in polyethylene is between 70 and 100 monomer units. These experimental results support Sohma's model for the fracture of amorphous polymers. Large shearing of single polymer molecules, which leads to a macroscopic fracture, requires a simultaneous motion of each constituent monomer in the chain. Such shearing movement of a polymer requires more energy than that of a substance of low molecular weight. For sufficiently long chains Sohma[1100, 2057] assumed the following relation:

$$n E_n \geqslant E_{C-C} \tag{8.4}$$

where n is the number of monomer units in the polymer molecule, E_n is the activation energy for moving one monomer unit ($E_n = 1.01$ kcal mol^{-1}) and E_{C-C} the bond energy ($E_{C-C} = 88$ kcal mol^{-1}). From this relation a critical value n = 83 is obtained, which agrees with the experimental results from measuring ESR spectra. When the number of monomers which move in a simultaneous and incorporated way is larger than 85, the C–C bond rupture consumes less energy than the shearing motion of the long chain. In such case the crack may proceed by breaking C–C bonds and not by shearing motion of polymer chains. A pair of radicals is formed for each main chain scission.

The mechanical destruction of polyethylene in the presence of air is followed by a rapid formation of peroxy radicals[510, 2057, 2505]. It was found that radicals produced in mechanical degradation are much more reactive towards oxygen than radicals formed by ionizing radiation. This difference in reactivity is presumably due to the difference in the trapping sites for the two kinds of radicals[2057]. During fracture, radicals are formed and trapped on freshly formed surfaces and, therefore, are easily accessible to oxygen, whereas free radicals, induced by ionizing radiation, are formed and trapped in the solid, where the presence and concentration of oxygen is limited by diffusion.

8.4. ESR Study of Free Radicals Formed during Fracture in Rubber

ESR spectroscopy was successfully applied for study of free radicals generated during fracture in rubber[45, 46, 306, 307, 320, 321, 371, 536, 537, 546, 1100, 1453, 1578, 1765, 1822, 1839, 1878, 2400] and especially in ozone atmosphere[537, 539, 544, 545, 548]. The results obtained were reviewed in detail by DeVries[537].

Rubber samples prestrained and fractured at low temperatures exhibited strong ESR signals (Fig. 8.9), while samples fractured at the same temperatures but without prestrain showed a low, if any, free radical concentration during deformation[46, 1453, 1578] (Fig. 8.10).

The large number of radicals produced during this deformation suggests that fracture is initiated in prestrained rubbers in many sites throughout the specimen volume. It was also found that the radical concentration varies with the rate of deformation, the crosslink density in the sample used, the degree of deformation, and the purity of the sample. When visible crazing occurs, the absorption of oxygen increases and results in the formation of peroxy radicals[1287, 1289].

The deformation of cross-linked natural rubber shows that three different kinds of deformation are possible below T_g, depending on temperature and strain rate at which the tensile test is performed. At temperatures immediately below T_g, the materials deform by shear yielding or cold drawing. At temperatures below 160 K and above 140 K, brittle behavior is obtained. Further decrease in test temperatures results in the formation of crazings. Figure 8.11 shows these three regions of deformation behavior in a strained rubber sample[46].

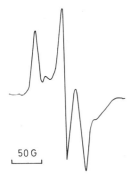

Fig. 8. 9. ESR spectrum resulting from deformation of natural rubber at 198 K and 100% prestrain[537]

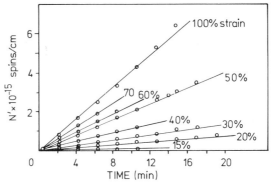

Fig. 8.10. Free radical concentration as a function of time for a acrylonitrile-butadiene rubber (NBR) (Hycar 1043) at different strains[537]

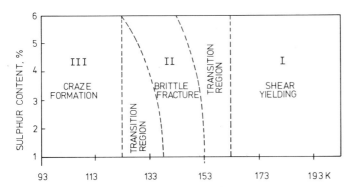

Fig. 8.11. Three regions I, II and III of different mechanical behaviour[46]

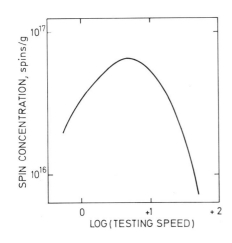

Fig. 8.12. Variation of spin concentration with strain rate for polychloroprene[46]

281

The strain rate also affects the spin concentration. Figure 8.12 shows the variation of spin concentration at a given overall strain for polychloroprene.

The number of radicals formed is related to the number of craze bands formed[46].

The formation of peroxy radicals in stretched rubber samples can be studied by three methods:

1. Direct observation of alkylperoxy radicals from ESR spectra.
2. Measuring of change in concentration of stable aryloxy radicals which are formed in reaction between alkylperoxy radicals and incorporated to the rubber sample stable radicals, e.g. galvinoxyl.
3. Study of aryloxy radicals from phenols, e.g. hydroxygalvinoxyl, incorporated in rubber sample.

Applications of these methods for the observation of radical formation in stretched peroxide or sulphur-cured rubbers are described in papers[46, 1578, 1765, 1822].

8.5. ESR Study of Fatigue Processes in Polymers

ESR spectroscopy was also applied to study of bond rupture under alternating tensile stresses[540]. Figure 8.13 shows the cumulative increase in free radical concentration when an alternating stress of $\pm 35\,000$ psi was superimposed upon a mean stress of 70000 psi.

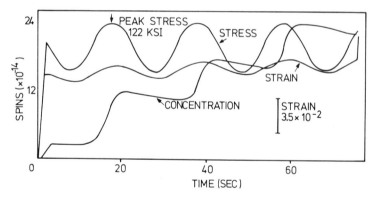

Fig. 8.13. Stress, strain and free radical concentration as function of time during cyclic loading of nylon 6 fibers[540]

It has been found that under these conditions the number of bonds that rupture before the final failure of the test sample is substantially lower than the value for fracture under a monotonically increasing stress.

Nagamura et al.[1557] applied ESR to study structural changes in high-density polyethylene occurring in the fatigue process. They suggested that the mosaic block crystals in the lamellae are first decomposed into smaller ones by the applied cyclic stress. This process probably occurs by heat generation in the specimen under the cyclic stress. During repeated loading, the molecular motion excited by the internally generated heat may tend to initiate more stable packing of deformed molecules.

8.6. ESR Study of Free Radicals Formed During Grinding and Machining of Polymers

Various operations such as machining, grinding(milling), sawing, and cutting may produce large amounts of new fracture surface in comparatively small quantities of polymeric material. Every polymer tested produces strong ESR signals after such treatment[122, 296, 303, 304, 306, 362, 364, 370, 372, 376, 540, 541, 543, 547, 576, 578, 602, 796, 1598, 1747, 1800, 1812, 1875, 1878, 2416, 2486, 2503]. Figures 8.14—8.19 show ESR spectra of different polymers obtained during the grinding.

In this case the ESR method allows an estimate of the number of bonds broken per unit area of new surface formed and the depth of damage at new surfaces as a function of fracture mode and fracture conditions. Results obtained indicate that during the grinding of nylon 6 at 77 K, 10^{13} to 10^{14} broken bonds/cm^2 of new surface are formed[540]. The majority of these broken bonds lie within a few μm below the surface. In semicrystalline polymers, such as nylon and polyethylene, the cracks propagate through amorphous regions surrounding the more ordered regions. Fractures take place along preferential routes, with the result that significantly fewer molecular chains per cm^2 of new surface area are ruptured than expected, on the basis of calculated number of chains passing through a unit cross section of bulk polymer[122, 547].

ESR spectra of polymer samples milled in vacuum differ appreciably from those milled in presence of air. Sohma et al.[2057] have shown that alkyl radicals formed during ball milling of polypropylene in vacuum (Fig. 8.19a), after contact with oxygen, give peroxy radicals (Fig. 8.19b).

Particles ground from elastomers at 77 K give a large number of radicals and differ from those ground at room temperature which give much fewer radicals. This indicates two distinct fracture mechanisms being operative above and below the glass transition temperature[543].

Free radicals have also been detected during milling and mastification of rubber, after trapping by freezing the samples in liquid nitrogen below the glass transition temperature (T_g)[371, 541], or by reaction with stable free radicals ("spin traps") previously introduced to the examined sample[1282, 1822, 1839] (see Chap. 11).

ESR has been used most successfully by Russian scientists to investigate bond rupture during grinding of biopolymers as cellulose[5, 7, 795, 796, 1146, 1414], starch[5], gelatine[4, 7], wool and silk[2425], albumines[3], and proteins[369].

ESR has also been used to study processes during the extraction of human teeth[535] and formation and breakdown of the structure in hard leather produced on an abrasive wear machine[1247, 1248].

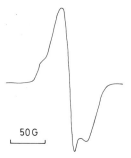

Fig. 8.14. ESR spectrum of
polystyrene ground at
77 K[543)

Fig. 8.15. ESR spectrum of
poly-butadiene ground at
77 K[543)

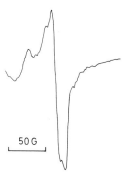

Fig. 8.16. ESR spectrum of
styrene-butadiene block
copolymer ground at
77 K[543)

Fig. 8.17. ESR spectrum of nylon
6 ground at 77 K[540)

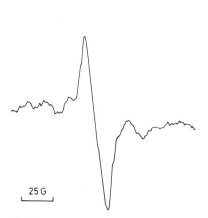

Fig. 8.18. ESR spectrum of polyurethane
ground at room temperature[543)

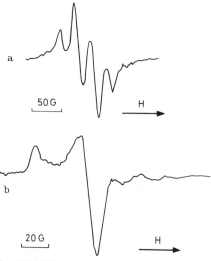

Fig. 8.19. ESR spectrum of polypropylene
milled: a. in vacuum and b. after contact with
oxygen[2057)

8.7. Anomalous Behavior of Free Radicals Obtained by Sawing Technique

It has been observed that decay curves of free radicals produced by sawing in liquid nitrogen show anomalous behavior[1100, 1879, 2051, 2057] (Fig. 8.20).

On raising the temperature to 150 K a rapid increase of radical concentration has been observed and interpreted as being caused by excess electric charge due to triboelectricity and oxygen dissolved in the liquid nitrogen used. It has been found that sawdust particles have a strong electrical charge. These experimental results suggest that the excess electric charges produced by friction may play some role in the production of new radicals in this temperature range. In the presence of a strong electron scarenger, e.g. tetra-cyano-ethylene, these anomalies in the decay curve of free radicals disappear (Fig. 8.21).

It has been shown by Theodorescu[2175, 2176] that the electric field effects the ESR spectra.

Oxygen plays an important role in the observed phenomena. No anomalous increase appears when mechanical degradation is made in vacuum (Fig. 8.22).

For explanation of these anomalies the following mechanism has been proposed by Sohma et al.[1879, 2057]:

1. During mechanical degradation of a polymer sample at 77 K in the presence of oxygen, free radicals (R·) and peroxy radicals (ROO·) are formed. At the same

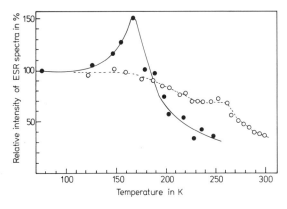

Fig. 8.20. Decay curves of the polypropylene radicals: (●) produced by sawing, (○) produced by γ-irradiation[1789]

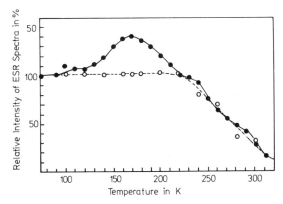

Fig. 8.21. Decay curves of the polypropylene radicals formed by ball-milling: (○) in the presence of tetra-cyano-ethylene, (●) without free radical scavenger[1789]

285

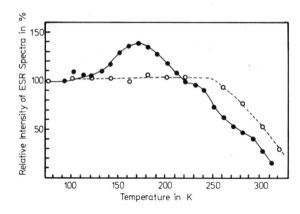

Fig. 8.22. Decay curves of polypropylene radicals formed by ball-
milling: (o) under high vacuum
10^{-5} mm Hg and (•) in the presence of air[1789]

time a great number of free radicals (R·) attract electrons from triboelectric charges
to form anions which have paired electrons and do not give ESR spectra:

$$R\cdot + e^- \longrightarrow R^- \tag{8.5}$$

2. During the warming in the temperature range of 100–173 K, formation occurs
of peroxy radicals ROO· from the anions and molecular oxygen:

$$R^- + O_2 \longrightarrow ROO\cdot + e^- \tag{8.6}$$

3. During further warming at temperatures above 173 K thermal decay of ROO·
radicals is observed. All excess electrons leak to earth and the sample becomes electri-
cally neutral. These results give an interpretation of the apparent role of excess charges
produced by friction in the reaction of radicals.

8.8. Mechanical Degradation of Polymers in Frozen Solution Matrix

Mechanically formed free radicals in frozen solvent matrix at low temperature may
abstract hydrogen by the following reaction:

$$R\cdot + HR_1 \longrightarrow RH + R_1\cdot \tag{8.7}$$

where R· is a macroradical and HR$_1$ hydrocarbon solvent. ESR studies of different
polymers such as polystyrene, poly(vinyl acetate), poly(vinylcyclohexane), and
poly(methyl methacrylate) degraded mechanically in different hydrogen-donor
solvents, e.g. toluene, ethylbenzene, isopropylbenzene, were reported[361, 372, 580].
Solvent radicals formed by hydrogen abstraction were found.

Chapter 9

ESR Studies of Graft Copolymerization

Under ionizing irradiation of solid polymers free radicals are formed which are trapped in a polymer matrix (see Chap. 5). Under certain conditions, these radicals can initiate graft polymerization. The grafting reaction is dependent on the physical state of the polymer and the properties of the free radicals formed in the polymer[166, 356, 409, 410, 1512, 2069]. For that reason ESR spectroscopy is a very important tool for the study of grafting mechanism. A sample of electron- or γ-irradiated polymer is brought into contact with monomer vapour, which can diffuse into the polymer phase and thus reach the trapped radical sites. This reaction can be examined very well by ESR spectroscopy and has been applied for the study of grafting of acrylic acid and methacrylic acid[386], methyl methacrylate[480, 1153, 1155, 1934, 1937], styrene[495, 496, 604, 1624, 1625, 1938, 2158, 2159, 2452], butadiene[1937], monochlorotrifluoroethylene[2130], tetrafluoroethylene[2129] to polyolefines, styrene to polytetrafluoroethylene[561, 2157] and poly(chloro-trifluoroethylene)[895], tetrafluoroethylene to polytetrafluoroethylene[121], and butadiene to poly(vinyl chloride)[857, 2099].

9.1. Graft Copolymerization to Polyethylene

When polyethylene is irradiated in air, three types of radicals are observed: alkyl, allyl, and peroxy (see Chap. 5 and Chap. 7). Among these, peroxy radicals are unreactive and do not participate in grafting reactions. The initially observed alkyl radicals disappear rapidly as the grafting reactions proceed. Most of the allyl radicals are trapped on surfaces of the crystallites, and only 10% of the radicals are trapped inside crystallites[1938]. The reactivity of allyl radicals depends on their contact with reactive monomers. The rate of reaction is related to the concentration of vapour pressure of the monomers (Figure 9.1).

Investigation of the grafting of methyl methacrylate[1935, 1938] and styrene[2138, 2159, 2452] shows that the grafting reactions are preferentially initiated by allyl radicals formed by the conversion of alkyl radicals. Seguchi and Tamura[1934, 1936, 1937] reported that alkyl radicals also initiate the grafting reactions, and the course of grafting reaction is different for allyl and alkyl radicals. In the atmosphere of a grafting monomer, alkyl radicals trapped in the crystalline region migrate to reach the surface of the crystallite and then react with the surrounding monomer to produce propagating radicals.

The effect of various solvents on the decay of alkyl radicals (Figure 9.2) and allyl radicals (Fig. 9.3) were investigated in detail[1938, 2452].

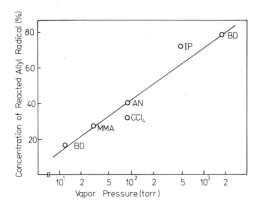

Fig. 9.1. Concentration of allyl radicals decaying after 3 min reactions as a function of the vapor pressure of reacted materials at 293 K: AN − acrylonitrile, BD − 1,3-butadiene, IP − isoprene, MMA − methyl methacrylate[1938)

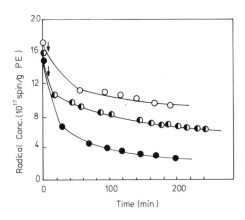

Fig. 9.2. Effects of solvents on the decay of alkyl radicals at 293 K in γ-irradiated polyethylene: (o)-benzene, (◐)-toluene, (●)-styrene. Arrows show the addition of solvents[2452)

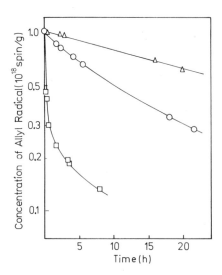

Fig. 9.3. Effects of solvents on the decay of allyl radicals: (△) m-xylene, (o) tetrachloroethane and (□) tetrachloromethane[1938)

The acceleration of the decay of radicals by solvents may be ascribed to:

1. The swelling of polyethylene, which facilitates the molecular motion of polymer chains and enhances the probability of coupling, and
2. Chain transfer in solvents.

Styrene is the most effective in reducing alkyl radicals, probably because it reacts rapidly with polyethylene radicals and also because grafted polymer chains loosen the crystal structure mechanically, and enhance the mobility of polyethylene radicals trapped in the crystals, accelerating thus the coupling reactions.

9.2. Cellulose Graft Copolymers

Graft and block copolymers of cotton may be prepared by free radical-initiated grafting of vinyl monomers to cellulose. In this way, the macromolecular, morphological, and textile properties of cotton may be modified. ESR studies of free radicals in graft polymerization of methacrylic acid[386], ethyl acrylate[1562], and styrene[391, 552, 839, 1220, 1225, 1804] on cotton cellulose were made.

The ESR study of the grafting of vinyl monomers onto γ-irradiated cellulose shows the formation of living polymer radicals[72, 928]. A typical ESR spectrum generated by trapped cellulosic radicals in γ-irradiated cotton cellulose is shown in Figure 9.4a.

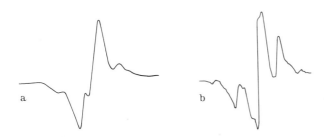

Fig. 9.4. ESR spectra of free radicals formed in: a. γ-irradiated cellulose, b. cellulose-poly(methacrylic acid) copolymer[72]

After treatment with aqueous solution of methacrylic acid (30 vol. %) for 3 min at 298 K, the ESR spectrum is changed (Fig. 9.4b). This spectrum is composed of a spectral component generated by propagating polymer radicals, and a component generated by trapped cellulose radicals located within the highly ordered regions of the macromolecules. The living polymer radical (9.1) can be generated as follows:

$$\text{cell} \cdot + n \begin{array}{c} H \ CH_3 \\ | \ \ | \\ C{=}C{-}COOH \\ | \\ H \end{array} \longrightarrow \text{cell} \left[\begin{array}{c} H \ CH_3 \\ | \ \ | \\ C{-}C \\ | \ \ | \\ H \ COOH \end{array} \right]_{n-1} \begin{array}{c} H \ CH_3 \\ | \ \ | \\ C{-}C{-}COOH \\ | \\ H \end{array} \quad (9.1)$$

An interesting result was reported in the ESR study of grafting of acrylamide on cellulose[72]. Direct photolysis of the solution of this monomer did not yield ESR-

289

detectable radicals. However, when solutions of monomers were photolysed in the presence of cotton cellulose, ESR spectra of propagating monomer radicals were recorded (Fig. 9.5).

ESR study of grafting different monomers on cellulosic materials using redox systems was described[76, 124]. Hydroxy radicals generated by Fe^{2+}–H_2O_2 in a flow system[76] (See Chap. 5.4) react with cellulose, giving free radicals which initiate graft polymerization of acrylonitrile. The triplet spectrum (Fig. 5.86) disappears during grafting and is replaced by two strong singlet spectra (Fig. 9.6).

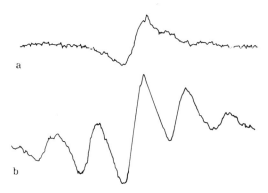

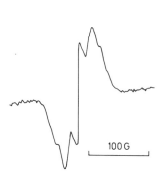

Fig. 9.5. ESR spectra of free radicals formed in: a. UV-irradiated cellulose at 333 K, b. in the presence of methacrylamide[72]

Fig. 9.6. ESR spectra of free radicals generated in the Fe^{2+}-H_2O_2 system on cellulose with acrylonitrile added, recorded at 163 K[76]

The broader of the two singlet spectra was probably due to a radical generated on carbon C_1 or C_4 (5.112) by degradation of the cellulose molecule, and the other probably to one generated at the end of the growing polyacrylonitrile molecular chain. If the interpretation is correct, the absence of a triplet spectrum provides a direct proof of which way the acrylonitrile monomer is being grafted onto the cellulose molecule.

ESR spectroscopy has also been applied for study of photo-grafting of methyl methacrylate[1237, 1635, 2009], acrylamide, methacrylamide and diacetone and acrylamide[1240, 1828] and of ceric ion initiated grafting of acrylonitrile[74] onto cellulose.

9.3. Miscellaneous Problems

ESR spectroscopy was also used for the study of: photochemical grafting of methylacrylamide on the surface of anhydroglucose triacetate film[1240], ferric ion (Fe^{3+}) sensitization of photo-initiated grafting of methyl methacrylate to poly(vinyl alcohol)[1636].

It has been found that commercial thermostabilizers such as dibutyl-tin-methy-lenebis(thioglycolate), dibutyl-tin-octylidenebis(thioglycolate), and dibutyl-tin-dilaurate lower the extent of radiation-induced grafting of acrylonitrile and styrene on poly(vinyl chloride). ESR studies show that the tin compounds retard the decay of the free radicals produced by radiation in poly(vinyl chloride)[1303].

Chapter 10

ESR Studies of Crosslinking

10.1. ESR Study of Enhanced Crosslinking of Polymers

Enhanced crosslinking occurs via a chain reaction involving both a polymer and a monomer. An example of this reaction is the curing of an unsaturated polyester-styrene mixture. The initiation of a chain reaction via a single radical allows the formation of a sequence of crosslinks. An ESR study of enhanced crosslinking of polyolefins with acrylic acid[438] and allyl ethers of Novolac with vinyl monomers in the presence of cationic catalysts was made[1531]. Using ESR Charlesby[1188] studied the role of S, Se and Te during radiation-induced crosslinking of polyethylene.

10.2. ESR Study of Free Radicals Observed During the Curing of Unsaturated Polyester Resins

The curing process of unsaturated polyester resins (free-radical copolymerization reactions between polyester fumarate and styrene) in a $TiCl_3-H_2O_2$ flow system was described by Lehmus[1333]. The copolymerization proceeds with extensive crosslinking and results in three-dimensional networks. Demmler and Schlag[532] have reported the importance of free radicals observed by ESR in the curing and stabilization of polyester resins.

10.3. ESR Study of the Vulcanization of Rubber

The reaction mechanism is an important problem in rubber vulcanization. Extensive ESR studies were made on the interaction of tetramethylthiuram disulphide with rubber at 373–433 K in the vulcanization process[232, 289, 1416]. The dissociation of tetramethylthiuran disulphide into radicals results in the formation of polymeric radicals. Their recombination produces three-dimensional structure (vulcanizates).

Other ESR studies were also presented, namely: γ-induced vulcanization of rubber in the presence of sulphur[1284, 1293, 2164, 2165], retardation of prevulcanization of rubber by N-nitrosodiphenylamine[1398], decomposition of vulcanization accelators[2514], vulcanization mechanism of liquid thiocols by sodium bichromate[1500], and reaction of hydroperoxide groups in rubber with phenothiazine[190].

292

10.4. ESR Spectra of Light-Irradiated Poly(vinyl cinnamate)

The ESR spectrum of light-irradiated poly(vinyl cinnamate) at 77 K consists of a broad singlet and a quartet (Fig. 10.1)[1561].

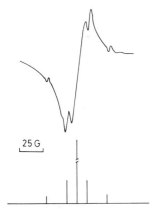

25 G

Fig. 10.1. ESR spectrum of poly(vinyl cinnamate)
1 hour light-irradiated at 77 K[1561]

The quartet spectrum decayed rapidly when resin was kept in the dark, while the singlet remained almost constant and decayed only at higher temperature. The quartet was attributed to a main chain radical and the singlet to a crosslinked cinnamoyl radical (*10.1*) produced by the light irradiation:

$$(10.1)$$

When crosslinking occurs, the polymer becomes insoluble in any solvent.

293

Chapter 11

Application of Stable Free Radicals in Polymer Research

Stable free radicals have been widely applied in the study of polymers. Three methods are commonly used:

1. Spin-probe technique,
2. Spin-trapping technique,
3. Spin-labeling technique.

These methods are mainly based on nitroxide radicals, which are highly stable in solid, and stable for many months in solution.

11.1. Nitroxide radicals

Nitroxide radicals as a class of stable organic free radicals were discovered in 1956[1059, 1846]. From that time on the synthesis and properties of these radicals received increasing attention[329, 351, 423, 524, 605, 650, 690, 818, 1045, 1114, 1140, 1259, 1296, 1336, 1337, 1402, 1547, 1548, 1587, 1586, 1675, 1689, 1856–1859, 1861, 2036, 2076, 2173, 2310, 2352].

The most stable nitroxide radicals contain tertiary carbons bonded to the nitrogen, e.g. di-tert-butylnitroxide (11.1)[636, 953], 2,2-dimethyl-oxazoline derivatives (11.2)[1105], pyrroline (11.3), and piperidine (11.4) nitroxides[309, 1857]:

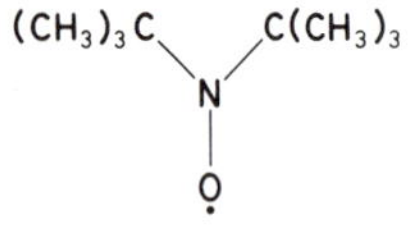

(11.1) 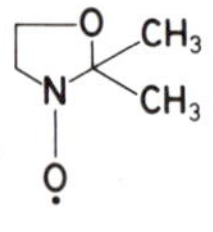 (11.2)

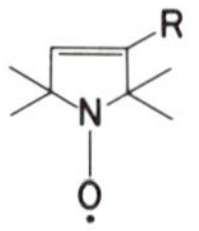

(11.3) 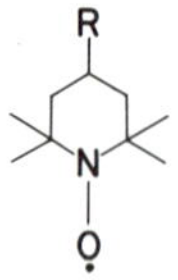 (11.4)

At sufficiently high temperatures the nitroxide radicals show triplet ESR spectra with equal components of equal intensities, which were interpreted as being due to the coupling to the nuclear spins of the nitrogen atom[479]. Hyperfine splitting constants (α_N) of some nitroxide radicals are given in Table 11.1.

When the nitroxide radicals are in solution of low viscosity, e.g. common organic solvents, as benzene at room temperature, rapid Brownian rotational motion occurs.

Table 11.1. Hyperfine splitting constants (a_N) of nitroxyl free radicals at room temperature[1259]

Compound	Color	Solvent	a_N (G)
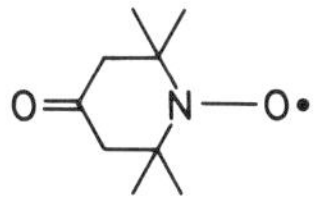	Orange	Benzene	14.5
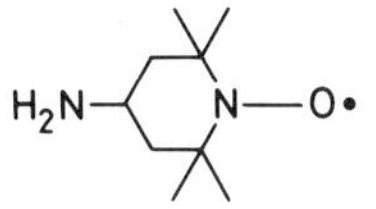	Orange	Methanol	16.4
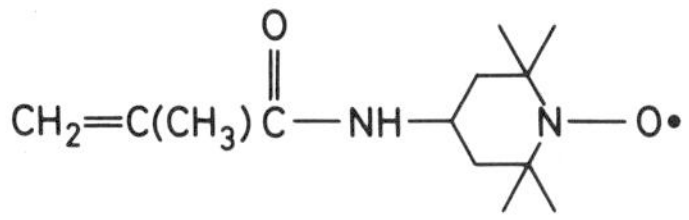	Orange	Methanol	16.4
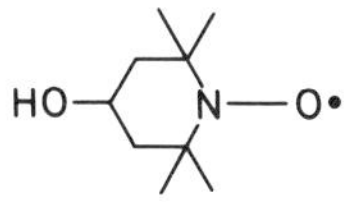	Orange	Benzene	17.0
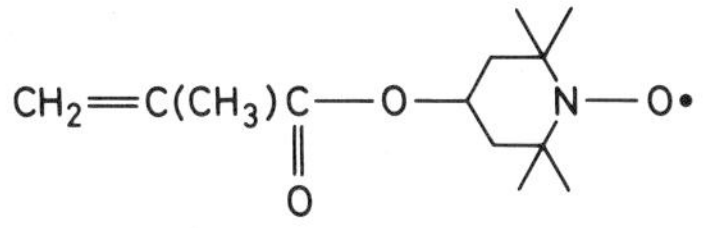	Orange	Benzene	17.0
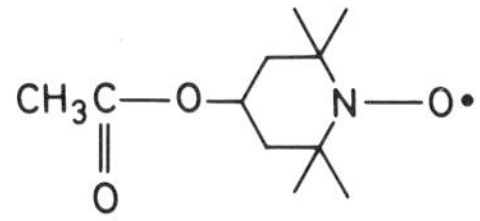	Orange	Methanol	16.3
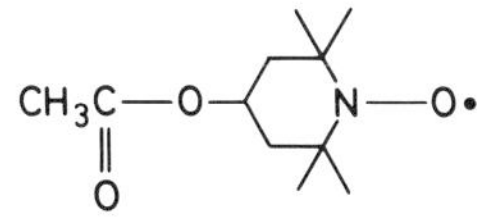	Orange	Benzene	15.6

This results in the averaging of the isotropic g-value and the isotropic hyperfine coupling constant (a). With increasing temperature, the ESR spectrum shows a triplet line with equal components of equal intensities (Fig. 11.1).

The time required for complete rotation of a nitroxide radical about its axis is called "the rotational correlation time" (τ) and is $10^{-11} < \tau < 10^{-6}$ s. The value

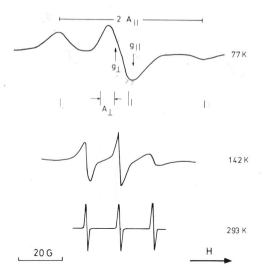

Fig. 11.1. ESR spectrum of di-tert-butyl-nitroxide (11.1)[2332]

(τ) can be defined as the time required for the nitroxide to rotate through an angle of one radian. It is calculated from one of the equations:

$$\tau = \Delta H_{pp} \left[\left(\frac{I_{pp0}}{I_{pp+1}} \right)^{\frac{1}{2}} - \left(\frac{I_{pp0}}{I_{pp-1}} \right)^{\frac{1}{2}} \right] C \tag{11.1}$$

$$\tau = \Delta H_{pp} \left[\left(\frac{I_{pp0}}{I_{pp+1}} \right)^{\frac{1}{2}} + \left(\frac{I_{pp+1}}{I_{pp-1}} \right)^{\frac{1}{2}} \right] - 2\ C' \tag{11.2}$$

where: ΔH_{pp} is peak-to-peak width, and I_{pp-1}, I_{pp0}, I_{pp+1} are the measured peak-to-peak amplitudes of three lines of ESR spectrum of nitroxide radical (Fig. 11.2). The values of C and C' are characteristic parameters for the nitroxide and the particular experimental conditions employed.

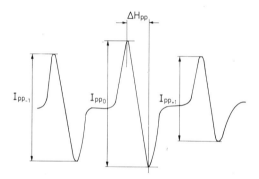

Fig. 11.2. Notation of the symbols used for determination of the rotational correlation time (τ) from ESR spectrum

An analysis of the width of hyperfine lines provides information about the correlation time for rotational diffusion when the solution is sufficiently diluted[2076] or, alternatively, about translational diffusion coefficient for more concentrated solutions ($> 10^{-3}$ M)[1494]. In viscous media the rotational motion is hindered and the tumbling rates of the molecules decrease. This gives an increase of the hyperfine coupling constants and a broadening of observed spectra. There are several methods for the calculation of rotational correlation time (τ) and the tumbling rate from ESR spectra[31, 155, 317, 338, 717–719, 787, 991, 1139, 1140, 1197, 1440, 1627, 1751, 2224, 2374].

11.2. Spin-Probe Technique

The spin-probe technique has been applied to the study of interaction between several low molecular organic nitroxides and host polymers in which they are dissolved. By this method it is possible to study, using ESR spectroscopy, the mobility of the dissolved molecule in its microscopic environment. Several studies on the motions of nitroxide radicals in polymer matrix are known[1133, 1208, 1209, 1212, 1276, 1347, 1792, 1793, 1904, 2082–2086, 2220, 2224, 2233, 2353, 2484].

Some of nitroxide radicals easily diffuse into various polymers[1212, 2083, 2085]. It was found that spin-probe radicals are located in the amorphous phase of solid polymers[2220, 2228, 2233]. Since the glass-transition temperature (T_g) corresponds to the on-set of liquid-like translational motions of long segments of molecules in the amorphous phase, and the rotations of nitroxyl radicals are sensitive to changes of geometry of their surroundings, the rotational relaxations of probe radicals in polymers at temperatures near T_g can be measured. ESR spectra of polyethylene (Fig. 11.3) and poly(vinyl chloride) (Fig. 11.4) above the glass transition temperature (T_g) containing nitroxide radicals (e.g. 2,2,6,6-tetramethyl-4-hydroxy-piperidine-1-oxyl benzoate (11.5)) show well-resolved triplet spectra.

$$C_6H_5COO$$

(11.5)

The triplet-line spectrum shows a high degree of rotational motion of nitroxide radicals in the amorphous part of the polymer. During the freezing of polymer matrix a decrease of the mobility of nitroxide radicals and change of ESR spectra are observed. From these spectra it is possible to calculate the rotational correlation time of free radical (τ) (11.1).

The rotational correlation time (τ) obey the Arrhenius law:

$$\tau = \tau_0 e^{\frac{E_a}{RT}} \quad \text{valid for } T > T_g \tag{11.3}$$

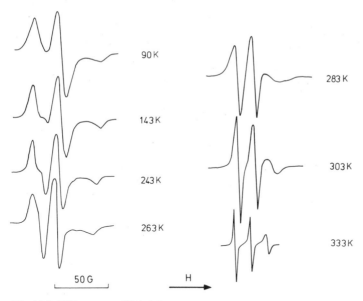

Fig. 11.3. ESR spectra of 2,2,6,6-tetramethyl-4-hydroxy-piperidine-1-oxyl benzoate (*11.5*) in polyethylene at different temperatures[1793]

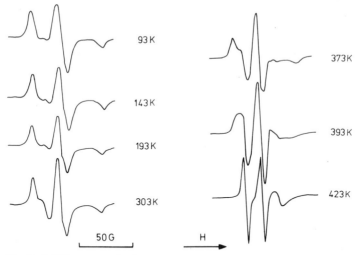

Fig. 11.4. ESR spectra of 2,2,6,6-tetramethyl-4-hydroxy-piperidine-1-oxyl benzoate (*11.5*) in poly(vinyl chloride) at different temperatures[1793]

where E_a is the rotational activation energy of the free radical, T is the absolute temperature (K) and R is the gas constant. For log τ = f(1/T) a straight line is obtained. It has been found that the rotational correlation time (τ) is dependent on the molecular volume (V) and follows the empirical equations:

$$\tau = A\,e^{-kV_g} \tag{11.4}$$

298

$$\tau = B \, e^{-kV_1} \qquad\qquad (11.5)$$

where V_g and V_1 are the molar volumes ($cm^3 \, mol^{-1}$), when $T < T_g$ and $T > T_g$, respectively, $k \approx 1$, and A and B are constants. In Table 11.2 the rotational activation energies of free radicals (E_a) for different polymers are listed.

Table 11.2. The rotational activation energy (E_a) of some spin probe radicals in linear polymers[2224]

Polymer	Radical (11.4) with R	E_a kcal mol^{-1}	References
Polyethylene (0.918 g cm^{-3})	H	10.7	2085)
Polyethylene (0.950 g cm^{-3})	H	10.4	2085)
Polyisobutylene	H	11.8	1857)
Poly(butyl methacrylate)	H	12	1857)
Polystyrene	H	22.6	1857)
Polyisoprene	H	5.35	1855)
Polystyrene		18.2	2353)
Atactic polypropylene		18.7	2353)
Isotactic polypropylene		10.5	2353)
Butadiene rubber		11.5	2353)
Natural rubber		8.5	2353)
Polyethylene (low density)	C_6H_5-COO	11.0	1793)
Polyethylene (high density)	C_6H_5-COO	10.0	1793)
Polypropylene	C_6H_5-COO	12.3	1793)
Polystyrene	C_6H_5-COO	18.3	1793)
	Radical (11.3) with R		
Poly(ethylene glycol)	CH_3COO	9.5	2223)

The rotational activation energy of free radical rotation (E_a) does not depend on the size of the radical, but only on the properties of the polymer matrix[1212]. This means that the radical rotation frequency is determined by the mobility of the polymer segments.

Kusumoto et al.[1276] studied, with the aid of the spin-probe method, the chain motions on the surface of grown and annealed polyethylene single crystals in relation to the surface structure. Figure 11.5 shows changes of ESR spectrum of 2,2,6,6-tetramethyl-4-hydroxy-piperidine-1-oxyl (11.6) obtained as grown polyethylene crystals at 349.5 K:

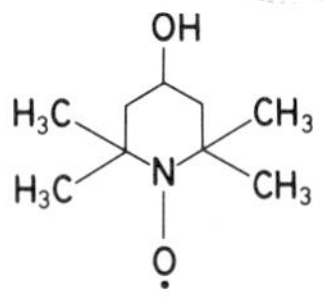

(11.6)

299

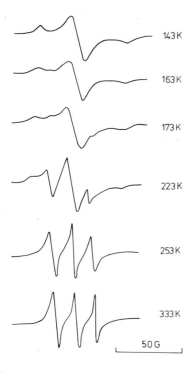

143 K

163 K

173 K

223 K

253 K

333 K

50 G

Fig. 11.5. ESR spectra of 2,2,6,6-tetramethyl-4-hydroxy-piperidine-1-oxyl (*11.6*) obtained as grown polyethylene crystals at 348.5 K[1276]

At low temperatures the spectra have a broad line-shape suggesting that the polymer matrix is in a rigid state. When the temperature is raised, a small subsplitting appears, because a part of the chain in the probe begins to rotate sufficiently to cause a change in the hyperfine pattern.

Spin-probe was also applied using polybutadiene and polyisoprene for the study of Overhauser effect, i.e. the dipole-dipole interaction between unpaired free radical electrons and protons[1346, 1348].

Spin-probe technique has also been applied for a study of polymer-plasticizer interactions[2219]. The rotational correlation time (τ) of probe radicals in plasticized poly(vinyl chloride) at 298 K decreases strongly when the amount of plasticizer is increased.

Rabold[1792] measured the interaction of sodium dodecyl sulphate with styrene-butadiene latices by using the spin-probe technique. By this method, it is possible to detect quantitatively the presence of micelles in a latex. A spin-probe method was also applied to study the critical micelle concentration of potassium palmitate[2221] and the dynamic structure of dimethyl sulphoxide-water mixture[339].

Nitrogen dioxide was used as a paramagnetic probe for examination of the molecular motion in rigid polymers at temperatures below the glass-transition temperature[2351].

Several papers[260, 1246, 1793, 2220] described determinations of the glass-transition temperatures by the spin-probe technique.

Readers may find several other applications of spin probe in polymer chemistry in original papers[1173, 1210, 1211, 1292, 1949, 2353].

11.3. Spin-Trapping Technique

This technique has been used for the detection and identification of short-lived free radicals by the ESR spectra of nitroxyl addition products (spin adducts) of free radicals (R·) to nitroso or nitrone compounds ("spin traps")[673, 1046, 1049, 1294, 1728]. This technique was mostly applied to liquid solutions using 2-nitroso-2-methyl propane (t-butyl-nitroxide) (11.7), nitrosobenzene (11.8), or phenyl-N-tert-butyl-nitrone (11.9) as spin trap for radical (R·):

$$R\cdot + CH_3-\underset{\underset{CH_3}{|}}{\overset{\overset{CH_3}{|}}{C}}-N{=}O \longrightarrow CH_3-\underset{\underset{CH_3}{|}}{\overset{\overset{CH_3}{|}}{C}}-\underset{\underset{R}{|}}{\overset{\overset{O^-}{|}}{N^{\cdot+}}} \qquad (11.7) \qquad\qquad (11.6)$$

$$R\cdot + \langle\!\!\!\bigcirc\!\!\!\rangle{-}N{=}O \longrightarrow \langle\!\!\!\bigcirc\!\!\!\rangle{-}\underset{\underset{R}{|}}{\overset{\overset{O^-}{|}}{N^{\cdot+}}} \qquad\qquad (11.7)$$

$$(11.8)$$

$$R\cdot + \langle\!\!\!\bigcirc\!\!\!\rangle{-}\underset{\underset{\underset{H_3C\quad CH_3}{\diagdown\quad\diagup}}{CH}}{\overset{\overset{H}{|}}{C}}{=}\overset{\overset{O^-}{|}}{N^+} \longrightarrow \langle\!\!\!\bigcirc\!\!\!\rangle{-}\underset{\underset{\underset{H_3C\quad CH_3}{\diagdown\quad\diagup}}{R\quad CH}}{\overset{\overset{H}{|}}{C}}{-}\overset{\overset{O^-}{|}}{N^{\cdot+}}$$

$$(11.9) \qquad\qquad\qquad\qquad\qquad\qquad (11.8)$$

Nitrogen and proton hyperfine splittings of the spin adducts are diagnostic parameters for the identification of the radical spin-adduct formed.

Possible application of the spin-trapping technique to quantitative studies was discussed by Janzen et al.[1048] and Perkins et al.[1729] who have reported a successful kinetic analysis of the decomposition of benzoyl peroxide in solution in the presence of phenyl-N-tert-butylnitrone (11.9).

The spin-trapping technique was applied to study radical polymerization[424, 1165, 1249, 1250, 1900]. Kunitake and Murakami[1249, 1250] have presented a detailed study of the spin-trapping method. They used 2-nitroso-2-methyl propane (11.7) and phenyl-N-tert-butylnitrone (11.9) for an ESR study of radical polymerization of different monomers, e.g. methyl methacrylate, styrene, vinyl acetate, acrylonitrile, butadiene, and isoprene. The reaction was initiated by thermal decomposition of azo-bis(isobutyronitrile)(AIBN), di-tert-butyl peroxalate or bis(4-tert-butylcyclohexyl)-peroxy dicarbonate.

Figure 11.6a shows an ESR spectrum of a polymerization mixture: AIBN-methyl methacrylate-2-nitroso-2-methyl-propane[1249].

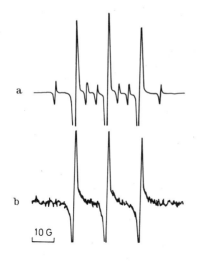

a

b

10 G

Fig. 11.6. ESR spectra of: a. polymerization mixture: AIBN-methyl
methacrylate-2-nitroso-2-methyl-
propane in benzene at 333 K,
b. Trapped poly(methyl methacrylate)
radical dissolved in benzene[1249]

This is principally a three-line spectrum formed by the exact overlapping of the spectra of trapped initiating radicals (*11.10*) and polymer-propagating radicals (*11.11*):

$$(11.9)$$

(*11.10*)

$$(11.10)$$

(*11.11*)

A clean triplet (a_N = 15.1 G) of diminished intensity, consistent with the structure characteristic for propagating polymer radical (*11.11*) is observed (Figure 11.6b).

The ESR spectrum obtained for the polymerization mixture AIBN-styrene-2-nitroso-2-methyl-propane at 343 K[1250] is shown in Figure 11.7.

This spectrum consists of the three-line spectrum and the six-line line spectrum which were attributed to the trapped initiating (*11.10*) and propagating radicals (*11.12*), respectively:

302

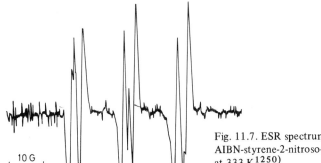

$$H_3C-\underset{\underset{CH_3}{|}}{\overset{\overset{CH_3}{|}}{C}}-N=O \;+\; -CH_2-\overset{\cdot}{C}H \;\longrightarrow\; H_3C-\underset{\underset{H_3C}{|}}{\overset{\overset{H_3C}{|}}{C}}-\underset{\underset{\underset{\underset{\underset{CH_2}{|}}{\overset{|}{\underset{}{C}}}}{|}}{\overset{\overset{O^-}{|}}{N}}{\cdot^+}$$

$$(11.11)$$

$$(11.12)$$

Fig. 11.7. ESR spectrum of polymerization mixture: AIBN-styrene-2-nitroso-2-methylpropane in benzene at 333 K[1250]

10 G

The main problem in the spin-trapping method applied to polymerization is the considerable change of the ESR spectrum of the examined system with time. For example, in the system discussed above, after 30 min the presence of radicals (11.12) was only indicated. After further heating at 343 K, the peak intensity decreased, and after 90 min the peak could no more be detected.

The application of the spin-trapping method to a living anionic polymerization system was reported by Forrester and Hepburn[709].

Especially interesting is the application of the spin-trapping method to the study of alternating copolymerization[2453]. The spin-trapping reagent used is 2,3,6-tri-t-butylnitrosobenzene (11.13)[2173], which can react with small-sized radicals giving radicals (11.14), while the bulky radicals are trapped in the form of radicals (11.15):

(11.13)

(11.14)

(11.15)

$$(11.12)$$

$$(11.13)$$

303

The radicals generated in the triethylaluminium-γ-butyrolactone-diacetyl peroxide system were trapped by 2,3,6-tri-t-butylnitrosobenzene (*11.13*) and are shown as ESR spectra in Figure 11.8.

Both types of spin-adducts (*11.14*) and (*11.15*) of methyl radicals derived from triethylaluminium were detected (Fig. 11.8a). The trapping of methyl radical as (*11.15*) type adduct shows the existence of some steric hindrance in the surroundings of methyl radicals by the aluminium atom or other components in this catalytic

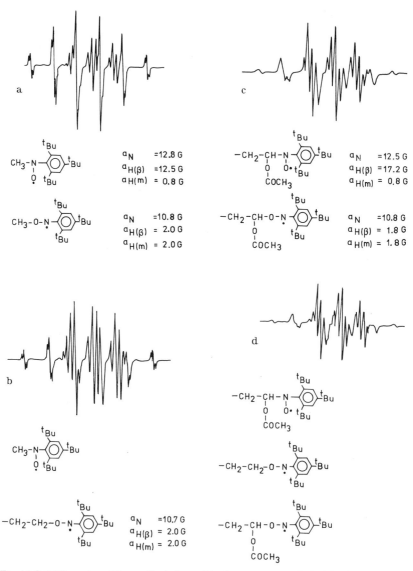

Fig. 11.8. ESR spectra of free radicals formed in the system: a. triethylaluminium-γ-butyrolactone-diacetyl peroxide, b. system a + ethylene, c. system a + vinyl acetate, d. system a + ethylene + vinyl acetate[245 3)]

system. The detection of "sterically hindered methyl radicals" shows the possibility of coordinated methyl radical formation.

In the presence of ethylene, the ESR spectra change (Fig. 11.8b). This ESR spectrum was assigned to (11.15) type adduct of primary radical derived from propagation ethylene, while (11.14) type adduct of free radical shows no change at all. When vinyl acetate monomer was introduced to the system, the new ESR spectrum was obtained (Fig. 11.8c), in which neither (11.14) nor (11.15) type adducts of methyl radical were observed. Instead, radical adducts (11.14) and (11.15), derived from vinyl acetate propagating radicals, were detected. ESR spectra of spin-trapped compounds in the presence of ethylene and vinyl acetate are as shown in Figure 11.8d, and were attributed to free propagation vinyl acetate radical and hindered ethylene and vinyl acetate radicals. This result shows the participation of "hindered radicals" in the ethylene-vinyl acetate copolymerization by coordination radicals[2453].

11.4. Spin-Labeling Technique

The spin-labeling technique is a simple way to extend the application of ESR spectroscopy to diamagnetic polymeric systems which do not contain unpaired electrons. Organic chemistry provides methods of synthetizing different nitroxide radicals which may be bound to natural and synthetic polymers:

1. By weak intermolecular forces such as hydrogen bonding or hydrophilic interaction.
2. By formation of chemical (covalent) bonds in the reaction of a polymer group with an R-group in the nitroxide radicals. Such groups are hydroxyl, carboxyl, cyano, isocyanate, meleimide, hydroxy-imino groups, etc.

Line-width measurements on the ESR spectra of macromolecules carrying stable nitroxide radicals may give information on the dynamics of the polymer chain[343, 348, 2224]. It has been found that in pure polymers there is no significant difference between rotational rates of free radicals (spin-probes) and bonded radicals (spin-labels)[2223]. This fact indicates that bonded radicals rotate quite freely around the C–O bond and that in viscous polymer matrices the rotational frequencies of the whole polymer molecule are very slow as compared to the local mode relaxations. The situation changes rapidly in dilute solutions. Theory of the rotational relaxation process in the range of rapid rotations ($\tau = 10^{-11} - 10^{-9}$ s) was presented in papers[94, 1140]. Comparison of rotational frequencies of spin-labeled macromolecules in different polymer matrices shows appreciable effects of the surrounding physical environment[348, 585, 2223].

The spin-labeling technique was originally applied to biological macromolecules[214, 314, 739, 820, 859, 906, 1065, 1075, 1443, 1537, 1716, 1907] and then extended to the synthetic polymers such as:

polyethylene[351, 1801, 1947],
poly(methyl vinyl ketone)[350],
poly(methyl methacrylate)[343, 819, 820, 1996],
polystyrene[32, 342, 345–349, 573],

305

poly(vinyl chloride)[2217, 2232],
poly(4-vinylpyridine)[1174],
pyrrolidone-allylamine copolymer[710, 1843],
maleic anhydride-dimethyl vinyl ether copolymer[820],
poly(ethylene oxide)[1902, 1963],
poly(ethylene glycol)[2217, 2218, 2222, 2223, 2229, 2230, 2232, 2233],
polyamides[2217, 2231, 2232],
polyurethane[2386],
poly(γ-benzyl glutamate)[1429],
polysiloxanes[1185, 1963],
cellulose and lignin[32, 1374, 2225–2227], and
polynucleic acids[235, 2089].

Low density polyethylene was spin-labeled with an oxazolidine-N-oxyl group, by the following method[351]:

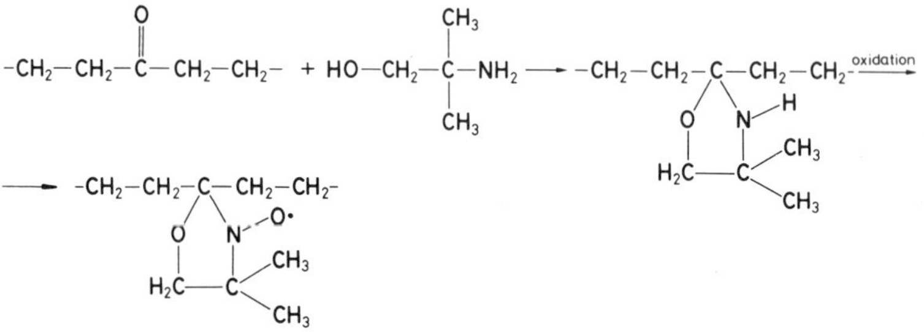

$$(11.14)$$

$$(11.16)$$

The dimethyl oxazolidine nitroxide *(11.16)* is rigidly held in the direction of z-axis of the nitrogen nuclear hyperfine tensor parallel to the extended chain direction (Fig. 11.9).

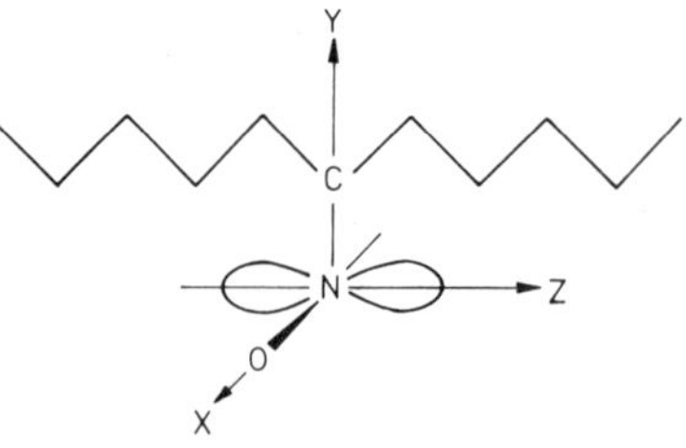

Fig. 11.9. Axis system for oxazolidine nitroxide spin-label[351]

The ESR spectrum of the solid amorphous spin-labeled polymer is shown in Figure 11.10.

This spectrum changes with temperature and it is analyzed to determine long correlation times $\tau > 3 \times 10^{-9}$ s. As the temperature of the solid polymer approached the melting point (about 383 K for low density polyethylene) the spectrum shows three narrow lines (Fig. 11.11).

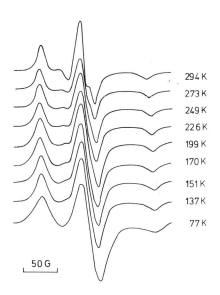

Fig. 11.10. Solid-state ESR spectra of spin-labelled polyethylene at different temperatures in the range 77–294 K[351]

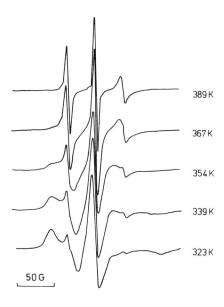

Fig. 11.11. Solid-state ESR spectra of spin-labelled polyethylene at different temperatures in the range 323–389 K[351]

In solution at 370 K, where anisotropic effects are largely averaged out, the three-line spectrum characteristic for nitroxide radicals is formed (Fig. 11.12).

As it was shown in Chapter 11.1 the rotational correlation time (τ) may be calculated from these ESR spectra[348].

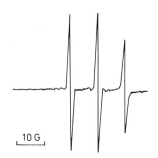

Fig. 11.12. ESR spectrum of spin-labelled polyethylene in xylene (3 w.t.%) at 370 K[351]

For solid polyethylene the ESR spectra revealed the following general relaxation processes:

1. α-relaxation in the range 300–308 K is attributed to large-scale reorientations of polymer chains within the crystalline regions.
2. β-relaxation is attributed to cooperative segmental motions in the amorphous phase in the range 220–300 K
3. γ-relaxation in the range 140–180 K is attributed to small-scale processes involving crankshaft-type motions of methylene segments.

The activation energies for these processes, determined by the study of dielectric and mechanical relaxation, are >100, 65–100 and 45–60 kJ mol^{-1}, respectively[1444].

The synthesis of spin-labeled polystyrene is described by Drefhal et al.[573] and Bullock et al.[342], and occurs by the following mechanism:

$$-CH_2-CH- \xrightarrow[HgOAc]{NOCl} -CH_2-CH- \xrightarrow{C_6H_5MgBr} -CH_2-CH- \xrightarrow{Ag_2O} -CH_2-CH-$$

(11.15)

(11.17)

The polymer contains approximately one nitroxide radical centre (11.17) per 1000 monomer units. The synthesis (11.15) is inconvenient for several reasons, e.g. the solubility of the mercurated polystyrene is low and the intermediate nitroso polystyrene is unstable. A better method for synthesis of spin-labeled polystyrene with t-butyl nitroxide is described by Bullock[345]:

$$-CH_2-CH- + BuLi (in\ C_6H_6) \longrightarrow -CH_2-CH- \xrightarrow{+(CH_3)_3C-N=O} -CH_2-CH- \xrightarrow[in\ air]{MeOH} -CH_2\ CH$$

(11.16)

(11.18)

By this method polystyrene is labeled exclusively in the para-position. Bullock et al.[347] also described a method for preparing meta-labeled polystyrene.

The ESR spectrum of the spin-labeled polystyrene (11.17) in solution is shown in Figure 11.13.

A triplet-line spectrum with an intensity ratio of approximately 1:1:1 is observed. A simple analysis of the ESR spectra shows that:

1. The low field and central components of the triplet have further splittings due to interaction with the aromatic protons.

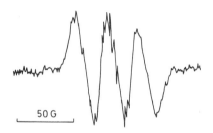

50 G

Fig. 11.13. ESR spectrum of spin-labelled polystyrene in toluene solution at 303 K[342]

2. The line-width depends on the magnetic quantum number M_I of the ^{14}N nucleus.

Figure 11.14 shows an ESR spectrum of spin-labeled polystyrene (*11.17*) in the solid state.

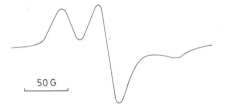

50 G

Fig. 11.14. ESR spectrum of the solid spin-labelled polystyrene at room temperature[342]

The outermost peaks arise from radicals oriented in such a way that the axis of π-orbital is parallel to the applied magnetic field.

From line-width measurements of randomly labeled polystyrene in toluene solution, the rotational correlation time (τ) and activation energy (E_a) for segmental rotation have been calculated[342, 348]. As the molecular weight decreases, the rotational frequency of the whole macromolecule increases[348].

The dependence of the rotational correlation times on solvent properties and polymer concentrations have been examined[349]. On account of the higher viscosity of α-chloronaphthalene and the poor solvent properties of cyclohexane, the rotational correlation time (τ) is greater in cyclohexane and α-chloronaphthalene than in toluene. In high molecular weight polymer samples, segmental reorientation is the main relaxation process. Its activation energy is 25.1 ± 0.8 kJ mol^{-1} in cyclohexane and 26.4 ± 1.3 kJ mol^{-1} in α-chloronaphthalene, as compared with 18.0 ± 0.8 kJ mol^{-1} in toluene. The internal energy barriers to rotation were calculated to be 8.8 kJ mol^{-1} in toluene and α-chloronaphthalene and 12.6 kJ mol^{-1} in cyclohexane. The higher internal barrier to segmental rotation in cyclohexane is probably due to a more tightly coiled conformation of the polymer chains in this poor solvent. The rotational correlation time (τ) increases slowly with polymer concentration up to a critical value, beyond which τ rises steeply. The critical concentration is close to the "entanglement" concentration as obtained by viscosity measurements.

End-labeled addition polymers can be prepared by reacting "living" polymeric anions with 2-nitroso-2-methylpropane (*11.7*) and then oxidizing the resulting hydroxylamine[343, 709]:

$$-CH_2-\underset{\underset{COOCH_3}{|}}{\overset{\overset{CH_3}{|}}{C^-}} \; + \; CH_3-\underset{\underset{CH_3}{|}}{\overset{\overset{CH_3}{|}}{C}}-N{=}O \longrightarrow -CH_2-\underset{\underset{COOCH_3}{|}}{\overset{\overset{CH_3}{|}}{C}}{-}\underset{\underset{C(CH_3)_3}{|}}{N}{-}O^- \xrightarrow{\quad MeOH \quad}$$

$$-CH_2-\underset{\underset{COOCH_3}{|}}{\overset{\overset{CH_3}{|}}{C}}{-}\underset{\underset{C(CH_3)_3}{|}}{N}{-}OH \xrightarrow{\quad oxidation \quad} -CH_2-\underset{\underset{COOCH_3}{|}}{\overset{\overset{CH_3}{|}}{C}}{-}\underset{\underset{C(CH_3)_3}{|}}{N}{-}O\cdot \qquad (11.17)$$

$$(11.19)$$

309

The ESR spectrum of end-labeled poly(methyl methacrylate) (*11.19*) is shown in Figure 11.15.

At 273 K, or after brief storage at room temperature, the end-labeled poly(methyl methacrylate) (*11.19*) gave a seven-line spectrum (a superposition of the spectra Figs. 11.15a and 11.15b), in which the outermost lines of the two spectra exactly

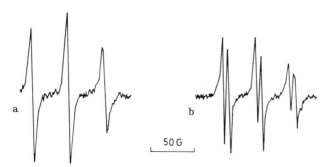

Fig. 11.15. ESR spectrum of end-labelled poly(methyl methacrylate): a. Polymerized anionically with n-buthyl litium in tetrahydrofuran at 203 K, b. Polymerized radically at 203 K[343]

overlapped. It was tentatively suggested that Figure 11.15b is the spectrum of radical (*11.20*), following the rearrangement of the

$$
\begin{array}{c}
\qquad\quad CH_3 \\
\qquad\quad | \\
-CH-C-H \qquad\quad (11.20)\\
\ \ |\qquad | \\
\ \ |\quad\ COOCH_3 \\
\ \ | \\
N-O\cdot \\
| \\
C(CH_3)_3
\end{array}
$$

poly(methyl methacrylate anion). The resulting ESR spectrum (Fig. 11.15b) shows the dependence of line-width upon M_N, the ^{14}N nucleus. This permits the evaluation of the rotational correlation time in end nitroxide-labeled poly(methyl methacrylate) (*11.19*), e.g. $\tau = 2.8 \times 10^{-10}$ s for poly(methyl methacrylate ($\overline{M}_v = 80000$) in ethyl acetate solution (5% wt.) at 298 K[343].

Spin-labeling technique has also been applied to investigation of other problems, such as

1. Study of the role of surface dispersion of bronzes in contact with polyisobutylene and polyurethanes during mechanical degradation in the friction process[794].

2. Study of micelle solutions, in particular evaluating the critical micelle concentration[2374].

3. Determining the ratio of the number of segments in loops to the number in trains along the surface for polymers absorbed at the solid/liquid interface[710, 1843].

Spin-labels attached to looped portions of the polymer are more mobile than those attached to trains, and the ESR spectra from the two sites are different.

4. Study of bimolecular recombination, which is a diffusion-controlled process. It is well known that the rate constants of termination reactions depend upon such factors as solution viscosity, chain flexibility, and molecular interactions between polymer and solvent. For investigation of these processes the spin-labeled method was successfully applied. The dependence of exchange rate constants and diffusion coefficients of stable macroradicals on the length of polymer chains in various solvents was investigated, e.g. for end nitroxide modified poly(dimethyl siloxane) and poly(ethylene oxide)[1963].

Chapter 12

ESR Spectroscopy of Stable Polymer Radicals and their Low Molecular Analogues

New types of stable polymer radicals and polyradical anions have been studied with increasing attention during the last few years. These polymers have interesting electrical properties and radical reactions, and are, in practise, useful as antioxidants and semiconductors. These types of polymers were reviewed in detail by Braun[262, 281]. The majority of stable polymer radicals have precursors consisting of low-molecular stable free radicals. In this chapter ESR spectra for low-molecular and polymer stable radicals are compared.

12.1. Poly(triphenylmethyl) Radicals

The precursor for this type of polyradicals is the well known and important triphenylmethyl free radical, which was investigated in detail by ESR spectroscopy[1512, 1516, 1517]. Poly-carbon radicals of the triphenyl type were synthesized by polymerization of vinyltriphenylmethylcarbinol (12.1). The poly(p-vinyltriphenylmethylcarbinol) (12.2) reacts with acetyl chloride in benzene chloride and forms poly(p-vinyltriphenylmethyl chloride) (12.3), which in a subsequent reaction with zinc-potassium gives poly(p-vinyltriphenylmethyl) (12.4) radicals[262, 265, 267, 276].

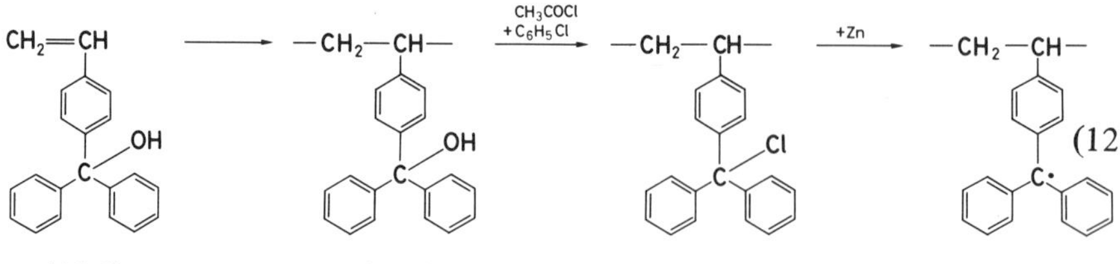

$$(12.1) \qquad (12.2) \qquad (12.3) \qquad (12.4)$$

In contrast to low-molecular weight triphenylmethyl radicals, the corresponding polyradicals give an ESR spectrum with a weakly resolved hyperfine singlet structure (Fig. 12.1)[267].

This was attributed to the high local concentration of the radicals in macromolecular coils and to their interaction, as well as to very small coupling parameters of triphenylmethyl radicals.

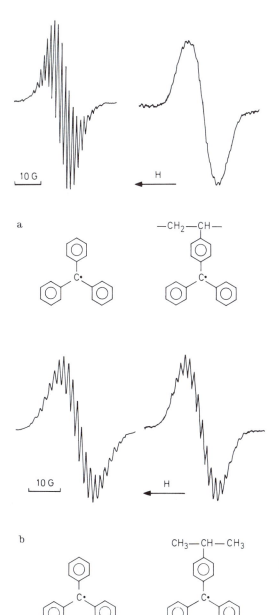

Fig. 12.1. ESR spectra of: a. triphenylmethyl and poly(vinyltriphenyl) radicals (0.0004 mole 1^{-1}) and b. 4-phenyltriphenylmethyl and 4-isopropyl-4′-phenyltriphenylmethyl radicals (0.003 mole 1^{-1}) in benzene solution at 296 K[267)]

In the presence of oxygen, carbon polyradicals form unstable peroxyradicals (Fig. 12.2) which react easily with carbon polyradicals to give crosslinked polymer free radicals.

313

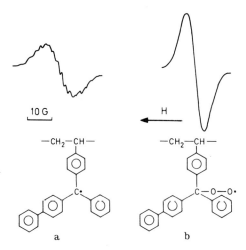

Fig. 12.2. ESR spectra of: a.poly-4-vinyl-4'-phenyltriphenylmethyl radical and b. poly-4-vinyl-4'-phenyltriphenylmethyl-peroxy radical in tetrahydrofurane $(0.0004$ mole $1^{-1})[267]$

12.2. Poly(α,α'-diphenyl-p-xylene) Radicals

The synthesis of these polyradicals was described by Braun et al.[282] and it occurs by the following mechanism:

$$(12.5) \qquad (12.6) \qquad (12.2)$$

A comparison of the ESR spectra of polymer radical with their analogue diphenyl[4-(2,2,2-triphenyl-ethyl)phenyl]methyl radical is shown in Figure 12.3. Both free radicals have an analogous structure.

12.3. Polyphenoxy Radicals

Various stable phenoxy radicals are the precursors for this type of polyradicals. They are obtained from Ag_2O or PbO_2 oxidation of 2,4,6-tri-substituted phenols and belong to the stable radicals which have well-determined ESR spectra[240, 273, 982—984, 1544, 1837, 1908, 2077]

Galvinoxyl (2,6-di-tert-butyl-α'-(3,5-di-tert-butyl-4-oxo-2,5-cyclohexadien-1-ylidene)-p-tolyloxy) (12.7) is one of the most stable free radicals in solid state and is also relatively stable in solution. The synthesis and ESR spectrum were described

314

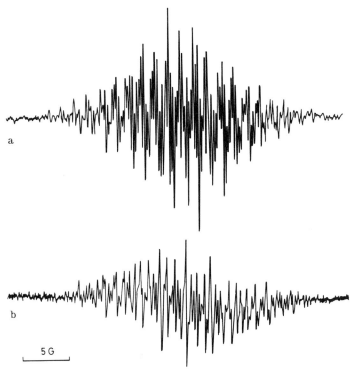

Fig. 12.3. ESR spectra of: a. diphenyl-p-tolylmethyl radicals and b. their polymer radical analog[282]

in several papers[177, 219, 478, 767, 1122, 1543, 1765, 2424]. Its ESR spectrum is shown in Figure 12.4.

(12.7)

Fig. 12.4. ESR spectrum of galvinoxyl radical in benzene solution[219]

It is an extremely effective scavenger for alkyl and alkylperoxy radicals[164] and has been applied in study of polypropylene oxidation[623].

Polymeric phenoxy[1190] and 2,6-disubstituted phenoxy radicals (12.8)[1903] were studied by ESR spectroscopy. The very small metha-hydrogen coupling constants indicate a linear 1,4-polyether structure:

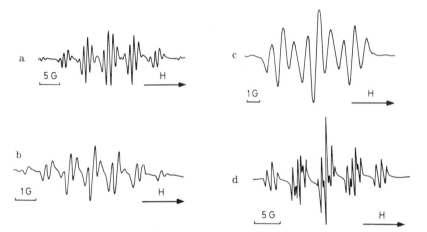

$$R : -CH_3 , -OCH_3$$
$$-C_2H_5 , -CH(CH_3)_2$$

$$(12.8)$$

Van den Hoek et al.[951] observed that the ESR spectra of most 2,6-disubstituted polyphenoxy radicals show hyperfine coupling with ortho-substituted groups in the ring at the end of the polymer chain. The ESR spectra of different polyphenoxy radicals are shown in Figure 12.5.

Fig. 12.5. ESR spectra of different polyphenoxy radicals: a. poly-2,6-dimethoxy-phenoxy radical, b. poly-2,6-dimethoxy-phenoxy radical, c. poly-2,6-di-t-butyl-phenoxy radical, d. poly-2,6-diethyl-phenoxy radical[1903]

Polyphenoxy radicals are also formed on the surface of crystalline copper(II) complexes: $CuCl(OCH_3)\cdot 2$ pyridine and $CuCl_2\cdot$cyclohexylamine from 2,6-xylenol[2200–2202, 2205].

Synthesis of polyalkenylphenoxy radicals (12.9) was described by Braun and Meier[273],

$$R :-H , -CH_3 \qquad (12.9)$$

$$(12.3)$$

ESR studies of polyalkenylphenoxyls show singlet-line spectra without hyperfine structure.

316

12.4. Poly(triphenylhydrazyl) Radicals

The commercially available (Eastman) free radical α,α'-diphenyl-β-picrylhydrazyl (DPPH) (12.10), whose molecular structure and approximate splittings (G) are shown below, is one of the most commonly used reagents in free-radical chemistry.

(12.10)

The complete synthesis of this compound is described by Poirier et al.[1750] and Tüdos et al.[2309]. DPPH crystallized from CS_2 or $CHCl_3$ interacts with O_2 giving broadening of the absorption line[40]. This reaction does not occur in benzene solution. For that reason DPPH should be crystallized from benzene and sealed in vacuum or in an inert gas. DPPH is photolyzed easily by UV-light[1750, 2097, 2098, 2403], and it should, therefore, be stored in the dark. Under ionizing irradiation DPPH is decomposed at the same rate both in vacuum and in presence of air[2471]. At low initial concentration of DPPH in solution (solid or liquid), the decomposition is accelerated in the presence of air. DPPH has been the subject of many ESR studies[60, 114, 197, 311, 449, 459, 516, 523, 654, 744, 746, 748, 831, 860, 955, 957, 981, 994, 1125, 1178, 1198, 1213, 1332, 1386, 1413, 1430, 1555, 1838, 1871, 2235, 2315, 2345]. ESR study of the water-soluble stable free radical p,p-disulpho-α,α'-diphenyl-β-picrylhydrazyl (SDPPH) (a sulphonated analogue of DPPH) has also been reported[1773].

DPPH stable radicals in solution produce spectra changing from a single sharp line for solids, to a broad line for concentrated solutions. Hyperfine structure appears in diluted solution[831, 994] (Fig. 12.6).

DPPH in benzene solution has a quintet-line spectrum with an overall line width of approximately 60 G (Fig. 12.7).

The ESR measurements in solution are considerably affected by dissolved oxygen. The majority of air-saturated organic solvents contain oxygen in the order of 10^{-3} mole l^{-1}, and this concentration is equivalent to 10^{18} spins cm^{-3}[879, 2346].

DPPH shows differences in observed line-width of ESR spectra, when measured in different solvents[523, 2318] and polymer solutions[733]. Differences in ESR spectra are also observed for DPPH crystallized from different solvents[40] (Table 12.1).

The DPPH in polymer matrix, e.g. in poly(methyl methacrylate), shows a poorly resolved five-line spectrum with the width of approximately 120 G, as shown in Figure 12.8.

The broadening is caused by the anisotropic hyperfine structure, in addition to the isotropic hyperfine structure due to the rigidity of the polymer[2471]. DPPH in polystyrene matrix shows an exponential, rather than a Gaussian shape, for the wings of the exchange-narrowed lines[493]. In Table 12.2, g-values for DPPH determined by various authors are given.

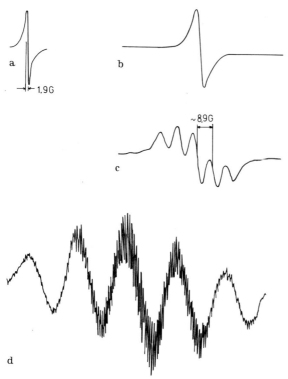

Fig. 12.6. ESR spectra of α,α'-diphenyl-β-picrylhydrazyl (DPPH): a. solid powdered sample[197], b. saturated solution of DPPH in tetrahydrofuran, c. saturated solution diluted by the factor 10^3, d. carefully dehydrated and degassed solution of 10^{-3} mole 1^{-1} DPPH in tetrahydro-furan

Fig. 12.7. ESR spectrum of α,α'-diphenyl-β-picrylhydrazyl (DPPH) in benzene solution $(5 \times 10^{-3}$ mole $1^{-1})$[2471]

Fig. 12.8. ESR spectrum of α,α'-diphenyl-β-picrylhydrazyl (DPPH) in poly(methyl methacrylate) film (matrix) $(7 \times 10^{-3}$ mole $1^{-1})$[2471]

318

Table 12.1. Line-width in DPPH samples crystallized from various solvents[40]

Solvent	ΔH, G		
	ν = 300 Mc		ν = 9400 Mc
	295 K	90 K	295 K
Benzene	6.8	4.6	4.7
Toluene	2.9	2.6	2.6
Xylene (mixture)	2.5	2.2	2.3
Pyridine	5.3	5.0	5.0
Bromoform	2.2	2.5	2.5
Carbon tetrachloride	1.9	2.7	2.3
Chloroform	1.7	2.1	2.0
Carbon disulfide	1.3	1.3	1.5

Table 12.2. Variation in g-values for DPPH at room temperature and 77 K[a]

Sample	g-Value	References
Single crystal[a]	2.0027–2.0039	2347)
Single crystal	2.0035–2.0041	2403)
Single crystal	2.0028–2.0038	1004)
Single crystal	2.0030–2.0040	40)
Powder[a]	2.0037	2347)
Polycrystalline	2.0036 ± 0.0003	1004)
Polycrystalline	2.0037	1760)

Polycrystalline DPPH sample has a g-value of 2.0036 ($g_{\parallel}$ = 2.0028 and $g_{\perp}$ = 2.0039)[2454]. Differences in the g-value for single crystals may occur, because the number of spins in crystal samples depends on the solvent used for the growing of the crystal[1505]. The ESR lines of DPPH in the whole temperature range from 293 K to 1.7 K have Lorentzian shape[2347]. ESR spectra and their characteristics for different analogues of DPPH have been collected by Braun et al.[263] and are shown in Table 12.3.

Hyde et al.[994] used the ELDOR technique to study solutions of DPPH, and obtained precise values for the hyperfine coupling of the main nitrogens N_{α} and N_{β}. The high resolution of the ENDOR technique was utilized to obtain precise values for the 12 proton hyperfine couplings (Fig. 12.9) and to estimate the various N hyperfine couplings in solutions of DPPH[502].

Application of ENDOR spectra helps to explain the rather complex ESR spectra of DPPH.

Osugi and Sasaki[1678] reported an ESR study of the photochemical reaction between DPPH and methyl methacrylate.

Table 12.3. Comparison of the different types of α,α-diphenylhydrazylen (I) and aminocarbazylen (II) with different R groups[263].

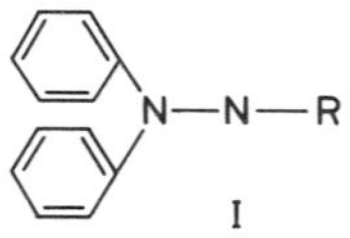	ESR Spectrum	number of lines	Coupling constant (G)	$a_{N_1}:a_{N_2}$ and g-Value	Stability	References
I (N–N–R, diphenyl)						
I R = Triphenylhydrazyl		4	$a_{N_1} = 4{,}28$ $a_{N_2} = 9{,}05$	0,47	$C_6H_6 = 10$ min; $CCl_4 = 1{,}5$ h	263)
I R 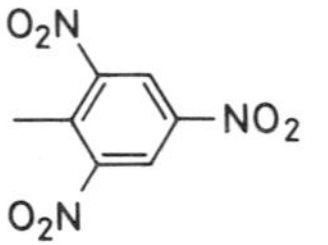 = DPPH	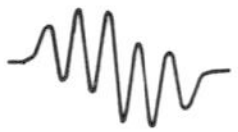	5	$a_{N_1} = 7{,}69$ $= 8{,}03$ $a_{N_2} = 9{,}16\,[45]$ $= 9{,}65\,[50]$	0,84; 0,83 $g = 2{,}00364 \pm 0{,}0001$	stable	59) 1386)
I R 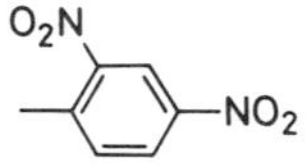 =		5	$a_{N_1} = 6{,}74$ $a_{N_2} = 9{,}49$	0,71 $g = 2{,}0037$	$t_{1/2} \sim 87$ h	59, 132) 1219) 2309)
I R 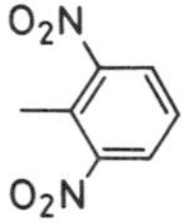 =		5	$a_{N_1} = 7{,}16$ $a_{N_2} = 10{,}38$	0,69 $g = 2{,}0038$	stable	2309)
I R 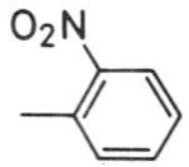 =		4	$a_{N_1} = 5{,}97$ $a_{N_2} = 9{,}63$	0,62	unstable	59)
I R 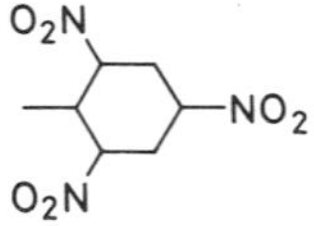 =		4	—	—	unstable $t_{1/2} \sim 1$ h	263)

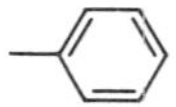

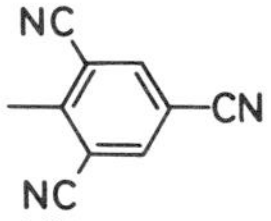 I R= (NC-C₆H₃-CN, NC) **DPTCH**	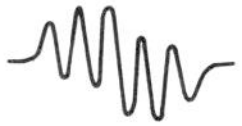 5	$a_{N1} = 7{,}58$ $a_{N2} = 9{,}39$	$0{,}81$ $g = 2{,}00362 \pm 0{,}0001$	stable	308)
I R= 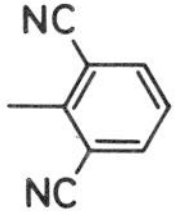 (NC, NC)	 5	$a_{N1} = 7{,}125$ $a_{N2} = 9{,}92$	$0{,}72$	stable	263)
I R= 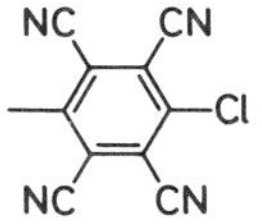 (NC, Cl, CN, NC)	 5	line distance 8,3	—	stable	308)
I R= 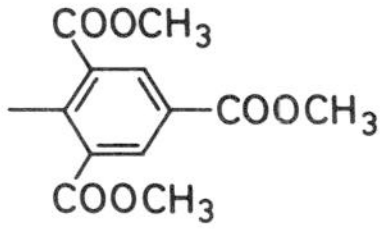 (COOCH₃, COOCH₃, COOCH₃) **DPTH**	 6	$a_{N1} = 6{,}07$ $a_{N2} = 10{,}97$	$0{,}57$ $g = 2{,}00369 \pm 0{,}0001$	stable	263) 280)
I R= 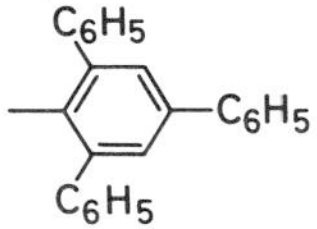 (C_6H_5, C_6H_5, C_6H_5)	 6	—	—	unstable $t_{1/2} \sim 2$ h	263)

Table 12.3. (continued)

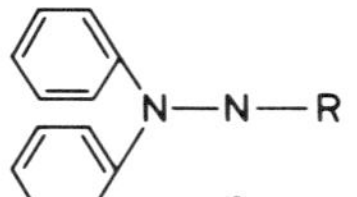	ESR Spectrum	number of lines	Coupling constant (G)	$a_{N_1}:a_{N_2}$ and g-Value	Stability	References
I $R =$ 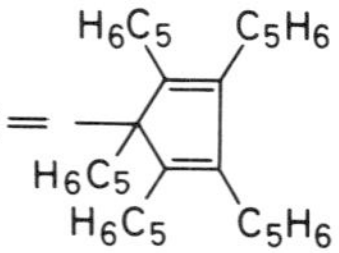 DPCPDH (H_6C_5, C_5H_6, C_5H_6, H_6C_5, H_6C_5)		7	$a_{N1} = 5,8$ $a_{N2} = 11,6$	0,5 $g = 2,00385 \pm 0,0001$	stable $t_{1/2} \sim$ 3 weeks	61) 263)
I $R = -C(C_6H_5)(C_6H_5)(C_6H_5)$ 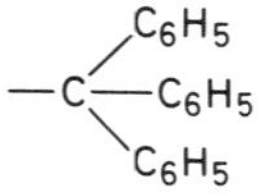		7	$a_{N1} = 5,68$ $a_{N2} = 12,02$	0,47	unstable	997)
I $R = -C(C_6H_4-NO_2)(C_6H_2-NO_2)(C_6H_4-NO_2)$ 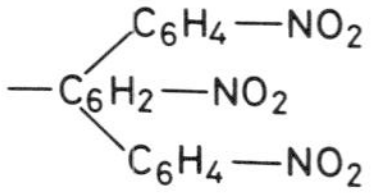	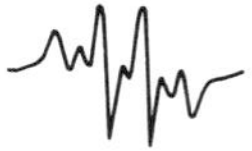	7	$a_{N1} = 6,70$ $a_{N2} = 11,50$	0,58	unstable	997)
I $R = -C(C_6H_4-C_6H_5)(C_6H_5)(C_6H_5)$ 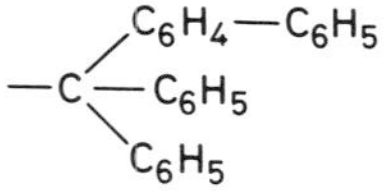		7	—	$\sim 0,50$	unstable $t_{1/2} \sim$ 3,5 h	263)
I $R = -C(C_6H_4-C_6H_5)(C_6H_4-C_6H_5)(C_6H_5)$ 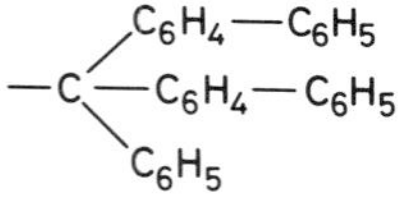		7	—	$\sim 0,50$	unstable $t_{1/2} \sim$ 3,5 h	263)

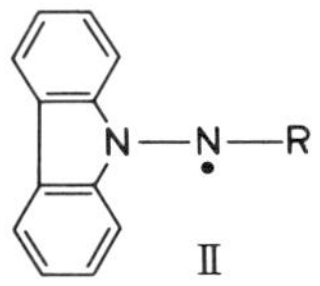

Base structure: carbazole–N–N•–R (II)

R	(number)	a_N	g-value	stability	ref.
$R =$ 2,4,6-trinitrophenyl (O_2N, NO_2, O_2N) — II	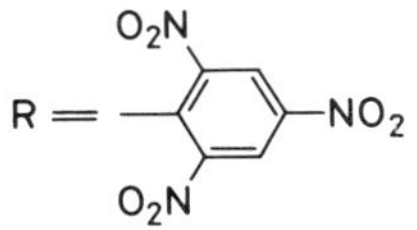 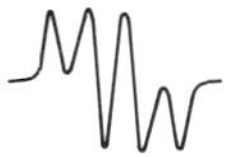 4	$a_{N1} = 5,8$ $= 5,8$ $a_{N2} = 10,2$ [44, 45] $= 11,2\cdots$	0,57 [44, 45]; 0,51 $g = 2,00353 \pm 0,0001$ $g = 2,0032\cdots$	rel. stable	263) 1050) 1386) 2309)
$R = -C-(C_6H_5)_3$ — II	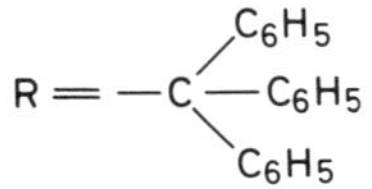  3	$a_{N1} = 0$ $a_{N2} = 12,10$	$g = 2,00383 \pm 0,0001$	$t_{1/2} \sim 2,5$ h	263)
$R =$ pentaphenylcyclopentadienyl (H_5C_6, C_6H_5, C_6H_5, H_5C_6, H_5C_6) — II	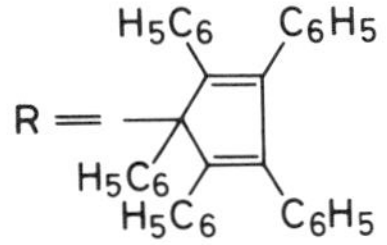  3	$a_{N1} = 0$ $a_{N2} = 12,10$	$g = 2,00383 \pm 0,0001$	very unstable $t_{1/2} \sim 0,75$ h	263)
$R =$ tricyanophenyl (CN, NC, NC) — II		—	—	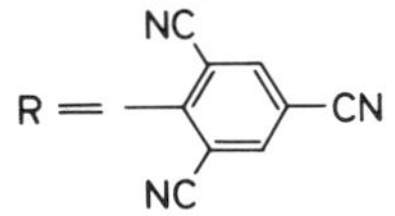 Hydrazine not oxydizable	263)
octahydrocarbazole–N–N•–picryl (NO_2, O_2N, O_2N)	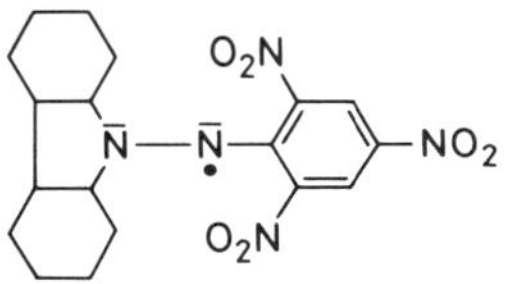 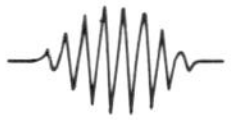 8	—	$g = 2,00382 \pm 0,0001$	unstable $t_{1/2} \sim 3,5$ h	263)

323

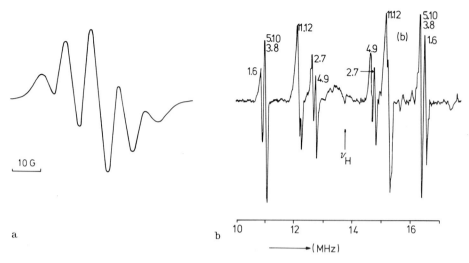

Fig. 12.9. a. ESR spectrum of α,α'-diphenyl-β-picrylhydrazyl (DPPH) in mineral oil used for ENDOR, b. ENDOR spectrum of DPPH in mineral oil at room temperature[502]

Polyvinyl-DPPH was first obtained by Henglein and Boysen[908] by γ-irradiation of high polymers in the presence of DPPH. In this way groups were incorporated into the main chain, being connected by one of the two phenyl groups (*12.11*) or the picryl group (*12.12*):

(*12.11*)   (*12.12*)

ESR spectra of polyvinyl-DPPH in tetrahydrofuran are shown in Figure 12.10.

Ovenall[1864] prepared the polymeric hydrazyl radical by polymerization of methyl methacrylate in the presence of DPPH. The polymer contained oxidizable N–H bonds, and upon oxidation gave polymer molecules containing free radicals with the ESR spectra identical to those of DPPH. Polystyrene radicals do not react with DPPH by this mechanism.

324

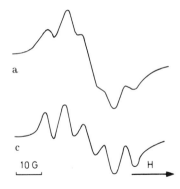

c

10 G

H

Fig. 12.10. ESR spectra of polyvinyl-DPPH in tetrahydrofuran: a. $c = 3 \times 10^4$ g cm^{-3} at 300 K (20.5%
radical groups), b. $c = 4.3 \times 10^{-4}$ g cm^{-3} at 300 K
(7% radical groups), c. $= 4.3 \times 10^{-4}$ g cm^{-3} at 360 K
(7% radical groups)[270]

A new method for synthesis of polyvinyl-DPPH was described by Braun et
al.[270, 277]. The particular reaction steps can be summarized in the following scheme:

$$(12.4)$$

$$(12.12)$$

Other polymeric hydrazyls (12.14) and (12.16) have been obtained by the following reactions[262, 277]:

1. Reaction of poly(vinyl-triphenyl chloride) (12.3) with diphenylhydrazine
   (12.13) and further oxidation in benzene by PbO_2:

$$(12.5)$$

$$(12.14)$$

325

2. Reaction of poly(vinylphenyl-tetraphenylcyclopentadienyl chloride) (*12.15*) with diphenylhydrazine (*12.13*) and further oxidation in benzene by PbO_2:

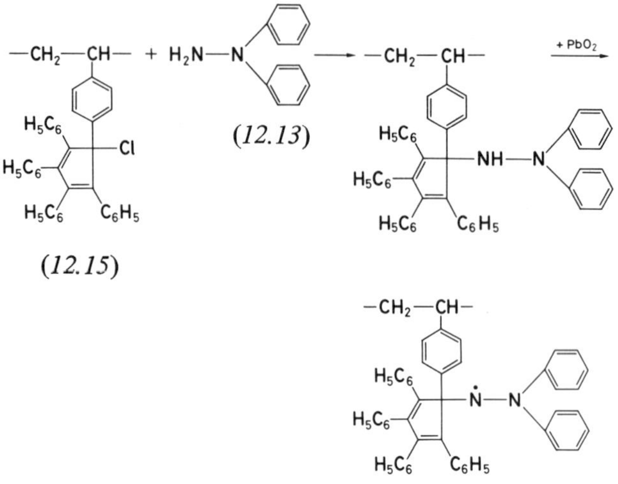

(*12.13*)

(*12.15*)

+ PbO₂ → ... (12.6)

(*12.16*)

Polymeric hydrazyl (*12.14*) and (*12.16*) obtained in this way are yellow-brown solid paramagnetic crystals, soluble in benzene and tetrahydrofuran.

12.5. Polymers Containing Stable Free Radicals of the Nitroxide Type

Formation of nitroxide stable free polymer radicals (*12.17*) have been observed during thermal oxidation of amine-terminated polyamides (nylon 6)[451]. The ESR spectra of these nitroxide radicals are shown in Figure 12.11.

$$-CH_2-N-CH_2- \qquad (12.17)$$
$$\underset{\overset{|}{O}}{}$$

A one-line ESR spectrum can be observed for nonterminated, and acid-terminate, polyamides heated in the presence of air. This singlet was attributed to conjugated systems with unpaired electrons[451, 803].

Several other polymers containing nitroxide groups were synthesized by polymer-analogous reactions and a few were synthesized by polycondensation of polyaddition reactions[573, 549, 819, 1691, 1860].

By oxidation of the poly-N-(4-diphenylamino)acryl with PbO_2, a stable poly-N-(4-diphenylnitroxyl) radical (*12.18*) is obtained[268], and its ESR spectra are shown in Figure 12.12.

326

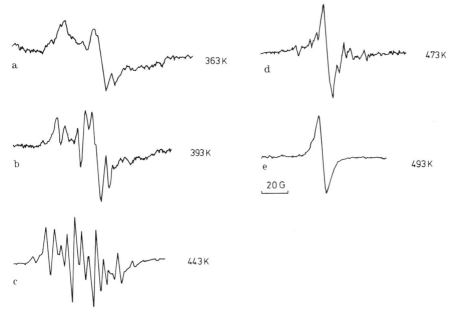

Fig. 12.11. ESR spectra of nitroxide radicals from nylon 6 thermal oxidation at: a. 363 K, b. 393 K, c. 443 K, d. 473 K, e. 493 K[451]

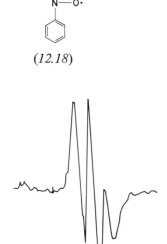

(12.18)

Fig. 12.12. ESR spectrum of poly-N-(4-diphenylnitroxide) radical in tetrahydrofuran solution[268]

Polymers containing stable nitroxyl free radical groups, poly-4-methacryloamino-2,2,6,6-tetramethylpiperidine-1-oxyl (*12.19*) and poly-4-methacrylooxy-2,2,6,6-tetramethylpiperidine-1-oxyl (*12.20*) have been prepared by the following method[1259, 1260]:

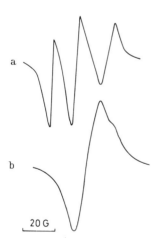

$$X: -NH- \qquad (12.19)$$

$$X: -O- \qquad (12.20)$$

Compared with monomeric nitroxyl compounds, the ESR spectra of nitroxyl polymers show different behavior. In Figure 12.13 spectra are shown for nitroxyl polymers of the amide type, which were prepared by the oxidation of polymeric precursor for 2 hr and 30 hr respectively.

a

b

20 G

Fig. 12.13. ESR spectra of poly-4-methacryloamino-2,2,6,6-tetramethylpiperidine-1-oxyl in methanol at room temperature, a. oxidation for 2 hr and b. for 30 hr[1259)

The characteristic feature of these spectra is that the triplet fine structure of N—O· radical, formed at the initial stage of the oxidation, becomes broadened and is finally merged into a singlet as the oxidation proceeds. The reason for the broadening and change to a singlet may be considered to be either the rate or the anisotropy of the spin-spin coupling. As the free radical density along the polymer chains in-

328

creases on oxidation, the increased possibility of spin-spin interaction tends to reduce the transition time from the excited to the ground state of spins. This effect may cause the broadening of the triplet lines to a singlet.

In the case of copolymers of (12.19) with styrene or methyl methacrylate, the ESR spectra are retained as triplets and do not merge into broad singlets during oxidation (Fig. 12.14).

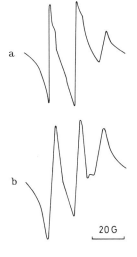

Fig. 12.14. ESR spectra of copolymer of poly-4-meth-acryloamino-2,2,6,6-tetra-methylpiperidine-1-oxyl with a: styrene, and b. methyl methacrylate. ESR spectra measured in methanol at room temperature[1259)]

In these copolymers, the nitroxyl radicals formed by oxidation are incapable of interacting with each other in long sequences, because the distribution density of the radicals along the polymer chains is lowered by the presence of nonradical co-monomer units. Consequently, the triplet signals were also observed after prolonged oxidation (60 hr).

A polymer containing nitronyl and nitroxyl groups, poly[4-(4',4',5',5'-tetra-methyl-1',3'-dihydroimidazol-2'-yl-3'-oxide-1'-oxyl)phenylethylene] (12.21) was prepared according to the following scheme[1511)]:

$$(12.8)$$

$$(12.21)$$

The nitronyl polymer is soluble in dimethylformamide and tetrahydrofuran giving a blue solution, but it is insoluble in benzene and methanol. The ESR spectrum of the model compound 2-(p-cumenyl)-4,4,5,5-tetramethylimidazoline-3-oxide-1-oxyl (*12.22*) showed a hyperfine structure with 5 lines due to the two nitrogen atoms, but that of the polymer showed a broad line (Fig. 12.15).

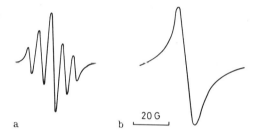

(*12.22*)

Fig. 12.15. ESR spectra of:
a. 2-(p-cumenyl)-4,4,5,5-tetramethyl-tetrahydroimidazole, and
b. poly[4-4',5',5',-tetramethyl-4,5-dihydroimidazol-2-yl-3-oxide-1'-oxyl)-phenylethylene] in tetrahydrofuran at room temperature in the air[1511]

a   b   20 G

12.6. Poly(pyridinyl) Radicals

Poly(pyridinyl) radical (*12.25*) has been prepared from poly(vinyl acetate) (*12.23*) and methyl ester of isonicotinic acid (*12.24*) according to the following mechanism[16]:

(*12.23*)   (*12.24*)   (12.9)

(*12.25*)

330

ESR spectra of poly(pyridinyl) radical and its low molecular analog are shown in Figure 12.16.

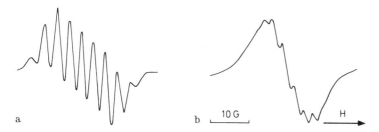

a   b |___10 G___| H

Fig. 12.16. ESR spectra of: a. pyridinyl radical, and b. polypyridinyl radical in tetrahydrofuran solution[16]

12.7. Poly(verdazyl) Radicals

Several authors reported the syntheses of polymers containing verdazyl free radical groups (12.26), which are very stable free radicals within a wide range of conditions[1242, 1261, 1262, 1507, 1592, 1923].

ESR spectrum of verdazyl (12.26) is shown in Figure 12.17.

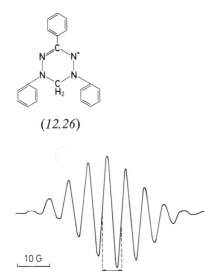

(12.26)

|___10 G___|
|¯6 G¯|

Fig. 12.17. ESR spectrum of 1,3,5-triphenyl-verdazyl radical[1242]

Polyverdazyl radical (12.28) was obtained by the reaction of polyformazon (12.27) with methanolic formaldehyde in the presence of HCl. The synthesis occurs by the following mechanism[1263]:

331

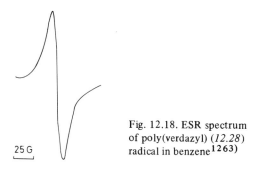

$$(12.10)$$

(12.27)   (12.28)

The green-colored polyverdazyl radical is completely insoluble. Its ESR spectrum is a singlet (Fig. 12.18).

Fig. 12.18. ESR spectrum
of poly(verdazyl) (12.28)
radical in benzene[1263]

25 G

Miura et al.[1508] recorded the anionic polymerization of 1,5-diphenyl-3-(p-vinylphenyl) verdazyl (12.29) and copolymerization of this monomer with styrene and methyl methacrylate. The polymerization of verdazyl monomer needs an excess of anionic initiators (n-butyllithium or sodium naphthalene), suggesting the formation of an intermediate organometallic compound (12.30) and radical (R·):

$$(12.11)$$

(12.29)   (12.30)   (12.31)   (12.32)   (12.33)

The polymer (12.31) formed is hydrolyzed (12.32) and oxidized to polyverdazyl radical (12.33). The polymeric products (12.33) have green color and molecular weights of about 2000, with radical contents up to 60% of the theoretically possible. ESR spectra of verdazyl copolymers are shown in Figure 12.19.

In a later paper, Miura et al.[1509] reported the preparation and the anionic polymerization of methyl and acryloyl derivatives of verdazyl, 1,5-diphenyl-3-(p-metha-

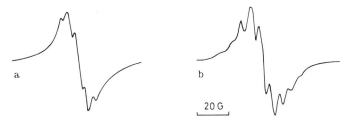

Fig. 12.19. ESR spectrum of copolymers of polyverdazyl with: a. styrene in benzene, and b. methyl methacrylate in dimethylformamide at room temperature[1508]

cryloyloxymethylphenyl)verdazyl (*12.34*), 1,5-diphenyl-3-(p-methacrylo-yloxy-phenyl)verdazyl, and 1,5-diphenyl-3-(p-acryloyloxymethylphenyl)verdazyl (*12.36*) by the following reactions:

$$(12.12)$$

(*12.34*)

(*12.35*)   (12.13)

333

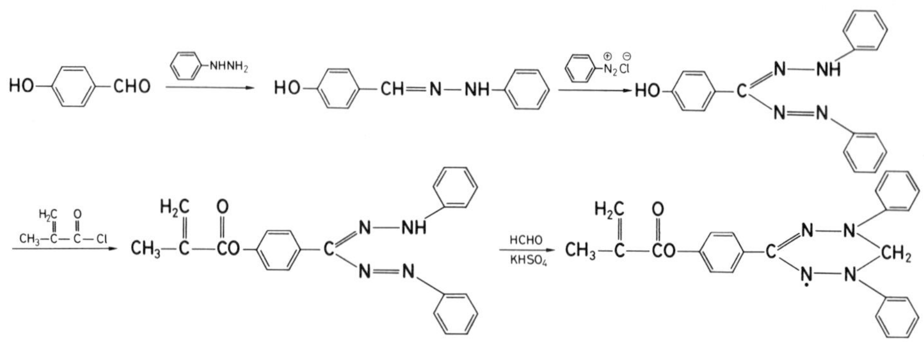

$$(12.36) \qquad (12.14)$$

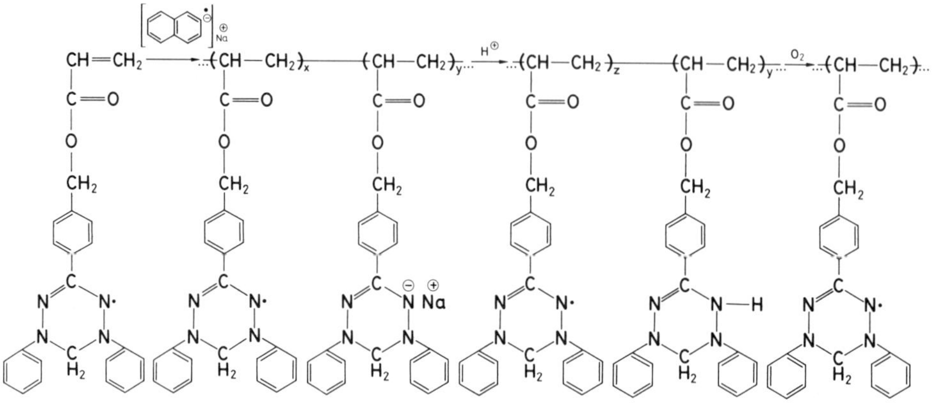

$$(12.37) \qquad (12.15)$$

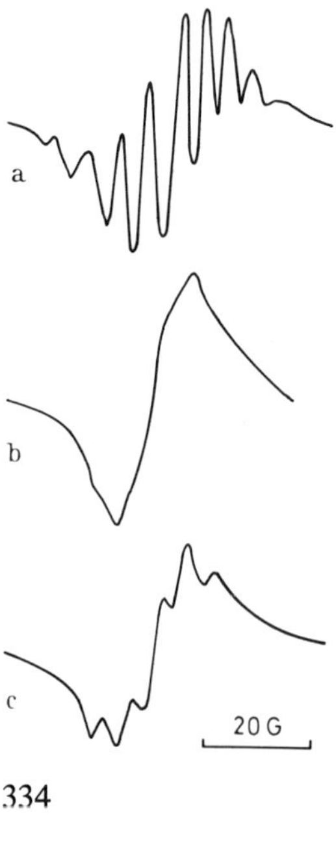

a

b

c

$\overline{}$ 20 G $\overline{}$

Fig. 12.20. ESR spectra of: a. verdazyl, b. poly[1,5-di-phenyl-3-(p-methacryloylmethylphenyl)verdazyl] and c. poly[1,5-diphenyl-3-(p-acryloyloxymethyl-phenyl)verdazyl] in dimethylformamide at room tempera-ture[1509]

ESR spectra of polymeric products (*12.35*) and (*12.37*) are shown in Figure 12.20.

All the polymeric products showed ESR spectra of less-resolved hyperfine structure than their monomers.

Miura et al.[1510] reviewed a new synthesis of verdazyl polymer (*12.40*) by coupling of aminoverdazyl (*12.38*) to a copolymer with maleic anhydride (*12.39*).

Figure 12.21 shows ESR spectrum of the two verdazyl copolymers (*12.40*) and a model compound (*12.41*).

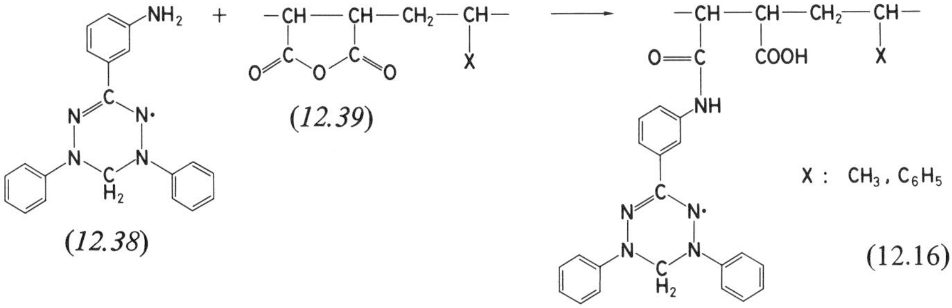

(*12.38*)   (*12.39*)   →   (*12.40*)   (12.16)

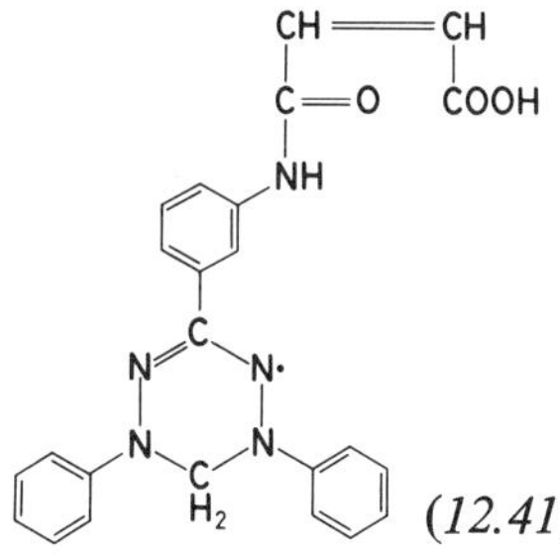

X : CH$_3$, C$_6$H$_5$

(*12.41*)

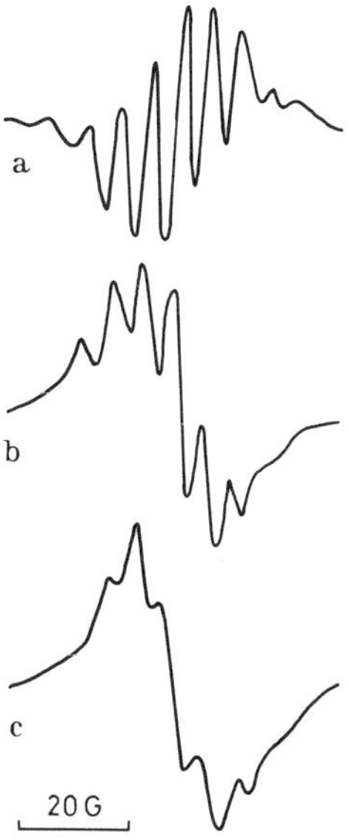

Fig. 12.21. ESR spectra of verdazyl monomer (*12.29*) and copolymers of this monomer with: a. styrene and b. propylene in dimethylformamide[1510]

20 G

335

12.8. Polymers Containing Stable Free Radicals of the Tetrazine Type

2,4-Diphenyl-6-(3-maleoyliminophenyl)-3,4-dihydro-s-tetrazine-1(2H)-yl (*12.42*) was prepared according to the scheme[1506]:

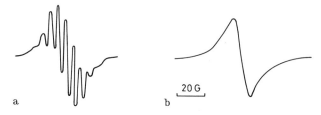

(12.17)

(*12.42*)

and anionically polymerized with a catalytic amount of sodium(1,4-dihydro-1,4-naphthalene). The polymer is a dark green powder, soluble in dimethylformamide and tetrahydrofuran but insoluble in benzene, acetone, and methanol. The ESR spectrum of the monomer shows a hyperfine structure of 9 lines, whereas the spectrum of the polymeric product shows one broad absorption (Fig. 12.22).

a   20 G   b

Fig. 12.22. ESR spectra of: a. 2,4-diphenyl-1,6-(3-maleoyliminophenyl)-3,4-dihydro-2-tetrazine-2(2H)-yl and b. its polymer in dimethylformamide at room temperature in air[1506]

12.9. Polyradical Anions

Radical anions are formed by addition of alkali metal to compounds containing C=C, C=O, C=N, N=N, N=NO or N=O double bonds:

$$A=B + Me \rightarrow \dot{A}-B^- \; Me^+$$ (12.18)

Polymeric ketyls (*12.43*) have been prepared by Braun and Löflund[269] and Greber and Egle[812-814]:

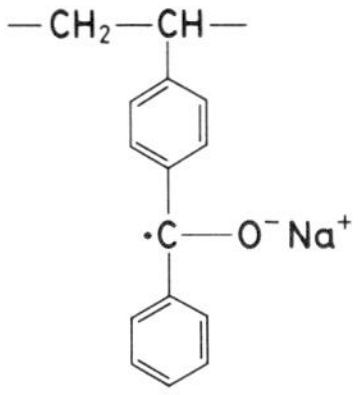

$$-CH_2-CH-$$

(*12.43*)

Polymeric ketyls (*12.43*) may initiate the polymerization of acrylonitrile, methyl methacrylate, and 4-vinylpyridine. Braun et al.[275] have reported graft copolymerization with these polyradical anions.

With metals, Schiff bases give polyradical anions (*12.44*)[1842] which may initiate the polymerization of acrylonitrile:

$$-CH_2-CH-$$

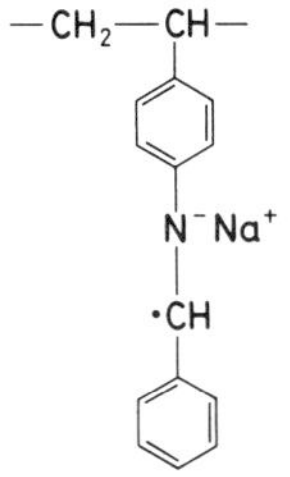

(*12.44*)

Polymeric azocompounds with alkali metals also give polyradical anions of the type (*12.45*)[264].

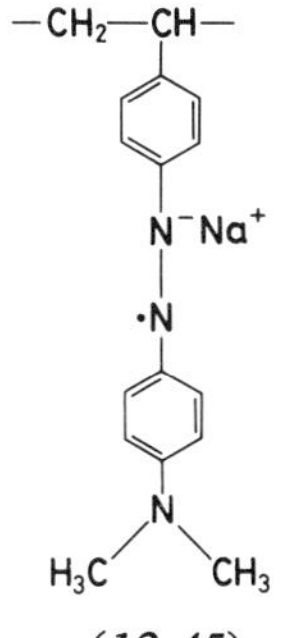

$$-CH_2-CH-$$

(*12.45*)

Poly(p-vinyltriphenylmethyl) radicals (*12.4*) form polyradical anions in reaction with alkali metals in tetrahydrofuran[267]. ESR spectra of these polyradical anions are presented in Figure 12.23.

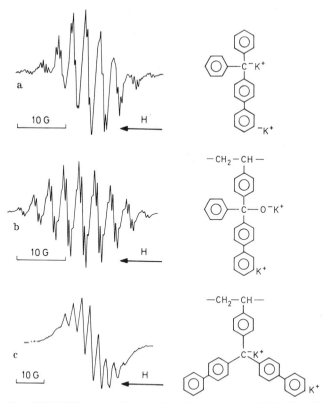

Fig. 12.23. ESR spectra of: a. 4-phenyltriphenylmethylchloride with K/Na in tetrahydrofuran, b. polyradical anion of poly-4-vinyl-4′-phenyltriphenylmethylacid K/Na salt in tetrahydrofuran, c. polyradical anion of poly-4-vinyl-4′-4″diphenyltriphenylmethylchlorid K/Na salt in tetrahydrofuran[267)]

Copolymers of p-vinyl-trans-stilbene and styrene in reaction with sodium in tetrahydrofuran gives deeply colored polyradical anion (12.46)[271)].

$$-CH_2-CH-$$

HC⁻Na⁺

HC·

(12.46)

This polyradical anion initiates the polymerization of acrylonitrile, methyl methacrylate, and styrene by an electron transfer mechanism. When styrene is used, "living polymers" are formed.

338

An ESR study of radical anion of acenaphthylene, obtained at low temperature from copolymers of styrene and acenaphthylene in tetrahydrofuran solution in the presence of sodium, was reported[922].

Polyquinoxaline polymers show a seven-line ESR spectrum (Fig. 12.24) which was attributed to the following polyradical anion (12.47)[1366].

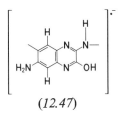

(12.47)

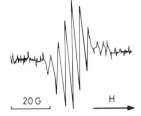

20 G   H

Fig. 12.24. ESR spectrum of polyquinoxaline in hexamethylphosphoramide at 298 K[1366)

After metal reduction, polyphenyl ethers form anion radicals of various length, e.g. (12.48), of which the ESR spectrum is shown in Figure 12.25[596].

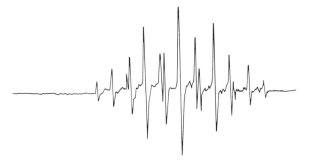

(12.48)

Fig. 12.25. ESR spectrum of polyphenyl ether radical (12.48)[596]

ESR was also used for detecting the presence of poly(phenylmethylsiloxane) radical anion (*12.49*) prepared by metal reduction[597]:

$$CH_3$$
$$\cdot\ Si—O^-K^+$$

(*12.49*)

The interpretation of ESR spectra of the radical anion (*12.49*) encountered several substantial difficulties; temperature and concentration dependences, inability to obtain good spectrum, overlapping spectra, and other gross effects that might be expected from multiple reduction sites within a single polymer molecule.

12.10. ESR Study of Donor-Acceptor Polymer Complexes

Polyradical ions are also formed in electron donor-acceptor complexes (EDA-complexes)[1555]:

$$(D\cdots A \;\rightarrow D^{\sigma+}\cdots A^{\sigma-}) \qquad (D^+\cdots A^-) \qquad D^{\cdot+} + A^{\cdot-} \qquad\qquad (12.19)$$

D-donor A-acceptor   solvated ion pair   free radical

The following EDA complexes have been examined by ESR spectroscopy:
copolymer of styrene and p-dimethylaminostyrene (D) – 2,3-dichloro-5,6-dicyano-p-quinone or chloranile (A)[263, 283],
poly(diphenylamine)(D) – tetracyanoethylene, chloranil or maleic anhydride (A)[1901],
poly(phenylacetylene)(D) – tetracyanoethylene, chloranil or 1,3,5-trinitrobenzene (A)[1724],
poly(2-vinylpyridine)(D) – tetracyanoethylene (A)[208],
poly(N-vinylcarbazole) (D) – antimony(V) chloride or iodide (A)[2443],
poly(N-vinylcarbazole)(D) – maleic anhydride (A)[2132],
poly(4-vinylpyridine)(D) – 2,4-dinitrobromobenzene (A)[1568].

12.11. Organic Polymeric Semiconductors

Polymers with high electrical conductivity and semiconducting properties were first prepared in 1958–1960 in the USSR. Polymeric semiconductors have a different structure from other polymers. The macromolecules of these polymers usually contain alternating systems of double, triple, or cumulative bonds, which bind two carbon atoms: C=C, C≡C, C=C=C=C, carbon and nitrogen: C=N, two nitrogen atoms N=N, benzene rings, etc. (see Table 12.4)[1720, 1746, 2412].

340

Table 12.4. Typical structures present in organic polymeric semiconductors

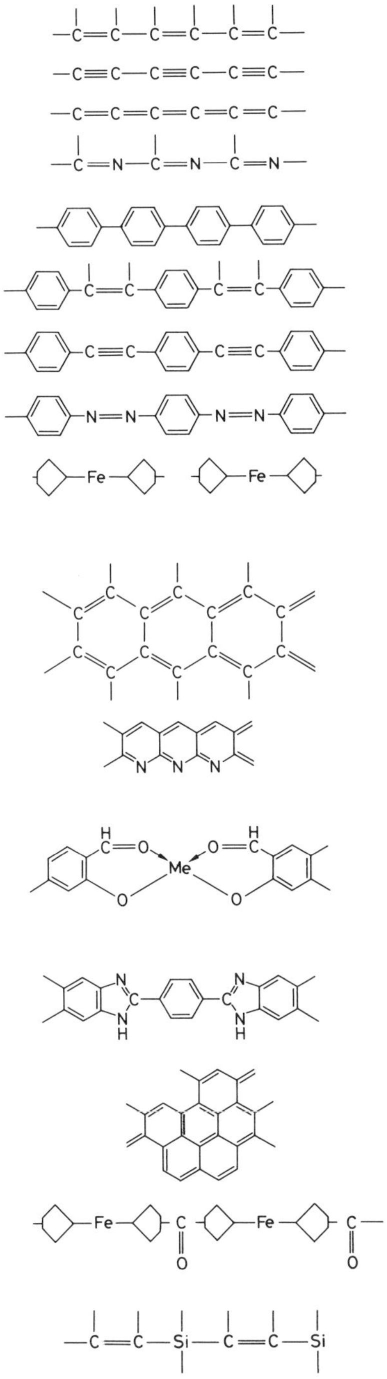

All of these semiconductor polymers contain unpaired electrons which give a narrow ESR signal, characteristic for polyconjugated polymer radicals. Figure 12.26 shows, for example, the ESR spectrum of poly(phenylacetylene) in xylene with a well-resolved structure.

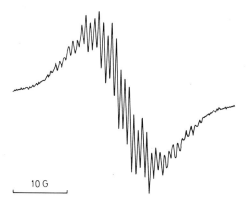

10 G

Fig. 12.26. ESR spectrum of polyphenyl-acetylene in xylene (11.1 wt.%) at 393 K[614]

The spectrum shows over 40 resolvable lines, which are approximately equally spaced about 1 G apart. The g-value is 2.0061 ± 0.00010[427, 614]. The symmetry of the hyperfine structure indicates the presence of a single paramagnetic species, and the even number of lines at least one odd-numbered set of equivalent protons. The large number of lines suggests that the unpaired electron interacts with more than one phenyl ring and the g-value agrees with that of hydrocarbon π-electron radical, in which the spin is delocalized over three aromatic rings[1930]. The g-value is distinctly different from that of an oxygen-centered free radical, indicating that the paramagnetism arises from the pure polymer, and not from its interaction with oxygen.

Different theories of the origin of ESR spectra in semiconductor polymers have been reviewed in detail by Kryszewski[1232], whereas the relation between ESR spectra and electric conductivity of polymers has been reviewed by Ehrlich[612] and Russian scientists[1355]. Readers may find the synthesis, properties, and ESR spectra of semiconducting polymers in original papers devoted to the following polymers:
poly(acetylenes)[82, 588, 688, 878, 1228, 1241, 1755, 1893, 2008, 2010],
poly(butoxyacetylenes)[182],
poly(cyanoacetylenes)[182, 188, 189],
poly(phenylacetylenes)[173, 182, 187, 198, 210–212, 378, 428, 446, 611, 613, 835, 959, 1230, 1582, 1583, 1674, 1724–1726, 1946, 2008, 2011, 2344],
poly(methines)[1304, 1411],
poly(cyclopentadienes)[1306, 1307],
poly(n-heptynes)[182],
poly(nitroacrylovinylenes)[209],
poly(diphenylbutadienes)[1401],
poly(bis(butadienyl)arenes)[2001],
poly(azopolyarylenes)[174, 2174],
poly(phenylenes)[200, 1230, 1718],

poly(hydroxyphenylenes)[1940],
poly(oxyphenylenes)[1189, 1227],
poly(azophenylenes)[1102],
poly(azothiophenylenes)[1759],
poly(phenylene-vinylenes)[1177],
poly(pyridines)[182, 207, 213, 1723, 2368],
poly(quinones)[233, 586, 2023],
poly(semiquinones)[204, 1376],
poly(sulphophenylenequinones)[206],
poly(phenyleneaminochloraniles)[1233],
poly(acenaphthylenes)[2000],
poly(amides)[185, 1278, 1305, 1412, 2059],
poly(phenylimines)[1600],
poly(nitriles)[1020, 1940],
poly(imidazoles)[205, 743, 1222, 1755],
poly(azines)[514],
poly(phenyleneisoxasoles)[968],
poly(pyromellitimide)[743, 1583]
poly(ferrocenes)[183, 203, 587, 1078, 1588, 1719, 1940].

A paramagnetic species was obtained in the reaction of polybenzyls with sulphur at 493 K and examined by ESR spectroscopy[780-782].

12.12. Free Radicals Formed by Pyrolysis of Polymers

ESR studies of stable free radicals obtained during thermolysis (373–573 K) and pyrolysis ($>$673 K) of different polymers such as:
polyethylene[131, 825, 827],
poly(vinyl chloride)[202, 825, 1679, 1680, 2184],
poly(vinylidene chloride)[2184, 2426],
polyacrylonitrile[201, 515, 591, 653, 875, 876, 896, 897, 1076, 1264, 1581, 2343],
poly(vinyl alcohol)[750],
poly(methyl vinyl ketone)[589],
poly(diphenylbutadiene)[445],
fluoroelastomers[522],
poly(divinylbenzene)(crosslinked)[1454, 2426],
polynaphthalenes[2380],
polyarylates[1397],
polypyrroles[242],
polypyromellitimides[223, 323],
polyacetylenes[1892, 2058],
poly(vinyl siloxanes)[2028],
polysaccharides[65, 191, 629, 1044, 1496],
lignin[1567],
and wool[570]
have been reported.

It has been found that the carbonizing temperature substantially affects the number and types of free radicals formed. For all polymers reported, the peak concentration develops at temperatures of 623–933 K, and the spin centres frequently disappear at 973–1173 K. Figure 12.27 shows the ESR spectrum of polyacrylonitrile pyrolized at 673 K.

The typical single line for pyrolized polymers has not yet been interpreted.

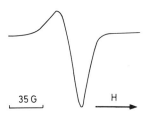

35 G   H

Fig. 12.27. ESR spectrum of polyacrylonitrile pyrolized at 673 K[896]

12.13. Free Radicals in Carbon Black

Studies of paramagnetic properties of carbon blacks have been reported by several authors[56, 66, 412, 442, 443, 526–530, 593, 747, 1538–1540, 1540, 1584, 1688]. In carbon blacks, the initially present radicals are fairly stable up to 773–873 K, but higher temperatures cause a rapid decrease of the spin concentration to a minimum at 1273–1673 K[468]. Above 1673 K to 3273 K radicals reappear[406, 407, 468, 910, 1081, 2016, 2060]. The ESR spectra of these carbons treated at high temperature differ little

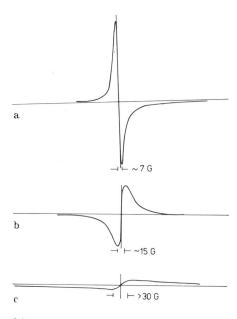

a

⊢⊢ ~7 G

b

⊢⊢ ~15 G

c

⊢ ⊢ >30 G

Fig. 12.28. The effect of the atmosphere on the linewidth of carbon black: a. in vacuum, b. in air and c. in oxygen[1743] in appearance from those treated at lower temperature. Treatment at very high temperatures (>2273 K) gives ESR spectra with increasing asymmetry.

Oxygen affects both the radical yield and the spectrum of observed free radicals[50, 868, 1136, 1743] (Figure 12.28).

The line-width of about 15 G for carbon black measured in air is reduced to about 7 G when the sample is evacuated, and broadens to at least twice the air value when oxygen is introduced to the evacuated sample. The concentration of radicals remains constant and only the line-width changes[571]. The width also varies considerably from one material to another and with sample preparation and environment.

A correlation between concentration of unpaired electrons in carbon black and the reinforcing properties of carbon black in the rubber vulcanization process were also investigated by ESR spectroscopy[618, 727, 1110, 2061]. An ESR study of stable macroradical complexes from polymers such as poly(n-butyl methacrylate) and poly(vinyl butyral) and carbon black has been made[444].

Chapter 13

ESR Study of Ion-Exchange Resins

ESR spectroscopy was also applied to the study of elementary processes in ion-exchangers[783, 784, 1601]. The exhaustive drying of sulphonated polystyrene-based ion-exchange resins produces free-radical centers (Fig. 13.1)[783]. The radicals disappear (a decrease by a factor of more than 100) upon wetting the resin and reappear on drying.

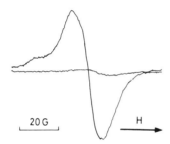

20 G   H

Fig. 13.1. ESR spectra at 298 K for Dowex 50 W – 8% XL ion exchange resin in: a. the ultra-dry state and b. under the saturation absorption conditions[783]

The reversibility of the appearance of the ESR signal upon drying and wetting may be interpreted as being due to the location of the spins. When the free spins appear at relatively inflexible sites, recombination may occur almost quantitatively on reabsorption of water. The role of water in this reversible process suggests further that the radicals may be located near ionic sites, where maximum stresses and local changes of dimensions are expected to arise as hydration of $-SO_3H$ functions occurs. When about four molecules of water per $-SO_3H$ group were absorbed in the resin, the free-spin concentration decreased from $10^{16}-10^{17}$ to about 10^{14} spins g^{-1} and remained on that level with further increase of water absorption.

The line-shapes of ESR spectra of Mn^{2+} ions on sulphonated polystyrene-based ion-exchange resin change also when the degree of hydration and the temperature are varied[784]. ESR spectroscopy has also been applied to study the state of copper ions in a sulphonic resin[2367], and the migration of Mn^{2+} ions in zeolites of the type NaA, NaX and NaY[1749].

ESR study of the effect of γ-irradiation on ion-exchange resins was made by Basinski et al.[1380]. Radiolysis of Zerolit FF-IP, a strongly basic polystyrene iso-porous anion exchange resin, reveals two types of free radicals. One of them was formed from the skeleton of the resin. It gave a narrow resonance line in the spectrum, which was probably due to an oxide or peroxide radical produced by oxidation of the polystyrene chain. The second kind of radical gave a wider resonance line,

346

and was probably formed in the functional group zone of the resin. The recombination rate of the second type was much faster than that of the first type. The radical formed in the functional group zone was less stable in the presence of air.

ESR study of free radicals formed in γ-irradiated and UV-irradiated cation-exchange resin was reported by Russian scientists[1079, 1281].

References

1. Aasa, R., Malmström, B., Saltman, P., Vänngård, T.: Biochem. Biophys. Acta *88*, 430 (1964).
2. Abagyan, G. V., Butyagin, P. Yu.: Biophysika *5*, 785 (1960).
3. Abagyan, G. V., Butyagin, P. Yu.: Dokl. Akad. Nauk SSSR *154*, 1444 (1964).
4. Abagyan, G. V., Butyagin, P. Yu.: Biophysika *9*, 180 (1964).
5. Abagyan, G. V., Butyagin, P. Yu.: Vysokomol. Soedin. *7*, 1410 (1965).
6. Abagyan, G. V., Butyagin, P. Yu.: Biophysika *10*, 763 (1965).
7. Abagyan, G. V., Butyagin, P. Yu.: Trans. Moscow Society of Naturalists *16*, 126 (1966).
8. Abagyan, G. V., Butyagin, P. Yu.: Studia Biophys. *15/16*, 291 (1969).
9. Abagyan, G. V., Krutova, Yu. I., Putilova, I. I., Butyagin, P. Yu.: Biophysika *12*, 820 (1967).
10. Abou-Kais, A., Vedrine, J. C., Massardier, J.: J. Chem. Soc. Faraday Trans. *71*, 1697 (1975).
11. Abragam, A., Bleaney, B.: Electron Paramagnetic Resonance of Transition Ions. London: Oxford Press 1970.
12. Abragam, A., Pryce, M. H. L.: Proc. Roy. Soc. *A205*, 135 (1961).
13. Abraham, R. J., Melville, H. W., Ovenall, D. W., Whiffen, D. H.: Trans. Faraday Soc. *54*, 1133 (1958).
14. Abraham, R. J., Whiffen, D. H.: Trans. Faraday Soc. *54*, 1291 (1958).
15. Abu-El'Khair, B. M., Lachinov, M. B., Zubov, V. P., Kabanov, V. A.: Vysokomol. Soedin. *17A*, 831 (1975).
16. Acosta, J. L., Schulz, R. C.: Makromol. Chem. *145*, 323 (1971).
17. Adam, F. C., Weissman, S. I.: J. Amer. Chem. Soc. *80*, 1518 (1958).
18. Adamic, K.: Stärke *20*, 3 (1968).
19. Adamic, K., Blinc, R.: Proc. Coloq. AMPERE *14*, 780 (1966).
20. Adams, R. F.: Electron Spin Resonance *1*, 81 (1972).
21. Adams, R. F., Staples, T. L., Szwarc, M.: Chem. Phys. Lett. *5*, 474 (1970).
22. Adema, E. H.: J. Polym. Sci. C, *16*, 3643 (1968).
23. Adema, E. H., Bartelnik, H. J. M., Smidt, J.: Rec. Trav. Chim. *80*, 173 (1961).
24. Adema, E. H., Bartelnik, H. J. M., Smidt, J.: Rec. Trav. Chim. *81*, 73 (1962).
25. Adler, G., Ballantine, D., Baysal, B.: J. Polym. Sci. *48*, 195 (1960).
26. Adler, G., Baysal, B.: Mol. Cryst. *6*, 361 (1969).
27. Adler, G., Petropolous, J. M.: J. Phys. Chem. *69*, 3712 (1965).
28. Ahmed, A. U., Rapson, W. H.: J. Polym. Sci. A1, *10*, 1945 (1972).
29. Akhmed-Zade, K. A., Baptizmanskii, V. V., Zakrevskii, V. A., Misrov, S. A., Tomashevskii, E. E.: Vysokomol. Soedin. *14*, 1360 (1972).
30. Akhvlediani, I. G., Shneiker, A. P., Nekhoroshev, V. G., Muromtsev, V. I., Abkin, A. D.: Vysokomol. Soedin. B, *16*, 777 (1974).
31. Aleksandrov, I. V., Ivanova, A. N., Korst, N. N., Lazarev, A. V., Prikhozenko, A. J., Stryukov, U. B.: Mol. Phys. *18*, 681 (1970).
32. Aleksandrova, T. A., Kovarskii, A. L.: Khim. Vysokomol. Soedin., Neftokhim. *1973*, 80.
33. Alger, R. S.: Electron Paramagnetic Resonance, Technique and Application. New York: Interscience-Wiley 1968.
34. Alger, R. S., Anderson, T. H., Webb, L. A.: J. Chem. Phys. *30*, 695 (1959).
35. Alifimov, M. V., Nikolskii, V. G., Buben, N. Ya.: Kinet. Kataliz. *5*, 268 (1964).
36. Allen, P. E. M., Brown, J. K., Obaid, R. M. S.: Trans. Faraday Soc. *59*, 1808 (1963).

37. Allendoerfer, R. D., Eustace, D. J.: J. Phys. Chem. *75*, 2765 (1971).
38. Almanov, G. A., Dzhavakhishvili, L. Ya., Ketiladze, G. D.: Optika i Spektr. *12*, 789 (1962).
39. Alt, L. Ya., Anufrienko, U. F., Tyulikova, T. Ya., Ermakov, Yu. I.: Kinet. Catal. *9*, 1031 (1968).
40. Altshuler, S. A., Kozyrev, B. M.: Electron Paramagnetic Resonance. New York: Academic Press 1964.
41. Amelin, A. V., Pozdnyakov, O. F., Regel, V. R., Sanfirova, T. P.: Soviet Physics-Solid State *12*, 2034 (1971).
42. Andersson, L. O.: Operating Technique in ESR Spectroscopy. Paper presented on 14th Varian EPR Workshop, Zürich, 1975.
43. Andrews, E. H.: Physical Basis of Yield and Fracture. Conf. Proc. Oxford, Sept. 1966, ed. Institute of Physics and Physical Society, London, 1966, p. 127.
44. Andrews, E. H.: Fracture in Polymers. New York: American Elsevier 1968.
45. Andrews, E. H., Reed, P. E.: J. Polym. Sci. B, *5*, 317 (1967).
46. Andrews, E. H., Reed, P. E.: in: Deformation and Fracture of High Polymers, (ed. Kausch, H. H., Hassell, J. A., and Jaffee, R. L.). New York: Plenum Press 1974, p. 259.
47. Andrianov, K. A., Gul, V. E., Chananashivili, L. M., Bulghakov, V. Ya.: in: Kinetics and Mechanism of Polyreactions. Budapest: Akademiai Kiado 1969, vol. 4, p. 217.
48. Angelescu, E., Nicolescu, I. V.: J. Polym. Sci. C, *22*, 203 (1967).
49. Angelescu, E., Nicolau, C., Simon, Z.: J. Amer. Chem. Soc. *88*, 3910 (1966).
50. Antonowicz, K.: J. Chem. Phys. *36*, 2046 (1962).
51. Antonowicz, K.: Polimery *11*, 450 (1966).
52. Antufev, V. V., Dokukina, A. F., Votinov, M. P., Suntsov, E. V., Boldyrev, A. G.: Vysokomol. Soedin. *7*, 380 (1965).
53. Aoki, A., Sawada, S.: Osaka Furitsu Kogyo-Shoreikan Hokoku, No. 40, 68, 1966.
54. Aono, S., Ono, I.: Kogyo Kagaku Zasshi. *68*, 1530 (1965).
55. Arai, S., Shida, S., Yamaguchi, Y., Kuri, Z.: J. Chem. Phys. *37*, 1885 (1962).
56. Araki, K.: Nippon Gomu Kyokaishi *34*, 616 (1961).
57. Araki, K., Campbell, D., Turner, D. T.: J. Polym. Sci. B, *3*, 993 (1965).
58. Araki, K., Campbell, D., Turner, D. T.: J. Polym. Sci. A1, *4*, 2597 (1966).
59. Arbuzov, A. E., Baiglidina, S. Yu., Valitova, F. G., Ryzhmanov, Yu. M., Yablokov, Yu. V.: Izv. Akad. Nauk SSSR *1966*, 1547.
60. Arbuzov, A. E., Valitova, F. G., Garifyanov, N. S., Kozyrev, B. M.: Dokl. Akad. Nauk SSSR *126*, 774 (1959).
61. Arbuzov, A. E., Valitova, F. G.: Dokl. Akad. Nauk SSSR *147*, 99 (1962).
62. Ard, W. B., Shields, H., Gordy, W.: J. Chem. Phys. *23*, 1727 (1955).
63. Arest-Yakubovich, A. A.: Vysokomol. Soedin *6*, 247 (1964).
64. Arlman, E. J.: J. Polym. Sci. *62*, S30 (1962).
65. Armstrong, J. W., Jackson, C., Marsh, H.: Carbon *2*, 227 (1964).
66. Arnold, G. M.: Carbon *5*, 33 (1967).
67. Arrington, C. A., Jr., Falick, A. M., Meyers, R. J.: J. Chem. Phys. *55*, 909 (1971).
68. Arthur, J. C., Jr.: Adv. Macromol. Chem. *2*, 1 (1970).
69. Arthur, J. C., Jr.: High Polym. *5*, 977 (1971).
70. Arthur, J. C., Jr.: J. Polym. Sci., C (36), 53 (1971).
71. Arthur, J. C., Jr.: in: Cellulose and Cellulose Derivatives (ed. Bikales, N. M. and Segal, L.), 2nd ed. vol. 5. New York: Wiley-Interscience 1971, p. 986.
72. Arthur, J. C., Jr.: in: Block and Graft Copolymers. Syracuse, N. Y., Syracuse University Press 1973, p. 295.
73. Arthur, J. C., Jr.: Polym. Prepr. Amer. Chem. Soc., Div. Polym. Chem. *16*, 419 (1975).
74. Arthur, J. C., Jr., Baugh, P. J., Hinojosa, O.: J. Appl. Polym. Sci. *10*, 1591 (1966).
75. Arthur, J. C., Jr., Hinojosa, O.: J. Polym. Sci. C, *36*, 53 (1971).
76. Arthur, J. C., Jr., Hinojosa, O., Bains, M. S.: J. Appl. Polym. Sci. *12*, 1411 (1968).
77. Arthur, J. C., Jr., Hinojosa, O., Tripp, V. W.: J. Appl. Polym. Sci. *13*, 1497 (1969).
78. Arthur, J. C., Jr., Mares, J. T.: J. Appl. Polym. Sci. *9*, 2581 (1965).
79. Arthur, J. C., Mares, J. T., Hinojosa, O.: Text Res. J. *36*, 630 (1966).

80. Arthur, J. C., Jr., Stannois, D. J., Mares, J. T., Hinojosa, O.: J. Appl. Polym. Sci. *11*, 1129 (1967).
81. Asai, M., Tazuke, S., Okamura, S.: J. Polym. Sci. A1, *12*, 45 (1974).
82. Aseev, Yu. G., Kasatochkin, V. I., Nedoshivin, Yu. N., Sladkov, A. M., Barkin, E. I.: Radiospektr. Tverd. Tela *1967*, 408.
83. Ashkinadze, B. M., Likachev, V. A., Ryvkin, S. M., Salamonov, V. M., Yaroshetskii, I. D.: Fiz. Tverd. Tela *8*, 2735 (1966).
84. Assenheim, H. M.: Introduction to Electron Spin Resonance. London: Hilger and Watts 1966.
85. Atchinson, G. J.: J. Polym. Sci. *49*, 385 (1961).
86. Atchinson, G. J.: J. Appl. Polym. Sci. *7*, 1471 (1963).
87. Atherton, N. M.: Lab. Pract. *13*, 1089 (1964).
88. Atherton, N. M.: Electron Spin Resonance *1*, 32 (1972).
89. Atherton, N. M.: Electron Spin Resonance, Theory and Application. New York: Halsted 1973.
90. Atherton, N. M.: Chem. Phys. Lett. *23* 454 (1973).
91. Atherton, N. M.: Electron Spin Resonance *2*, 36 (1974).
92. Atherton, N. M., Henshaw, P. A.: J. Chem. Soc. Perkin Trans. *2*, 258 (1975).
93. Atherton, N. M., Weissman, S. I.: J. Amer. Soc. *83*, 1330 (1961).
94. Atkins, P. W.: Adv. Mol. Relaxation Processes *2*, 121 (1972).
95. Atkins, P. W.: Electron Spin Resonance *1*, 47 (1972).
96. Atkins, P. W., Symons, M. C. R.: The Structure of Inorganic Radicals. Amsterdam: Elsevier 1967.
97. Atwater, H. A.: Amer. Soc. Testing Mater. Spec. Techn. Publ. *384*, 32 (1964).
98. Atwater, H. A.: J. Appl. Phys. *36*, 2220 (1965).
99. Auerbach, I.: Polym. Prepr. Amer. Chem. Soc., Div. Polym. Chem. *3*, 203 (1962).
100. Auerbach, I.: Polym. Prepr. Amer. Chem. Soc., Div. Polym. Chem. *4*, 351 (1963).
101. Auerbach, I.: Polymer *7*, 283 (1966).
102. Auerbach, I.: Polymer *8*, 63 (1967).
103. Auerbach, I.: Polymer *9*, 1 (1968).
104. Auerbach, I., Sanders, L. H.: Polymer *10*, 579 (1969).
105. Ausloos, P.: Ann. Rev. Phys. Chem. *17*, 205 (1966).
106. Ausloos, P.: Fundamental Processes in Radiation Chemistry. New York: Wiley 1968.
107. Auerbuch, P.: Magnetic Resonance and Radiofrequency Spectroscopy. Amsterdam: North-Holland 1969.
108. Azori, M., Tüdös, F., Mohos, B.: Proc. Third Tihany Symp. on Radiation Chem. Budapest: Akademiai Kiado 1972, p. 617.
109. Ayscough, P. B.: Electron Spin Resonance in Chemistry. London: Methuen 1967.
110. Ayscough, P. B., Brooks, B. R., Evans, H. E.: J. Phys. Chem. *68*, 3889 (1964).
111. Ayscough, P. B., Collins, R. G., Dainton, F. S.: Nature *205*, 965 (1965).
112. Ayscough, P. B., Eden, C., Steiner, H.: J. Catal. *4*, 278 (1965).
113. Ayscough, P. B., Evans, E. H.: Trans. Faraday Soc. *60*, 801 (1964).
114. Ayscough, P. B., Evans, H. E., McCann, A. P.: Nature *203*, 1378 (1964).
115. Ayscough, P. B., Ivin, K. J., O'Donnell, J. H.: Proc. Chem. Soc. *1961*, 71.
116. Ayscough, P. B., Ivin, K. J., O'Donnell, J. H.: Trans. Faraday Soc. *61*, 1110 (1965).
117. Ayscough, P. B., Munari, S.: J. Polym. Sci. B, *4*, 503 (1966).
118. Ayscough, P. B., Roy, A. K., Croce, R. G., Munari, S.: J. Polym. Sci. A1, *6*, 1307 (1968).
119. Ayscough, P. B., Thomson, C.: Trans. Faraday Soc. *58*, 1477 (1962).
120. Babenko, V. P., Ryabehenko, S. M., Salkova, E. N.: Cryogenics *7*, 44 (1967).
121. Baberkin, A. S., Pechnikov, V. G., Volkova, E. V.: Kinet. Katal. *7*, 1084 (1966).
122. Backman, D. K., DeVries, K. L.: J. Polym. Sci. A1, *7*, 2125 (1969).
123. Baider, L. M., Voevodskaya, M. V., Fok, N. V.: Khim. Vys. Energ. *5*, 422 (1971).
124. Bains, M. S.: J. Polym. Sci. C, (37), 125 (1972).
125. Bains, M. S., Arthur, J. C., Jr., Hinojosa, O.: J. Phys. Chem. *72*, 2250 (1968).
126. Bains, M. S., Arthur, J. C., Jr., Hinojosa, O.: Prepr. 155th Nation. Meeting Am. Chem. Soc., San Francisco, Calif. April 1968, No. 163.

127. Bains, M. S., Arthur, J. C., Jr., Hinojosa, O.: J. Amer. Chem. Soc. *91*, 4673 (1969).
128. Bains, M. S., Arthur, J. C., Jr., Hinojosa, O.: Inorg. Chem. *9*, 1570 (1970).
129. Bains, M. S., Hinojosa, O., Arthur, J. C., Jr.: Carbohyd. Res. *6*, 233 (1968).
130. Bains, M. S., Hinojosa, O., Arthur, J. C., Jr.: Carbohyd. Res. *20*, 436 (1971).
131. Bakh, N. A., Bityukov, V. D., Vannikov, A. V., Grishina, N. A.: Dokl. Akad. Nauk SSSR *144*, 135 (1962).
132. Balaban, A. T., Frangpol, P. T., Marculescu, M., Bally, J.: Tetrahedron *13*, 258 (1961).
133. Balaban, L., and Kucerovsky, Z.: Chem. Prumysl *13*, 74 (1963).
134. Bamford, C. H., Biddy, A., Eastmond, G. C.: J. Polym. Sci., C, (16), 2417 (1967).
135. Bamford, C. H., Biddy, A., Eastmond, G. C.: Polymer *9*, 653 (1968).
136. Bamford, C. H., Crighton, J. S., Ward, J. C.: Soc. Chem. Ind. (London) Monograph No. 17, 284 (1963).
137. Bamford, C. H., Eastmond, G. C., Sakai, Y.: Nature *200*, 1284 (1963).
138. Bamford, C. H., Eastmond, G. C., Ward, J. C.: Proc. Roy. Soc. (London) A271 (1963).
139. Bamford, C. H., Jenkins, A. D.: in: Formation and Trapping of Free Radicals (ed. Bass, A. M., and Broida, H. P.). New York: Academic Press 1960.
140. Bamford, C. H., Jenkins, A. D., Ingram, D. J. E., Symons, M. C. R. Nature *175*, 894 (1955).
141. Bamford, C. H., Jenkins, A. D., Symons, M. C. R., Townsend, M. G.: J. Polym. Sci. *34*, 181 (1959).
142. Bamford, C. H., Jenkins, A. D., Ward, J. C.: Nature *186*, 713 (1960).
143. Bamford, C. H., Symons, M. C. R., Townsend, M. G.: J. Polym. Sci. *34*, 181 (1959).
144. Bamford, C. H., Ward, J. C.: Polymer *2*, 277 (1961).
145. Bamford, C. H., Ward, J. C.: Trans. Faraday Soc. *58*, 971 (1962).
146. Baraboym, N. K.: Mechanochemistry of Polymers. Moscow: Khimia 1971 (in Russ.).
147. Barabanov, V. S.: Yadiern. Magnet. Rezonans, *1969*, 7.
148. Barchuk, V. I., Dubinsky, A. A., Grinberg, O. Ya., Lebedev, Ya. S.: Chem. Phys. Lett. *34*, 476 (1975).
149. Bargon, J.: J. Polym. Sci. B, *9*, 681 (1971).
150. Bargon, J., Fischer, H.: Z. Naturforsch. *22a*, 1556 (1967).
151. Bargon, J., Fischer, H.: Z. Naturforsch. *23a*, 2109 (1968).
152. Bargon, J., Fischer, H., Johnsen, U.: Z. Naturforsch. *22a*, 1551 (1967).
153. Barkalov, I. M., Goldanskii, V. I., Enikolopyan, N. S., Terekhova, S. F., Trofimova, G. M.: Dokl. Akad. Nauk SSSR *147*, 395 (1962).
154. Barkalov, I. M., Goldanskii, V. I., Enikolopyan, N. S., Terekhova, S. F., Trofimova, G. M.: J. Polym. Sci. C, (4), 897 and 909 (1964).
— 155. Barratt, M. D., Green, D. K., Chapman, D.: Biochem. Biophys. Acta *152*, 20 (1968).
156. Bartelnik, H. J. M., Bos, H., Smidt, J., Vrinssen, C. H., Adema, E. H.: Rec. Trav. Chim. *81*, 225 (1962).
157. Bartenev, G. M., and Lukyanov, I. A.: Plast. Massy *1963*, 46.
158. Bartenev, G. M., Zuyev, Y. S.: Strength and Failure of Viscoelastic Materials. New York: Pergamon Press 1969.
159. Bartoň, J., Horanská, V.: Europ. Polym. J., Suppl. 1969, p. 261.
160. Bartoň, J., Horanská, V.: Makromol. Chem. *157*, 87 (1972).
161. Bartoň, J., Horanská, V.: J. Polym. Sci., Symp. (40), 157 (1973).
162. Bartoň, J. Szöcs, F., Nemček, J.: Makromol. Chem. *124*, 28 (1969).
163. Bartoň, J., Tino, J., Horanská, V.: Makromol. Chem. *164*, 215 (1973).
164. Bartlett, P. D., Funahashi, T.: J. Amer. Chem. Soc. *84*, 2596 (1962).
165. Bass, A. M., Broida, H. P.: Formation and Trapping of Free Radicals. New York: Academic Press 1960.
166. Battaerd, H. A., Tregear, G. W.: Graft Copolymers, Polymer Rev., (16). New York: Interscience 1967.
167. Baudet, J., Berthier, G., Pullman, B.: C. R. Acad. Sci. *254*, 762 (1962).
168. Baugh, P. J., Hinojosa, O., Arthur, J. C., Jr.: J. Phys. Chem. *71*, 1135 (1967).
169. Baugh, P. J., Hinojosa, O., Arthur, J. C., Jr.: J. Appl. Polym. Sci. *11*, 1139 (1967).
170. Baugh, P. J., Hinojosa, O., Arthur, J. C., Jr., Mares, T.: J. Appl. Polym. Sci. *12*, 249 (1968).

171. Baugh, P. J., Hinojosa, O., Mares, T., Hoffman, M. J., Arthur, J. C., Jr.: Text. Res. J. *37*, 942 (1967).
172. Bawn, C. E. H., Ledwith, A., Sambhi, M.: Polymer *12*, 209 (1971).
173. Bazhin, N. M., Chibrikin, V. M., Voevodskii, V. V.: Vysokomol. Soedin. *6*, 1478 (1964).
174. Bazhin, N. M., Terpugova, M. P., Kuznetsov, E. V., Kotlyarevskii, I. L.: Izv. Akad. Nauk SSSR, *1966*, 1154.
175. Baysal, B., Adler, G., Ballantine, D., Colombo, P.: J. Polym. Sci. *44*, 117 (1960).
176. Beachell, H. C., Chang, I. L.: J. Polym. Sci. A1, *10*, 503 (1972).
177. Becconsall, J. K., Clough, S., Scott, G.: Trans. Faraday Soc. *56*, 459 (1960).
178. Becht, J., Fischer, H.: Kolloid. Z. u. Polymere *240*, 766 (1970).
179. Becht, J., Fischer, H.: Angew. Makromol. Chem. *18*, 81 (1971).
180. Beerbower, A., von Rosenberg, A., Cross, N. O.: in: Kirk-Othmer Encyclopedia of Chemical Technology. New York: Wiley-Interscience 1968, vol. 17, p. 65.
181. Beinert, H., Orme-Johnson, W. H.: in: Magnetic Resonance in Biological Systems (ed. Ehrenberg, A., Malmström, B., Vänngård, T.). London: Pergamon Press 1967.
182. Belov, V. F., Oganesov, S. S.: Tr. Moskov. Inst. Neftokhim. Gaz. Promysl. (58), 136 (1965).
183. Below, V. F., Sokolinskaya, T. A., Paushkin, Ya. M., Vishnyakova, T. P.: Dokl. Akad. Nauk SSSR *169*, 831 (1964).
184. Belyakov, V. K., Berlin, V. K., Bukin, I. I., Orlov, V. A., Tarakanov, O. G.: Vysokomol. Soedin. *A 10*, 599 (1968).
185. Belyakov, V. K., Kagan, G. I., Kosobutskii, V. A., Kuznetsov, G. A., Sokolov, L. B.: Vysokomol. Soedin. *B 14*, 657 (1972).
186. Benderskii, V. A., Blumenfeld, L. A.: Dokl. Akad. Nauk SSSR *144*, 813 (1962).
187. Benderskii, V. A., Blumenfeld, L. A., Pristupa, A. I.: Vysokomol. Soedin. *A 9*, 171 (1967).
188. Benes, M., Peska, J., Wichterle, O.: Chem. Ind. (London) *1962*, 562.
189. Benes, M., Peska, J., Wichterle, O.: J. Polym. Sci. C (5), 1377 (1964).
190. Beniska, I., Kavun, S. M., Tarasova, Z. N.: Vysokomol. Soedin. *8*, 893 (1966).
191. Bennett, J. E., Ingram, D. J. E., Tapley, J. G.: J. Chem. Phys. *23*, 215 (1955).
192. Bennett, J. E., Howard, J. A.: Chem. Phys. Lett. *9*, 460 (1971).
193. Bensasson, R. V., Bernas, A., Bodard, M., Marx, R.: J. Chim. Phys. *1963*, 950.
194. Bensasson, R. V., Bernas, A., Bodard, M., Marx, R.: Proc. Tihany Symp. on Radiation Chem. Budapest: Akademiai Kiado 1964, p. 153.
195. Bensasson, R., Durup, M., Dworkin, A., Magat, M., Marx, R., Szwarc, H.: Discussions Faraday Soc. (36), 177 (1963).
196. Bensasson, R., Dworkin, A., Marx, R.: J. Polym. Sci. C (4), 881 (1964).
197. Beringer, R., Castle, J. G.: Phys. Rev. *81*, 82 (1951).
198. Berlin, A. A., Belova, R. N.: Vysokomol. Soedin. *B 9*, 718 (1967).
199. Berlin, A. A., Belyakov, V. K., Nevskii, L. V., Tarakanov, O. G.: Vysokomol. Soedin. *A 9*, 1677 (1967).
200. Berlin, A. A., Cherkashin, M. I.: Izv. Akad. Nauk SSSR *1964*, 568.
201. Berlin, A. A., Dubinskaya, A. M., Moshkovskii, Yu. Sh.: Vysokomol. Soedin. *6*, 1938 (1964).
202. Berlin, A. A., Kasatochkin, V. I., Aseeva, R. M., Finkelshtein, S. B.: Vysokomol. Soedin. *5*, 1303 (1963).
203. Berlin, A. A., Liogonskii, B. I., Parini, V. P.: Vysokomol. Soedin. *5*, 330 (1963).
204. Berlin, A. A., Liogonskii, B. I., Ragimov, A. V., Vonsyatskii, A. V.: Izv. Akad. Nauk SSSR *1963*, 1351.
205. Berlin, A. A., Liogonskii, B. I., Shamraev, G. M., Belova, G. V.: Vysokomol. Soedin. *A 9*, 1936 (1967).
206. Berlin, A. A., Ragimov, A. V., Liogonskii, B. I.: Izv. Akad. Nauk SSSR *1962*, 1863.
207. Berlin, A. A., Razvadovskii, E. F.: Dokl. Akad. Nauk SSSR *140*, 598 (1961).
208. Berlin, A. A., Sherle, A. I., Markova, N. A.: Vysokomol. Soedin. *B 11*, 21 (1969).
209. Berlin, A. A., Vakulskaya, T. N., Zadoncev, B. G., Chauser, M. G., Cherkashin, M. I., Chibirskii, W. M., Chigir, A. N.: Dokl. Akad. Nauk SSSR *182*, 582 (1968).

352

210. Berlin, A. A., Vinogradov, G. A., Kobryanskii, V. M.: Izv. Akad. Nauk SSSR *1970*, 1192.
211. Berlin, A. A., Vinogradov, G. A., Ovchinnikov, A. A.: Chem. Phys. *23*, 406 (1973).
212. Berlin, A. A., Vinogradov, G. A., Ovchinnikov, A. A.: Opt. Spektrosk. *36*, 611 (1974).
213. Berlin, A. A., Zherebtsova, L. V., Razvadovskii, E. F.: Vysokomol. Soedin. *6*, 58 (1964).
214. Berliner, L. J.: Spin Labeling, Theory and Application. New York: Academic Press 1974.
215. Bernard, O. R., Gagnaire, D.: Proc. Colloq. AMPERE *12*, 269 (1963).
216. Bernard, O. R., Gagnaire, D., Servoz-Gavin, P.: J. Chim. Phys. *60*, 1348 (1963).
217. Bersohn, M., Baird, J. C.: An Introduction to Electron Paramagnetic Resonance. New York: Benjamin 1966.
218. Bersohn, M., Thomas, J. R.: J. Amer. Chem. Soc. *86*, 959 (1964).
219. Besev, C., Lund, A., Vanngard, T., Kakansson, R.: Acta Chem. Scand. *17*, 2281 (1963).
220. Beuche, F.: J. Appl. Phys. *281*, 784 (1957).
221. Bichutinskii, A. A., Prokofiev, A. I., Shabalkin, V. A.: Zh. Fiz. Khim. *38*, 983 (1964).
222. Bielski, B. H. J., Gebicki, J. M.: Atlas of Electron Spin Resonance Spectra. New York: Academic Press 1967.
223. Bikbulatova, L. A., Messerle, P. E.: Vysokomol. Soedin. *B14*, 469 (1972).
224. Biktimirov, R. S., Kondratiev, Yu. A.: Khim. Vysok. Energ. *4*, 169 (1970).
225. Birks, J. B.: Photophysics of Aromatic Compounds. New York: Wiley-Interscience 1970.
226. Blasi, F.: Int. J. Radiat. Biol. *11*, 393 (1966).
227. Blayden, H. E., Westcott, D. J.: Proc. Fifth Carbon Conf. Pennsylvania State Univ. London: ed. Pergamon Press 1963, p. 97.
228. Blinc, R., Samec, M.: Stärke *15*, 245 (1963).
229. Blinder, S. M.: J. Chem. Phys. *33*, 1074 (1960).
230. Bloembergen, N., Purcell, E. M., Pound, R. V.: Phys. Rev. *73*, 679 (1948).
231. Blois, M. S., Brown, H. W., Lemmon, R. M., Lindblom, R. O., Weissbluth, M.: Free Radicals in Biological Systems. New York: Academic Press 1961.
232. Blokh, G. A.: Dokl. Akad. Nauk SSSR *129*, 361 (1959).
233. Blumenfeld, L. A., Berlin, A. A., Matveeva, N. G., Kalmanson, A. E.: Vysokomol. Soedin. *1*, 1647 (1959).
234. Blumenfeld, L. A., Voyevodskii, V. V., Siemionov, A. G.: Application of the Electron Spin Resonance in Chemistry. Novosibirsk: Izd. Sybirskoyi Akademii Nauk SSR 1962 (in Russ.).
235. Bobst, A. M.: in: XXIII Intern. Congress of Pure and Applied Chemistry, Macromol. Prepr., vol. 2, p. 1229 (1971).
236. Bobyleva, A. V., Berlyant, S. M., Klinshpont, E. R.: Vysokomol. Soedin. *B15*, 527 (1973).
237. Bobyleva, A. V., Berlyant, S. M., Milinchuk, V. K., Karpov, V. L.: Vysokomol. Soedin. *A13*, 185 (1971).
238. Bodard-Gauthier, M., Marx, R.: J. Polym. Sci. C (16), 4241 (1968).
239. Bodily, D. M., Dole, M.: J. Chem. Phys. *45*, 1428 (1966).
240. Bolon, D. A.: J. Amer. Chem. Soc. *88*, 3148 (1966).
241. Bolt, R. O., Carroll, J. G.: Radiation Effects in Organic Materials. New York: Academic Press 1963.
242. Bolto, B. A., McNeill, R., Weiss, D. E.: Australian J. Chem. *16*, 1090 (1963).
243. Bolton, J. R., Fraenkel, G. K.: J. Chem. Phys. *40*, 3307 (1964).
244. Bolton, J. R., Borg, D. C., Szwartz, H. M.: in: Biological Application of Electron Spin Resonance (ed. Swartz, H. M., Bolton, J. R., and Borg, D. C.). New York: Wiley 1972, p. 63.
245. Bonazzola, L., Fenistein, S., Marx, R.: Mol. Phys. *22*, 689 (1971).
246. Borg, D. C.: in: Biological Applications of Electron Spin Resonance (ed. Swartz, H. M., Bolton, J. R., and Borg, D. C.). New York: Wiley 1972, p. 265.
247. Borsig, E., Klimova, M., Szöcs, F., Tino, J.: Makromol. Chem. *176*, 3401 (1975).
248. Borvkova, L. Ya., Bagdasaryan, Kh. S.: Khim. Vysok. Energii *1*, 340 (1967).
249. Borvkova, L. Ya., Bagdasaryan, Kh. S.: Khim. Vysok. Energii *5*, 337 (1971).
250. Bovey, F. A.: The Effect of Ionizing Radiation on Natural and Synthetic High Polymers. New York: Wiley-Interscience 1958.
251. Bowden, M. J., O'Donnell, J. H.: Macromolecules *1*, 499 (1968).

252. Bowden, M. J., O'Donnell, J. H.: J. Polym. Sci. A1, 7, 1665 (1969).
253. Bowden, M. J., O'Donnell, J. H., Sothman, R. D.: Makromol. Chem. 122, 185 (1969).
254. Bowden, M. J., O'Donnell, J. H., Sothman, R. D.: Macromolecules 5, 269 (1972).
255. Bowen, E. J.: Chemical Aspects of Light. Oxford: Oxford University Press 1946.
256. Bowen, E. J.: in: Advances in Photochemistry (ed. Noyes, W. A., Jr., Hammond, G. S., Pitts, J. N., Jr.). New York: Interscience-Wiley 1963, vol. 1, p. 23.
257. Bower, H. J., McRae, J. A., Symons, M. C. R.: Chem. Commun. 1967, 542.
258. Bowers, K. D., Kamper, R. A., Lustig, C. D.: Proc. Roy. Soc. A 251, 565 (1959).
259. Bowers, K. W.: Adv. Magn. Resonance 1, 317 (1966).
260. Boyer, R. F.: Macromolecules 6, 288 (1973).
261. Box, H. C., Freund, H. G., Budzinski, E.: J. Chem. Phys. 49, 3974 (1968).
262. Braun, D.: J. Polym. Sci. C (24), 7 (1968).
263. Braun, D.: Pure Appl. Chem. 30, 41 (1972).
264. Braun, D., Arcache, G.: Makromol. Chem. 148, 119 (1971).
265. Braun, D., Faust, R. J.: Angew. Chem. 5, 838 (1966).
266. Braun, D., Faust, R. J.: Angew. Chem. 78, 905 (1966).
267. Braun, D., Faust, R. J.: Makromol. Chem. 121, 205 (1969).
268. Braun, D., Hauge, S.: Makromol. Chem. 150, 57 (1971).
269. Braun, D., Löflund, I.: Makromol. Chem. 53, 219 (1962).
270. Braun, D., Löflund, I., Fischer, H.: J. Polym. Sci. 58, 667 (1962).
271. Braun, D., Lucas, F. J. Q., Neumann, W.: Makromol. Chem. 127, 253 (1969).
272. Braun, D., Meier, B.: Makromol. Chem. 175, 791 (1974).
273. Braun, D., Meier, B.: Makromol. Chem.: Suppl. 1975, p. 111.
274. Braun, D., Neumann, W.: Makromol. Chem. 92, 180 (1966).
275. Braun, D., Neumann, W., Arcache, G.: Makromol. Chem. 112, 97 (1968).
276. Braun, D., Neumann, W., Faust, J.: Makromol. Chem. 85, 143 (1965).
277. Braun, D., Peschk, G.: Angew. Chem. 7, 945 (1968).
278. Braun, D., Peschk, G.: Makromol. Chem. 164, 61 (1973).
279. Braun, D., Peschk, G.: Makromol. Chem. 164, 75 (1973).
280. Braun, D., Peschk, G., Hechler, E.: Chimia 21, 536 (1967).
281. Braun, D., Peschk, G., Hechler, E.: Chem. Z. 94, 703 (1970).
282. Braun, D., Platzek, U., Hefter, H. J.: Chem. Ber. 104, 2581 (1971).
283. Braun, D., Sterzel, H. J.: Ber. Bunsen Ges. physik. Chem. 76, 551 (1972).
284. Breck, A. K., Taylor, C. L., Russell, K. E., Wan, J. K. S.: J. Polym. Sci. A1, 12, 1505 (1974).
285. Breitenbach, J. W., Burgmann, H., Olaj, O. F., Vana, N.: Monatsh. Chem. 97, 1479 (1966).
286. Breitenbach, J. W., Campbell, D., Schindler, A.: J. Polym. Sci. B, 3, 1017 (1965).
287. Breitenbach, J. W., Goldenberg, H., Olaj, O. F.: J. Polym. Sci. B, 10, 911 (1972).
288. Breslav, Yu. A., and Kotov, A. G.: Khim. Vysok. Energ. 4, 149 (1970).
289. Bresler, S. E., Dogadkin, B. A., Kazbekov, E. N., Saminskii, E. M., Shershenev, V. A.: Vysokomol. Soedin. 2, 174 (1960).
290. Bresler, S. E., Kazbekov, E. N.: Fortschr. Hochpolym. Forsch. 3, 688 (1964).
291. Bresler, S. E., Kazbekov, E. N.: in: Encyclopedia of Polymer Science and Technology. New York: Wiley-Interscience 1966, vol. 5, p. 669.
292. Bresler, S. E., Kazbekov, E. N.: Uspek. Khim. 36, 720 (1967).
293. Bresler, S. E., Kazbekov, E. N., Fomichev, V. N.: Kinet. Kataliz. 6, 820 (1965).
294. Bresler, S. E., Kazbekov, E. N., Fomichev, V. N., Shadrin, V. N.: Vysokomol. Soedin. B12, 678 (1970).
295. Bresler, S. E., Kazbekov, E. N., Fomichev, V. N., Shadrin, V. N.: Makromol. Chem. 157, 167 (1972).
296. Bresler, S. E., Kazbekov, E. N., Fomichev, V. N., Szöcs, F., Smejtek, P.: Fiz. Tverd. Tela 5, 675 (1963).
297. Bresler, S. E., Kazbekov, E. N., Saminskii, E. M.: Vysokomol. Soedin. 1, 132 (1959).
298. Bresler, S. E., Kazbekov, E. N., Saminskii, E. M.: Vysokomol. Soedin. 1, 1374 (1959).
299. Bresler, S. E., Kazbekov, E. N., Saminskii, E. M.: Rubber Chem. Technol. 33, 469 (1960).

300. Bresler, S. E., Kazbekov, E. N., Saminskii, E. M.: J. Polym. Sci. *52*, 119 (1961).
301. Bresler, S. E., Kazbekov, E. N., Shadrin, V. N.: Makromol. Chem. *175*, 2875 (1974).
302. Bresler, S. E., Kazbekov, E. N., Shadrin, V. N.: Vysokomol. Soedin. *A17*, 507 (1975).
303. Bresler, S. E., Osminskaya, A., Popov, A.: Kolloid Z. *20*, 403 (1958).
304. Bresler, S. E., Osminskaya, A., Popov, A.: Zh. Tekhn. Fiz. *29*, 358 (1959).
305. Bresler, S. E., Poddubny, I. Ya., Tsereteli, Yu.: Vysokomol. Soedin. *B11*, 151 (1969).
306. Bresler, S. E., Zhurkov, S. N., Kazbekov, E. N., Saminskii, E. M., Tomashevskii, E. E.: Zh. Tekh. Fiz. *29*, 358 (1959).
307. Bresler, S. E., Zhurkov, S. N., Saminskii, E. M., Kazbekov, E. N., Tomashevskii, E. E.: Rubber Chem. Technol. *33*, 462 (1960).
308. Bretschneider, J., Wallenfels, K.: Tetrahedron *24*, 1063 (1968).
309. Briere, R., Lemaire, H., Rassat, A.: Bull. Soc. Chim. Fr. *1965*, 3273.
310. Brintzinger, H. H.: J. Amer. Chem. Soc. *89*, 6871 (1967).
311. Brivati, J. A., Gross, J. M., Symons, M. C. R., Tinling, D. J. A.: J. Chem. Soc. *1965*, 6504.
312. Brodskii, A. I., Fomenko, A. S., Abramova, T. M., Dareva, E. P., Galina, A. A., Furman, E. G., Kotorlenko, L. A., Gardenina, A. P.: Vysokomol. Soedin. *7*, 116 (1965).
313. Brodskii, A. I., Fomenko, A. S., Abramova, T. M., Furman, E. G., Dareva, E. P., Kukhtenko, I. I., Galina, A. A.: Dokl. Akad. Nauk SSSR *156*, 1147 (1964).
314. Brotherus, J., Törmälä, P.: Finska Kemistsamf. Medd. *81*, 49 (1972).
315. Brown, D. M., Sainton, F. S.: Rad. Res. Rev. *1*, 241 (1968).
316. Brown, D. W., Florin, R. E., Wall, L. A.: J. Phys. Chem. *66*, 2602 (1962).
317. Brown, I. M.: Chem. Phys. Lett. *17*, 404 (1972).
318. Brown, I. M., Thrush, B. A., Tuck, A. F.: Proc. Roy. Soc. *A 302*, 311 (1968).
319. Brown, J. R., O'Donnell, J. H.: Macromolecules *5*, 109 (1972).
320. Brown, R. T., DeVries, K. L., Williams, M. L.: in: Polymer Networks: Structural and Mechanical Properties (ed. Chompff, A. J.) Plenum Press 1971, p. 409.
321. Brown, R. T., DeVries, K. L., Williams, M. L.: J. Polym. Sci. B, *10*, 327 (1972).
322. Browning, H. L., Jr., Ackerman, H. D., Patton, H. W.: J. Polym. Sci. A1, *4*, 1433 (1966).
323. Bruck, S. D.: Polymer *6*, 319 (1965).
324. Bruk, M. A., Abkin, A. D., Khomikovskii, P. M.: Dokl. Akad. Nauk SSSR *149*, 1322 (1963).
325. Bruk, M. A., Chuiko, K. K., Eroshina, L. V., Aulov, V. A., Abkin, A. D.: Vysokomol. Soedin. *A14*, 794 (1972).
326. Brumby, S.: Z. Naturforsch. *25a*, 12 (1970).
327. Bryant, W. M. D.: J. Polym. Sci. *6*, 359 (1951).
328. Bubonov, N. N., Prokofiev, A. I., Volodkin, A. A., Belostotskaya, I. S., Ershov, V. V.: Dokl. Akad. Nauk SSSR *210*, 354 (1873).
329. Buchachenko, A. L.: Optika i Spektroskopia *13*, 910 (1962).
330. Buchachenko, A. L.: Stable Radicals. New York: Consultant Bureau 1965.
331. Buchachenko, A. L.: Zh. Fiz. Khim. *41*, 2586 (1967).
332. Buchachenko, A. L., Khloplyankina, M. S., Neiman, M. B.: Dokl. Akad. Nauk SSSR *143*, 146 (1962).
333. Buchachenko, A. L., Lebedev, Ya. S., Neiman, M. B.: Advances in Physics and Chemistry of Polymers (ed. Rogovin, Z. A.). Moscow: Khimia 1970, p. 407 (in Russ.).
334. Buckmaster, H. A., Chatterjee, R., Dering, J. C., Fry, D. J. I., Shing, Y. H., Skirrow, J. D., Venkatesan, B.: J. Magn. Resonance *4*, 113 (1971).
335. Budanova, G. P., Mazurek, V. V.: Vysokomol. Soedin. *A12*, 1062 (1970).
336. Bueche, F.: Physical Properties of Polymers. New York: Wiley 1961.
337. Bukanaeva, F. M., Pecherskaya, Ya. I., Kazanskii, V. B., Dzisko, V. A.: Kinet. Kataliz. *3*, 358 (1962).
338. Bulla, I., Törmälä, P., Lindberg, J. J.: Finn. Chem. Lett. *1974*, 129.
339. Bulla, I., Törmälä, P., Lindberg, J. J.: Acta Chem. Scand. *A29*, 89 (1975).
340. Bullock, A. T., Burnett, G. M., Kerr, C. M. L.: Europ. Polym. J. *7*, 791 (1971).
341. Bullock, A. T., Burnett, G. M., Kerr, C. M. L.: Europ. Polym. J. *7*, 1011 (1971).
342. Bullock, A. T., Butterworths, J. H., Cameron, G. G.: Europ. Polym. J. *7*, 445 (1971).
343. Bullock, A. T., Cameron, G. G., Elsom, J. M.: Polymer *15*, 74 (1974).

344. Bullock, A. T., Cameron, G. G., Nicol, B. D.: Chem. Zvesti 26, 193 (1972).
345. Bullock, A. T., Cameron, G. G., Smith, O.: Polymer 13, 89 (1972).
346. Bullock, A. T., Cameron, G. G., Smith, P. M.: J. Polym. Sci. A2, 11, 1263 (1973).
347. Bullock, A. T., Cameron, G. G., Smith, P. M.: Polymer 14, 525 (1973).
348. Bullock, A. T., Cameron, G. G., Smith, P. M.: J. Phys. Chem. 77, 1635 (1973).
349. Bullock, A. T., Cameron, G. G., Smith, P. M.: J. Chem. Soc. Faraday Transactions 70, 1202 (1974).
350. Bullock, A. T., Cameron, G. G., Smith, P. M.: Makromol. Chem. 176, 2153 (1975).
351. Bullock, A. T., Cameron, G. G., Smith, P. M.: Europ. Polym. J. 11, 617 (1975).
352. Bullock, A. T., Griffiths, W. E.: J. Polym. Sci. A2, 6, 1451 (1968).
353. Bullock, A. T., Griffiths, W. E., Sutcliffe, L. H.: Trans. Faraday Soc. 63, 1846 (1967).
354. Bullock, A. T., Sutcliffe, L. H.: Trans. Faraday Soc. 60, 625 (1964).
355. Bullock, A. T., Sutcliffe, L. H.: Trans. Faraday Soc. 60, 2112 (1964).
356. Burlant, W. B., Hoffman, A.: Block and Graft Copolymers. New York: Reinhold 1960.
357. Burrell, E. J., Jr.: J. Amer. Chem. Soc. 83, 574 (1961).
358. Burton, M.: Adv. Chem. Ser. No. 80, p.
359. Burton, M., Smith, J. S., Magee, J. L.: Comparative Effects of Radiation. New York: Wiley 1960.
360. Butyagin, P. Yu.: Dokl. Akad. Nauk SSSR 140, 145 (1961).
361. Butyagin, P. Yu.: Dokl. Akad. Nauk SSSR 148, 129 (1963).
362. Butyagin, P. Yu.: Vysokomol. Soedin. 5, 1829 (1963).
363. Butyagin, P. Yu.: Dokl. Akad. Nauk SSSR 165, 103 (1965).
364. Butyagin, P. Yu.: Vysokomol. Soedin. A9, 136 (1967).
365. Butyagin, P. Yu.: Usepkh. Khim. 40, 1935 (1971).
366. Butyagin, P. Yu.: Pure Appl. Chem. 30, 57 (1972).
367. Butyagin, P. Yu.: Chem. Techn. 25, 258 (1973).
368. Butyagin, P. Yu.: Vysokomol. Soedin. A16, 63 (1974).
369. Butyagin, P. Yu., Abagyan, G. V.: in: Physikalische Chemie biogener Makromolekule, II, Jenaer Symp., 1963. Berlin: Akademie Verlag 1964.
370. Butyagin, P. Yu., Berlin, A. A., Kalmanson, A. E., Blumenfeld, L. A.: Vysokomol. Soedin. 1, 865 (1959).
371. Butyagin, P. Yu., Drozdovskii, V. F., Razgon, D. R., Kolbanev, I. V.: Fiz. Tverd. Tela 7, 941 (1965).
372. Butyagin, P. Yu., Dubinskaya, A. M.: Vysokomol. Soedin. B9, 103 (1967).
373. Butyagin, P. Yu., Dubinskaya, A. M., Kolbanev, I. V., Radstig, V. A.: Radiospektrosk. Tverd. Tela Dokl. Vses. Soveshch. Krasnoyarsk, USSR, 1964, 414.
374. Butyagin, P. Yu., Dubinskaya, A. M., Radstig, V. A.: Usepkh. Khim. 37, 593 (1969).
375. Butyagin, P. Yu., Kolbanev, I. V., Dubinskaya, A. M., Kisluk, M. U.: Vysokomol. Soedin. A10, 2265 (1968).
376. Butyagin, P. Yu., Kolbanev, I. V., Radstig, V. A.: Fiz. Tverd. Tela 5, 2257 (1963).
377. Butyagin, P. Yu., Radstig, V. A.: Plaste Kautschuk 1972, 82.
378. Byrd, N. R., Kleist, F. K., Stamires, D. N.: J. Polym. Sci. A2, 10, 957 (1972).
379. Cadena, D. G., Rowlands, J. R.: J. Chem. Soc. B, 1968, 488.
380. Cafasso, F., Sundheim, B. R.: J. Chem. Phys. 31, 809 (1959).
381. Calvert, J. G., Pitts, N. J., Jr.: Photochemistry. New York: Wiley 1966.
382. Cambrisson, J., Uebersfeld, J.: C. R. Acad. Sci. 238, 1397 (1954).
383. Campbell, D.: Macromol. Rev. 4, 91 (1970).
384. Campbell, D.: J. Polym. Sci. B, 8, 313 (1970).
385. Campbell, D., Araki, K., Turner, D. T.: J. Polym. Sci. A1, 4, 2597 (1966).
386. Campbell, D., Charlesby, A.: Europ. Polym. J. 9, 301 (1973).
387. Campbell, D., Loney, F. D.: Austral. J. Chem. 15, 642 (1962).
388. Campbell, D., Peterlin, A.: J. Polym. Sci. B, 6, 481 (1968).
389. Campbell, D., Turner, D. T.: J. Polym. Sci. A1, 5, 2199 (1967).
390. Campbell, D., Turner, D. T.: J. Polym. Sci. B, 6, 1 (1968).
391. Campbell, D., Williams, J. L., Stannett, V.: J. Polym. Sci. A1, 7, 429 (1969).
392. Canbäck, G., Rånby, B.: Macromolecules (in press).

393. Carrington, A.: Proc. Roy. Soc. *A 302,* 291 (1968).
394. Carrington, A., McLachlan, A. D.: Introduction to Magnetic Resonance. New York: Harper and Row 1969, IInd. ed.
395. Carrington, A., Stein, G.: Nature *193,* 976 (1962).
396. Carrington, R. A.: Computers for Spectroscopists. New York: Halsted 1974.
397. Carstensen, P.: Dansk Kemi *49,* 97 (1968).
398. Carstensen, P.: Acta Polytechn. Scand. Chem. No. 93 (1970).
399. Carstensen, P.: Makromol. Chem. *135,* 219 (1970).
400. Carstensen, P.: Makromol. Chem. *142,* 131 (1971).
401. Carstensen, P.: Makromol. Chem. *142,* 145 (1971).
402. Carstensen, P.: in: ESR Application to Polymer Research (Ed. Kinell, P. O., Rånby, B., Runnström-Reio, V.). Stockholm: Almqvist-Wiksell 1973, p. 159.
403. Carstensen, P., Rånby, B.: in: Radiation Research. Amsterdam: North-Holland 1967, p. 297.
404. Caspers, W. J.: Theory of Spin Relaxation. New York: Interscience-Wiley 1964.
405. Casteleijn, G., Bosh, J. J., Shmidt, J.: J. Appl. Phys. *39,* 4375 (1968).
406. Castle, J. G., Jr.: Phys. Rev. *92,* 1063 (1953).
407. Castle, J. G., Jr.: Phys. Rev. *94,* 1410 (1954).
408. Castle, J. G., Jr., Chester, P. F., Wagner, P. E.: Phys. Rev. *119,* 953 (1960).
409. Ceresa, R. J.: Block and Graft Copolymerization. New York: Wiley, vol. 1, 1973 and vol. 2, 1974.
410. Ceresa, R. J.: in: Encyclopedia of Polymer Science and Technology. New York: Wiley-Interscience 1965, vol. 2, p. 485.
411. Cernia, E., Mantovani, E., Marconi, W., Mazzei, M., Palladino, N., Zanobi, A.: J. Appl. Polym. Sci. *19,* 15 (1975).
412. Cerutti, M., Theobald, J., Uebersfeld, J.: C. R. Acad. Sci. *256,* 3029 (1963).
413. Chachaty, C.: J. Chim. Phys. *64,* 608 (1967).
414. Chachaty, C., Forchioni, A.: C. R. Acad. Sci. *C264,* 1421 (1967).
415. Chachaty, C., Forchioni, A.: J. Polym. Sci. A1, *10,* 1905 (1972).
416. Chachaty, C., Hayon, E.: Nature *200,* 59 (1963).
417. Chachaty, C., Hayon, E.: J. Chim. Phys. *61,* 1115 (1964).
418. Chachaty, C., Latimer, M., Forchioni, A.: J. Polym. Sci. A1, *13,* 189 (1975).
419. Chachaty, C., Marx, R.: J. Chim. Phys. *1961,* 787.
420. Chachaty, C., Marx, R.: J. Chim. Phys. *1962,* 792.
421. Chachaty, C., Shmidt, M. C.: J. Chim. Phys. Chim. Biol. *62,* 527 (1965).
422. Chadwick, K. H.: Versl. Lanbouwk. Onderz. (Neth.), No. 763 (1971).
423. Chalfont, G. R., Perkins, M. J.: J. Chem. Soc., B *1970,* 401.
424. Chalfont, G. R., Perkins, M. J., Horsfield, A.: J. Amer. Chem. Soc. *90,* 7141 (1968).
425. Chambers, K. W., Collinson, E., Dainton, F. S.: Trans. Faraday Soc. *66,* 142 (1970).
426. Chance, B., Bicking, L., Legallais, V.: in: Free Radical in Biological Systems (Ed. Blois, M. S., Brown, H. W., Lemmon, R. M., Lindblom, R. O., Weissbluth, M.). New York: Academic Press 1961, p. 101.
427. Chang, S. H. C., Mertzluft, E. C., Ehrlich, P., Allendoerfer, R. D.: Macromolecules (in press).
428. Chang, S. H. C., Mertzluft, E. C., Ehrlich, P., Allendoerfer, R. D.: Polym. Prepr. Amer. Chem. Soc., Div. Polym. Chem. *16,* 723 (1975).
429. Chapiro, A.: Polymerizations en phase solide, actions chimiques et biologiques des radiations *10,* 189 (1966).
430. Chapiro, A.: Radiation Chemistry of Polymeric Systems. New York: Wiley 1962.
431. Chapiro, A.: Adv. Chem. Ser. No. 66, 22 (1967).
432. Chapiro, A.: in: Encyclopedia of Polymer Science and Technology. New York: Wiley-Interscience 1969, vol. 11, 702.
433. Chapman, T. M.: in: Encyclopedia of Polymer Science and Technology. New York: Wiley-Interscience 1968, vol. 9, p. 275.
434. Charles, M. W.: J. Polym. Sci. A2, *10,* 1609 (1972).
435. Charlesby, A.: Radiation Sources. New York: MacMillan 1964.

436. Charlesby, A.: Proc. Second Tihany Symp. on Radiation Chem. Budapest: Akademiai Kiado 1962, p. 175.
437. Charlesby, A.: Atomic Radiation and Polymers. London: Pergamon Press 1960.
438. Charlesby, A., Campbell, D.: in: ESR Application to Polymer Research (Ed. Kinell, P. O., Rånby, B., Runnström-Reio, V.). Stockholm: Almqvist-Wiksell 1973, p. 147.
439. Charlesby, A., Libby, D., Ormerod, M. G.: Proc. Roy. Soc. *A262*, 207 (1961).
440. Charlesby, A., Patridge, R. H.: Proc. Roy. Soc. *A283*, 329 (1965).
441. Charlesby, A., Thomas, D. K.: Proc. Roy. Soc. *A269*, 104 (1962).
442. Charlier, A., Tagland, P., Donnet, J. B., Lahaye, J., Metzger, J., Papier, E.: Carbon *3*, 321 (1965).
443. Charlier, M. F., Kappel, G., Malan, F., Charlier, A., Tagland, P.: Carbon *8*, 692 (1970).
444. Chatterji, A. K.: Polym. Prepr. Amer. Chem. Soc., Div. Polym. Chem. *7*, 535 (1966).
445. Chauser, M. G., Cherkashin, M. I., Kushneryov, M. Ya., Vakulskaya, T. I., Protsuk, T. F., Berlin, A. A.: Vysokomol. Soedin. *A10*, 916 (1968).
446. Chauser, M. G., Vakulskaya, T. I., Vinogradov, G. A., Cherkashin, M. I., Berlin, A. A.: Izv. Akad. Nauk SSSR *1971*, 1591.
447. Che, M., Tench, A. J.: J. Polym. Sci. B, *13*, 345 (1975).
448. Chen, K. S., Hirota, N.: Tech. Chem. (N.Y.), *6*, 565 (1974).
449. Chen, M. M., Sane, K. V., Walter, R. I., Weil, J. A.: J. Amer. Chem. Soc. *65*, 713 (1961).
450. Chiang, Ren-Tai, Tseng, Der-Ling: Ho Tzu K'o Hsueh (Taiwan) *11*, 41 (1974).
451. Chiang, T. C., Sibilia, J. P.: J. Polym. Sci. A1, *10*, 605 (1972).
452. Chiang, T. C., Sibilia, J. P.: J. Polym. Sci. A2, *10*, 2249 (1972).
453. Chiang, Y. S., Craddock, J., Mickewich, D. M., Turkevich, J.: J. Phys. Chem. *70*, 3509 (1966).
454. Chidambareswaran, P. K., Sundaraman, V., Prakesh, J., Verma, N. C., Singh, B. B.: J. Polym. Sci. A1, *9*, 2651 (1971).
455. Chidambareswaran, P. K., Sundaraman, V., Singh, B. B.: J. Polym. Sci. A1, *10*, 2655 (1972).
456. Chien, J. C. W., Boss, C. R.: J. Amer. Chem. Soc. *89*, 571 (1967).
457. Chien, J. C. W., Boss, C. R.: J. Polym. Sci. A1, *5*, 1683 (1967).
458. Chien, J. C. W., Boss, C. R.: J. Polym. Sci. A1, *5*, 3091 (1967).
459. Chirkov, A. K., Matevosyan, R. O.: Zh. Eksper. Teor. Fiz. *33*, 1053 (1957).
460. Chung, Y. J., Squire, D. R., Stannett, V.: J. Macromol. Sci. Chem. *A8*, 1401 (1974).
461. Chung, Y. J., Takeda, K., Williams, Ff.: Macromolecules *3*, 264 (1970).
462. Chung, Y. J., Yamakawa, S., Stannett, V.: Macromolecules *7*, 204 (1974).
463. Claxon, T. A., Gough, T. E., Symons, C. R.: Trans. Faraday Soc. *62*, 279 (1962).
464. Clay, M. R., Charlesby, A.: Europ. Polym. J. *11*, 187 (1975).
465. Cole, T., Heller, H. C.: J. Chem. Phys. *42*, 1668 (1965).
466. Coles, R. B., Nicholls, C. H.: J. Soc. Dyers Colour. *91*, 19 (1975).
467. Collins, M. A.: Nature *193*, 1061 (1962).
468. Collins, R. L., Bell, M. D., Kraus, G.: J. Appl. Phys. *30*, 56 (1959).
469. Compton, R. N., Huebner, R. H.: in: Advances in Radiation Chemistry (Ed. Burton, M., and Magee, J. L.). New York: Wiley 1969, vol. 1, p. 245.
470. Connor, H. D., Shimada, K., Szwarc, M.: Macromolecules *5*, 801 (1972).
471. Connor, H. D., Shimada, K., Szwarc, M.: Chem. Phys. Lett. *14*, 402 (1972).
472. Coogan, C. K., Ham, N. S., Stuart, S. N., Pilbrow, J. R., Wilson, G.: Magnetic Resonance. New York: Plenum 1970.
473. Cook, J. B., Wyard, S. J.: Nature, London *210*, 526 (1966).
474. Cook, J. B., Wyard, S. J.: Int. J. Rad. Biol. *11*, 357 (1966).
475. Cooper, J. W., Griller, D., Ingold, K. U.: J. Amer. Chem. Soc. *97*, 233 (1975).
476. Cooper, W., Vaugham, G.: Progr. Polym. Sci. *1*, 91 (1967).
477. Copeland, E. S.: in: Biological Application of Electron Spin Resonance (Ed. Swartz, H. M., Bolton, J. R., Borg, D. C.). New York: Wiley 1972, p. 449.
478. Coopinger, G.: J. Amer. Chem. Soc. *79*, 501 (1957).
479. Coopinger, G., Swallen, J.: J. Amer. Chem. Soc. *83*, 4900 (1961).
480. Correia, P., Bandeira, M. H., Novais, H.: XXIII Inter. Congress of Pure and Applied Chemistry, Marcomol. Prepr., vol. 1, p. 384 (1971).

481. Corvaja, C., Fischer, H., Giacometti, G.: Z. physik. Chem. NF, 45, 1 (1965).
482. Costaschuk, F. M., Gilson, D. F. R., Pierre, L. E. St.: Macromolecules 4, 333 (1971).
483. Cowley, A. H., Huoosh, M. H.: J. Amer. Chem. Soc. 88, 2595 (1966).
484. Cozzens, R. F., Moniz, W. B., Fox, R. B.: J. Chem. Phys. 48, 581 (1968).
485. Cox, A., Kemp, T. J.: Introduction to Photochemistry. New York: McGraw-Hill 1971.
486. Cracco, F. A., Arvia, A. J., Dole, M.: J. Chem. Phys. 37, 2449 (1962).
487. Crick, F. H. C., Watson, J. D.: Proc. Roy. Soc. London A223, 80 (1954).
488. Crist, B., Peterlin, A.: Makromol. Chem. 171, 211 (1973).
489. Cross, P. E.: Nature 208, 892 (1965).
490. Crouzet, C., Marchal, J.: in: Kinetics and Mechanism of Polyreactions. Budapest: Akademiai Kiado 1969, vol. 5, 259.
491. Cucinella, S., Mazzer, A., Marconi, W., Busetto, C.: J. Macromol. Sci. Chem.: A4, 1549 (1970).
492. Cundall, R. B., Gilbert, A.: Photochemistry. New York: Appleton 1970.
493. Cusumoto, C., Troupp, G. J.: Phys. Lett. 44, 441 (1973).
494. Czapski, G.: J. Phys. Chem. 75, 2957 (1971).
495. Czvikovszki, T., Marx, R.: in: Kinetics and Mechanism of Polyreactions. Budapest: Akademiai Kiado 1969, vol. 4, 237.
496. Czvikovszki, T., Marx, R.: J. Chim. Phys. Physicochim. Biol. 68, 1660 (1971).
497. Dainton, F. S., Keene, J. P., Kemp, T. J., Salmon, G. A., Teply, J.: Proc. Chem. Soc. 1964, 265.
498. Dainton, F. S., Salmon, G. A.: Proc. Roy. Soc. A285, 319 (1965).
499. Dainton, F. S., Salmon, G. A., Teply, J.: Proc. Roy. Soc. A286, 27 (1965).
500. Dainton, F. S., Salmon, G. A., Wardman, P.: Proc. Roy. Soc. A313, 1 (1969).
501. Dainton, F. S., Salmon, G. A., Wardman, P., Zucker, U.: Proc. Second Tihany Symp. on Radiation Chem. Budapest: Akademiai Kiado 1967, p. 247.
502. Dalal, N. S., Kennedy, D. E., McDowell, C. A.: J. Chem. Phys. 59, 3403 (1973).
503. Dalton, L. R.: Magn. Resonance Rev. 1, 301 (1972).
504. Damerau, W., Lassmann, G., Thom, H. G.: Z. physik. Chem. (Leipzig) 233, 59 (1963).
505. David, C.: in: Degradation of Polymers (Ed. Bamford, C. H., Tripper, C. F. H.) Comprehensive Chemical Kinetics. Amsterdam: Elsevier 1975, vol. 14, p. 175.
506. David, C., Geuskens, G., Verhasselt, A., Jung, P., Oth, J. F. M.: Mol. Phys. 11, 257 and 599 (1966).
507. David, C., Jansen, P., Geuskens, G.: Int. J. Radiat. Phys. Chem. 4, 51 (1972).
508. Davis, L. A.: J. Polym. Sci. A2, 12, 75 (1974).
509. Davis, L. A., Baugham, R. H., Pampillo, C. A.: J. Polym. Sci. A1, 11, 2441 (1973).
510. Davis, L. A., Pampillo, C. A., Chiang, T. C.: J. Polym. Sci. A2, 11, 841 (1973).
511. Davis, P., Evans, M. G., Higginson, W. C. E.: J. Chem. Soc. (London) 1961, 2563.
512. Davydov, E. Ya., Ledneva, O. A., Mikheev, Yu. A., Pariskii, G. B., Toptygin, D. Ya.: Dokl. Akad. Nauk SSSR 195, 875 (1970).
513. Davydov, E. Ya., Pariskii, G. B., Toptygin, D. Ya.: Izv. Akad. Nauk SSSR 1974, 1717.
514. Davydov, B. E., Zakharyan, R. Z., Karpacheva, G. P., Krentsel, B. A., Lapitskii, G. A., Khutareva, G. V.: Dokl. Akad. Nauk SSSR 160, 650 (1965).
515. Dawans, F., Gallard, J., Teyssie, P., Traynard, P.: J. Polym. Sci., C, (4), 1385 (1964).
516. Deal, R. M., Koski, W. S.: J. Chem. Phys. 31, 1138 (1959).
517. Deas, T. M., Hofer, H. H., Dole, M.: Macromolecules 5, 223 (1972).
518. Decker, C., Mayo, F. R.: J. Polym. Sci. A1, 11, 2847 (1973).
519. Decker, C., Mayo, F. R., Richardson, H.: J. Polym. Sci. A1, 11, 2879 (1973).
520. Deffner, U.: Kolloid. Z. 201, 65 (1965).
521. Deffner, U., Paretzke, H.: Radiat. Res. 49, 272 (1972).
522. Degeteva, T. G., Sedova, I. M., Kuzminskii, A. S.: Vysokomol. Soedin. 5, 1485 (1963).
523. Deguchi, Y.: J. Chem. Phys. 32, 1584 (1960).
524. Deguchi, Y.: Bull. Chem. Soc. Japan 35, 260 (1962).
525. Dehl, R., Fraenkel, G. K.: J. Chem. Phys. 39, 1793 (1963).
526. Delhaes, P.: C. R. Acad. Sci. 265, 575 (1967).

527. Delhaes, P.: Carbon 6, 925 (1968).
528. Delhaes, P., Carmona, F.: Carbon 10, 677 (1972).
529. Delhaes, P., Marchand, A.: J. Phys. 28, 67 (1967).
530. Delhaes, P., Marchand, A.: Carbon 6, 257 (1968).
531. Delmelle, M., Duchesne, J.: C. R. Acad. Sci. 264, 138 (1967).
532. Demmler, K., Schlag, J.: Kunststoffe 57, 566 (1967).
533. Deren, J. Haber,
533. Deren, J., Haber, J., Kosek, K.: Bull. Acad. Polon. Sci. Chem. 13, 21 (1965).
534. Derouane, E. G.: Ind. Chim. Belge 34, 183 (1969).
535. Despain, R. R., DeVries, K. L., Luntz, R. D., Williams, M. L.: J. Dental Res. 49, 876 (1970).
536. DeVries, K. L.: Appl. Polym. Symp. No. 32, 325 (1971).
537. DeVries, K. L.: Rubber Chem. Technol. 48, 445 (1975).
538. DeVries, K. L., Lloyd, B. A., Williams, M. L.: J. Appl. Polym. Sci. 42, 4644 (1971).
539. DeVries, K. L., Moore, N. B., Williams, M. L.: J. Appl. Polym. Sci. 16, 1377 (1972).
540. DeVries, K. L., Roylance, D. K., Williams, M. L.: J. Polym. Sci. A1, 8, 237 (1970).
541. DeVries, K. L., Roylance, D. K., Williams, M. L.: J. Polym. Sci. B, 9, 605 (1971).
542. DeVries, K. L., Roylance, D. K., Williams, M. L.: Int. J. Fracture Mechanics 7, 197 (1971).
543. DeVries, K. L., Roylance, D. K., Williams, M. L.: J. Polym. Sci. A2, 10, 599 (1972).
544. DeVries, K. L., Simonson, E. R., Williams, M. L.: J. Basic Eng. 1969, 587.
545. DeVries, K. L., Simonson, E. R., Williams, M. L.: J. Appl. Polym. Sci. 14, 3049 (1970).
546. DeVries, K. L., Wilde, T. B., Williams, M. L.: J. Macromol. Sci. Phys. B, 7, 633 (1973).
547. DeVries, K. L., Williams, M. L.: Surface and Interfaces 2, 139 (1968).
548. DeVries, K. L., Williams, M. L.: Int. J. Nondestruct. Test. 2, 363 (1971).
549. DeVries, K. L., Williams, M. L.: J. Macromol. Sci. Phys. B, 8, 691 (1973).
550. Dewing, J., Longster, G. F., Myatt, J., Todd, P. F.: Chem. Commun. 1965, 391.
551. Dilli, S., Ernst, I. T., Garnett, J. L.: Austral. J. Chem. 20, 911 (1967).
552. Dilli, S., Garnet, J. L., Martin, E. C., Phuoc, D. H.: J. Polym. Sci. C, (37), 57 (1972).
553. Dillon, M. A.: Creation and Detection of the Excited State (Ed. Lamola, A. A.). New York: Dekker 1971, p. 375.
554. Dixon, W. T.: Ann. Rep. Progr. Chem. Sect., B, 69, 189 (1972).
555. Dixon, W. T., Norman, R. O. C.: Nature 196, 891 (1962).
556. Dixon, W. T., Norman, R. O. C.: J. Chem. Soc. 1963, 3119.
557. Dixon, W. T., Norman, R. O. C., Buley, A. L.: J. Chem. Soc. 1963, 3625.
558. Djabiev, T. S., Sabirova, R. D., Shilov, A. E.: Kinet. Kataliz. 5, 441 (1964).
559. Dobbs, A. J.: Electron Spin Resonance 2, 281 (1974).
560. Dobo, J., Hedvig, P.: Makromol. Chem. 82, 298 (1965).
561. Dobo, J., Hedvig, P.: J. Polym. Sci. C, (16), 2577 (1967).
562. Dodonov, V. A., Petukhov, G. G., Razuev, G. A.: Izv. Akad. Nauk SSSR 1965, 1109.
563. Doetschman, D. C.: Rev. Sci. Instr. 43, 143 (1972).
564. Doi, Y., Rånby, B.: J. Polym. Sci., C, (31), 231 (1970).
565. Dolan, H. E., Jr.: Ann. Chem. 40, 303R (1968).
566. Dole, M.: Radiation Chemistry of Macromolecules. New York: Academic Press, vol. 1 (1972) and vol. 2 (1973).
567. Dole, M.: in: Radiation Chemistry of Macromolecules. New York: Academic Press 1972, vol. 1, p. 265.
568. Dole, M., Cracco, F.: J. Phys. Chem. 66, 193 (1962).
569. Dole, M., Keeling, C. D., Rose, D. G.: J. Amer. Chem. Soc. 76, 4304 (1954).
570. Domburga, G., Sergeeva, V. N., Koshik, M. F., Salna, L.: Latv. PSR Zinat. Akad. Vestis. Khim. Ser. 1968, 497.
571. Donnet, J. B., Rigaut, M., Furstenberger, R., Ehrburger, P.: Carbon 11, 68 (1973).
572. Draghicescu, M., Draghicescu, P., Grosescu, R., Ianculovici, B., Balla, D.: Rev. Roum. Phys. 14, 567 (1969).
573. Drefahl, G., Hörhold, H. H., Hofmann, K.: J. Prakt. Chem. 37, 137 (1968).
574. Dubinskaya, A. M.: Vysokomol. Soedin. A14, 783 (1972).

575. Dubinskaya, A. M.: Chem. Zvesti 26, 224 (1972).
576. Dubinskaya, A. M., Butyagin, P. Yu.: Vysokomol. Soedin. B9, 525 (1967).
577. Dubinskaya, A. M., Butyagin, I. Yu.: A10, 240 (1968).
578. Dubinskaya, A. M., Butyagin, P. Yu.: Kinet. Kataliz. 9, 1016 (1968).
579. Dubinskaya, A. M., Butyagin, P. Yu.: Dokl. Akad. Nauk SSSR 211, 141 (1973).
580. Dubinskaya, A. M., Butyagin, P. Yu., Berlin, A. A.: Dokl. Akad. Nauk SSSR 159, 595 (1964).
581. Dubinskaya, A. M., Butyagin, P. Yu., Odinskaya, R. R., Berlin, A. A.: Vysokomol. Soedin. A10, 410 (1968).
582. Dubinskaya, A. M., Dushkina, L. I.: Vysokomol. Soedin. A14, 1467 (1972).
583. Dubinskaya, A. M., Yusobov, I. I.: Izv. Akad. Nauk SSSR 1974, 1484.
584. Dubinskii, A. A., Gribnerg, O. Ya., Tabachnik, A. A., Lebedev, Ya. S.: Dokl. Akad. Nauk SSSR 215, 631 (1974).
585. Dugas, H.: Can. J. Spectrosc. 18, 110 (1973).
586. Dulov, A. A., Liogonskii, B. I., Ragimov, A. V., Slinkin, A. A., Berlin, A. A.: Izv. Akad. Nauk SSSR 1964, 909.
587. Dulov, A. A., Slinkin, A. A., Rubinshtein, A. M.: Vysokomol. Soedin. 5, 1441 (1963).
588. Dulov, A. A., Slinkin, A. A., Rubinshtein, A. M., Kotlyarevskii, I. L.: Izv. Akad. Nauk SSSR 1963, 1910.
589. Dulov, A. A., Slinkin, A. A., Rubinshtein, A. M.: Izv. Akad. Nauk SSSR 1964, 26.
590. Duncan, C. K., Kearns, D. R.: Chem. Phys. Lett. 12, 306 (1971).
591. Du-Roure, H. C., Nechtschein, M.: C. R. Acad. Sci. 260, 880 (1965).
592. Dzhavakhishvilli, L. Ya., Ketiladze, G. D.: Fiz. Tverd. Tela 10, 3715 (1968).
593. Dyson, F. J.: Phys. Rev. 98, 349 (1955).
594. Eargle, D. H.: J. Amer. Chem. Soc. 86, 349 (1964).
595. Eargle, D. H.: Anal. Chem. 38, 371 R (1966).
596. Eargle, D. H., Moniz, W. B.: J. Org. Chem. 32, 2227 (1967).
597. Eargle, D. H., Moniz, W. B.: J. Polym. Sci. A1, 6, 1153 (1968).
598. Eastman, M. P., Kooser, P. G., Das, M. R., Freed, J. H.: J. Chem. Phys. 51, 2690 (1969).
599. Eastmond, G. C.: Mol. Cryst. Liquid Cryst. 9, 383 (1969).
600. Eaton, W. C., Keighley, J. H.: J. Text. Inst. 60, 556 (1969).
601. Eaton, W. C., Keighley, J. H.: Appl. Polym. Symp. No. 18, 263 (1971).
602. Eckert, R. E., Maykantz, T. R.: J. Polym. Sci. B, 6, 213 (1968).
603. Eda, B., Nunome, K., Iwasaki, M.: J. Polym. Sci. B, 7, 91 (1969).
604. Eda, B., Nunome, K., Iwasaki, M.: J. Polym. Sci. A1, 8, 1831 (1970).
605. Edelstain, N., Kwok, A., Maki, A. H.: J. Chem. Phys. 41, 179 (1964).
606. Edge, D. J., Kochi, J. K.: J. Amer. Chem. Soc. 94, 2635 (1972).
607. Edlund, O., Kinell, P. O., Lund, A., Shimazu, A.: J. Polym. Sci. B, 6, 133 (1968).
608. Ehrenberg, A., Ehrenberg, L., Löfroth, G.: Nature 200, 376 (1963).
609. Ehrenberg, A., Malmström, B., Vänngård, T.: Magnetic Resonance in Biological Systems. London: Pergamon 1967.
610. Ehrenberg, A., Rupprecht, A., Ström, G.: Science 157, 1317 (1967).
611. Ehrlich, P.: J. Macromol. Sci. B, 2, 153 (1968).
612. Ehrlich, P.: in: Electric Properties of Polymers (Ed. Frisch, K. C.). Westport, Conn.: Technomic 1972.
613. Ehrlich, P., Kern, R. J., Pierron, E. D., Provder, T.: J. Polym. Sci. B, 5, 911 (1967).
614. Ehrlich, P., Mertzlufft, E. C., Allendoerfer, R. D.: J. Polym. Sci. B, 12, 125 (1974).
615. Eisinger, J., Shulman, R. G.: Proc. Natl. Acad. Sci., U.S. 50, 694 (1963).
616. Ekström, A., Suenram, R., Willard, J. E.: J. Phys. Chem. 74, 1888 (1970).
617. Elliot, J. P.: Proc. Roy. Soc. A302, 361 (1968).
618. Ellis, B., and Baugher, J. F.: J. Polym. Sci. A2, 11, 1461 (1973).
619. Ellison, A., Oubridge, J. O. V., Sing, K. S. W.: Trans. Faraday Soc. 66, 1004 (1970).
620. Ellison, A., Sing, K. S. W.: Discuss. Faraday Soc. 41, 315 (1966).
621. El-Sayed, M. A.: Ann. Rev. Phys. Chem. 26, 235 (1975).
622. Emanuel, N. M., Sharpatyi, V. A., Nadzhimiddinova, M. T., Kudryashov, L. I., Yarovaya, S. M., Kochetkov, N. K.: Dokl. Akad. Nauk SSSR 177, 1142 (1967).

623. Emanuel, N.M., Zaikov, G.E.: Vysokomol. Soedin. *A 17*, 2122 (1975).
624. Engelmann, H.D., Von Harpe, H.: Fressenius Y. Anal. Chem. *267*, 37 (1973).
625. Erbeia, A.: Resonance Magnetique. Paris: Masson, 1969.
626. Erlich, R.H., Chignell, C.F.: Chem. Instrum. *5*, 65 (1973).
627. Ermolaev, V.K., Voevodskii, V.V.: Proc. Second Tihany Symp. on Radiation Chem.. Budapest: Akademiai Kiado, 1967, p. 211.
628. Ermolenko, I.N., Potapovich, A.K., Makatun, V.N.: Vesti Akad. Nauk Belarusk. SSR, Ser. Fiz. Tekhn. Nauk *1963*, 65.
629. Ermolenko, I.N., Sviridova, R.N., Potapovich, A.K.: Vesti Akad. Nauk Belarusk. SSR, Ser. Khim. Nauk *1966*, 111.
630. Ermolenko, I.N., Sviridova, R.N., Potapovich, A.K.: Zh. Anal. Khim. *22*, 260 (1967).
631. Ershov, B.G., Pikaev, A.K., Rad. Res. Rev. *2*, 1 (1969).
632. Ershov, Yu.A., Gak, Yu.V.: Izv. Akad. Nauk SSSR *1967*, 778.
633. Evans, A.G., Evans, J.C., Moon, E.H.: J. Chem. Soc. *1974*, 2390.
634. Evenson, K.M., Burch, D.S.: J. Chem. Phys. *44*, 1715 (1966).
635. Faber, R.J., Fraenkel, G.K.: J. Chem. Phys. *47*, 2462 (1967).
636. Faber, R.J., Markley, F.W., Weil, J.A.: J. Chem. Phys. *46*, 1652 (1967).
637. Fadner, T.A., Morawetz, H.: J. Polym. Sci. *45*, 475 (1960).
638. Fallgatter, M. B., Dole, M.: J. Phys. Chem. *68*, 1988 (1964).
639. Fallick, A. M.: U.S. At. Energy Comm. *1967*, UCRL-17 543.
640. Fallick, A. M., Mahan, B. H., Meyers, R. J.: J. Chem. Phys. *42*, 1837 (1965).
641. Faucitano, A., Adler, G.: J. Macromol. Sci. Chem. *A 4*, 1471 (1970).
642. Faucitano, A., Facucitano-Martionotti, F.: Europ. Polym. J. *10*, 489 (1974).
643. Fedorov, V. B.: Cryogenics *5*, 12 (1965).
644. Fedoseeva, T. S., Kuzminskii, A. S., Neiman, M. B., Buchachenko, A. L., Lebedev, Ya. S., Chertkova, V. F.: Vysokomol. Soedin. *6*, 241 (1964).
645. Feher, G.: Phys. Rev. *103*, 834 (1956).
646. Feichmayer, F., Schlag, J.: Melliand Textil. *45*, 526 (1964).
647. Feichmayer, F., Schlag, J., Wuerstlin, F.: Kunststoffe *64*, 405 (1974).
648. Fenner, H.: Mitt. Deut. Pharm. Ges. *41*, 165 (1971).
649. Ferruti, P.: J. Amer. Chem. Soc. *92*, 3704 (1970).
650. Ferruti, P., Gill, D., Klein, M. R., Wang, H. H., Entine, G., Calvin, M.: J. Amer. Chem. Soc. *92*, 3704 (1970).
651. Fessenden, R. W., Schuler, R. H.: J. Chem. Phys. *38*, 773 (1962).
652. Fessenden, R. W., Schuler, R. H.: J. Chem. Phys. *39*, 2147 (1963).
653. Fialkov, A. S., Kabardina, V. A., Samoilov, V. S., Galeev, G. S.: Vysokomol. Soedin. *B9*, 505 (1967).
654. Filipov, A. I., Ryzhmanov, Yu. A., Kozyrev, B. M.: Dokl. Akad. Nauk SSSR. *215*, 144 (1974).
655. Findeisen, E., Roth, H. K., Haser, H., Hellbrand, J.: Plaste. Kauth. *15*, 881 (1968).
656. Firth, E. W., Ingram, D. J. E.: J. Sci. Instrum. *44*, 821 (1967).
657. Fischer, H.: Kolloid-Z. u. Z. Polymere *180*, 64 (1962).
658. Fischer, H.: J. Chem. Phys. *37*, 1094 (1962).
659. Fischer, H.: Z. Naturforsch. *18a*, 1142 (1963).
660. Fischer, H.: Z. Naturforsch. *19a*, 267 (1964).
661. Fischer, H.: J. Polym. Sci. B, *2*, 529 (1964).
662. Fischer, H.: Z. Naturforsch. *19a*, 866 (1964).
663. Fischer, H.: Z. Naturforsch. *20a*, 428 (1965).
664. Fischer, H.: Kolloid-Z. u. Z. Polym. *206*, 131 (1965).
665. Fischer, H.: Mol. Phys. *9*, 149 (1965).
666. Fischer, H.: Kunststoffe *55*, 344 (1965).
667. Fischer, H.: in: Magnetic Properties of Free Radicals. Landolt-Börnstein, New Series, Group II, vol. 1, Berlin, Heidelberg, New York: Springer, 1965.
668. Fischer, H.: Makromol. Chem. *98*, 179 (1966).
669. Fischer, H.: Chimie et Industrie, Gen. Chim. *97*, 8 (1967).

670. Fischer, H.: Z. Bunsenges. Physik Chem. *71*, 685 (1967).
671. Fischer, H.: Proc. Roy. Soc. *A 302*, 321 (1968).
672. Fischer, H.: Fortsch. Hochpolymeren-Forsch. *5*, 463 (1968).
673. Fischer, H.: Accounts Chem. Res. *4*, 110 (1971).
674. Fischer, H.: J. Mol. Spectrosc. *40*, 414 (1971).
675. Fischer, H.: Nucl. Magn. Res. *4*, 301 (1971).
676. Fischer, H.: in: Free Radicals, (Ed. Kochi, J. K.) New York: Wiley, 1973, vol. II, p. 435.
677. Fischer, H., Bargon, J.: Accounts Chem. Res. *2*, 210 (1969).
678. Fischer, H., Giacometti, G.: J. Polym. Sci. C (16), 2763 (1967).
679. Fischer, H., Hellwege, K. H.: J. Polym. Sci. *56*, 33 (1962).
680. Fischer, H., Hellwege, K. H.: Z. Naturforsch. *18 a*, 994 (1963).
681. Fischer, H., Hellwege, K. H.: Makromol. Chem. *91*, 107 (1966).
682. Fischer, H., Hellwege, K. H., Johnsen, U.: Kolloid-Z. *170*, 61 (1960).
683. Fischer, H., Hellwege, K. H., Johnsen, U., Neudörfl, P.: Kolloid-Z. u. Z. Polymere *195*, 129 (1964).
684. Fischer, H., Hellwege, K. H., Neudörfl, P.: J. Polym. Sci., A1, *1*, 2109 (1963).
685. Fischer, H., Hummel, D. O.: in: Polymer Spectroscopy (Ed. Hummel, D. O.) Weinheim: Verlag Chemie 1974, p. 289.
686. Fischer, L.: Acta Fac. Rerum Natur. Univ. Comenianae Phys., No. 11, 163 (1971).
687. Fitzgerald, J. M.: Analytical Photochemistry and Photochemical Analysis: Solids, Solutions and Polymers. New York: Dekker, 1971.
688. Flandrois, S.: C. R. Acad. Sci. *C264*, 1244 (1967).
689. Fleischer, G., Hellebrand, J.: Wiss. Z. Karl-Marx-Univ. Leipzig. Math.-Naturwiss. Reihe *21*, 653 (1972).
690. Flesia, E., Surzur, J. M.: Tetrahedron Lett. *1975*, 2893.
691. Flockhart, B. D.: Compr. Anal. Chem. *1971*, 206.
692. Flockhart, B. D., Ivin, K. J., Pink, R. C., Sharma, B. D.: Chem. Comm. *1971*, 339.
693. Flockhart, B. D., Ivin, K. J., Pink, R. C., Sharma, B. D., XXIII Intern. Congress of Pure and Applied Chemistry, Macromol. Prepr., vol. 1, p. 319 (1971).
694. Flockhart, B. D., Ivin, K. J., Pink, R. C., Sharma, B. D.: J. Chem. Soc. *1971*, 339.
695. Flockhart, B. D., Ivin, K. J., Pink, R. C., Sharma, B. D.: in: ESR Application to Polymer Research (Ed. Kinell, P. O., Rånby, B., and Runnström-Reio, V.), Stockholm: Almqvist-Wiksell, 1973, p. 17.
696. Florin, R. E., Sicilio, F., Wall, L. A.: J. Research Nat. Bur. Stand., Phys. Chem. *72A*, 49 (1968).
697. Florin, R. E., Sicilio, F., Wall, L. A.: J. Phys. Chem. *72*, 3154 (1968).
698. Florin, R. E., Wall, L. A.: Trans, Faraday Soc. *56*, 1305 (1960).
699. Florin, R. E., Wall, L. A.: J. Research Nat. Bur. Stand. Phys. Chem. *65A*, 375 (1961).
700. Florin, R. E., Wall, L. A.: J. Polym. Sci. A1, *1*, 1163 (1963).
701. Florin, R. E., Wall, L. A.: J. Chem. Phys. *57*, 1791 (1972).
702. Florin, R. E., Wall, L. A., Brown, D. W.: Trans. Faraday Soc. *56*, 1304 (1960).
703. Florin, R. E., Wall, L. A., Brown, D. W.: Polym. Prepr. Amer. Chem. Soc., Div. Polym. Chem. *2*, 5 (1961).
704. Florin, R. E., Wall, L. A., Brown, D. W.: J. Polym. Sci. A1, *1*, 1521 (1963).
705. Flynn, J. H., Wall, L. A., Morrow, W. L., J. Research Nat. Bur. Stand., Phys. Chem. *71A*, 25 (1967).
706. Földes-Berehunskh, T., Tüdös, F., Jokay, L., Mohos, B.: Bull. Inst. Polit. Iasi *16*, 221 (1970).
707. Forchioni, A., Chachaty, C.: J. Polym. Sci. A1, *10*, 1923 (1972).
708. Forrestal, L. J., Hodgson, W. G.: J. Polym. Sci. A1, *2*, 1275 (1972).
708. Forrestal, L. J., Hodgson, W. G.: J. Polym. Sci. A1, *2*, 1275 (1964).
709. Forrestal, A. R., Hepburn, S. P.: J. Chem. Soc. *1971*, 701.
710. Fox, K.K., Robb, O. D., Smith, R.: J. Chem. Soc. Faraday Trans. *70*, 1186 (1974).
711. Fraenkel, G. K., Hirshon, J. M., Walling, Ch.: J. Amer. Chem. Soc. *76*, 3606 (1954).
712. Fraenz, J.: Atomkernenergie *8*, 63 (1963).

References

713. Franconi, C.: in: Magnetic Resonance in Biology (Ed. Franconi, C.,) New York: Gordon and Breach, 1971, p. 397.
714. Freed, J. H.: J. Chem. Phys. *43*, 2312 (1965).
715. Freed, J. H.: J. Phys. Chem. *71*, 38 (1967).
716. Freed, J. H.: Ann. Rev. Phys. Chem. *23*, 265 (1972).
717. Freed, J. H., Bruno, G. U., Polanaszek, C. F.: J. Phys. Chem. *75*, 3385 (1971).
718. Freed, J. H., Fraenkel, G. K.: J. Chem. Phys. *39*, 326 (1963).
719. Freed, J. H., Fraenkel, G. K.: J. Chem. Phys. *40*, 1815 (1964).
720. Freed, J. H., Leniart, D. S., Hyde, J.S.: J. Chem. Phys. *47*, 2762 (1967).
721. Freeman, A. J., Frankel, R. B.: Hyperfine Interactions: New York: Academic Press 1967.
722. Friedlander, H. Z.: Polym. Prepr., Amer. Chem. Soc., Div. Polym. Chem. *4*, 300 (1963).
723. Fritzche, C., Fischer, E. W.: Kolloid-Z. u. Z. Polym. *251*, 721 (1973).
724. Froix, M. F., Beatty, C. L., Pochan, J. M., Hinman, D. D.: J. Polym. Sci. A2, *13*, 1269 (1975).
725. Frunze, N. K., Berlin, A. A.: Vysokomol. Soedin. *A11*, 1444 (1969).
726. Fueki, K., Kuri, Z.: J. Amer. Chem. Soc. *87*, 923 (1965).
727. Fujimoto, K., Inomata, I., Fujiwara, S.: Nippon Gomu Kyokaishi *46*, 232 (1973).
728. Fujimura, T., Haykawa, N., Tamura, N.: Rep. Prog. Polym. Phys. Japan *14*, 557 (1971).
729. Fujimura, T., Haykawa, N., Tamura, N.: Rep. Prog. Polym. Phys. Japan *14*, 561 (1971).
730. Fujimura, T., Tamura, N.: J. Polym. Sci. B, *10*, 469 (1972).
731. Fujimura, T., Tamura, N.: Rep. Progr. Polym. Phys. Japan *17*, 541 (1974).
732. Fujimura, T., Tamura, N.: J. Phys. Chem. *79*, 1859 (1975).
733. Fujita, N., Matsumura, O.: Mem. Fac. Sci. Kyushu Univ. Ser. B, *3*, 65 (1962).
734. Fujiwara, S., Nagashima, K., Codell, M.: Bull. Chem. Soc. Japan *37*, 49 (1964).
735. Fujiwara, S., Nagashima, K., Codell, M.: Bull. Chem. Soc. Japan *37*, 773 (1964).
736. Fukuda, K.: Mem. Fac. Sci. Kyushu Univ. Ser. B, *3*, 141 (1960).
737. Fukuda, K., Kusumoto, N., Kawano, G., and Takayagi, K.: Kogyo Kagaku Zasshi *67*, 2163 (1964).
738. Furukawa, J., Hirai, R.: J. Polym. Sci. A1, *10*, 2139 (1972).
739. Gaffney, B. J., McNamee, C. M.: Methods Enzymol. *32*, 161 (1974).
740. Gager, W. B., Sliemers, F. A., Kircher, J. F., and Leininger, B. I.: Polym. Prepr., Amer. Chem. Soc., Div. Polym. Chem. *3*, 67 (1962).
741. Gak, Yu. V., Gogolev, R. S., Sukhov, V. A., and Lukovnikov, A. F.: Khim. Vysokh. Energ. *5*, 45 (1970).
742. Gak, Yu. V., Lukovnikov, A. F.: Vysokomol. Soedin. *A12*, 2415 (1970).
743. Gallard-Nechtschein, J., Pecher-Reboul, A., Traynard, P.: J. Catal. *13*, 261 (1969).
744. Gomo, K., Masuda, K., Yamaguchi, J., Kakitani, T.: J. Phys. Soc. Japan *20*, 1730 (1965).
745. Gaponova, I. S., Ershov, B. G.: Izv. Akad. Nauk SSSR *1974*, 2453.
746. Garifyanov, N. S., Kozyrev, B. M.: Dokl. Akad. Nauk SSSR *118*, 738 (1958).
747. Garifyanov, N. S., Illyasov, A. V., Ryzhmanov, Yu. M.: Zh. Tekhn. Fiz. *31*, 694 (1961).
748. Garifyanov, N. S., Illyasov, A. V., Yablokov, Yu. V.: Dokl. Akad. Nauk. SSSR *149*, 876 (1963).
749. Garst, J. F.: in: Free Radicals (Ed. Kochi, J. K.), New York: Wiley, 1973, vol. 1, p. 503.
750. Gelfman, A. Ya., Bidnaya, D. S., Sigalova, L. V., Buravleva, M. G., Koba, V. S.: Dokl. Akad. Nauk SSSR *154*, 894 (1964).
751. Gendell, J., Freed, J. H., Fraenkel, G. K.: J. Chem. Phys. *41*, 949 (1964).
752. Georgiev, G. S., Apollonova, I. P., Golubev, V. B., Zubov, V. I., Kabanov, V. A.: Vysokomol. Soedin. *A14*, 2714, (1973).
753. Georgiev, G. S., Kaplan, A. M., Zubov, V. P., Golubev. V. B., Barkalov, I. M., Goldanskii, V. I., Kabanov, V. A.: Vysokomol. Soedin. *A14*, 177 (1972).
754. Georgiev, G. S., Panasenko, A. A., Yun, E., Golubev, V. B., Zubov, V. I., Kargin, V.A.: Vysokomol. Soedin. *B11*, 479 (1969).
755. Georgiev, G. S., Pergushov, V. I., Golubev, V. B., Zubov, V. I., Kabanov, V. A.: Vysokomol. Soedin. *A15*, 2008 (1973).

364

756. Gerasimov, G. N., Khomikovskii, P. M., Abkin, A. D.: Dokl. Akad. Nauk SSSR *156*, 1150 (1964).
757. Gerasinov, G. N., Khomikovskii, P. M., Abkin, A. D.: Vysokomol. Soedin. *A 9*, 877, (1967).
758. Gerson, F.: Hochauflösende ESR-Spektroskopie Weinheim: Verlag Chemie, 1967.
759. Gerson, F.: High Resolution Electron Spin Theory. New York: Wiley-Interscience 1971.
760. Geschwind, S.: Electron Paramagnetic Resonance. New York: Plenum, 1972 p. 575.
761. Geuskens, G.: in: Degradation of Polymers Comprehensive Chemical Kinetics (Ed. Bamford, C. H., Tipper, C. F. H.), Amsterdam: Elsevier, 1975, vol. 14 p. 333.
762. Geuskens, G.: Degradation and Stabilization of Polymers. New York: Wiley, 1975.
763. Geuskens, G., David, C.: Makromol. Chem. *165*, 273 (1973).
764. Geuskens, G., Fuld, D., David, C.: Makromol. Chem. *160*, 135 (1972).
765. Gibson, J. F., Roy. Inst. Chem. *86*, 37 (1962).
766. Gibson, J. F., Symons, M. C. R., Townsend, M. G.: J. Chem. Soc. *1959*, 269.
767. Gierke, W., Harrer, W., Kurreck, H., Reusch, J.: Tetrahedron Lett. *1973*, 3681.
768. Gilbert, B. C.: Essays in Chem. *4*, 61 (1972).
769. Gilbert, B. C.: Electron Spin Resonance *1*, 205 (1972).
770. Gill, E. K., Laidler, K. J.: Can. J. Chem. *36*, 79 (1958).
771. Gillbro, T.: J. Polym. Sci. B, *11*, 309 (1973).
772. Gillbro, T., Kinell, P. O.: in: ESR Applications in Polymer Research (Ed. Kinell, P. O., Rånby, B., and Runnström-Reio, V.,) Stockholm: Almqvist-Wiksell 1973, p. 83
773. Gillbro, T., Kinell, P. O., Lund, A.: J. Phys. Chem. *73*, 4167 (1969).
774. Gillbro, T., Kinell, P. O., Lund, A.: J. Polym. Sci. A2, *9*, 1495 (1971).
775. Gillbro, T., Lund, A.: Chem. Phys. *5*, 283 (1974).
776. Gillbro, T., Lund, A.: Chem. Phys. Lett. *34*, 375 (1975).
777. Gillbro, T., Yamaoka, H., Okamura, S.: J. Phys. Chem. *77*, 1165 (1973).
778. Girard, Y., Che. M., Guyot, A., Chanzy, H.: Makromol. Chem. *162*, 119 (1972).
779. Glegg, R. E., Kertesz, Z. I.: J. Polym. Sci. *26*, 289 (1957).
780. Glukhovskii, V. S., Kostin, E. S., Yukelson, I. I.: Vysokomol. Soedin. *B12*, 136 (1970).
781. Glukhovskii, V. S., Kostin, E. S., Yukelson, I. I.: Vysokomol. Soedin. *B13*, 801 (1971).
782. Goldammer, von B. E., Conway, D. H., Paskovich, D. H., Reddoch, A. H.: J. Polym. Sci. A1, *11*, 2767 (1973).
784. Goldammer, von B. E., Müller, A., Conway, B. E.: Ber. Bunsenges. physik. Chem. *78*, 35 (1974).
785. Goldin, S. I., Markevich, S. V.: Vesti Akad. Nauk Belorus. SSR *1970*, 47.
786. Goldin, S. I., Markevich, S. V.: Khim. Vysok. Energ. *5*, 463 (1971).
787. Goldman, S. A., Bruno, G. V., Freed, J. H.: J. Phys. Chem. *76*, 1858 (1972).
788. Golubev, V. B., Zubov, V. P., Georgiev, G. S. Stoyachenko, I. L., Kabanov, V. A.: J. Polym. Sci. A1, *11*, 2463 (1973).
789. Golubev, V. B., Zubov, V. P., Valuev, L. I., Naumov, G. S., Kabanov, V. A., Kargin, V. A.: Vysokomol. Soedin. *A 11*, 2689 (1969).
790. Gölz, W. L. F., Zachmann, H. G.: Makromol. Chem. *176*, 2721 (1975).
791. Gonet, B.: Postepy Hig. Med. Doswiad. *23*, 733 (1969).
792. Goodhead, D. T.: J. Polym. Sci. A2, *9*, 999 (1971).
793. Gordy, W., Morehause, R.: Phys. Rev. *151*, 207 (1966).
794. Gorkhovskii, G. A., Chernenko, P. A., Vonsyatskii, V. A., Popov, I. A.: Dokl. Akad. Nauk SSSR *204*, 126 (1972).
795. Gorokhov, G. I.: Gidroliz. i Lesokhim. Prom. *18*, 13 (1965).
796. Gorokhov, G. I., Abagyan, G. V., Butyagin, P. Yu.: Izv. Vysokikh Uchebn. Zavedenii Lesn. Zh. *8*, 135 (1965).
797. Gorokhov, G. I., Radstig, V. A., Butyagin, P. Yu.: Vysokomol. Soedin. *B14*, 568 (1972).
798. Gothe, K. H.: Wiss. Z. Techn. Hochsch. Illmenau *16*, 67 (1970).
799. Gough, T. E., Hindle, P. R.: Can. J. Chem. *47*, 1698 (1969).
800. Gough, T. E., Hindle, P. R.: Can. J. Chem. *47*, 3393 (1969).

References

801. Gough, T. E., Symons, M. C. R.: Trans, Faraday Soc. *62*, 269 (1962).
802. Grassie, N.: Pure Appl. Chem. *16*, 389 (1968) .
803. Graves, C. T., Ormerod, M. G.: Polymer *4*, 81 (1963).
804. Gräslund, A.: Report of Department of Biophysic, Stockholm University, Sweden 1970.
805. Gräslund, A., Ehrenberg, A., Ruppert, A., Ström, G.: Biochem. Biophys. Acta *254*, 172 (1971).
806. Gräslund, A., Ehrenberg, A., Rupprecht, A., Tjälldin, B., Ström, G.: Radiat. Res. *61*, 488 (1975).
807. Gräslund, A., Ehrenberg, A., Rupprecht, A.: Ann. N. Y. Acad. Sci. *222*, 374 (1973)
808. Gräslund, A., Lööfroth, G.: Acta Chem. Scand. B *29*, 475 (1975).
809. Gräslund, A., Rigler, R., Ehrenberg, A.: Febs Lett. *4*, 227 (1969).
810. Gräslund, A., Rupprecht, A., Ström, G.: Photochem. Photobiol. *21*, 153 (1975).
811. Graves, C. T., Ormerod, M. G.: Polymer *4*, 81 (1963).
812. Greber, G., Egle, G.: Makromol. Chem. *53*, 206 (1962).
813. Greber, G., Egle, G.: Makromol. Chem. *54*, 136 (1962).
814. Greber, G., Egle, G.: Makromol. Chem. *59*, 174 (1963).
815. Gregoli, S., Bertinchamps, A.: Int. J. Radiat. Biol. *21*, 65 (1972).
816. Gregoli, S., Meelhuysen, R., Bertinchamps, A.: EUR-4689 Report, 1971.
817. Greyson, J., Ingalls, R. B., Keen, R. T.: J. Chem. Phys. *45*, 3755 (1966).
818. Griffith, O. H.: J. Amer. Chem. Soc. *89*, 5072 (1957).
819. Griffith, O. H., Keana, J. F. W., Rottschaeffer, S., Warlick, T. A.: J. Amer. Chem. Soc. *89*, 5072 (1967).
820. Griffith, O. H., Waggoner, A. S.: Accounts Chem. Res. *2*, 17 (1969).
821. Griffiths, W. E., Congster, G. H., Myatt, J., Todd, P. H.: J. Chem. Phys. B, *1969*, 530.
822. Griffiths, W. E., Sutcliffe, L. H.: Trans. Faraday Soc. *62*, 2837 (1966).
823. Grinberg, O. Ya., Dubinskii, A. A., Ozherelev, B. V., Chistota, A. A., Lebedov, Ya. S.: Soviet Physic-Solid State *14*, 2886 (1973).
824. Grishina, A. D.: Dokl. Akad. Nauk SSSR *150*, 809 (1963).
825. Grishina, A. D.: Kinet. Kataliz. *8*, 685 (1967).
826. Grishina, A. D., Bakh, N. A.: Vysokomol. Soedin. *7*, 1968 (1965).
827. Grishina, A. D., Chervonenko, V. S., Bakh, N. A.: Elektrokhimiya *4*, 1184 (1968).
828. Grosescu, R., Constantinescu, O., Balla, D.: Rev. Roum. Phys. *14*, 1271 (1970).
829. Gruber, K., Forrer, J., Schweiger, A., Gunthard, H. H.: J. Phys. E. Sci. Instrum. *7*, 569 (1974).
830. Gruber, K., Forrer, J., Zopfi, E.: J. Phys. Ed. *6*, 666 (1973).
831. Gubanov, V. A., Chirkov, A. K.: Acta Phys. Pol. *A 43*, 361 (1973).
832. Gubanov, V. A., Inishev, V. D., Chirkov, A. K.: Zh. Strukt. Khim. *13*, 349 (1972).
833. Gueron, M.: in: Creation and Detection of the Excited States (Ed. Lamola, A. A.), New York: Dekker 1971, vol. 1A, p.303.
834. Gueron, M., Eisinger, J., Shulman, R. G.: Mol. Phys. *14*, 111 (1968).
835. Guk, A. F., Karpukhin, O. N., Kisilitsa, P. P., Charkashin, M. I., Tsepalov, V. F.: Izv. Akad. Nauk SSSR *1967*, 1394.
836. Guk, A. F., Kozlova, Z. G., Tsepalov, V. F.: Vysokomol. Soedin. *B 15*, 41 (1973).
837. Gupta, R. P.: J. Phys. Chem. *66*, 849 (1962).
838. Gupta, R. P.: J. Phys. Chem. *68*, 1229 (1964).
839. Guthrie, J. T., Huglin, M. B., Phillips, G. O.: Europ. Polym. J. *8*, 747 (1972).
840. Guthrie, J. C., Huglin, M. B., Phillips, G. O.: J. Polym. Sci. C (37), 205 (1972).
841. Gutowsky, H. S., Pake, G. E.: J. Chem. Phys. *18*, 162 (1950).
842. Guyot, A., Mordini, J., Spitz, R.: C. R. Acad. Sci. *C 269*, 483 (1969).
843. Guyot, A., Bert, M., Michel, A., Spitz, R.: J. Polym. Sci. A 1, *8*, 1596 (1970).
844. Haissinsky, M.: Nuclear Chemistry and its Applications. Reading, Mass., USA: Addison-Wesley 1964.
845. Hajimoto, Y., Tamura, N., Okamato, S.: J. Polym. Sci. A 1, *3*, 255 (1965).
846. Hales, B. J.: J. Amer. Chem. Soc. *97*, 5993 (1975).
847. Hales, B. J., Bolton, J. R.: Photochem. Photobiol. *12*, 239 (1970).

366

848. Ham. J. S., Davis, M. K., Song, H.H.: J. Polym. Sci. A2, *11*, 217 (1973).
849. Hama, Y., Furui, Y., Hosono, K., Shinohara, K.: Rep. Progr. Polym. Phys. Japan *12*, 1433 (1966).
850. Hama, Y., Hoscono, K., Furui, Y., Shinohara, K.: J. Polym. Sci. A1, *9*, 1411 (1971).
851. Hama, Y., Miura, K., Ooi, Y., Shonohara, K.: Rep. Progr. Polym. Phys. Japan *16*, 561 (1973).
852. Hama, Y., Okamoto, S., Tamura, N.: Rep. Progr. Polym. Phys. Japan *7*, 351 (1964).
853. Hama, Y.,Shinohara, K.: J. Polym. Sci. A1, *8*, 651 (1970).
854. Hama, Y., Ooi, T., Shiotsubo, M., Shinohara, K.: Rep. Progr. Polym. Phys. Japan *15*, 537 (1972).
855. Hama, Y., Ooi, T., Shiotsubo, M., Shinohara, K.: Polymer *15*, 787 (1974).
856. Hamanoue, K., Kamantauskas, V., Tabata, Y., Silverman, J.: J. Chem. Phys. *61*, 3439 (1974).
857. Hamanoue, K., Shimizu, M., Higaki, H., Block, I., Silverman, J.: J. Polym. Sci. A2, *12*, 1189 (1974).
858. Hamill, W. H.: in: Radical Ions (Ed. Kaiser, E. T., and Kevan, L.), New York: Wiley 1968, p. 321.
859. Hamilton, C. L., McConnell, H. M.: in: Structural Chemistry and Molecular Biology (Ed. Rich., A., Davidson, N.) San Francisco: Freeman, 1968, p. 115.
860. Heniotis, Z., Guenthard, H. H.: Hel. Chim. Acta *51*, 561 (1968).
861. Hanna, M. W.,McLachan, A. D., Dearman, H. H., McConnell, H. M.: J. Chem. Phys. *37*, 361 (1962).
862. Hansen, R. H.: in: Thermal Stability of Polymers (Ed. Conley, R. T.), New York: Dekker 1970, p. 153.
863. Hansen, R. H., Pascale, J. V., DeBenedicts, T., Rentzepis, P. M.: J. Polym. Sci. A1, *3*, 2205 (1965).
864. Hardham, W. M.: in: Creation and Detection of the Excited State (Ed. Lamola, A. A.,) New York: Dekker 1971, vol. 1B, p. 615.
865. Hardy, G., Varga, J., Nagy, G., Cser, F., Ero, J.: J. Polym. Sci. C, (16), 2583 (1967).
866. Haret, J. C. R., Combier, A. L., Chachaty, C.: J. Phys. Chem. *78*, 899 (1974).
867. Harker, H., Gallagher, J. T., Parkin, A.: Carbon *4*, 401 (1966).
868. Harker, H., Jackson, C., Wynne-Jones, W. F. K.: Proc. Roy. Soc. *A262*, 328 (1961).
869. Harrah, L. A.: Mol. Cryst. Liquid Cryst. *9*, 197 (1969).
870. Harris, J. A., Hinojosa, O., Arthur, J. C. Jr.: Polym. Prepr., Amer. Chem. Soc. Div. Polym. Chem. *13*, 479 (1972).
871. Harris, J. A., Hinojosa, O., Arthur, J. C. Jr.: J. Polym. Sci. A1, *11*, 3215 (1973).
872. Harris, J. A., Hinojosa, O., Arthur, J. C. Jr.: J. Polym. Sci. A1, *12*, 679 (1974).
873. Harris, J. A., Hinojosa, O., Arthur, J. C. Jr.: Polym. Prepr. Amer. Chem. Soc., Div. Polym. Chem. *15*, 491 (1974).
874. Hase, H., Yamaoka, H.: Radiation Effects *19*, 195 (1973).
875. Hasegawa, S., Shimizu, T.: Oyo Butsuri *37*, 669 (1968).
876. Hasegawa, S., Shimazu, T.: Jap. J. App. Phys. *9*, 958 (1970).
877. Hatakeyama, H., Nakano, J.: Cellul. Chem. Technol. *4*, 281 (1970).
878. Hatano, M.: Kogyo Kagaku Zasshi *65*, 723 (1962).
879. Hausser, K. H.: High Resolution ESR, Proc. Colloq. Spectr. Intern., 10th, College Park, USA, 1962.
880. Hausser, K. H.: Method Chim. *1*, 227 (1974).
881. Hausser, K. H.: Method Chim. *1*, 379 (1974).
882. Hay, J. N.: J. Polym. Sci. A1, *8*, 1201 (1970).
883. Hayashi, K.: IUPAC Intern. Symp. on Macromol. Chem., Budapest, 1969, Budapest: Akademiai Kiado 1969, p. 507.
884. Hayashi, K., Okamura, S.: J. Polym. Sci., C (22) 15 (1968).
885. Hayashi, K., Yamaoka, H., Okamura, S.: Proc. Second Tihany Symp. on Radiation Chemistry, Budapest: Akademiai Kiado 1967, p. 451.
886. Hecht, H. G.: Magnetic Resonance Spectroscopy. New York: Wiley 1967.

367

887. Heckly, R. J.: in: Biological Applications of Electron Spin Resonance (Ed. Swartz, H. M., Bolton, J. R., Borg, D. C.) New York: Wiley 1972, p. 197.
888. Hedvig, P.: Europ. Polym. J. Suppl. *1969*, p. 285.
889. Hedvig, P.: J. Polym. Sci. A1, *7*, 1145 (1969).
890. Hedvig, P.: Kinetics and Mechanism of Polyreactions. Budapest: Akademiai Kiado 1969, vol. 5, p. 277.
891. Hedvig, P.: Enka Biniiru To Porima *10*, 46 (1970).
892. Hedvig, P.: in: Radiation Chemistry of Macromolecules (Ed. Dole, M.) New York: Academic Press 1972, vol. 1, p. 55.
893. Hedvig, P.: in: ESR Application to Polymer Research (Ed. Kinell, P. O., Rånby, B., Runnström: Reio, V.) Stockholm: Almqvist-Wiksell 1973, p. 215.
894. Hedvig, P.: Experimental Quantum Chemistry. Budapest: Akademiai Kiado 1975.
895. Hedvig, P., Dobo, J., Proc. Second Tihany Symp. on Radiation Chemistry. Budapest: Akademiai Kiado 1967, 685.
896. Hedvig, P., Kiss, L.: in: Kinetics and Mechanism of Polyreactions. Budapest: Akademiai Kiado 1969, vol. 5, p. 291.
897. Hedvig, P., Kulcsár, S., Kiss, L.: Europ. Polym. J. *4*, 601 (1968).
898. Hedvig, P., Zentai, G.: Microwave Study of Chemical Structure and Reactions, Ilffe. Budapest: London-Akademiai Kiado 1969.
899. Hefter, H., Fischer, H.: Ber. Bunsenges. physik. Chem. *73*, 633 (1969).
900. Helbert, J. N., Bales, B., Kevan, L.: J. Chem. Phys. *57*, 723 (1972).
901. Helbert, J. N., Wagner, B. E., Poindexter, E. H., Kevan, L.: J. Polym. Sci. A 2, *13*, 825 (1975).
902. Helene, C., Santus, R., Douzou, P.: Photochem. Photobiol. *5*, 127 (1966).
903. Hellebrand, J., Fleischer, G.: Wiss. Z. Karl-Marx-Univ. Leipzig, Math. Naturwiss. Reihe *21*, 649 (1972).
904. Heller, C., McConnell, H. M.: J. Chem. Phys. *32*, 1535 (1960).
905. Hellwege, K. H., Lehnig, M.: (Unpubl. results).
906. Hemmerich, P.: Proc. Roy. Soc. *A 302*, 335 (1968).
907. Henglein, A.: in: Radiation Research (Ed. Silini G.), Amsterdam: North-Holland 1967, p. 316.
908. Henglein, A., Boysen, M.: Makromol. Chem. *20*, 83 (1956).
909. Hennessy, J. J., Moore, W. H.: Rev. Sci.,Instr. *37*, 55 (1966).
910. Henning, G. R., Smaller, B., Yasaitis, E. L.: Phys. Rev. *95*, 1088 (1954).
911. Henning, J. C. M.: J. Chem. Soc. *44*, 2139 (1966).
912. Henrici-Olivé, G., Olivé, S.: Angew. Chem. *6*, 790 (1967).
913. Henrici-Olivé, G., Olivé, S.: Angew. Chem. *7*, 821 (1968).
914. Henrici-Olivé, G., Olivé, S.: J. Polym. Sci. C (22). 965 (1969).
915. Henrici-Olivé, G., Olivé, S.: Adv. Polym. Sci. *6*, 421 (1969).
916. Henriksen, T.: in: ESR Application to Polymer Research (Ed. Kinell, P. O., Rånby, B., Runnström-Reio, V.), Stockholm: Almqvist-Wiksell 1973, p. 249.
917. Henriksen, T., Snipes, W.: Radiat. Res. *42*, 255 (1970.
918. Herak, J. N., Galogaza, V.: J. Chem. Phys. *50*, 3101 (1969).
919. Herak, J. N., Gordy, W.: Proc. Natl. Acad. Sci. US, *55*, 698 (1966).
920. Herak, J. N., Gordy, W.: Proc. Natl. Acad. Sci. US, *55*, 1373 (1966).
921. Herman, M.: Postepy Fiz. *17*, 185 (1966).
922. Hertz, J. E., Andre, J. J.: Europ. Polym. J. *6*, 1505 (1970).
923. Hesse, P., Heusinger, H.: Int. J. Radiat. Phys. Chem., (in press).
924. Hesse, P., Rosenberg, A., Heusinger, H.: Europ. Polym. J. *9*, 581 (1973).
925. Heuvel, H. M., Lind, K. C.: J. Polym. Sci. A2, *8*, 401 (1970).
926. Higuchi, J.: J. Chem. Phys. *39*, 2366 (1963).
927. Hill, D. A., Haster, B. A., Hwang, C. F.: Phys. Lett. *23*, 63 (1966).
928. Hinojosa, O., Arthur, J. C. Jr.: J. Polym. Sci. B, *10*, 161 (1972).
929. Hinojosa, O., Arthur, J. C. Jr., Mares, M.: J. Appl. Polym. Sci. *18*, 2509 (1974).
930. Hinojosa, O., Harris, J. A., Arthur, J. C. Jr.: Carbohydrate Res. *41*, 31 (1975).

931. Hinojosa, O., Nakamura, Y., Arthur, J. C. Jr.: J. Polym. Sci. C, (37) 27 (1972).
932. Hinschberger, A., Marchal, J., Vacherot, M.: J. Polym. Sci. C, (16), 4437 (1969).
933. Hirahara, T., Inoue, H., Minoura, Y.: Europ. Polym. J. *10*, 109 (1974).
934. Hirahara, T., Sigimura, T., Minoura, Y.: J. Polym. Sci. A1, *8*, 2827 (1970).
935. Hirai, H., Fujiwara, M.: Nippon Kagaku-kaishi *1972*, 968.
936. Hirai, H., Hiraki, K., Noguchi, I., Makishima, S.: J. Polym. Sci. A1, *8*, 147 (1970).
937. Hirai, H., Hiraki, K., Noguchi, I., Inoue, T., Makishina, S.: J. Polym. Sci. A1, *8*, 2393 (1970).
938. Hirai, H., Ikegami, T.: J. Polym. Sci. A1, *8*, 2407 (1970).
939. Hirai, H., Ikegami, T., Makishima, S.: J. Polym. Sci. A1, *7*, 2059 (1969).
940. Hirai, H., Okuzawa, S., Ikegami, T., Makishima, S.: J. Faculty of Engeneer.,

940. Hirai, H., Okuzawa, S., Ikegami, T., Makishima, S.: J. Faculty of Engineer., Univ. Tokyo *29*, 115 (1967).
941. Hiraki, K., Hirai, H.: J. Polym. Sci. A1, *11*, 901 (1973).
942. Hiraki, K., Ionue, T., Hirai, H.: J. Polym. Sci. A1, *8*, 2543 (1970).
943. Hiraki, K., Kaneko, S., Hirai, H.: J. Polym. Sci. B, *10*, 199 (1972).
944. Hiraki, K., Kaneko, S., Hirai, H.: Nagasaki Daigaku Kogakubu Kenkyo Hokoku *4*, 106 (1973).
945. Hirano, T., Saito, Y., Hirai, H., Maki s
945. Hirano, T., Saito, Y., Hirai, H., Makishina, S.: Kogyo Kagaku Zasshi *66*, 1158 (1963).
946. Hirokawa, S.: J. Phys. Soc. Japan *37*, 897 (1974).
947. Hirota, N.: J. Amer. Chem. Soc. *90*, 3603 (1968).
948. Ho, S. K., Siegel, S.: J. Chem. Phys. *50*, 1142 (1969).
950. Hoare, F. E., Jackson, L. C., Kurti, N.: Experimental Cryophysics. London: Butterworths 1961.
951. Hoek, van den W. J., de Winter, J. F., Smidt, J.: J. Magn. Resonance *6*, 15 (1972).
952. Höfelmann, K., Jagur-Grodzinski, J., Szwarc, M.: J. Amer. Chem. Soc. *91*, 4645 (1969).
953. Hoffman, A. K.: J. Amer. Chem. Soc.: *83*, 4671 (1961).
954. Hoffman, W., Sauer, W.: Plaste Kaut. *19*, 8 (1972).
955. Holden, A. N., Kittel, C., Merritt, F. R., Yager, W. A.: Phys. Rev. *75*, 1614 (1949).
956. Hollahan, J. R., J. Chem. Educ.: *43*, A401 (1966).
957. Holmberg, R. W., Livingston, R., Smith, W. T.: J. Chem. Phys. *33*, 541 (1960).
958. Holman, D. E., Weiss, J.J.: Int. J. Radiat. Biol. *14*, 187 (1968).
959. Holob, G. M., Ehrlich, P., Allendoerfer, R. D.: Macromolecules *5*, 569 (1972).
960. Holroyd, R. A., Glass, J. W.: Int. J. Radiat. Biol. *14*, 445 (1969).
961. Holzmueller, W.: Plast. Kaut. *13*, 321 (1966).
962. Holzmueller, W.: Polimery *11*, 455 (1966).
963. Hon, N. S.: J. Polym. Sci. A1, *13*, 955 (1975).
964. Hon, N. S.: J. Polym. Sci. A1, *13*, 1347 (1975).
965. Hon, N. S.: J. Polym. Sci. A1, *13*, 1933 (1975).
966. Hon, N. S.: J. Polym. Sci. A1, *13*, 2362 (1975).
967. Hon, N. S.: J. Appl. Polym. Sci. *19*, 2789 (1975).
968. Hong, S. J., Iwakura, Y., Uno, K.: Polymer *12*, 521 (1971).
969. Horanská, V., Bartoň, J., Maňásek, Z.: J. Polym. Sci. A1, *10*, 2701 (1972).
970. Horanská, V., Bartoň, J.: J. Polym. Sci. A1, *12*, 513 (1974).
971. Horne, S. E. Jr., Carman, C. J.: J. Polym. Sci. A1, *9*, 3039 (1971).
972. Horsfield, A.: Chimia *17*, 42 (1963).
973. Hoskins, R. H., Pastor, R. C.: J. Appl. Phys. *31*, 1506 (1960).
974. Hoyer, H., Fitzky, H. G.: Makromol. Chem. *161*, 49 (1972).
975. Hudson, A.: Electron Spin Resonance *1*, 69 (1972).
976. Hughes, O. R., Coard, L. C.: J. Polym. Sci. A1, *7*, 1861 (1969).
977. Huisjen, M., Hyde, J. S.: J. Chem. Phys. *60*, 1682 (1974).
978. Hukuda, K., Kusumoto, N., Kawano, I., Takayanagi, M.: Kogyo Kagaku Zasshi *67*, 2163 (1964).

369

979. Hukuda, K., Kusumoto, N., Kawano, I., Takayanagi, M.: J. Polym. Sci. B, *3*, 743 (1965).
980. Hurst, G. S.: Elementary Radiation Physics. New York: Wiley 1970.
981. Hutchison, C. A., Pastor, R. C., Kowalsky, A. G.: J. Chem. Phys. *20*, 534 (1952).
982. Huysmans, W. G. B., Waters, W. A.: J. Chem. Soc. *1966B*, 1047.
983. Huysmans, W. G. B., Waters, W. A.: J. Chem. Soc. *1967B*, 1163.
984. Huysmans, W. G. B., Waters, W. A.: Proc. Roy. Soc. *A 302*, 329 (1968).
985. Hyde, J. S.: Principles on ESR Instrumentation, paper presented on 5th Varian NMR-ESR Workshop, Palo Alto, Calif., 1961.
986. Hyde, J. S.: Experimental Techniques in EPR, paper presented on 6th Varian NMR-EPR Workshop, Palo Alto, Calif., 1962.
987. Hyde, J. S.: J. Chem. Phys. *43*, 1806 (1965).
988. Hyde, J. S.: in: Free Radicals in Chemistry (Ed. Blumenfeld L. A.) Novosybirsk, USSR: Nauka 1972, p. 24 (in Russ.).
989. Hyde, J. S.: Ann. Rev. Phys. Chem. *25*, 407 (1974).
990. Hyde, J. S., Chien, J. C. W., Freed, J. H.: J. Chem. Phys. *48*, 4211 (1968).
991. Hyde, J. S., Dalton, L.: Chem. Phys. Lett. *16*, 568 (1972).
992. Hyde, J. S., Maki, A. H.: J. Chem. Phys. *40*, 3117 (1964).
993. Hyde, J. S., Rist, G. H., Eriksson, L. E. G.: J. Phys. Chem. *72*, 4269 (1968).
994. Hyde, J. S., Sneed, R. C. Jr., Rist, G. H.: J. Chem. Phys. *51*, 1404 (1969).
995. Ichikawa, T., Iwasaki, M., Kuwata, K.: J. Chem. Phys. *44*, 2979 (1966).
996. Ichikawa, T., Yoshida, H., Hayashi, K.: Bull. Chem. Soc. Japan *46*, 812 (1973).
997. Ikarina, M. A., Illyasov, A. V.: Dokl. Akad. Nauk SSSR *147*, 618 (1962).
998. Ikegami, T., Hirai, H.: J. Polym. Sci. A1, *8*, 463 (1970).
999. Inaki, Y., Ishyama, M., Hibino, K., Takemoto, K.: Makromol. Chem. *176*, 3135 (1973).
1000. Ingalls, R. B., Kivelson, D.: J. Chem. Phys. *38*, 1907 (1963).
1001. Ingalls, R. B., Wall, L. A.: J. Chem. Phys. *35*, 370 (1961).
1002. Ingalls, R. B., Young, W. A.: J. Chem. Phys. *43*, 1759 (1965).
1003. Ingold, K. U., Morton, J. R.: J. Amer. Chem. Soc. *86*, 3400 (1964).
1004. Ingram, D. J. E.: Free Radicals as Studied by Electron Spin Resonance. London: Butterworths 1958.
1005. Ingram, D. J. E.: Lab. Pract. *12*, 518 (1963).
1006. Ingram, D. J. E.; Symons, M. C. R., Townsend, M. G.: Trans. Faraday Soc. *54*, 409 (1958).
1007. Irie, M., Hayashi, K., Okamura, S., Yoshida, H.: J. Chem. Phys. *48*, 922 (1968).
1008. Irie, M., Hayashi, K., Okamura, S., Yoshida, H.: Int. J. Radiat. Phys. Chem.*1*, 297 (1969).
1009. Irie, M., Hayashi, K., Okamura, S., Yoshida, H.: J. Phys. Chem. *75*, 476 (1971).
1010. Irie, M., Nakahigashi, N.: Ann. Rep. Res. Reactor Inst. Kyoto Univ. *5*, 20 (1972).
1011. Irie, M., Tomimoto, S., Hayashi, K.: J. Phys. Chem. *76*, 1419 (1972).
1012. Irie, M., Yoshida, H., Hayashi, K., Okamura, S.: Bull. Chem. Soc. Japan *45*,2347 (1972).
1013. Irie, M., Tomimoto, S., Hayashi, K.: J. Polym. Sci. A1, *10*, 3235 (1972).
1014. Irie, M., Tomimoto, S., Hayashi, K.: J. Polym. Sci. A1, *10*, 3243 (1972).
1015. Isoya, J.: Kagaku No Jikken *24*, 501 (1973).
1016. Ito, M., Kuri, Z.: Kogyo Kagaku Zasshi *69*, 531 (1966).
1017. Ito, M., Kuri, Z.: Kogyo Kagaku Zasshi *70*, 109 (1967).
1018. Itzkowits, M.: J. Chem. Phys. *46*, 3048 (1967).
1019. Ivanchev, S. S., Yurzhenko, A. I., Lukovnikov, A. F., Peredereeva, S. I., Gak, Yu. V.: Dokl. Akad. Nauk SSSR *171*, 894 (1966).
1020. Ivanov, V. F., Grishina, A. D., Lunin, A. F.: Dokl. Akad. Nauk SSSR *222*, 107 (1975).
1021. Ivin, K. J., Sharma, B. D.: J. Indian Chem. Soc. *49*, 1239 (1972).
1022. Iwakura, Y., Takeda, K., Nakazawa, T.: J. Polym. Sci. B, *6*, 115 (1968).
1023. Iwamoto, T., Hayashi, K., Okamura, S., Yoshida, H.: Int. J. Radiat. Phys. Chem. *1*, 1 (1969).
1024. Iwasaki, M.: J. Chem. Phys. *45*, 990 (1966).
1025. Iwasaki, M.: Kobunshi *18*, 18 (1969).
1026. Iwasaki, M.: Kobunshi *18*, 131 (1969).
1027. Iwasaki, M., Fukaya, M., Fujii, S., Muto, H.: J. Phys. Chem. *77*, 2739 (1973).

1028. Iwasaki, M., Ichikawa, T., J. Chem. Phys. *46*, 2851 (1967).
1029. Iwasaki, M., Ichikawa, T., Ohmori, T.: J. Chem. Phys. *50*, 1984 (1969).
1030. Iwasaki, M., Ichikawa, T., Ohmori, T.: J. Chem. Phys. *50*, 1991 (1969).
1031. Iwasaki, M., Ichikawa, T., Toriyama, K.: J. Polym. Sci. B, *5*, 423 (1967).
1032. Iwasaki, M., Nunome, K., Ichikawa, T., Toriyama, K.: Bull. Chem. Soc. Japan *44*, 1522 (1971).
1033. Iwasaki, M., Nunome, K., Muto, H., Toriyama, K.: J. Chem. Phys. *54*, 1839 (1971).
1034. Iwasaki, M., Sakai, Y.: J. Polym. Sci. A2, *6*, 265 (1968).
1035. Iwasaki, M., Sakai, Y.: J. Polym. Sci. A1, *7*, 1537 (1969).
1036. Iwasaki, M., Toriyama, K.: J. Chem. Phys. *47*, 559 (1967).
1037. Iwasaki, M., Toriyama, K., Ohmori, T.: J. Phys. Chem. *72*, 4347 (1968).
1038. Iwasaki, M., Toriyama, K., Sawaki, T., Inoue, M.: J. Chem. Phys. *47*, 554 (1967).
1039. Izumi, Z., Rånby, B.: IUPAC Macromol. Chem., Helsinki, 1972, Pure Appl. Chem., London: Butterworth 1972, vol. 8, p. 107.
1040. Izumi, Z., Rånby, B.: Polym. J. Japan *5*, 208 (1973).
1041. Izumi, Z., Rånby, B.: J. Polym. Sci. A1, *11*, 1903 (1973).
1042. Izumi, Z., Rånby, B.: in: ESR Application to Polymer Research (Ed. Kinell P. O., Rånby, B., Runnström-Reio, V.) Stockholm: Almqvist-Wiksell 1973, p. 43.
1043. Izumi, Z., Rånby, B.: Macromolecules *8*, 151 (1975).
1044. Jackson, I. C., Wynne-Jones, W. F. K.: Carbon *2*, 227 (1964).
1045. Janzen, E. G.: Accounts Chem. Res. *2*, 279 (1969).
1046. Janzen, E. G.: Accounts Chem. Res. *4*, 31 (1971).
1047. Janzen, E. G., DuBose, C. M. Jr.: J. Phys. Chem. *70*, 3372 (1966).
1048. Janzen, E. G., Evans, C. A., Nishi, Y.: J. Amer. Chem. Soc. *94*, 8236 (1972).
1049. Janzen, E. G., Lopp, I. E.: J. Phys. Chem. *76*, 2056 (1972).
1050. Jarret, H. S.: J. Chem. Phys. *21*, 761 (1963).
1051. Jefcoate, C. R. E., Norman, R. O. C.: J. Chem. Soc. *1967B*, 48.
1052. Jenkins, A. D., Ledwith, A.: Reactivity, Mechanism and Structure in Polymer Chemistry. New York: Wiley 1974.
1053. Joffe, Z., Rånby, B.: in: ESR Application to Polymer Research (Ed. Kinell P. O., Rånby, B., Runnström-Reio, V.) Stockholm: Almqvist-Wiksell 1973, p. 171.
1054. Johnsen, R. H.: J. Phys. Chem. *65*, 2144 (1961).
1055. Johnsen, R. H.: in: Creation and Detection of the Excited State (Lamola A. A.,) New York: Dekker 1971, p. 429.
1056. Johnsen, U.: Kolloid-Z. *178*, 161 (1961).
1057. Johnsen, U.: J. Polym. Sci. *54*, S6 (1961).
1058. Johnsen, U., Klinkenberg, D.: Kolloid-Z. e. Z-Polym. *251*, 843 (1973).
1059. Johnson, D. R., Rogers, M., Trappe, G.: J. Chem. Soc. *1956*, 1093.
1060. Johnson, D. R., Wen, W. Y., Dole, M.: Macromolecules *5*, 223 (1972).
1061. Johnson, D. R., Wen, W. Y., Dole, M.: J. Phys. Chem. *77*, 2174 (1973).
1062. Johnsen, P. M., Albrecht, A. C.: J. Chem. Phys. *44*, 1845 (1966).
1063. Jones, M. T.: J. Phys. Chem. *38*, 2892 (1963).
1064. Jones, M. T., Weissman, S. I.: J. Chem. Phys. *84*, 4269 (1962).
1065. Jost, P., Griffith, O. H.: Methods Pharmacol. *2*, 233 (1972).
1065. Kabanov, V. A., Sergeev, G. B., Zubov, V. P., Kargin, V. A.: Vysokomol. Soedin. *1*, 1859 (1959).
1068. Kaeriyama, K.: Bull. Chem. Soc. Japan *47*, 753 (1974).
1069. Kaeriyama, K., Shimura, Y.: Makromol. Chem. *167*, 129 (1973).
1070. Kaeriyama, K., Shimura, Y., Yamaguchi, T.: J. Appl. Polym. Sci. *16*, 3035 (1972).
1071. Kagiya, T., Izu, M., Kawai, S., Fukui, K.: J. Polym. Sci., B, *4*, 387 (1966).
1072. Kaiser, E. T., Kevan, L.: Radical Ions. New York: Interscience 1968.
1073. Kalinowski, M. K., Sadlej, A. J.: Wiad. Chem. *17*, 91 (1963).
1074. Kalinowski, M. K., Sadlej, A. J.: Wiad. Chem. *17*, 171 (1963).
1075. Kalmanson, A. E., Grigoryan, G. L.: in: Experimental Methods in Biophysical Chemistry (Ed. Nicolau, C.) New York: Wiley 1972, p. 589.

1076. Kambara, S., Hatano, M., Kubushiro, K.: Bull. Toyko Inst. Technol. *39*, 29 (1964).
1077. Kan, R. O.: Organic Photochemistry. New York: McGraw-Hill 1966.
1078. Karimov, Yu. S., Schegolev, I. F.: Dokl. Akad. Nauk SSSR *146*, 1370 (1962).
1079. Karpukhina, T. A., Kiseleva, E. D., Chmutkov, K. V., Glazunov, M. P.:
      Zh. Fiz. Khim. *44*, 1003 (1970).
1080. Karra, J., Turkevich, J.: Discuss. Faraday Soc. *41*, 310 (1966).
1081. Kasatochkin, V. I., Nedoshivin, Yu. N.: Zh. Fiz. Khim. *39*, 359 (1965).
1082. Kashiwabara, H.: J. Phys. Soc. Japan *16*, 2493 (1961).
1083. Kashiwabara, H.: J. Phys. Soc. Japan *17*, 567 (1962).
1084. Kashiwabara, H.: J. Appl. Phys. Jap. *3*, 384 (1964).
1085. Kashiwabara, H.: Kobunshi *15*, 979 (1966).
1086. Kashiwabara, H., Shimada, S., Sohma, J.: in: ESR Application to Polymer Research
      (Ed. Kinell, P. O., Rånby, B., Runnström-Reio, V.) Stockholm: Almqvist-Wiksell
      1973, p. 275.
1087. Kashiwabara, H., Shinohara, K.: J. Phys. Soc. Japan *15*, 1129 (1960).
1088. Kashiwagi, M.: J. Chem. Phys. *36*, 575 (1962).
1089. Kashiwagi, M.: J. Polym. Sci. A1, *1*, 189 (1963).
1090. Kato, K., Yoshizaki, O., Nagai, E.: Kogyo Kagaku Zasshi *68*, 2495 (1965).
1091. Kato, Y., Nishioka, A.: Rep. Progr. Polym. Phys. Japan *9*, 477 (1966).
1092. Katzer, H., Heusinger, H.: Makromol. Chem. *163*, 195 (1973).
1093. Kausch, H. H.: J. Macromol. Sci. Rev. Makromol. Chem. C, *4*, 243 (1970).
1094. Kausch, H. H.: J. Polym. Sci. C, (32), 1 (1971).
1095. Kausch, H. H., Becht, J.: Rheol. Acta *9*, 137 (1970).
1096. Kausch, H. H., Becht, J.: Kolloid-Z. u. Z-Polym. *250*, 1048 (1972).
1097. Kausch, H. H., Hassell, J. A., Jaffee, R. L.: Deformation and Fracture of High Polymers
      Plenum 1974.
1098. Kausch, H. H., Moghe, S. R., Hsiao, C. C.: J. Appl. Phys. *38*, 201 (1967).
1099. Kawashima, T., Nakamura, M., Shimada, S., Kashiwabara, H., Sohma, J.: Rep. Progr.
      Polym. Phys. Japan *12*, 469 (1969).
1100. Kawashima, T., Shimada, S., Kashiwabara, H., Sohma, J.: Polym. J. Japan *5*, 135 (1973).
1101. Kawai, K., Shirota, Y., Tsubomura, H., Mikawa, H.: Bull. Chem. Soc. Japan *45*, 77 (1972).
1102. Kazanova, Z. S., Parini, V. P., Liogonskii, B. I.: Izv. Akad. Nauk SSSR *1965*, 419.
1103. Kazanskii, V. B., Pecherskaya, Yu. N.: Kinet. Katliz. *4*, 244 (1963).
1104. Kazanskii, V. B., Turkevich, J.: J. Catal. *8*, 231 (1967).
1105. Keana, J. F. W., Keana, S. B., Beetham, D.: J. Amer. Chem. Soc. *89*, 3055 (1967).
1106. Kearney, J. J., Clark, H. G., Stannett, V., Campbell, D.: J. Polym. Sci. A1, *9*, 1197 (1971).
1107. Kearns, D. R.: Chem. Rev. *71*, 395 (1971).
1108. Kearns, D. R., Khan, A. U.: Photochem. Photobiol. *10*, 193 (1969).
1109. Kearns, R. D., Khan, A. U., Duncan, C. K., Maki, A. H.: J. Amer. Chem. Soc. *91*,
      1039 (1969).
1110. Kecki, Z., Lancman, L.: Przem. Chem. *46*, 94 (1967).
1111. Keighley, J. H.: J. Textl. Inst. *59*, 470 (1968).
1112. Keii, T.: Kinetics of Ziegler-Natta Polymerization. New York-Tokyo: Chapman-Kodansha
      1972.
1113. Kelleher, P. G., Jassie, L. B., Gasner, B. D.: J. Appl. Polym. Sci. *11*, 137 (1967).
1114. Kende, I., Tüdös, F., Sümegi, L.: Acta Chim. Acad. Sci. Hungaricae *54*, 315 (1967).
1115. Kessenikh. A. V.: Tr. Nauchn.-Issled. Fiz. Khim. Inst. *1963*, 187.
1116. Kevan, L.: Actions chim. biol. Radiat. *13*, 57 (1969).
1117. Keyser, R. M., Tsuji, K., Williams, Ff.: Macromolecules *1*, 289 (1968).
1118. Keyser, R. M., Tsuji, K., Williams, Ff.: in: The Radiation Chemistry of Macromolecules
      (Ed. Dole M.,) New York: Academic Press 1972, vol. 1, p. 145.
1119. Keyser, R. M., Williams, Ff.: J. Phys. Chem. *73*, 1623 (1969).
1120. Khamidov, D. S., Azizov, U. A., Milinchuk, V. K.: Uzb. Khim. Zh. *14*, 39 (1970).
1121. Khamidov, D. S., Azizov, U. A., Milinchuk, V. K.: Vysokomol. Soedin. *14*, 838 (1972).
1122. Kharash, M. S., Joshi, B. S.: J. Org. Chem. *22*, 1439 (1957).

1123. Kharitonenkov, I. G.: Biofizyka *11*, 905 (1966).
1124. Kholmogorov, V. E.: Usepkh. Khim. *37*, 1492 (1968).
1125. Kim, Y. W., Chalmers, J. S.: J. Chem. Phys. *44*, 3591 (1966).
1126. Kimmer, W., Voekel, G., Wartewig, S., Windsch, W.: Plaste Kautsch. *10*, 345 (1963).
1127. Kinell, P. O., Komatsu, T., Lund, A., Shiga, T., Shimazu, A.: Acta Chem. Scand. *24*, 3265 (1970).
1128. Kinell, P. O., Lund, A.: Kem. Tidskr. *83*, 36 (1971).
1129. Kinell, P. O., Lund. A.: Kem. Tidskr. *83*, 44 (1971).
1130. Kinell, P. O., Lund, A., Vänngård, T.: Acta Chem. Scand. *19*, 2113 (1967).
1131. Kircher, J. F., Sliemers, F. A., Markle, R. A., Gager, W. G., Leininger, R. I.: J. Phys. Chem. *69*, 189 (1965).
1132. Kirsh, Yu. E., Kovner, V. Ya., Kokorin, A. I., Zamaraev, K. I., Chernyak, V. Ya., Kabanov, V. A.: Europ. Polym. J. *10*, 671 (1974).
1133. Kirsh, Yu. E., Staradupcev, C. G., Grebenshikov, Yu. B., Lichtensthein, G. I., Kabanov, V. A.: Dokl. Akad. Nauk SSSR *194*, 1357 (1970).
1134. Kirukhin, V. P., Milinichuk, V. K.: Vysokomol. Soedin. *A 16*, 816 (1974).
1135. Kirukhin, V. P., Milinchuk, V. K.: Khim. Vysok. Energ. *3*, 451 (1969).
1136. Kiselev, A. V., Kozlov, G. A., Lygin, V. I.: Zh. Fiz. Khim. *40*, 1959 (1966).
1137. Kiselev, A. G., Mokulskii, M. A., Lazurkin, Yu. S.: Vysokomol. Soedin. *2*, 1678 (1960).
1138. Kiss, F., Bagdasaryan, K. S.: Proc. Second Tihany Symp. on Radiation Chemistry. Budapest: Ed. Akademia Kiado 1967, p. 257.
1139. Kivelson, D.: J. Chem. Phys. *27*, 1087 (1957).
1140. Kievelson, D.: J. Chem. Phys. *33*, 1094 (1960).
1141. Klein, E., Möbius, K., Winterhoff, H.: Z. Naturforsch. *22a*, 1704 (1967).
1142. Kleinert, T. N.: Monatsh. Chem. *95*, 387 (1964).
1143. Kleinert, T. N.: Holtzforsch. *18*, 24 (1964).
1144. Kleinert, T. N.: Textil Rundschau *20*, 336 (1965).
1145. Kleinert, T. N.: Monatsh. Chem. *96*, 1925 (1965).
1146. Kleinert, T. N., Morton, J. R.: Nature *196*, 334 (1962).
1147. Klinshpont, E. R., Gilyazitdinov, D. G., Kirukhin, V. P., Milinchuk, V. K.: Khim. Vysokh. Energ. *7*, 188 (1973).
1148. Klinshpont, E. R., Milinchuk, V. K.: Khim. Vysokh. Energ. *1*, 242 (1967).
1149. Klinshpont, E. R., Milinchuk, V. K.: Khim. Vysokh. Energ. *3*, 81 (1969).
1150. Klinshpont, E. R., Milinchuk, V. K.: Khim. Vysokh. Energ. *4*, 84 (1970).
1151. Klinshpont, E. R., Milinchuk, V. K.: Khim. Vysokh. Energ. *5*, 16 (1971).
1152. Klinshpont, E. R., Milinchuk, V. K.: Vysokomol. Soedin. *B 15*, 332 (1973).
1153. Klinshpont, E. R., Milinchuk, V. K.: Vysokomol. Soedin. *B 16*, 35 (1974).
1154. Klinshpont, E. R., Milinchuk, V. K.: Vysokomol. Soedin. *B 17*, 358 (1975).
1155. Klinshpont, E. R., Milinchuk, V. K., Dmitriev, S. M., Kurilenko, A. I.: Vysokomol. Soedin. *A 14*, 1596 (1972).
1156. Klinshpont, E. R., Milinchuk, V. K., Paschenko, V. I., Gilyazitdinov, D. G.: Vysokomol. Soedin. *A 16*, 49 (1974).
1157. Klinshpont, E. R., Milinchuk, V. K., Pshezhetskii, S. Ya.: Khim. Vysokh. Energ. *3*, 74 (1969).
1158. Klinshpont, E. R., Milinchuk, V. K., Pshezhetskii, S. Ya.: Khim. Vysokh. Energ. *3*, 357 (1969).
1159. Klinshpont, E. R., Milinchuk, V. K., Pshezhetskii, S. Ya.: Vysokomol. Soedin. *B 12*, 86 (1970).
1160. Klinshpont, E. R., Milinchuk, V. K., Pshezhetskii, S. Ya.: Vysokomol. Soedin. *A 12*, 1509 (1970).
1161. Klinshpont, E. R., Milinchuk, V. K., Pshezhetskii, S. Ya.: Vysokomol. Soedin. *A 15*, 1963 (1973).
1162. Klopfenstein, C., Jost, P., Griffith, O. H.: Comput. Chem. Biochem. Res. *1*, 175 (1972).
1163. Kloza, M.: Pomiary Autom. Kontr. *17*, 337 (1971).
1164. Kneübuhl, F. K.: J. Chem. Phys. *33*, 1074 (1960).

1165. Ko, M., Sato, T., Otsu, T.: Makromol. Chem. *176*, 643 (1975).
1166. Kochi, J. K.: in: Free Radicals (Ed. Kochi, J. K.) New York: Wiley 1973, vol. 2, p. 665.
1167. Kochi, J. K., Krusic, P. J.: J. Amer. Chem. Soc. *91*, 3940 (1969).
1168. Kodaira, T., Aoyama, F.: J. Polym. Sci. A1, *12*, 897 (1974).
1169. Kodaira, T., Morishita, K., Yamaoka, H., Aida, H.: J. Polym. Sci. B, *11*, 347 (1973).
1170. Kodratoff, Y., Girard, Y., Kibler, M. R., Guyot, A.: Makromol. Chem. *162*, 135 (1972).
1171. Köhnlein, W., Müller, A.: in: Free Radical in Biological Systems (Ed. Blois, M. S., Brown, H. W., Lemmon, R. M., Lindblom, R. O., Weissbluth, M.) New York: Academic Press 1961, p. 113.
1172. Kokes, R. J.: in: Experimental Methods in Catalysis Research (Ed. Andersson, R. D.,) New York: Academic Press 1968, p. 11.
1173. Kokorin, A. I., Kirsh, Yu. E., Zamaryev, K. I.: Vysokomol. Soedin. *A 17*, 1618 (1975).
1174. Kokorin, A. I., Kirsh, Yu. E., Zamaryev, K. I., Kabanov, V. A.: Dokl. Akad. Nauk SSSR *208*, 1391 (1973).
1175. Kokoszka, G. F., Gordon, G.: Techn. Inorg. Chem. *7*, 151 (1968).
1176. Kolek, A., Rabek, J. F.: Problems in Industrial Photochemistry. Proceedings of the Institute of Organic Technology and Plastic of the Wroclaw Polytechnic University, No. 4, 1971.
1177. Kolesnikov, G. S., Vainer, A. Yu., Zhuravlev, I. V., Rode, V. V., Sidnev, A. I.: Vysokomol. Soedin. *A 13*, 923 (1971).
1178. Kolker, P. L., Waters, W. A.: J. Chem. Soc. *1964*, 1136.
1179. Koller, L. R.: Ultraviolet Radiation. New York: Wiley 1965, 2 nd ed.
1180. Komarynski, M. A., Weissman, S. I.: J. Amer. Chem. Soc. *97*, 1589 (1975).
1181. Komatsu, T., Seguchi, T., Kashiwabara, H., Sohma, J.: J. Polym. Sci. C, (16), 535 (1967).
1182. Komatsu, T., Sohma, J., Kashiwabara, H., Seguchi, T.: Rep. Progr. Polym. Phys. Japan *7*, 327 (1964).
1183. Komitami, S.: J. Phys. Chem. *76*, 1729 (1972).
1184. Komitami, S., Akasaka, K., Umegaki, H., Hatano, H.: Chem. Phys. Lett. *9*, 510 (1971).
1185. Konobeevskii, K. S., Finkelshtein, E. S., Nametkin, N. S., Kdovin, V. M., Ivanov, V. P.: Vysokomol. Soedin. *B 14*, 811 (1972).
1186. Konobeevskii, K. S., Guselnikov, L. E., Namietkin, N. S., Polak, L. E., Chernysheva, T. I.: Vysokomol. Soedin. *8*, 533 (1966).
1187. Konzelmann, U.: Dissertation Universität Stuttgart, Germany, 1974.
1188. Kopp, P. M., Charlesby, A.: Intern. J. Appl. Radiation Isotopes *17*, 352 (1966).
1189. Kopylev, V. V., Pravednikov, A. N.: Vysokomol. Soedin. *B 1o*, 254 (1968).
1190. Kopylev, V. V., Pravednikov, A. N.: Vysokomol. Soedin. *A 10*, 1170 (1968).
1191. Koritskii, A. T., Molin, Yu. N., Shamshev, V. N., Buben, N. N., Voevodskii, V. V.: Vysokomol. Soedin. *1*, 1182 (1959).
1192. Koritskii, A. T., Zubov, A. V.: Khim. Vyokh. Energ. *1*, 123 (1967).
1193. Koritskii, A. T., Zubov, A. V.: Vysokomol. Soedin. *A 9*, 789 (1967).
1194. Koritskii, A. T., Zubov, A. V., Tochin, V. A.: Khim. Vyokh. Energ. *2*, 552 (1968).
1195. Korolev, G. V., Smirnov, B. R.: Vysokomol. Soedin. *6*, 1140 (1964).
1196. Korolev, G. V., Smirnov, B. R., Makhonina, L. I.: Vysokomol. Soedin. *7*, 1417 (1965).
1197. Korst, N. N., Lazarev, A. V.: Mol. Phys. *17*, 481 (1969).
1198. Koryakov, V. I., Gubanov, V. A., Belyakov, Yu. M., Chirkov, A. K., Pankratov, V. N., Matevosyan, R. O.: Izv. Akad. Nauk SSSR *1971*, 2468.
1199. Kosek, S., Zielinski, W.: Polimery *13*, 260 (1968).
1200. Kosek, S., Zielinski, W.: Polimery *13*, 297 (1968).
1201. Kössel, H.: in: Encyclopedia of Polymer Science and Technology. New York: Wiley-Interscience 1969, vol. 11, p. 364.
1202. Kotomin, E. A., Plotnikov, O. V., Rajavcc, E., Janson, J.: Zh. Prikl. Spektrosk. *22*, 363 (1975).
1203. Koton, M. M., Andreeva, I. V., Getmanchuk, Yu. P., Madorskaya, L. Ya., Pokrovskii, E. I., Koltsov, A. I., Filatova, V. A.: Vysokomol. Soedin. *7*, 2039 (1965).
1204. Kotov, B. V., Pravednikov, A. N.: Proc. Second Tihany Symp., on Radiation Chemistry. Budapest: Ed. Akademiai Kiado 1967, p. 775.

1205. Kotov, B. V., Zhaludkova, I.: Dokl. Akad. Nauk SSSR *159*, 640 (1964).
1206. Koǔrim, P., Vacek, K.: Tetrahedron Lett. *23*, 1051 (1962).
1207. Koǔrim, P., Vacek, K.: Trans. Faraday Soc. *61*, 415 (1965).
1208. Kovarskii, A. L., Arkina, S. N., Vasserman, A. M.: Vysokomol. Soedin. *B12*, 35 (1970).
1209. Kovarskii, A. L., Vasserman, A. M., Aleksandrova, T. A., Tager, A. A.: Dokl. Akad. Nauk SSSR *210*, 1372 (1973).
1210. Kovarskii, A. L., Vasserman, A. M., Buchachenko, A. L.: Vysokomol. Soedin. *B12*, 211 (1970).
1211. Kovarskii, A. L., Vasserman, A. M., Buchachenko, A. L.: Vysokomol. Soedin. *A13*, 1647 (1971).
1212. Kovarskii, A. L., Vasserman, A. M., Buchachenko, A. L.: J. Magn. Reson. *7*, 225 (1972).
1213. Kozerev, B. M., Yablokov, Yu. V., Matevosyan, R. O., Ikarina, M. R., Illyasov, A. V., Stashkov, L. I., Shatrukov, P. A.: Opt. Spektrosk. *15*, 625 (1963).
1214. Kozlov, V. T., Evseev, A. G.: Vysokomol. Soedin. *A11*, 467 (1969).
1215. Kozlov, V. T., Evseev, A. G., Karlin, A. V., Zubov, P. I.: Vysokomol. Soedin. *A11*, 2223 (1969).
1216. Kozlov, V. T., Kashevskaya, N. G.: Vysokomol. Soedin. *B14*, 315 (1972).
1217. Kozlov, V. T., Lanin, S. N., Kashevskaya, N. G., Khan, A. A., Soldatenko A. I.: Vysokomol. Soedin. *B17*, 3 (1975).
1218. Kozlov, V. T., Tarasova, Z. N.: Vysokomol. Soedin. *8*, 943 (1966).
1219. Kozyrev, B. M., Yablokov, Yu. V., Matevosyan, R. O.: Opt. Spektrosk. *15*, 625 (1963).
1220. Kraessig, H. A.: Tappi *46*, 654 (1963).
1221. Krauss, H. L., Huttmann, H., Deffiner, V.: Z. Anorg. Allgem. Chem. *341*, 164 (1965).
1222. Krieg, B., Manecke, G.: Makromol. Chem. *108*, 210 (1967).
1223. Kroh, J.: Free Radicals in Radiation Chemistry. Warszawa: PWN 1967 (in Polish).
1224. Kroh, J.: Radiation Chemistry. Warszawa: PWN 1970 (in Polish).
1225. Kroh, J., Walicki, M.: Bull. Acad. Polon. Sci., Ser. Sci. Chem. *13*, 635 (1955).
1226. Krongelb, S., Strandberg, M. W. P.: J. Chem. Phys. *31*, 1196 (1959).
1227. Kryazhev, Yu. G., Cherkashin, M. I., Semenova, E. F., Salurev, V. N., Berlin A. A., Shostakovskii, M. F.: Dokl. Akad. Nauk SSSR *177*, 846 (1967).
1228. Kryazhev, Yu. G., Petrinska, V. B., Brodskaya, E. N., Shostakovskii, S. M.: Khim. Vysokomol. Soedin. Neftokhim. *1973*, 104.
1229. Kryazhev, Yu. G., Vakulskaya, T. I., Brodskaya, E. N., Yushmanova, T. I.: Izv. Akad. Nauk SSSR *1973*, 2600.
1230. Kryazhev. Yu. G., Vakulskaya, T. I., Cherkashin, M. I., Vysokomol. Soedin. *A15*, 1011 (1973).
1231. Krylova, Z. L., Dolin, P. I.: Proc. Second Tihany Symp. on Radiation Chemistry. Budapest: Akademiai Kiado 1967, p. 133.
1232. Kryszewski, M.: Polimery *11*, 459 (1966).
1233. Kryszewski, M., Wojciechowski, P.: in: Kinetic and Mechanism of Polyreactions, Akademiai Kiado, Budapest, 1969, vol. 1, p. 221.
1234. Kubarev, A. V., Gasilov, A. L., Telminov, M. M., Slutskin, M. A.: Izmiertitel. Tekhn. *1966*, 71.
1235. Kubo, S., Dole, M.: Macromolecules *6*, 774 (1973).
1236. Kubo, S., Dole, M.: Macromolecules *7*, 190 (1974).
1237. Kubota, N., Ogiwara, Y., Matsuzaki, K.: J. Polym. Sci. A1, *12*, 2809 (1974).
1238. Kubota, H., Ogiwara, Y., Matsuzaki, K.: J. Appl. Polym. Sci. *19*, 1291 (1975).
1239. Kubota, Y., Miura, M.: Bull. Chem. Soc. Japan *40*, 2989 (1967).
1240. Kudrna, S. K.: Vysokomol. Soedin. *8*, 1828 (1966).
1241. Kudryatsev, Yu. P., Sladkov, A. M., Aseev, Yu. G., Nedoshivin, Yu. N., Kasatochkin, V., Korshak, V. V.: Dokl. Akad. Nauk SSSR *158*, 389 (1964).
1242. Kuhn, R., Trischmann, H.: Monatsh. Chem. *95*, 457 (1964).
1243. Kuksenko, V. S., Gezalov, M. A., Slutsker, A. I., Yastrebinsky, A. A., Zakrevskii, V. A.: J. Polym. Sci. C, (38), 357 (1972).
1244. Kuksenko, V. S., Slutsker, A. I.: Fiz. Tverd. Tela *11*, 405 (1969).

References

1245. Kumanotani, J., Yoshikawa, T., Kubo, E., Kawanishi, I., Okamura, T.: J. Polym. Sci. C, (16), 1705 (1967).
1246. Kumler, P. L., Boyer, R. F.: Polym. Prepr. Amer. Chem. Soc. Div. Polym. Chem. *16*, 572 (1975).
1247. Kumpikas, P., Majauskiene, N.: Izv. Vyssh. Ucheb. Zaved. Tekhnol. Legk. Prom. *1971*, 8.
1248. Kumpikas, P., Majauskiene, N.: Izv. Vyssh. Ucheb. Zaved. Tekhnol. Legk. Prom. *1971*, 22.
1249. Kunitake, T., Murakami, S.: Polymer J. Japan *3*, 249 (1972).
1250. Kunitake, T., Murakami, N.: J. Polym. Sci. A1, *12*, 67 (1974).
1251. Kuran, W., Pasynkiewicz, S., Florjanczyk, Z., Kowalski, A.: Makromol. Chem. *175*, 3411 (1974).
1252. Kuri, Z., Ueda, H.: J. Polym. Sci. *50*, 349 (1961).
1253. Kuri, Z., Ueda, H., Shida, S.: Izotopes and Radiation *2*, 496 (1959).
1254. Kuri, Z., Ueda, H., Shida, S.: J. Chem. Phys. *32*, 371 (1960).
1255. Kurita, Y.: J. Chem. Soc. Japan *85*, 833 (1964).
1256. Kurita, Y.: J. Chem. Phys. *41*, 3926 (1964).
1257. Kurita, Y.: Nippon Kagaku Zasshi *85*, 833 (1964).
1258. Kurita, Y., Ohigashi, H., Kashiwagi, M.: Bussei *9*, 87 (1968).
1259. Kurosaki, T., Lee, K. W., Okawara, M.: J. Polym. Sci. A1, *10*, 3295 (1972).
1260. Kurosaki, T., Takahashi, O., Okawara, M.: J. Polym. Sci. A1, *12*, 1407 (1974).
1261. Kurusu, Y., Yoshida, H., Okawara, M.: Tetrahedron Lett. *1967*, 3595.
1262. Kurusu, Y., Yoshida, H., Okawara, M.: J. Chem. Soc. Japan, Ind. Chem. Sec. *72*, 1402 (1969).
1263. Kurusu, Y., Yoshida, H., Ogiwara, M.: Makromol. Chem. *143*, 73 (1971).
1264. Kustanovich, I. M., Patalakh, I. I., Polak, L. S.: Kinet. Kataliz. *4*, 167 (1963).
1265. Kusumoto, N.: J. Polym. Sci. C, (23), 837 (1968).
1266. Kusumoto, N., Hukuda, K.: Rep. Progr. Polym. Phys. Japan *10*, 517 (1967).
1267. Kusumoto, N., Hukuda, K., Takayanagi, M.: Rep. Progr. Polym. Phys. Japan *8*, 315 (1965).
1268. Kusumoto, N., Hukuda, K., Takayanagi, M.: Rep. Progr. Polym. Phys. Japan *8*, 317 (1965).
1269. Kusumoto, N., Kawaho, I., Hukuda, K., Takayanagi, M.: Kogyo Kagaku Zasshi *68*, 825 (1965).
1270. Kusumoto, N., Matsumoto, K., Takayanagi, M.: J. Polym. Sci. A1, *7*, 1773 (1969).
1271. Kusumoto, N., Nakahara, T., Takayanagi, M.: Rep. Progr. Polym. Phys. Japan *10*, 523 (1967).
1272. Kusumoto, N., Shirano, K., Takayanagi, M.: Kogyo Kagaku Zasshi *68*, 1553 (1965).
1273. Kusumoto, N., Shirano, K., Takayanagi, M.: Rep. Progr. Polym. Phys. Japan *10*, 519 (1967).
1274. Kusumoto, N., Suehiro, K., Takayanagi, M.: J. Polym. Sci. B, *10*, 81 (1972).
1275. Kusumoto, N., Yamamoto, T., Takayanagi, M.: J. Polym. Sci. A2, *9*, 1173 (1971).
1276. Kusumoto, N., Yonezawa, M., Motozato, Y.: Polymer *15*, 793 (1974).
1277. Kuwata, K.: Kagaku *19*, 1077 (1964).
1278. Kuwata, K., Kageyama, Y., Hirota, K.: Bull. Chem. Soc. Japan *38*, 510 (1965).
1279. Kuwata, K., Kotake, Y., Inada, K., Ono, M.: J. Phys. Chem. *76*, 2061 (1972).
1280. Kuwata, K., Nishikida, K., Kawazura, H., Hirota, H.: Bull. Chem. Soc. Japan *36*, 925 (1963).
1281. Kuzin, I. A., Semushin, A. M., Antufiev, V. V., Votinov, M. P., Evdokimov, V. F.: Zh. Fiz. Khim. *36*, 395 (1962).
1282. Kuzminskii, A. S., Angert, L. G., Buchachenko, A. L., Mikhailova, G. N., Shlvakhova, L. P.: Dokl. Akad. Nauk SSSR *167*, 586 (1966).
1283. Kuzminskii, A. S., Fedoseeva, T. S. Buchachenko, A. L.: Kauchuk i Rezina *24*, 10 (1965).
1284. Kuzminskii, A. S., Fedoseeva, T. S., Chartkova, V. F.: Ind. Use Large Radiation Sources. Proc. Conf. Salzburg, Austria, 1963, vol. 1, p. 345.
1285. Kuzminskii, A. S., Fedoseeva, T. S., Lebedev, Ya. S., Buchachenko, A. L., Zhuravskaya, E. V.: Vysokomol. Soedin. *6*, 1308 (1964).
1286. Kuzminskii, A. S., Kvashenko, J. J., Fedoseeva, T. S.: Proc. Second Tihany Symp. on Radiation Chemistry. Budapest: Akademiai Kiado 1967, p. 741.

376

1287. Kuzminskii, A. S., Lyubchanskaya, L. I.: Dokl. Akad. Nauk SSSR 93, 519 (1953).
1288. Kuzminskii, A. S., Lyubchanskaya, L. I.: Rubb. Chem. Technol. 29, 770 (1956).
1289. Kuzminskii, A. S., Manzels, M. G., Lezhnev, N. W.: Dokl. Akad. Nauk SSSR 71, 319 (1960).
1290. Kuzminskii, A. S., Neiman, M. B., Fedoseeva, T. S., Lebedev, Ya. S., Buchachenko, A. L., Chertkova, V. F.: Dokl. Akad. Nauk SSSR 146, 611 (1962).
1291. Kuzminskii, A. S., Nikitina, T. S., Fedoseeba, T. S., Lebedev, Ya. S., Buchachenko, A. L., Chertkova, V. F.: Dokl. Akad. Nauk SSSR 146, 611 (1962).
1292. Kuznetsov, A. N., Talroze, R. V., Tenchev, B. G., Shibaev. V. P.: Vysokomol. Soedin. A 17, 1332 (1975).
1293. Kvashenko, J. J., Kuzminskii, A. S., Fedoseeva, T. S.: Vysokomol. Soedin. 8, 2150 (1966).
1294. Lagercrantz, C.: J. Phys. Chem. 75, 3466 (1971).
1295. Laidler, K. J.: The Chemical Kinetics of Excited States. Oxford: Clarendon 1955.
1296. Lakatos, B., Turscányi, B., Tüdös, F.: Acta Chim. Acad. Sci. Hungaricae 70, 225 (1971).
1297. Lambe, J., Kikuchi, C.: Phys. Rev. 11, 71 (1960).
1298. Lamola, A. A.: Creation and Detection of the excited States. New York: Dekker 1971.
1299. Lancaster, G.: J. Mater. Sci. 2, 489 (1967).
1300. Landgraf, W. C.: EPR's Role in Free Radical Chemistry. Palo Alto, Calif, USA: Varian Associates 1964.
1301. Lando, J. B., Morawetz, H.: J. Poym. Sci. C, (4), 789 (1964).
1302. Landolt-Börnstein: Neue Serie Gruppe II. Berlin, Heidelberg, New York: Springer 1966, vol. 2, p. 4.
1303. Langner, H., Zeppenfeld, G., Bartl, A.: Plaste Kautsch. 13, 76 (1966).
1304. Lapitskii, G. A., Makin, S. M., Berlin, A. A.: Vysokomol. Soedin. A 9, 1274 (1967).
1305. Lapitskii, G. A., Makin, S. M., Presnov, A. E., Karpacheva, G. P.: Vysokomol. Soedin. B 9, 508 (1967).
1306. Lapshin, N. M., Khidekel, M. L.: Zh. Strukt. Khim. 3, 713 (1962).
1307. Lapshin, N. M., Ovcharenko, N. I., Khidekel, M. L.: Zh. Strukt. Khim. 5. 305 (1964).
1308. Lawton, E. J., Balwit, J. S.: J. Phys. Chem. 65, 815 (1961).
1309. Lawton, E. J., Balwit, J. S., Powell, R. S.: J. Chem. Phys. 33, 395 (1960).
1310. Lawton, E. J., Balwit, J. S., Powell, R. S.: J. Chem. Phys. 33, 405 (1960).
1311. Lawton, E. J., Powell, R. S., Balwit, J. S.: J. Polym. Sci. 32, 277 (1958).
1312. Lawton, E. L., J. Polym. Sci. A 1, 10, 1857 (1972).
1313. Lazár, M., Szöcs, F.: Coll. Szechoslav. Chem. Commun. 31, 1902 (1966).
1314. Lazár, M., Szöcs, F.: J. Polym. Sci. C, (16), 461 (1967).
1315. Leaver, I. H.: Text. Res. J. 38, 729 (1968).
1316. Leaver, I. H., Ramsay, G. C.: Text. Res. J. 39, 722 (1969).
1317. Lebedev, Ya. S.: Kinet. Kataliz. 3, 615 (1962).
1318. Lebedev, Ya. S.: Zh. Strukt. Khim. 4, 22 (1963).
1319. Lebedev, Ya. S.: Rad. Effects 1, 213 (1964).
1320. Lebedev, Ya. S.: Dokl. Akad. Nauk. SSSR 171, 378 (1966).
1321. Lebedev, Ya. S.: Kinet. Kataliz. 8, 245 (1967).
1322. Lebedev, Ya. S., Chernikova, D. M., Tikhomirova, N. N.: Zh. Strukt. Khim. 2, 690 (1961).
1323. Lebedev, Ya. S., Chernikova, D. M., Tikhomirova, N. N., Voevodskii, N. N.: Atlas of Electron Spin Resonance Spectra. Consultants New York: Bureau 1963.
1324. Lebedev, Ya. S., Muromtsev, V. I.: EPR and Relaxation of Stabilized Radicals. Moscow, USSR: Khimia 1972.
1325. Lebedev, Ya. S., Tsepalov, V. F., Shlyapintokh, V. Ya.: Dokl. Akad. Nauk SSSR 139, 1409 (1961).
1326. Lebedev, Ya. S., Tsvetkov, Yu. D.: Strukt. Khim. 2, 607 (1961).
1327. Lebedev, Ya. S., Tsvetkov, Yu. D., Voevodskii, V. V.: Opt. Spektr. 8, 811 (1960).
1328. Lebedev, Ya., S., Tsvetkov, Yu. D., Voevodskii, V. V.: Vysokomol. Soedin. 5, 1500 (1963).
1329. Lebedev, Ya. S., Tsvetkov, Yu. D., Voevodskii, V. V.: Vysokomol. Soedin. 5, 1608 (1963).

References

1330. Lebedev, Ya. S., Tsvetkov, Yu. D., Zhidomirov, G. M.: Zh. Strukt. Khim. *3*, 21 (1962).
1331. Leca, M., Fulea, A. O.: Mater. Plast. (Bucharest) *9*, 482 (1972).
1332. Lefebvre, R., Maruani, J., Marx, R.: J. Chem. Phys. *41*, 585 (1964).
1333. Lehmus, P.: in: ESR Application to Polymer Research (Ed. Kinell, P. O., Rånby, B., Runnström-Reio, V.). Stockholm: Almqvist-Wiksell 1973, p. 141.
1334. Lehr, M. H., Carman, C. J.: Macromolecules *2*, 217 (1969).
1335. Leighton, P. A.: Photochemistry of Air Pollution. New York: Academic Press 1961.
1336. Lemaire, H., Marchal, Y., Ramasseul, R., Rassat, A.: Bull. Chim. Soc. France *1965*, 372.
1337. Lemaire, H., Rassat, A., Ravet, A. M.: Bull. Chim. Soc. France *1963*, 1980.
1338. Lembke, R., Kevan, L.: Int. J. Radiat. Phys. Chem. *7*, 547 (1975).
1339. Lenherr, A. D., Ormerod, M. G.: Biochim. biophys. Acta *166*, 298 (1968).
1340. Lenherr, A. D., Ormerod, M. G.: Nature *225*, 546 (1970).
1341. Lenherr, A. D., Ormerod, M. G.: Proc. Roy. Soc. *A 325*, 81 (1971).
1342. Leniart, D. S.: Computer Application in ESR, paper presented on 14th Varian ESR Workshop, Zürich, Swiss, 1975.
1343. Leniart, D. S.: Powder ESR, paper presented on 14th Varian ESR Workshop, Zürich, Swiss, 1975.
1344. Lenk, R.: Czech. J. Phys. *11*, 876 (1961).
1345. Lenk, R.: Czech. J. Phys. *12*, 833 (1962).
1346. Lenk, R.: Mol. Phys. *21*, 57 (1971).
1347. Lenk, R.: Chimia *28*, 51 (1974).
1348. Lenk, R., Rousseau, A.: Proc. Coll. AMPERE (Ed. Averbuch, P.) Amsterdam: North-Holland 1968, p. 285.
1349. Lennox, F. G., King, M. G., Leaver, I. H., Ramsay, G. C., Savige, W. E.: Appl. Polym. Symp. No. 18, 353 (1971).
1350. Lentz, R. L.: Organic Chemistry of Synthetic High Polymers. New York: Wiley 1967.
1351. Lepley, A. R., Closs, G. L.: Chemically Induced Magnetic Polarization. New York: Wiley 1973.
1352. Lerman, L. S.: J. Mol. Biol. *3*, 18 (1961).
1353. Lerner, N. R.: J. Chem. Phys. *50*, 2902 (1969).
1354. Lerner, N. R.: J. Polym. Sci. A1, *12*, 2477 (1974).
1355. Levich, V. G., Markin, V. S., Chirkov, Yu. G.: Dokl. Akad. Nauk SSSR *149*, 894 (1963).
1356. Levitskii, M. B., Roginskii, V. A., Dzhagatspanyan, R. V., Pshezhetskii, S. Ya.: Vysokomol. Soedin. *B15*, 48 (1973).
1357. Levy, R. A.: Phys. Rev. *102*, 31 (1956).
1358. Levy, D. H., Myers, R. J.: J. Chem. Phys. *41*, 1062 (1964).
1359. Levy, D. H., Myers, R. J.: J. Chem. Phys. *44*, 4177 (1966).
1360. Levy, M., Szwarc, M.: J. Amer. Chem. Soc. *82*, 521 (1960).
1361. Lewis, F. M., Mayo, F. R.: Ind. Eng. Chem. Anal. Ed. *17*, 134 (1945).
1362. Libby, D., Ormerod, M. G.: Phys. Chem. Solids *18*, 316 (1961).
1363. Libby, D., Ormerod, M. G., Charlesby, A.: Polymer *1*, 212 (1960).
1364. Liebman, S. A., Ahlstrom, D. A., Quinn, E. J., Geigley, A. G., Meluskey, J. T.: J. Polym. Sci. A1, *9*, 1921 (1971).
1365. Liebman, S. A., Reuwer, J. F., Gollatz, K. A., Nauman, C. D.: J. Polym. Sci. A1, *9*, 1823 (1971).
1366. Liepins, R., Verma, G. S. P., Walker, C.: Macromolecules *2*, 419 (1969).
1367. Liming, F. G., Gorgy, W.: Proc. Nat. Acad. Sci. *60*, 794 (1968).
1368. Limura, D., Shimada, S., Phno, K., Kashiwabara, H., Sohma, J.: Rep. Progr. Polym. Phys. Japan *14*, 545 (1971).
1369. Lin, D. P.: J. Chin. Chem. Soc. (Taipei) *21*, 201 (1974).
1370. Lin, J., Tsuji, K., Williams, Ff.: Chem. Phys. Lett. *1*, 1279 (1967).
1371. Lin, J., Tsuji, K., Williams, Ff.: J. Chem. Phys. *46*, 4982 (1967).
1372. Lin, J., Tsuji, K., Williams, Ff.: Trans, Faraday Soc. *64*, 2896 (1968).
1373. Lin, J., Tsuji, K., Williams, Ff.: J. Amer. Chem. Soc. *90*, 2766 (1968).
1374. Lindberg, J. J., Bulla, I., Törmälä, P.: J. Poym. Sci. C, (51), 167 (1975).

1375. Linder, R. E., Ling, A. C.: Can. J. Chem. *50*, 3982 (1972).
1376. Liogonskii, B. I., Ragimov, A. V., Berlin, A. A.: Teor. i Esperim. Khim. Akad. Nauk Ukr. SSR *1*, 511 (1965).
1377. Lipatova, E., Siderko, V. M.: Vysokomol. Soedin. *7*, 1476 (1965).
1378. Lishnevskii, V. A.: Dokl. Akad. Nauk SSSR *182*, 596 (1968).
1379. Lishnevskii, V. A.: Vysokomol. Soedin. *B11*, 44 (1969).
1380. Litowska, M., Narebska, A., Basinski, A.: Bull. Acad. Pol. Sci., Ser. Sci. Chem. *19*, 439 (1971).
1381. Livingston, M.: High Energy Accelerators. New York: Wiley 1954.
1382. Livingston, R., Zeldes, H.: J. Chem. Phys. *44*, 1245 (1966).
1383. Livingston, R., Zeldes, H.: J. Amer. Chem. Soc. *88*, 4333 (1966).
1384. Livingston, R., Zeldes, H.: J. Mag. Resonance *1*, 169 (1969).
1385. Lloyd, B. A., DeVries, K. L., Williams, M. L.: J. Polym. Sci. A2, *10*, 1415 (1972).
1386. Lord, N. W., Blinder, S. M.: J. Chem. Phys. *34*, 1693 (1961).
1387. Low, W.: Paramagnetic Resonance. New York: Academic Press 1963, vol. 1 and vol. 2.
1388. Loy, B. R.: J. Polym. Sci. *44*, 341 (1960).
1389. Loy, B. R.: J. Polym. Sci. *50*, 145 (1961).
1390. Loy, B. R.: J. Phys. Chem. *65*, 58 (1961).
1391. Loy, B. R.: SPE Trans. *2*, 157 (1962).
1392. Loy, B. R.: J. Polym. Sci. A1, *1*, 2251 (1963).
1393. Lund, A.: J. Phys. Chem. *76*, 1411 (1972).
1394. Lund, A., Kevan, L.: J. Phys. Chem. *77*, 2180 (1973).
1395. Lund, A., Shiga, T., Kinell, P. O.: in: ESR Application to Poymer Research (Ed. Kinell, P. O., Rånby, B., Runnström-Reio, V.) Stockholm: Almqvist-Wiksell 1973, p. 95.
1396. Lund, T., Raynor, J. B.: Electron Spin Resonance *2*, 295 (1974).
1397. Lyamenkova, E. K., Zhuravleva, I. V., Ayupova, R. S., Papov, V. S., Matochkin, V. S., Valetskii, P. M., Vinogradova, S. V., Pavlova, S. A., Korshak, V. V.: Vysokomol. Soedin. *A17*, 698 (1975).
1398. Lyankina, S. P., Dobromyslova, A. V., Kazakova, V. M., Dontsov, A. A., Dogadkin, B. A.: Vysokomol. Soedin. *A15*, 2773 (1973).
1399. Lyons, A. R., Symons, M. C. R., Yandell, J. K.: Makromol. Chem. *157*, 103 (1972).
1400. Lyons, A. R., Symons, M. C. R., Yandell, J. K.:
1401. Lyubchenko, L. S., Livshits, V. A., Strigutskii, V. P.: Dokl. Akad. Nauk SSSR *186*, 860 (1969).
1402. Mackor, A., Wajer, A. J. W., deBoer, J., Tetrahedron Lett.: *1966*, 2115.
1403. Magaril, R. Z.: Mechanism and Kinetics of the Homogenous Thermal Transformations of Hydrocarbons. Moscow: Khimia 1970.
1404. Magat, M.: Polymer *3*, 449 (1962).
1405. Maguire, W. J., Pink, R. C.: Trans. Faraday Soc. *63*, 1097 (1967).
1406. Makatun, V. N., Potapovich, A. K., Ermolenko, I. N.: Vysokomol. Soedin. *5*, 467 (1963).
1407. Makhlis, F. A.: Radiation Physics and Chemistry of Polymers. New York: Wiley 1975.
1408. Maki, A. H.: Ann. Rev. Phys. Chem. *18*, 9, (1967).
1409. Maki, A. H., Allendoerfer, R. D., Danner, J. C., Keys, T.: J. Amer. Chem. Soc. *90*, 4225 (1968).
1410. Maki, A. H., Randall, E. W., J. Amer. Chem. Soc.: *82*, 4109 (1960).
1411. Makin, S. M., Lapitskii, G. A., Kolunova, A. M.: Zh. Vses. Khim. Obshestva im.D.I. Mendeleva *8*, 708 (1963).
1412. Maklakov, A. I., Maklakov, L. I., Nikitina, V. I., Balakirova, R. S., Shepelev, V. I., Kurzhunova, Z. Z.: Izv. Vyssh. Ucheb. Zaved. Khim. Khim. Tekhnol. *10*, 90 (1967).
1413. Makosa, A.: Acta Phys. Polon. *39*, 161 (1971).
1414. Malchevskii, V. A., Zakrevskii, V. A.: Izv. Vyssh. Ucheb. Zaved. Khim. Khim. Tcknol. *15*, 276 (1972).
1415. Mamunya, E., Vonsyatskii, V. A., Lebedev, Ya. S.: Teor. Eksp. Khim. *10*, 794 (1974).
1416. Manko, E. N., Kavun, S. M., Illina, E. A., Sherchev, V. A., Dogadkin, B. A.: Vysokomol. Soedin. *B11*, 447 (1069).

References

1417. Manausadzhyan, V. G., Babyan, G. V.: Izv. Akad. Nauk. Arm. SSR, Biol. Nauk *18*, 11(1965).
1418. Mao, S. W., Kevan, L.: Chem. Phys. Lett. *24*, 505 (1974).
1419. Marcotte, F. B., Campbell, D., Cleaveland, J. A., Turner, D. T.: J. Polym. Sci. A1, *5*, 481 (1967).
1420. Marek, M., Toman, L.: Appl. Polym. Symp. No. 42, 339 (1973).
1421. Marek, M., Toman, L., Pilar, J.: J. Polym. Sci. A1, *13*, 1565 (1975).
1422. Margomenou-Leonidopoulu, G.: Chem. Chron. Epistem. Ekdosis *35*, 76 (1970).
1423. Martinmaa, J.: Dissertation, Universtiy of Helsinki, 1974.
1424. Marx, R.: J. Chim. Phys. *62*, 767 (1965).
1425. Marx, R., Bensasson, M. R.: J. Chim. Phys. *57*, 673 (1960).
1426. Marx, R., Chachaty, M. C.: J. Chim. Phys. *58*, 527 (1961).
1427. Marx, R., Fenistein, S.: J. Chim. Phys. *64*, 1424 (1967).
1428. Marx, R., Fenistein, S., Bonazzola, L.: in: ESR Application to Polymer Research (Ed. Kinell, P. O., Ranby, R., Runnström-Reio, V.) Stockholm: Amqvist-Wiksell 1973, p. 77.
1429. Mason, R. P., Polnaszek, C. F., Freed, J. H.: J. Phys. Chem. *78*, 1324 (1974).
1430. Matsuda, K., Yamaguchi, J.: J. Phys. Soc. Japan *20*, 1340 (1965).
1431. Matsuda, T., Ono, Y., Keii, T.: J. Polym. Sci. A1, *4*, 730 (1966).
1432. Matsugashita, T., Shinohara, K.: J. Chem. Phys. *32*, 954 (1960).
1433. Matsugashita, T., Shinohara, K., J. Chem. Phys. *35*, 1652 (1961).
1434. Matsumoto, A., Tanaka, H., Goto, N., Bull. Chem. Soc. Japan *38*, 45 (1965).
1435. Matsuura, T., Nishinaga, A.: Chemistry of Free Radicals. Tokyo-Kyoto, Nankodo 1967.
1436. Matkovskii, P. E., Russiyan, L. N., Beikhold, G. A., Dyachovskii, F. S., Larina, T. I., Brikenshtein, K. M., Chirov, N. M.: Vysokomol. Soedin. *A 15*, 805 (1973).
1437. Matthies, P., Schlag, J., Schwartz, E.: Angew. Chem. *77*, 323 (1965).
1438. Mauclaire, G., Marx, R.: J. Chim. Phys. *65*, 213 (1968).
1439. Mazzolini, C., Patron, L., Moretti, A., Campanelli, M.: Ind. Eng. Chem. Prod. Res. Devel. *9*, 504 (1970).
1440. McCalley, R. C., Schimshick, E. J., McConell, H. M.: Chem. Phys. Lett. *3*, 115 (1972).
1441. McConnell, H.: J. Chem. Phys. *34*, 13 (1961).
1442. McConnell, H., McLachlan, A.: J. Chem. Phys. *34*, 33 (1961).
1443. McConnell, H. M., McFarland, B. G., Quart. Rev. Biophys. *3*, 91 (1970).
1444. McCrum, N. G., Read, B. E., Williams, G.: Anelastic and Dielectric Effects in Polymer Solids. New York: Wiley 1967, p. 353.
1445. McLaren, A. D., Shugar, D.: Photochemistry of Proteins and Nucleic Acids. Oxford: Pergamon Press 1964.
1446. McLauchlan, K. A.: Magnetic Resonance. Oxford: University Press 1975.
1447. McMillan, J. A.: Electron Paramagnetism. New York: Reinhold 1968.
1448. McMillan, J. A.: Syst. Mater. Anal. *1974*, 193.
1449. McNeil, A. A. C., Raynor, J. B., Symons, M. C. R.: Mol. Phys. *10*, 297 (1966).
1450. McNesby, J. R., Braun, W., Bell, J.: in: Creation and Detection of the Excited State (Ed. Lamola A. A.) New York: Dekker 1971, vol. 1, p. 503.
1451. McRae, J. A., Symons, M. C. R.: J. Chem. Soc. B, *1968*, 428.
1452. McTaggart, T. K.: Plasma Chemistry in Electrical Discharges. New York: Elsevier 1967.
1453. Mead, W. T., Reed, P. E.: Polym. Eng. Sci. *14*, 22 (1974).
1454. Meguro, K., Koischi, M., Hayashi, T.: Kogyo Kagaku Zasshi. *70*, 249 (1967).
1455. Meisel, D., Czapski, G., Samuni, A.: J. Chem. Soc. Perkin Trans., II, 1702 (1973).
1456. Melville, H.: Acad. Roy. Belg. Classe Sci. Mèm. *33*, 33 (1961).
1457. Memory, J. D.: Quantum Theory of Magnetic Resonance Parameters. New York: McGraw Hill 1968.
1458. Memory, J. D., Parker, G. W.: Methods Exper. Phys. *1974*, 465.
1459. Mercier, J., Smets, G.: J. Polym. Sci. *57*, 763 (1962).
1460. Metcalfe, A. R., Waters, W. A.: J. Chem. Soc. B *1967*, 34.
1461. Meyer, B.: Low Temperature Spectroscopy. New York: Elsevier 1971.
1462. Michel, A., Ceysson, M., Spitz, R., Vialle, J., Guyot, A.: C. R. Acad. Sci. *276C*, 1151 (1973).

1463. Michel, R. E.: J. Polym. Sci. A 2, *10*, 1841 (1972).
1464. Michel, R. E., Chapman, F. W.: J. Polym. Sci. A 2, *8*, 1159 (1970).
1465. Michel, R. E., Chapman, F. W., Mao, T. J.: J. Chem. Phys. *45*, 4604 (1966).
1466. Michel, R. E., Chapman, F. W., Mao, T. J.: J. Polym. Phys. A 1, *5*, 677 (1967).
1467. Mielnikov, M. Ya., Sklyarenkom V. I., Fok, N. V.: Dokl. Akad. Nauk SSSR *218*, 938 (1974).
1468. Mikhailov, V. I., Kuzina, S. I., Lukovnikov, A. F., Goldanskii, V. I.: Dokl. Akad. Nauk SSSR *204*, 383 (1972).
1469. Mikheev, Yu. A., Pariskii, G. B., Shubnyakov, V. F., Toptygin, D. Ya.: Khim. Vysokh. Energ. *5*, 77 (1971).
1470. Mikulasova, D., Horie, K., Tkač, A.: Europ. Polym. J. *10*, 1039 (1974).
1471. Milevskaya, I. S., Volkensthein, M. V.: Opt. Spektr. *11*, 349 (1961).
1472. Milevskaya, I. S., Volkenshtein, M. V.: Opt. Spektr. *12*, 381 (1962).
1473. Milinchuk, V. K.: Vysokomol. Soedin. *7*, 1293 (1965).
1474. Milinchuk, V. K.: Khim. Vysokh. Energ. *3*, 533 (1969).
1475. Milinchuk, V. K., Dudarev, V. Ya.: Khim. Vysokh. Energ. *3*, 133 (1969).
1476. Milinchuk, V. K., Klinshpont, E. R.: Khim. Vysokh. Energ. *1*, 352 (1967).
1477. Milinchuk, V. K., Klinshpont, E. R.: Khim. Vysokh. Energ. *3*, 366 (1969).
1478. Milinchuk, V. K., Klinshpont, E. R.: Polym. Symp., No. 40, p. 1 (1973).
1479. Milinchuk, V. K., Klinshpont, E. R., Pshezhetskii, S. Ya.: Radiat. Khim. Polim. Mater. Simp. Moscow, 1964, p. 211.
1480. Milinchuk, V. K., Klinshpont, E. R., Pshezhetskii, S. Ya.: Radiation Chemistry of Polymers. Akad. Nauk SSSR, Moscow, 1966.
1481. Milinchuk, V. K., Pshezhetskii, S. Ya.: Dokl. Akad. Nauk SSSR *152*, 665 (1963).
1482. Milinchuk, V. K., Pshezhetskii, S. Ya.: Vysokomol. Soedin. *5*, 946 (1963).
1483. Milinchuk, V. K., Pshezhetskii, S. Ya.: Vysokomol. Soedin. *6*, 666 (1964).
1484. Milinchuk, V. K., Pshezhetskii, S. Ya.: Vysokomol. Soedin. *6*, 1605 (1964).
1485. Milinchuk, V. K., Pshezhetskii, S. Ya.: Vysokomol. Soedin. Separate: Chemical Properties and Polymer Modification, 1964, p. 70.
1486. Milinchuk, V. K., Pshezhetskii, S. Ya.: Vysokomol. Soedin., Separate: Chemical Properties and Polymer Modification, 1964, p. 222.
1487. Milinchuk, V. K.: Pshezhetskii, S. Ya., Kotov, A. G., Tupikov, V. I., Tsivenko, V. I.: Vysokomol. Soedin. *5*, 71 (1963).
1488. Milinchuk, V. K., Zhdanov, G. S., Pshezhetskii, S. Ya.: Proc. Third Tihany Symp. on Radiation Chemistry. Budapest: Akademiai Kiado 1972, p. 1077.
1489. Milinchuk, V. K., Zhdanov, G. S., Pshezhetskii, S. Ya.: Vysokomol. Soedin. *12*, 658 (1970).
1490. Millard, M. M., Windle, J. J., Pavlath, A. E.: J. Appl. Polym. Sci. *17*, 2501 (1973).
1491. Miller, A. A.: J. Phys. Chem. *63*, 1755 (1959).
1492. Miller, G. H., Chock, D., Chock, E. P., J. Polym. Sci. A 1, *3*, 3353 (1965).
1493. Miller, J. H., White, F. H. Jr., Riesz, P., Kon, H.: Photobiol. *14*, 577 (1971).
1494. Miller, T. A., Adams, R. N., Richards, P. M.: J. Chem. Phys. *44*, 4022 (1966).
1495. Milov, A. D., Shchirov, M. D., Khmelinskii, V. E.: Dokl. Akad. Nauk SSSR *218*, 941 (1974).
1496. Milsch, B., Windisch, W., Heinzelman, H.: Carbon *6*, 807 (1968).
1497. Mims, W. B.: Rev. Sci. Instrum. *36*, 1472 (1965).
1498. Mims, W. B.: in: Electron Paramagnetic Resonance (Ed. Geschwind S.,). New York: Plenum 1972, p. 263.
1499. Minami, H., Hirano, S., Fujiwara, S., Araki, S.: Instrumental Analysis of Polymers. Hirokawa Publ., Japan, 1961, vol. 2, p. 203.
1500. Minkin, V. S., Averko-Antonovich, L. A., Kirpichnikov, P. A.: Vysokomol. Soedin. *B 17*, 26 (1975).
1501. Minkoff, G. J.: Frozen Free Radicals, Interscience, New York, 1960.
1502. Minoura, Y., Katano, M.: J. Appl. Polym. Sci. *13*, 2057 (1969).
1503. Minoura, Y., Tsuboi, S.: J. Polym. Sci. A 1, *8*, 125 (1970).

1504. Minoura, Y., Tsuboi, S.: J. Org. Chem. *37*, 2064 (1972).
1505. Misra, B. N., Gupta, S. K.: Bull. Chem. Soc. Japan *46*, 3067 (1973).
1506. Miura, Y., Kinoshita, M.: Makromol. Chem. *175*, 23 (1974).
1507. Miura, Y., Kinoshita, M., Imoto, M.: Mem. Fac. Eng. Osaka City Univ. *11*, 79 (1969/70).
1508. Miura, Y., Konoshita, M., Imoto, M.: Makromol. Chem. *146*, 69 (1971).
1509. Miura, Y., Kinoshita, M., Imoto, M.: Makromol. Chem. *157*, 51 (1972).
1510. Miura, Y., Makita, N., Kinoshita, M.: Makromol. Sci. Chem. A, *7*, 1007 (1973).
1511. Miura, Y., Nakai, K., Kinoshita, M.: Makromol. Chem. *172*, 233 (1973).
1512. Moacanin, J., Holden, G., Tschoegel, N. W.: Block Copolymers. New York: Wiley 1969
1513. Moacanin, J., Rembaum, A.: J. Polym. Sci. B, *2*, 979 (1964).
1514. Moan, J., Hovik, B.: J. Phys. Chem. *79*, 2220 (1975).
1515. Mohos, B., Tüdös, F., Jokay, L.: Phys. Lett. *24A*, 310 (1967).
1516. Mohos, B., Tüdös, F., Jokay, L.: Jeol News *5*, 5 (1967).
1517. Mohos, B., Tüdös, F., Jokay, L.: Acta Chim. Acad. Sci. Hunharicae *55*, 73 (1968).
1518. Molin, Yu. N., Koritskii, A. T., Semenov, A. G., Buben, N. Ya., Shamshev, V. N.: Pribory i Tekhn. Eksperim. *6*, 73 (1960).
1519. Molin, Yu. N., Koritskii, A. T., Shamshev, V. N., Buben, N. Ya.: Vysokomol. Soedin. *4*, 690 (1962).
1520. Molin, Yu. N., Tsvetkov, Yu. D.: Zh. Fiz. Khim. *33*, 1668 (1959).
1521. Mongini, L., Thonet, C.: Program for Theoretical Reconstruction of ESR Spectra. Luxemburg: Com. Eur. Communities 1972.
1522. Mönig, H., Ringsdorf, H.: Makromol. Chem. *127*, 204 (1969).
1523. Mönig, H., Ringsdorf, H.: Makromol. Chem. *175*, 811 (1974).
1524. Mönig, H., Ringsdorf, H., Spirk, E.: Int. J. Radiat, Phys. Chem. *5*, 67 (1973).
1525. Mönig, H., Ringsdorf, H., Wessel, S.: Makromol. Chem. *176*, 1323 (1975).
1526. Moore, W. S.: Phys. Educ. *3*, 11 (1968).
1527. Morawetz, H., Fadner, T. A.: Makromol. Chem. *34*, 162 (1959).
1528. Morawetz, H., Rubin, I. D.: J. Polym. Sci. *57*, 669 (1962).
1529. Mori, K., Tabata, Y., Oshima, K.: Kogyo Kagaku Zasshi. *73*, 2475 (1970).
1530. Morigaki, K., Murayama, K.: Nippon Butsuri Gakkaishi *30*, 356 (1975).
1531. Morinaga, M., Murayama, S., Nakatsuka, R.: Kobunshi Kagaku *22*, 618 (1965).
1532. Morita, S., Mizutani, T., Ieda, M.: Jap. J. Apll. Phys. *10*, 1275 (1971).
1533. Moriuchi, S., Kashiwabara, H., Sohma, J., J. Chem. Phys. *51*, 2981 (1969).
1534. Moriuchi, S., Nakamura, M., Shimada, S., Kashiwabara, H., Sohma, J.: Polymer *11*, 630 (1970).
1535. Moriuchi, S., Sohma, J.: Molecular Phys. *21*, 369 (1971).
1536. Moriuchi, S., Sohma, J.: Mem. Fac. Eng. Hokkaido Univ. *13*, 335 (1974).
1537. Morrisett, J. D., Wien, R. W., McConnell, H. M.: Ann. N. Y. Acad. Sci. *222*, 149 (1973).
1538. Mrozowski, S.: Carbon *3*, 205 (1965).
1539. Mrozowski, S.: Carbon *4*, 227 (1966).
1540. Mrozowski, S.: Carbon *9*, 97 (1971).
1541. Müller, A.: Progr. Biophys. Mol. Biol. *17*, 99 (1967).
1542. Müller, A.; Huetterman, J.: Ann. N. Y. Acad. Sci. *222*, 387 (1973).
1543. Müller, E., Ley, K., Scheffler, K., Mayer, E.: Chem. Ber. *91*, 2682 (1958).
1544. Müller, E., Mayer, E., Scheffler, K.: Ann. Chem. *645*, 75 (1961).
1545. Munari, S., Tealdo, G., Vigo, F., Bonta, G.: Proc. Second Tihany Symp., on Radiation Chemistry. Budapest: Akademiai Kiado 1967, p. 573.
1546. Murai, H., Obi, K.: J. Phys. Chem. *79*, 2446 (1975).
1547. Murayama, K., Morimura, S., Yoskioka, T.: Bull. Chem. Soc. Japan *42*, 1640 (1969).
1548. Murayama, K., Yoshika, T.: Bull. Chem. Soc. Japan *42*, 1942 (1969).
1549. Muromtsev, V. I., Akhvlediani, I. G., Asaturyan, R. A., Bruk, M. A., Slovokhtova, N. A.: Dokl. Akad. Nauk SSSR *171*, 389 (1966).
1550. Muromtsev, V. I., Asaturyan, R. A., Bruk, M. A., Abkin, A. D., Akhvlediani, I. G.: Khim. Vyokh. Energ. *3*, 252 (1969).

1551. Muromtsev, V. I., Bruk, M. A., Akhvlediani, I. G., Asaturyan, R. A., Zhidomirov, G. M., Abkin, A. D.: Teor. Eksp. Khim. *2*, 679 (1966).

1552. Myers, R. J.: Molecular Magnetism and Magnetic Resonance Spectroscopy. Englewood Cliff's, New York: Prentice-Hall 1973.

1553. Nagai, S., Ohnishi, S., Nitta, I.: Nippon Genshiryoku Kenkyusho Nempo, 1971, 128 JAERI 5027.

1554. Nagai, S., Ohnishi, S., Nitta, I.: Chem. Phys. Lett. *13*, 379 (1972).

1555. Nagakura, S.: Excited States *2*, 321 (1975).

1556. Nagamura, T., Fukitani, T., Takayanagi, M.: J. Polym. Sci. A 2, *13*, 1515 (1975).

1557. Nagamura, T., Kusumoto, N., Takayanagi, M.: J. Polym. Sci. A 2, *11*, 2357 (1973).

1558. Nagamura, T., Takayanagi, M.: J. Polym. Sci. A 2, *12*, 219 (1974).

1559. Nagamura, T., Takayanagi, M.: J. Polym. Sci. A 2, *13*, 567 (1975).

1560. Nakagawa, T., Hopfenberg, H. B., Stannett, V.: J. Appl. Polym. Sci. *15*, 747 (1971).

1561. Nakamura, K., Kikuchi, S.: Bull. Chem. Soc. Japan *40*, 2684 (1967).

1562. Nakamura, Y., Hinojosa, O., Arthur, J. C. Jr.: J. Appl. Polym. Sci. *15*, 391 (1971).

1563. Nakano, J.: Kami-pa Gikyoshi *20*, 167 (1966).

1564. Nakayama, Y., Hayashi, K., Okamura, S.: J. Appl. Polym. Sci. *18*, 3633 (1974).

1565. Nakayama, Y., Tsuruta, T., Furukawa, J.: Makromol. Chem. *40*, 79 (1960).

1566. Nanassy, A. J., Desai, R. L.: J. Appl. Polym. Sci. *15*, 2245 (1971).

1567. Nanassy, A. J., Vyas, A.: J. Macromol. Sci. Phys. *3*, 271 (1969).

1568. Nanov, V. F., Grishina, A. D., Richmond, D., Aliev, K.: Vysokomol. Soedin. *B 17*, 275 (1975).

1569. Nara, S., Kashiwabara, H., Sohma, J.: Rep. Progr. Polym. Phys. Japan *9*, 473 (1966).

1570. Nara, S., Kashiwabara, H., Sohma, J.: Rep. Progr. Polym. Phys. Japan *10*, 479 (1967).

1571. Nara, S., Kashiwabara, H., Sohma, J.: J. Polym. Sci. A 2, *5*, 929 (1967).

1572. Nara, S., Shimada, S., Kashiwabara, H., Sohma, J.: Rep. Progr. Polym. Phys. Japan *10*, 483 (1967).

1573. Nara, S., Shimada, S.: Kashiwabara, H., Sohma, J.: Rep. Progr. Polym. Phys. Japan *11*, 465 (1968).

1574. Nara, S., Shimada, S., Kashiwabara, H., Sohma, J.: J. Polym. Sci. A 2, *6*, 1435 (1968).

1575. Narasimhan, P. T.: J. Indian Chem. Soc. *52*, 275 (1975).

1576. Naruse, T., Kuri, Z.: Kobunshi Kagaku *21*, 431 (1964).

1577. Nasirov, F. M., Karpacheva, G. P., Davydov, B. E., Krentzel, B. A.: Izv. Akad. Nauk SSSR *1964*, 1697.

1578. Natarajan, R., Reed, P. E.: J. Polym. Sci. A 2, *10*, 585 (1972).

1579. Natsuume, T., Nishimura, M., Fujimatsu, M., Shimizu, Y., Shirota, H., Hirata, H., Kusabayashi, S., Mikawa, H.: Polym. J. Japan *1*, 181 (1970).

1580. Natta, G., Zambelli, A., Lanzi, G., Pasquon, I., Mognaschi, E., Segre, A.: Centola: P.: Makromol. Chem. *81*, 161 (1964).

1581. Nechtschein, M.: J. Polym. Sci. C (4), 1367 (1964).

1582. Nechtschein, M.: C. R. Acad. Sci. *260*, 6348 (1965).

1583. Nechtschein, M.: Commis. Energ. At. (France) Rapp. *1968*, CEA-R-3352.

1584. Nedoshivin, Yu. N., Grigoreve, Z. V.: Strukt. Khim. Ungleroda Uglei *1969*, 153.

1585. Neiman, M. B., Fedoseeva, T. S., Chubarova, G. V., Buchachenko, A. L., Lebedev, Ya. S.: Vysokomol. Soedin. *5*, 1339 (1963).

1586. Neiman, M. B., Rozantsev, E., Mamedova, J.: Nature *196*, 472 (1963).

1587. Nelsen, S. N.: in: Free Radicals (Ed. Kochi, J. K.) New York: Wiley 1973, vol. 2, p. 527.

1588. Nesmeyanov, A. N., Rubinshtein, A. M., Slonimskii, G. L., Slinkin, A. A., Kochetkova, N. S., Maternikova, R. B.: Dokl. Akad. Nauk SSSR *138*, 125 (1961).

1589. Neudörfl, P.: Kolloid-Z. u. Z-Polym. *204*, 38 (1965).

1590. Neudörfl, P.: Kolloid-Z. u. Z-Polym. *224*, 25 (1968).

1591. Neudörfl, P.: Kolloid-Z. u. Z-Polym. *224*, 132 (1968).

1592. Neugebauer, F. A., Trischmann, H.: J. Polym. Sci. B, *6*, 255 (1968).

1593. Nevskii, L. V., Tarakanov, O. G., Belyakov, V. K.: Plast. Massy *1966*, 20.

1594. Newman, L.: Comput. Chem. Instr. *1973*, 3.

1595. Nicolau, C.: Experimental Methods in Biophysical Chemistry. New York: Wiley 1973.
1596. Nielsen, L. E., Dahm. D. J., Berger, P. A., Murty, V. S.: J. Polym. Sci. A 2, *12*, 1239 (1974).
1597. Niki, E., Decker, C., Mayo, F. R.: J. Polym. Sci. A 1, *11*, 2813 (1973).
1598. Nikitina, I. I., Dubinskaya, A. M.: Vysokomol. Soedin. *A 16*, 1782 (1974).
1599. Nikitina, T. S., Zhuravskaya, E. V., Kuzminsky, A. S.: Effect of Ionizing Radiation on High Polymers. New York: Ed. Gordon 1963.
1600. Nikitina, V. I., Maklakov, A. I., Balakirieva, R. S., Pudovnik, A. N. : Vysokomol. Soediṇ., Geterotsepnye Vysokomol. Soedin. *1964*, 87.
1601. Nikolaev, N. I., Muromtsev, V. I., Chuvileva, G. G., Kalina, M. D.: Okislitelno-Vosstanov. Vysokomol. Soedin. *1967*, 47.
1602. Nikolskii, V. G., Alifimov, M. V., Buben, N. Ya.: Zh. Fiz. Khim. *37*, 2797 (1963).
1603. Nisbet, P. S.: J. Oil Colour Chem. Ass. *55*, 285 (1972).
1604. Nishii, M., Tsuji, K., Takakura, K., Hayashi, K., Okamura, S.: Nippon Hoshasen Kobunshi Kenkyu Kyokai Nenpo 6, 181 (1964—65).
1605. Nishikida, K., Hiramoto, Y., Sakata, S., Kubota, T., Oishi, H.: Kogyo Kagaku Zasshi *73*, 1220 (1970).
1606. Nishimura, H., Tamura, N., Tabata, Y., Oshima, K.: Kogyo Kagaku Zasshi *73*, 1220 (1970).
1607. Nishimura, H., Tamura, N., Tabata, Y., Oshima, K.: Kogyo Kagaku Zasshi *73*, 2692 (1970).
1608. Nishimura, H., Tamura, N., Tabata, Y., Oshima, K.: Kogyo Kagaku Zasshi *74*, 489 (1971).
1609. Nishimura, H., Tamura, N., Tabata, Y., Oshima, K.: Kogyo Kagaku Zasshi *74*, 2166 (1971).
1610. Nishimura, N.: J. Macromol. Chem. *1*, 257 (1966).
1611. Nistor, S. V.: Rev. Chim. (Bucharest) *19*, 117 (1968).
1612. Nitta, I., Ohnishi, S., Sugimoto, S.: Nippon Hoshasen Kobunshi Kenkyo Kyokai Nempo *4*, 273 (1962).
1613. Nitta, I., Ohnishi, S., Sugimoto, S.: Nippon Hoshasen Kobunshi Kenkyo Kyokai Nempo *4*, 277 (1962).
1614. Nitta, I., Ohnishi, S., Sugimoto, S., Hayashi, K.: Nippon Hoshasen Kobunshi Kenkyo Kyokai Nempo *4*, 275 (1962).
1615. Noble, G. A., Serway, R. A., O'Donnell, A., Freeman, E. S.: J. Phys. Chem. *71*, 4326 (1967).
1616. Norman, R. O. C.: Proc. Roy. Soc. *A 302*, 315 (1968).
1617. Norman, R. O. C.: Specialist Periodical Reports: Electron Spin Resonance. London: Ed. Chem. Soc. 1974, vol. 1 and vol. 2.
1618. Norman, R. O. C., Storey, P. M.: J. Chem. Soc. *B 1971*, 1009.
1619. Novikova, O. A., Kuznetsova, V. P., Kornev, K. A.: Ukr. Khim. Zh. *34*, 617 (1968).
1620. Nozawa, Y.: Bull. Chem. Soc. Japan *43*, 657 (1970).
1621. Nozawa, Y., Suzuki, M., Higashida, F.: Nippon Kagaku Kaishi *11*, 2043 (1973).
1622. Nozawa, Y., Takeda, M.: Bull. Chem. Soc. Japan *42*, 2431 (1969).
1623. Noyes, W. A. Jr., Leighton, P. A.: The Photochemistry of Gases. New York: Reinhold 1941.
1624. Nunome, K., Eda, B., Iwasaki, M.: J. Appl. Polym. Sci. *18*, 2711 (1974).
1625. Nunome, K., Eda, B., Iwasaki, M.: J. Appl. Polym. Sci. *18*, 2719 (1974).
1626. Ochiai, E., Hirai, H., Makishima, S.: J. Polym. Sci. B, *4*, 1003 (1966).
1627. Odgaard, E., Meloe, T. B., Henriksen, T.: J. Magn. Reson. *18*, 436 (1975).
1628. O'Donnell, J. H., McGarvey, B., Moravetz, H.: J. Amer. Chem. Soc. *86*, 2322 (1964).
1629. Ogasawara, M., Inaba, S., Yoshida, H., Hayashi, K.: Bull. Chem. Soc. Japan *47*, 1611 (1974).
1630. Ogasawara, M., Ohno, K., Hayashi, K., Sohma, J.: J. Phys. Chem. *74*, 3221 (1970).
1631. Ogasawara, M., Takaoka, H., Hayashi, K.: Bull. Chem. Soc. Japan *46*, 35 (1973).
1632. Ogawa, S.: J. Phys. Soc. Japan *16*, 1488 (1960).
1633. Ogiwara, Y., Hon, N. S., Kubota, H.: J. Appl. Polym. Sci. *18*, 2057 (1974).
1634. Ogiwara, Y., Kubota, H.: J. Polym. Sci. A 1, *11*, 3243 (1973).
1635. Ogiwara, Y., Kubota, H., Yasunaga, T.: J. Appl. Polym. Sci. *19*, 887 (1975).
1636. Ogiwara, Y., Yasunaga, T., Kubota, H.: J. Apll. Polym. Sci. *19*, 1119 (1975).
1637. Ohmori, T., Ichikawa, T., Iwasaki, M.: Bull. Chem. Soc. Japan *46*, 1383 (1973).

1638. Ohnishi, S., Bull. Chem. Soc. Japan 35, 254 (1962).
1639. Ohnishi, S., Ikeda, Y., Kashiwagi, N., Nitta, I.: Isotopes, Radiation 1, 210 (1958).
1640. Ohnishi, S., Ikeda, Y., Kashiwagi, M., Nitta, I.: Polymer 2, 119 (1961).
1641. Ohnishi, S., Ikeda, Y., Sugimoto, S., Nitta, I.: J. Polym. Sci. 47, 503 (1960).
1642. Ohnishi, S., Kashiwagi, M., Ikeda, Y., Nitta, I.: in: Ind. Use Large Radiation Sources. Warsaw: Proc. Conf. 1959, vol. 1, p. 291.
1643. Ohnishi, S., Morokuma, K.: Electron Spin Resonance, Its. Application to Chemistry. Kyoto: Kagakudojin 1964.
1644. Ohnishi, S., Nakajima, Y., Nitta, I.: J. Appl. Polym. Sci. 6, 629 (1962).
1645. Ohnishi, S., Nitta, I.: J. Polym. Sci. 38, 451 (1959).
1646. Ohnishi, S., Nitta, I.: J. Chem. Phys. 39, 2848 (1963).
1647. Ohnishi, S., Sugimoto, S., Hayashi, K., Nitta, I.: Bull. Chem. Soc. Japan 37, 524 (1964).
1648. Ohnishi, S., Sugimoto, S., Nitta, I.: Nippon Isotope Kaigi Hobunshu 4, 393 (1961).
1649. Ohnishi, S., Sugimoto, S., Nitta, I.: J. Chem. Phys. 37, 1283 (1962).
1650. Ohnishi, S., Sugimoto, S., Nitta, I.: J. Polym. Sci. A1, 1, 605 (1963).
1951. Ohnishi, S., Sugimoto, S., Nitta, I.: J. Polym. Sci. A1, 1, 625 (1963).
1652. Ohnishi, S., Sugimoto, S., Nitta, I.: J. Chem. Phys. 39, 2647 (1963).
1653. Ohnishi, S., Tanei, T., Nitta, I.: J. Chem. Phys. 37, 2402 (1962).
1654. Ohno, K., Sohma, J.: Chem. Instrum. 2, 121 (1969).
1655. Ohno, K., Takemura, I., Sohma, J.: J. Chem. Phys. 56, 1202 (1972).
1656. Okamura, S., Hayashi, K.: Kyoto Daigaku Nippon Kagakuseni Kenkyusho Koenshu 1967, 93.
1657. Okaya, T., Yoshida, H., Hayashi, K., Okamura, S.: Kobunshi Kagaku 21, 358 (1964).
1658. Okura, I., Keii, T.: Nippon Kagaku Kaishi 1973, 250 (1973).
1659. Okura, I., Sendoa, Y., Keii, T.: Kogyo Kagaku Zasshi 73, 276 (1970).
1660. Oleinik, A. V.: Practical Course in the Structure of Matter, Textbook No. 1 Electron Paramagnetic Resonance. Ed. Gork. Gos. Univ. SSSR, 1973.
1661. O'Malley, J. J., Yanus, J. F., Pearson, J. M.: Macromolecules 5, 158 (1972).
1662. O'Mera, J. P., Shaw, T. M.: Food Technol. 9, 132 (1957).
1663. Ono, Y., Keii, T.: J. Polym. Sci. A1, 4, 2441 (1966).
1664. Ooi, T., Shitsubo, M., Hama, Y., Shonohara, K.: Rep. Progr. Polym. Phys. Japan 17, 543 (1974).
1665. Ooi, T., Shitsubo, M., Hama, Y., Shinohara, K.: Polymer 16, 510 (1975).
1666. Ootani, S., Ishikawa, T.: Kogyo Kagaku Zasshi 66, 1012 (1963).
1667. O'Reilly, A.: Adv. Catal. 12, 31 (1960).
1668. Ormerod, M. G.: Polymer 4, 451 (1963).
1669. Ormerod, M. G.: Phil. Mag. 12, 657 (1965).
1670. Ormerod, M. G.: Int. J. Radiat. Biol. 9, 291 (1965).
1671. Ormerod, M. G.: Charlesby, A. Polymer 4, 459 (1963).
1672. Ormerod, M. G.: Charlesby, A. Polymer 5, 67 (1964).
1673. Osawa, Z., Cheu, E. L., Ogiwara, Y.: J. Polym. Sci. B, 13, 535 (1975).
1674. Oshima, A.: Kogyo Kagaku Zasshi 70, 1818 (1967).
1675. Osiecki, J. H., Ullman, E. F.: J. Amer. Chem. Soc. 90, 1078 (1968).
1676. Oster, G., Nan-Loh Yang: Chem. Rev. 68, 125 (1968).
1677. Oster, G., Oster, G. K., Kryszewski, M.: J. Polym. Sci. 57, 937 (1962).
1678. Osugi, J., Sasaki, M.: Rev. Phys. Chem. Japan 34, 65 (1964).
1679. Otani, S.: Kogyo Kagaku Zasshi 66, 1012 (1963).
1680. Ouchi, I.: J. Polym. Sci. A1, 3, 2685 (1965).
1681. Ovenall, D. W.: Nature 184, 181 (1959).
1682. Ovenall, D. W.: J. Polym. Sci. 41, 199 (1959).
1683. Ovenall, D. W.: J. Phys. Chem. Solids 21, 309 (1961).
1684. Ovenall, D. W.: J. Polym. Sci. 49, 225 (1961).
1685. Ovenall, D. W.: J. Chem. Phys. 38, 2448 (1963).
1686. Ovenall, D. W.: J. Phys. Chem. Solids 26, 81 (1965).
1687. Oversberger, C. G., Mulvaney, J. E., Schiller, A. M.: in: Encyclopedia Polymer Science and Technology. New York: Wiley 1965, vol. 2, p. 95.

References

1688. Pacault, A., Marchand, A., Bothorel, P., Zanchetta, J. V., Boy, A., Cherville, J., Oberlin, J.: J. Chim. Phys. 7, 892 (1962).
1689. Pacifici, J. G., Browning, H. L. Jr.: J. Amer. Chem. Soc. 42, 5231 (1970).
1690. Pake, G. E.: Paramagnetic Resonance. New York: Benjamin 1962.
1691. Pakhomova, I. E.: Makromol. Chem. 146, 69 (1971).
1692. Pampillo, C. A., Davis, L. A.: J. Appl. Phys. 43, 4277 (1972).
1693. Panasenko, A. A., Golubev, V. B., Zubov, V. I., Kabanov, V. A., Kargin, V. A.: Vysokomol. Soedin. A12, 865 (1970).
1694. Panasenko, A. A., Golubev, V. B., Zubov, V. I., Kargin, V. A.: Vysokomol. Soedin. B10, 139 (1968).
1695. Panayotov, A. A., Petrova, D. T., Tsvetanov, C. B.: Makromol. Chem. 176, 815 (1975).
1696. Panayotov, A. A., Rashkov, I. B.: Compt. rend. Acad. Bulgare Sci. 21, 885 (1968).
1697. Panayotov, I. M., Rashkov, I. B.: Compt. rend. Acad. Bulgare Sci. 22, 173 (1969).
1698. Panayotov, I. M., Rashkov, I. B.: Izv. Akad. Nauk Bulgaria 2, 213 (1969).
1699. Panayotov, I. M., Tsvetanov, Ch. B.: Monatshefte Chem. 101, 1672 (1970).
1700. Panayotov, I. M., Tsvetanov, Ch. B., Berlinova, I. V., Velichkova, R. S.: Makromol. Chem. 134, 313 (1970).
1701. Panayotov, I. M., Tsvetanov, Ch. B., Rashkov, I. B.: Vysokomol. Soedin. B10, 845 (1968).
1702. Panayotov, I. M., Tsvetanov, Ch. B., Velitschkova, R. St.: Monatshefte Chem. 103, 1119 (1972).
1703. Panayotov, I. M., Tsvetanov, Ch. B., Yokhnovski, J. N.: Europ. Polym. J. 6, 1625 (1970).
1704. Panfilov, V. N.: in: Free Radical Reactions in Chemistry. Novosibirsk: Nauka 1972, p. 171.
1705. Pariskii, G. B., Davydov, E. Ya., Zaitseva, N. I., Toptygin, D. Ya.: Izv. Akad. Nauk SSSR 1972, 281.
1706. Pariskii, G. B., Postnikov, L. M., Davydov, E. Ya., Toptygin, D. Ya.: Vysokomol. Soedin. A16, 482 (1974).
1707. Pariskii, G. B., Postnikov, L. M., Toptygin, D. Ya., Davydov, E. Ya.: J. Polym. Sci. C, (43), 1287 (1973).
1708. Pariskii, G. B., Toptygin, D. Ya., Davydov, E. Ya., Lednev, D. A., Mikheev, Yu. A., Karasov, V. M.: Vysokomol. Soedin. B14, 511 (1972).
1709. Park, G. S., Ward, J. C.: Nature 202, 389 (1964).
1710. Parr, R. G.: Quantum Theory of Molecular Electronic Structures. New York: Benjamin 1963.
1711. Parrish, C. F.: Mol. Cryst. Liquid Cryst. 9, 453 (1969).
1712. Partridge, R. H.: J. Chem. Phys. 52, 1277 (1970).
1713. Partridge, R. H.: J. Chem. Phys. 52, 2485 (1970).
1714. Partridge, R. H.: J. Chem. Phys. 52, 2492 (1970).
1715. Partridge, R. H.: J. Chem. Phys. 52, 2501 (1970).
1716. Paton, R. M., Kaiser, E. T.: J. Amer. Chem. Soc. 92, 4723 (1970).
1717. Patten, F., Gordy, W.: Proc. Natl. Acad. Sci. 46, 1137 (1960).
1718. Paushkin, Ya. M., Omarov, I. Yu.: Vysokomol. Soedin. 7, 710 (1965).
1719. Paushkin, Ya. M., Polak, L. S., Vishnyakova, T. P., Patalakh, I. I., Machus, F. F., Sokolinskaya, T. A.: J. Polym. Sci. C, (4), 1481 (1964).
1720. Paushkin, Ya. M., Vishnyakova, T. P., Lunin, A. F., Nizova, S. A.: Organic Polymeric Semiconductors. New York: Wiley 1974.
1721. Pearson, J. T., Smith, P., Smith, T. C.: Can. J. Chem. 42, 2022 (1964).
1722. Pecherskaya, Yu. N., Kazanskii, V. B.: Zh. Fiz. Khim. 34, 2617 (1960).
1723. Penkovski, V. V.: Vysokomol. Soedin. 6, 1755 (1964).
1724. Penkovskii, V. V.: Teor. Eksper. Khim. 3, 106 (1967).
1725. Penkovskii, V. V., Kuts, V. S.: Teor. Eksper. Khim. 1, 167 (1965).
1726. Penkovskii, V. V., Kuts, V. S.: Teor. Eksper. Khim. 1, 254 (1965).
1727. Perkins, H. K., Sienko, M. J.: J. Chem. Phys. 46, 2398 (1967).
1728. Perkins, M. J.: In Essays on Free Radical Chemistry. London: Chem. Soc. 1970, p. 97.
1729. Perkins, M. J., Ward, P., Horsfield, A.: J. Chem. Soc. B 1970, 395.
1730. Peterlin, A.: J. Polym. Sci. A2, 7, 1151 (1969).

1731. Peterlin, A.: J. Polym. Sci. C, (32), 297 (1971).
1732. Peterlin, A.: J. Phys. Chem. 75, 3921 (1971).
1733. Peterlin, A.: Int. J. Fracture Mech. 7, 496 (1971).
1734. Peterlin, A.: in: Structure and Properties of Polymer Fibers. (Ed. Lentz, R. W., Stein, R. S.) New York: Plenum Press 1972, p. 253.
1735. Peterlin, A.: Int. J. Fracture Mech. 8, 235 (1972).
1736. Peterlin, A.: J. Macromol. Sci. Phys. B, 6, 583 (1972).
1737. Peterlin, A.: in: ESR Application to Polymer Research (Ed. Kinell, P. O., Rånby, B., Runnström-Reio, V.) Stockholm: Almqvist-Wiksell 1973, p. 235.
1738. Peterlin, A., Corneliussen, R.: J. Polym. Sci. A2, 6, 1273 (1968).
1739. Peterlin, A., Meinel, G.: J. Polym. Sci. B, 5, 197 (1967).
1740. Peyroche, J., Girard, Y., Laputte, R., Guyot, A.: Makromol. Chem. 129, 215 (1969).
1741. Phillips, G. O., Hinojosa, O., Arthur, J. C. Jr., Mares, T.: Textile Res. J. 36, 822 (1966).
1742. Phillips, W. D., Rowell, J. C., Weissman, S. I.: J. Chem. Phys. 33, 626 (1960).
1743. Piette, L. H.: in: NMR and ESR Spectroscopy Ed. NMR-ESR Staff of Varian Associates. New York: Pergamon 1960, p. 207.
1744. Piette, L. H., Bulow, G., Loeffler, K.: Prepr. Div. Petrol. Chem. Amer. Chem. Soc., No. C–9, 1964.
1745. Piette, L. H., Landgraff, W. C.: J. Chem. Soc. 32, 1107 (1960).
1746. Pigon, K., Guminski, K., Ventulani, J.: Organic Semiconductors. PWN Warszawa, 1964 (in Polish).
1747. Pilar, J., Ulbert, K.: Polymer 16, 730 (1975).
1748. Pinkerton, D. M.. Whelan, D. J.: Austral. J. Chem. 24, 183 (1971).
1749. Piontovskaya, M. A., Taranukha, O. M., Bobonich, F. M., Neimark, I. E., Galich, P. N., Lebedev, Ya. S.: Neorg. Ionoobmen. Mater. 1, 140 (1974).
1750. Pitts, J. N., Jr., Scuck, E. A., Wan, J. K. S.: J. Amer. Chem. Soc. 86, 296 (1964).
1751. Poggi, G., Johnson, S.: J. Magn. Reson. 3, 436 (1970).
1752. Pogorelko, V. Z., Ryabov, A. V.: Tr. Khim. Khim. Tekhnol. 1963, 347.
1753. Pogorelko, V. Z., Ryabov, A. V.: Tr. Khim. Khim. Tekhnol. 1964, 299.
1754. Pogorelko, V. Z., Ryabov, A. V.: Tr. Khim. Khim. Tekhnol. 1972, 41.
1755. Pohl, H. A., Chartoff, R. P.: J. Polym. Sci. A2, 2, 2787 (1964).
1756. Poirer, R. H., Kahler, E. J., Benington, F.: J. Organ. Chem. 1952, 1437.
1757. Pomponiu, C., Balaban, A. T.: Rev. Roum. Chim. 18, 1173 (1973).
1758. Pomponiu, C., Balaban, A. T.: Inst. Fiz. Atom. (Roumania), 1973, Report CO-27.
1759. Poninski, M., Kryszewski, M.: J. Polym. Sci. C, (16), 3901 (1965).
1760. Poole, C. P., Jr.: Electron Spin Resonance – A Comprehensive Treatise on Experimental Technique. New York: Wiley-Interscience 1967.
1761. Poole, C. P., Jr.: in: Guide to Modern Methods of Instrumental Analysis (Ed. Gouw T. H.). New York: Wiley-Interscience 1972, p. 279.
1762. Poole, C. P., Jr., Anderson, R. S.: J. Chem. Phys. 31, 346 (1959).
1763. Poole, C. P., Jr., Farach, H. A.: The Theory of Magnetic Resonance. New York: Wiley-Interscience 1972.
1764. Poole, C. P., Jr., Kehl, W. L., MacIver, D. S.: J. Catalyst 1, 407 (1962).
1765. Potter, W. D., Scott, G.: Europ. Polym. J. 7, 489 (1971).
1766. Povich, M. J.: Anal. Chem. 47, 346 (1975).
1767. Přikryl, R., Tkač, A., and Staško, A.: Coll. Czechoslov. Chem. Commun. 37, 1295 (1972).
1768. Pruden, B., Snipes, W., Gordy, W.: Proc. Nat. Acad. Sci. US 53, 917 (1965).
1769. Pryor, W. A.: Free Radicals. New York: McGraw-Hill 1966.
1770. Pshezhetskii, S. Ya., Kotov, A. G., Milinchuk, V. K., Roginskii, V. A., Tupikov, V. I.: EPR of Free Radicals in Radiation Chemistry. New York: Halsted 1974.
1771. Pshezhetskii, S. Ya., Kotov, A. G., Milinchuk, V. K., Tupikov, V. I.: Int. J. Radiat. Phys. Chem. 6, 159 (1974).
1772. Ptak, M.: Biochemie 57, 483 (1975).
1773. Putirskaya, G., Matus, I.: Proc. Tihany Symp. on Radiation Chemistry. Budapest: Ed. Akademiai Kiado 1964, p. 385.

References

1774. Putzger, D.: Faserforsch. Textiltech. 26, 19 (1975).
1775. Rabek, J. F.: Polimery 9, 128 and 221 (1964).
1776. Rabek, J. F.: Polimery 10, 514 (1965).
1777. Rabek, J. F.: Wiadom. Chem. 20, 291, 355 and 435 (1966).
1778. Rabek, J. F.: Polimery 11, 1 (1966).
1779. Rabek, J. F.: Photochem. Photobiol. 7, 5 (1968).
1780. Rabek, J. F.: Polimery 16, 257 (1971).
1781. Rabek, J. F.: Wiadom. Chem. 25, 293, 365 and 435 (1971).
1782. Rabek, J. F.: in: XXIIIrd IUPAC Congress, Boston, USA, 1971, Pure Appl. Chem. 8, 29. London: Butterworths 1972.
1783. Rabek, J. F.: in 3rd Technical Conf. on Photopolymers, Soc. Plast. Eng. Ellenville: Mid-Hudson Sec. 1973, p. 27.
1784. Rabek, J. F.: in: Degradation of Polymers. Comprehensive Chemical Kinetics (Ed. Bamford, C. H., Tipper, C. F. H.). Amsterdam: Elsevier 1975, vol. 14, p. 425.
1785. Rabek, J. F.: in: Int. Symp. on "Ultraviolet Light Induced Reaction in Polymers Amer. Chem. Soc. Philadelphia: 1975, p. 190.
1786. Rabek, J. F., Canbäck, G., Rånby, B.: J. Appl. Polym. Sci. (in press).
1787. Rabek, J. F., Canbäck, G., Rånby, B.: in: Second IUPAC Symp. on Photochemical Processes in Polymer Chemistry. Leuven, Belgium, 1976.
1788. Rabek, J. F., Canbäck, G., Lucky, J., Rånby, B.: J. Polym. Sci. A 1 14, 1447 (1976).
1789. Rabek, J. F., Rånby, B.: in: ESR Application to Polymer Research (Ed. Kinell, P. O., Rånby, B., Runnström-Reio, V.). Stockholm: Almqvist-Wiksel 1973, p. 201.
1790. Rabek, J. F., Rånby, B.: J. Polym. Sci. A1, 12, 273 (1974).
1791. Rabek, J. F., Rånby, B.: Polym. Eng. Sci. 15, 40 (1975).
1772. Rabold, G. P.: J. Polym. Sci. A1, 7, 1187 (1969).
1793. Rabold, G. P.: J. Polym. Sci. A1, 7, 1203 (1969).
1794. Radford, H. E.: Phys. Rev. 122, 114 (1961).
1795. Radics, L., Neszmelyi, A.: Magy. Tud. Akad. Biol. Tud. Oszt. Kozlem 10, 329 (1967).
1796. Rado, R., Srocs, F., Lazar, M.: Coll. Czech. Chem. Commun. 30, 894 (1965).
1797. Radstig, V. A.: Zhurn. Strukt. Khim. 11, 235 (1970).
1798. Radstig, V. A.: Vysokomol. Soedin. A17, 154 (1975).
1799. Radstig, V. A., Butyagin, P. Yu.: Vysokomol. Soedin. 7, 922 (1965).
1800. Radstig, V. A., Butyagin, P. Yu.: Vysokomol. Soedin. A9, 2549 (1967).
1801. Radstig, V. A., Shapiro, A. B., Rozantsev, E. G.: Vysokomol. Soedin. B14, 685 (1972).
1802. Raevskii, A. B., Gainulin, I. F.: Kauczuk i Rezina 25, 5 (1966).
1803. Rafikov, S. R., Tsui-Tsi-Pin: Vysokomol. Soedin. 3, 56 (1961).
1804. Ramalingam, K. V., Werezak, G. N., Hodgins, J. W.: J. Polym. Sci. C, (2), 153 (1963).
1805. Rånby, B.: Tidskr. Vetenskaplig Forsk. 36, 226 (1965).
1806. Rånby, B.: in: Svensk Naturvetenskap-1967. Stockholm: Statens Naturvetenskapliga Forskningsråd 1967, p. 22.
1807. Rånby, B.: Svensk Kemisk Tidskrift 79, 6 (1967).
1808. Rånby, B.: Kemisk Tidskrift 1971, 32.
1809. Rånby, B.: in: ESR Application to Polymer Research (Ed. Kinell, P. O., Rånby, B., Runnström-Reio, V.). Stockholm: Almqvist-Wiksell 1973, p. 53.
1810. Rånby, B.: Kemia-Kemi 1974, 477.
1811. Rånby, B.: Appl. Polym. Symp. No 26, 327 (1975).
1812. Rånby, B., Carstensen, P.: Adv. Chem. Ser. No. 66, 256 (1967).
1813. Rånby, B., Kringstad, K., Cowling, E. B., Lin, S. Y.: Acta Chem. Scand. 23, 3257 (1969).
1814. Rånby, B., Rabek, J. F.: Photodegradation, Photo-oxidation and Photostabilization of Polymers. London: Wiley 1975.
1815. Rånby, B., Rabek, J. F.: in Int. Symp. on: Ultraviolett Light Induced Reaction in Polymers. Amer. Chem. Soc., Philadelphia, 1975, p. 148.
1816. Rånby, B., Rabek, J. F., Canbäck, G.: in: Second Intern. Symp. on: Polyvinylchloride, Lyon, France, 1976.
1817. Rånby, B., Yoshida, H.: J. Polym. Sci. C, (12), 263 (1966).

1818. Randolph, M. L.: in: Biological Applications of Electron Spin Resonance (Ed. Swartz, H. M., Bolton, J. R., and Borg, D. C.). New York: Wiley 1972, p. 119.
1819. Rashkov, I. B., Panayotov, I. M.: Compt. rend. Acad. Bulgare Sci. *24*, 889 (1971).
1820. Rataiczak, R. D., Jones, M. T.: J. Chem. Phys. *56*, 3898 (1972).
1821. Raven, A. V., Hesse, P., Heusinger, H.: Makromol. Chem. *163*, 215 (1973).
1822. Razgon, D. R., Drozdovskii, V. F.: Vysokomol. Soedin. *A12*, 1538 (1970).
1823. Razumovskii, S. D., Kefeli, A. A., Zaikov, G. E.: Europ. Polym. J. *7*, 275 (1971).
1824. Reed, R. I.: Ion Production by Electron Impact. New York: Academic Press 1962.
1825. Reeder, K. M., Evenson, K. M., Burch, D. S.: Rev. Sci. Instrum. *37*, 141 (1966).
1826. Regel, V. R., Muinov, T. M.: Fiz. Tverd. Tela *8*, 2364 (1966).
1827. Reich, L., Schindler, A.: Polymerization by Organometallic Compounds. New York: Wiley 1966.
1828. Reine, A. H., Hinojosa, O., Arthur, J. C., Jr.: J. Appl. Polym. Sci. *17*, 3337 (1973).
1829. Reinisch, G., Jaeger, W., Damerau, W., Lassmann, G.: Naturwiss. *52*, 55 (1965).
1830. Rembaum, A., Haack, R. F., Hermann, A. M.: J. Macromol. Chem. *1*, 673 (1966).
1831. Rembaum, A., Maocanin, J.: Polym. Prepr., Amer. Chem. Soc. Div. Polym. Chem. *3*, 251 (1962).
1832. Rembaum, A., Moacanin, J.: in: Exchange Reactions. Proc. Symp., Upton, N.Y., USA, 1965, p. 173.
1833. Rembaum, A., Moacanin, J., Cuddihy, E.: J. Polym. Sci. C, (4), 529 (1964).
1834. Rembaum, A., Moacanin, J., Haack, R.: J. Macromol. Chem. *1*, 657 (1966).
1835. Restaino, A. J., Mesorobian, R. B., Morawetz, H., Bellantine, D. S., Dienes, G. J., Metz, D. J.: J. Amer. Chem. Soc. *78*, 2939 (1956).
1836. Revillon, A., Couble, P., Spitz, R.: C. R. Acad. Sci. *270*, C, 791 (1970).
1837. Revillon, A., Couble, P., Spitz, R.: Europ. Polym. J. *11*, 735 (1975).
1838. Revina, A. A., Bakh, N. A.: Dokl. Akad. Nauk SSSR *141*, 409 (1961).
1839. Reztova, E. V., Chubarova, G. V.: Vysokomol. Soedin. *7*, 1335 (1965).
1840. Rexroad, H. N., Gordy, W.: J. Chem. Phys. *30*, 399 (1959).
1841. Rieger, R. H., Bernal, I., Reinmuth, W. H., Fraenkel, G. K.: J. Amer. Chem. Soc. *85*, 683 (1963).
1842. Ringsdorf, H.: J. Polym. Sci. C, (4), 987 (1964).
1843. Robb, I. D., Smith, R.: Europ. Polym. J. *10*, 1005 (1974).
1844. Rocaboy, F.: Bull. Soc. Sci. Bretagne *40*, 33 (1965).
1845. Rode, T. B., Pecherskaya, Yu. N., Kazanskii, V. B.: Zh. Fiz. Khim. *35*, 2370 (1961).
1846. Rogers, M.: J. Chem. Soc. *1956*, 2102.
1847. Rogowski, R. S.: J. Polym. Sci. A2, *9*, 1911 (1971).
1848. Rogowski, R. S., Pezdirtz, G. F.: J. Polym. Sci., A2, *9*, 2111 (1971).
1849. Romanko, J.: NASA Accession No. N66-13092, Rept. No. NASA-CR-68435 (1964).
1850. Ronfard, H. J. C., Lablache, C. A., Chachaty, C.: J. Phys. Chem. *78*, 899 (1974).
1851. Rosato, D., Schwartz, R. T.: Environmental Effect on Polymer Material. New York: Wiley-Interscience 1968, vol. 1 and vol. 2.
1852. Rosen, B.: Fracture Processes in Polymer Solids. New York: Interscience 1964.
1853. Rosenthal, J., Yarmus, L.: Rev. Sci. Instrum. *37*, 381 (1966).
1854. Rouge, M., Spitz, R., Guyot, A.: J. Polym. Sci. B, *12*, 407 (1974).
1855. Rousseau, A., Lenk, R.: Mol. Phys. *15*, 425 (1968).
1856. Rozantsev, E. G.: Izv. Akad. Nauk SSSR *1963*, 1669.
1857. Rozantsev, E. G.: Free Nitroxyl Radicals. New York: Plenum 1970.
1858. Rozantsev, E. G., Mamedova, Yu. G., Neiman, M. B.: Izv. Akad. Nauk SSSR *1962*, 2250.
1859. Rozantsev, E. G., Papko, R. A.: Izv. Akad. Nauk SSSR *1963*, 764.
1860. Rozantsev, E. G., Pavelko, G. F.: Vysokomol. Soedin. B, *9*, 866 (1967).
1861. Rozantsev, E. G., Sholle, V. D.: Synthesis *1971*, 190.
1862. Roylance, D. K., DeVries, K. L., Williams, M. L.: Fracture. London: Chapman 1969.
1863. Rubin, I. D., Huber, L. M.: J. Polym. Sci. B, *4*, 337 (1966).
1864. Rychly, J., Lazar, M.: J. Polym. Sci. B, *7*, 843 (1969).
1865. Rychly, J., Lazar, M.: Europ. Polym. J. *8*, 711 (1972).

References

1866. Rychly, J., Lazar, M., Hybl, C.: Chem. Zvesti 24, 245 (1970).
1867. Rychly, J., Lazar, M., Pavlinec, J.: Polym. Symp., No. 40, 133 (1973).
1868. Sadovskaya, G. K., Slo okhotova, N. A., Vasiliev, L. A., Kargin, V. A.: Tr. Komis. Spektr. Akad. Nauk SSSR 1964, 483.
1869. Sadykov, M. U., Azizov, U. A., and Usmeanov, Kh. U.: Strukt. Modif. Khlop. Tsellul. 1966, 93.
1870. Saegusa, T., Yatsu, T., Miyaji, S., Fujii, H.: Polym. J. 1, 7 (1970).
1871. Sagdeev, R. Z., Molin, Yu. N.: Zh. Strukt. Khim. 7, 38 (1966).
1872. Saito, E., Bielski, B. H. J.: J. Amer. Chem. Soc. 83, 4467 (1961).
1873. Saito, E., Cannava, C., Hayashi, K.: J. Polym. Sci. A1, 8, 2309 (1970).
1874. Saito, E., Saegusa, T.: Makromol. Chem. 117, 86 (1968).
1875. Sakaguchi, M., Kodama, S., Edlund, O., Sohma, J.: Rep. Progr. Polym. Phys. Japan 17, 539 (1974).
1876. Sakaguchi, M., Kodama, S., Edlund, O., Sohma, J.: J. Polym. Sci. B, 12, 609 (1974).
1877. Sakaguchi, M., Sohma, J.: Rep. Progr. Polym. Phys. Japan 16, 547 (1973).
1878. Sakaguchi, M., Sohma, J.: J. Polym. Sci. A2, 13, 1233 (1975).
1879. Sakaguchi, M., Sohma, J.: Polymer J. 7, 490 (1975).
1880. Sakaguchi, M., Sugimoto, T., Shiotani, M., Sohma, J.: Rep. Progr. Polym. Phys. Japan 17, 537 (1974).
1881. Sakaguchi, M., Yamakawa, H., Sohma, J.: J. Polym. Sci. B, 12, 193 (1974).
1882. Sakai, Y.: Nagoya Kogyo Gijutsu Shikensho Hokoku 13, 444 (1964).
1883. Sakai, Y.: J. Polym. Sci. A1, 7, 3177 (1969).
1884. Sakai, Y.: J. Polym. Sci. A1, 7, 3191 (1969).
1885. Sakai, Y., Iwasaki, M.: J. Polym. Sci. A1, 7, 1749 (1969).
1886. Sakai, Y., Iwasaki, M.: J. Polym. Sci. A1, 7, 3143 (1969).
1887. Salmon, A.: Discuss. Faraday Soc. 36, 284 (1963).
1888. Salovey, R., Luongo, J. P., Yager, W. A.: Macromolecules 2, 198 (1969).
1889. Salovey, R., Malm, D. L., Beach, A. L., Luongo, J. P.: J. Polym. Sci. A1, 2, 3067 (1964).
1890. Salovey, R., Shulman, R. G., Walsh, W. M., Jr.: J. Chem. Phys. 39, 839 (1963).
1891. Salovey, R., Yager, W. A.: J. Polym. Sci. A2, 2, 219 (1964).
1892. Salurov, V. N., Kryazhev, Yu. G., Brodskaya, E. I., Vakulskaya, T. I.: Vysokomol. Soedin. A15, 1029 (1973).
1893. Salurov, V. N., Kryazhev, Yu. G., Yushamanova, T. I. Vakulskaya, T. I., Voronkov, M. G.: Makromol. Chem. 175, 757 (1974).
1894. Samec, M., Blinc, R., Herak, K., Adamic, J.: Stärke 16, 181 (1964).
1895. Sands, R. H.: Phys. Rev. 99, 1222 (1955).
1896. Sargent, F. P., Gardy, E. M., Falle, H. R.: Chem. Phys. Lett. 24, 120 (1974).
1897. Sasakura, H., Takeuchi, N., Mizuno, T.: J. Phys. Soc. Japan 17, 572 (1962).
1898. Sato, T.: Kagaku 29, 910 (1974).
1899. Sato, T., Kashino, E., Fukumura, N., Otsu, T.: Makromol. Chem. 162, 9 (1972).
1900. Sato, T., Kita, S., Otsu, T.: Makromol. Chem. 176, 561 (1975).
1901. Sato, Y., Kuwata, K., Hirota, K.: Nippon Kagaku Zasshi 86, 1244 (1965).
1902. Savolainen, A.: Dissert. Univ. Helsinki, 1964.
1903. Savolainen, A., Saavalainen, A.: Europ. Polym. J. 10, 815 (1974).
1904. Savalainen, A., Törmala, P.: J. Polym. Sci. A2, 12, 1251 (1974).
1905. Savostin, A. Ya., Tomashevskii, E. E.: Fiz. Tverd. Tela 12, 2857 (1970).
1906. Sawai, T., Shinozaki, T., Meshitsuka, G.: Bull. Chem. Soc. Japan 45, 984 (1972).
1907. Sawaryn, A.: Postepy Biochem. 21, 21 (1975).
1908. Scheffler, K.: Z. Elektrochem. 65, 439 (1961).
1909. Scheffler, K., Stegmann, H. B.: Ber. Bunsenges. Phys. Chem. 67, 864 (1963).
1910. Scheffler, L., Stegmann, H. B.: Elektronenspinresonanz. Berlin, Heidelberg, New York: Springer 1970.
1911. Schleyer, H.: Ann. N. Y. Acad. Sci. 212, 57 (1973).
1912. Schmidberger, R.: Dissertation Univ. Stuttgart, Germany, 1974.
1913. Schmidt, D.: Low Temperature ESR Measurements. Paper presented on 14th Varian EPR Workshop, Zurich, 1975.

1914. Schneider, E. E.: Disc. Faraday Soc. *19*, 158 (1955).
1915. Schneider, E. E.: J. Chem. Phys. *23*, 978 (1955).
1916. Schneider, E. E., Day, E. J., Stein, G.: Nature *168*, 645 (1951).
1917. Schneider, F.: Z. Instrument-tekn. *71*, 315 (1963).
1918. Schneider, F., Moebius, K., Plato, M.: Angew. Chem. *77*, 888 (1965).
1919. Schneider, F., Plato, M.: Elektronenspin-Resonanz Experimentelle Technik. Munich: Thiemig 1971.
1920. Schoffa, G.: Elektronenspinresonanz in der Biologie. Karlsruhe: Braun 1963.
1921. Schuler, R. H., Fessenden, R. W.: in: Radiation Research (Ed. Silni, G.). Amsterdam: North-Holland 1967.
1922. Schulte-Frohlinde, D., Vacek, K.: Curr. Topics Rad. Res. *5*, 39 (1969).
1023. Schultz, R. C., Kinoshita, M.: Makromol. Chem. *111*, 137 (1968).
1924. Schumacher, R. T.: Introduction to Magnetic Resonance, Principles and Applications. New York: Benjamin 1970.
1925. Schuskus, A. J.: Phys. Rev. *127*, 1529 (1962).
1926. Schwoerer, M., Wolf, H. C.: Mol. Cryst. *3*, 177 (1967).
1927. Searl, J. W., Smith, R. C., Wyard, S. J.: Proc. Phys. Soc. (London) *74*, 1174 (1954).
1928. Searle, N. Z.: Analytical Photochemistry and Photochemical Analysis (Ed. Fitzgerald, J. M.). New York: Dekker 1971, p. 249.
1929. Seddon, W. A., Smith, D. R.: Can. J. Chem. *45*, 3085 (1967).
1930. Segal, B. G., Kaplan, M., Fraenkel, G. K.: J. Chem. Phys. *43*, 4191 (1965).
1931. Segal, B. G., Reymond, A., Fraenkel, G. K.: J. Chem. Phys. *51*, 1336 (1969).
1932. Seguchi, T., Makuuchi, K., Suwa, T., Tamura, N., Abe, T., Takeshita, M.: Rep. Progr. Polym. Phys. Japan *15*, 513 (1972).
1933. Seguchi, T., Makuuchi, K., Suwa, T., Tamura, N., Abe, T., Takeshita, M.: Nippon Kagaku Kaishi, 1974, 1309.
1934. Seguchi, T., Tamura, N.: in: Large Radiation Sources for Industrial Processes. Intern. Atomic Energy Agency, Vienna, 1969, p. 353.
1935. Seguchi, T., Tamura, N.: Rpe. Progr. Polym. Phys. Japan *12*, 489 (1969).
1936. Seguchi, T., Tamura, N.: J. Phys. Chem. *77*, 40 (1973).
1937. Seguchi, T., Tamura, N.: J. Polym. Sci. A1, *12*, 1671 (1974).
1938. Seguchi, T., Tamura, N.: J. Polym. Sci. A1, *12*, 1953 (1974).
1939. Seidel, H.: Z. Phys. *165*, 218 (1961).
1940. Selezneva, E. N., Nedoshivin, Yu. N., Grigorieva, Z. V., Kasatochkin, V. I.: Zh. Strukt. Khim. *10*, 533 (1969).
1941. Selivanov, P. I., Kirilova, E. I., Maksimov, V. L.: Vysokomol. Soedin. *8*, 1418 (1966).
1942. Sevilla, M. D.: J. Phys. Chem. *75*, 626 (1971).
1943. Sevilla, M. D., Van Paemel, C., Nichols, C.: J. Phys. Chem. *76*, 3571 (1972).
1944. Sevilla, M. D., Van Paemel, C., Zorman, G.: J. Phys. Chem. *76*, 3577 (1972).
1945. Shamonina, N. F., Kotov, A. G.: Vysokomol. Soedin. *B16*, 342 (1974).
1946. Shantorovich, P. S., Shlyapnikova, I. A.: Vysokomol. Soedin. *3*, 1495 (1961).
1947. Shapiro, A. B., Baimagambetov, K., Radtsig, V. A., Rozantsev, E. G.: Vysokomol. Soedin. *B15*, 300 (1973).
1948. Shapiro, A. B., Lebedeva, L. P., Suskina, V. I., Antipina, G. N., Smirnov, L. N., Levin, P. I., Rozantsev, E. G.: Vysokomol. Soedin. *A15*, 2673 (1973).
1949. Shapiro, A. M., Struykov, V. B., Rozenberg, B. A., Grygoryan, G. L., Rozantsev, E. G.: Vysokomol. Soedin. *B17*, 265 (1975).
1950. Shapkin, V. V.: Electron Paramagnetic Resonance. Leningrad Gosud.Pedagog. Inst., Leningrad, 1974.
1951. Sharnoff, M.: J. Chem. Phys. *51*, 451 (1969).
1952. Sharpatyi, V. A., Aptekar, E. L., Zakatsova, N. V., Pravednikov, A. N.: Dokl. Akad. Nauk SSSR *156*, 626 (1964).
1953. Sharpatyi, V. A., Nadzhimiddinova, M. T., Kruglyakova, K. E., Emanuel, H. M.: Dokl. Akad. Nauk SSSR *180*, 412 (1968).
1954. Sharpatyi, V. A., Pravednikov, A. N.: J. Polym. Sci. C, (16), 1599 (1967).

1955. Sharpatyi, V. A., Safarov, S. A., Ynova, K. G.: Dokl. Akad. Nauk SSSR *147*, 863 (1962).
1956. Shatky, A.: Text. Res. J. *41*, 975 (1971).
1957. Shatky, A.: Photochem. Photobiol. *19*, 299 (1974).
1958. Shatky, A., Michaeli, I.: Radiat. Res. *43*, 485 (1970).
1959. Shatky, A., Michaeli, I.: Text. Res. J. *41*, 269 (1971).
1960. Shatky, A., Michaeli, I.: Photochem. Photobiol. *12*, 119(1972).
1961. Shatrov, V. D., Chkeidze, I. I., Shamshiev, V. N., Buben, N. Ya.: Khim. Vysokh. Energ. *2*, 413 (1968).
1962. Shatrov, V. D., Chkeidze, I. I., Shamshiev, V. N., Buben, N. Ya.: Dokl. Akad. Nauk SSSR *181*, 376 (1968).
1963. Shaulov, A. Yu., Shapiro, A. B., Sklyarova, A. G., Wasserman, A. M., Buchachenko, A. L., Rozantsev, E. G.: Europ. Polym. J. *10*, 1077 (1974).
1964. Shevelev, V. A., Relaksationnye Yavleniya Polim. *1972*, 44.
1965. Shida, T., Hamill, W. H.: J. Amer. Chem. Soc. *88*, 5371 (1966).
1966. Shields, H.: Exp. Methods Biophys. Chem. *1973*, 417.
1967. Shields, H., Gordy, W.: Bull. Amer. Phys. Soc. *1*, 267 (1956).
1968. Shields, H., Gordy, W.: Proc. Nat. Acad. Sci. US, *45*, 269 (1959).
1969. Shields, H., Hamrick, P.: J. Chem. Phys. *37*, 202 (1962).
1970. Shiga, T.: J. Phys. Chem. *69*, 3805 (1965).
1971. Shiga, T., Lund, A.: Ber. Bunsenges. physik. Chem. *78*, 259 (1974).
1972. Shiga, T., Lund, A., Kinell, P. O.: Acta Chem. Scand. *25*, 1508 (1971).
1973. Shiga, T., Lund, A., Kinell, P. O.: Int. J. Radiat. Chem. *3*, 145 (1971).
1974. Shiga, T., Lund, A., Kinell, P. O.: Acta Chem. Scand. *26*, 383 (1972).
1975. Shiga, T., Matsuyama, T., Yamaoka, H., Okamura, S.: Makromol. Chem. *175*, 217 (1974).
1976. Shimada, K., Shimazato, Y., Szwarc, M.: J. Amer. Chem. Soc. *97*, 5834 (1975).
1977. Shimada, K., Szwarc, M.: Chem. Phys. Lett. *28*, 540 (1974).
1978. Shimada, K., Moshuk, G., Connor, H. D., Caluwe, P., Szwarc, M.: Chem. Phys. Lett. *14*, 396 (1972).
1979. Shimada, M., Nakamura, Y., Kusama, Y., Matsuda, O., Tamura, N., Kageyama, E.: J. Appl. Polym. Sci. *18*, 33 (1974).
1980. Shimada, M., Nakamura, Y., Kusama, Y., Matsuda, O., Tamura, N., Kageyama, E.: J. Appl. Polym. Sci. *18*, 3387 (1974).
1981. Shimada, S., Kashiwabara, H., Sohma, J.: Rep. Progr. Polym. Phys. Japan *11*, 467 (1968).
1982. Shimada, S., Kashiwabara, H., Sohma, J.: Rep. Progr. Polym. Phys. Japan *11*, 471 (1968).
1983. Shimada, S., Kashiwabara, H., Sohma, J.: Rep. Progr. Polym. Phys. Japan *12*, 465 (1969).
1984. Shimada, S., Kashiwabara, H., Sohma, J.: Rep. Progr. Polym. Phys. Japan *13*, 475 (1970).
1985. Shimada, S., Kashiwabara, H., Sohma, J.: J. Polym. Sci. A2, *8*, 1291 (1970).
1986. Shimada, S., Kashiwabara, H., Sohma, J.: Rep. Progr. Polym. Phys. Japan *14*, 547 (1971).
1987. Shimada, S., Kashiwabara, H., Sohma, J.: Rep. Progr. Polym. Phys. Japan *14*, 551 (1971).
1988. Shimada, S., Kashiwabara, H., Sohma, J., Nara, S.: Japan J. Appl. Phys. *8*, 145 (1969).
1989. Shimokawa, S., Ohno, Y., Sohma, J.: Nippon Kagaku Kaishi *1973*, 2016.
1990. Shimozato, Y., Shimada, K., Szwarc, M.: J. Amer. Chem. Soc. *97*, 5831 (1975).
1991. Shimshick, E. J., McConnell, H. M.: Biochem. Biophys. Res. Commun. *46*, 321 (1972).
1992. Shinohara, Y., Ballantine, D.: J. Chem. Phys. *36*, 3042 (1962).
1993. Shioji, Y., Ohnishi, S., Nitta, I.: J. Polym. Sci. A1, *1*, 3373 (1963).
1994. Shioji, Y., Ohnishi, S., Nitta, I.: Ann. Rep. Japanese Assoc. Radiat. Res. Polym. *2*, 253 (1960).
1995. Shiotani, M., Chachaty, C.: Bull. Chem. Soc. Japan *47*, 28 (1974).
1996. Shiotani, M., Sohma, J.: Rep. Progr. Polym. Phys. Japan *17*, 505 (1974).
1997. Shirom, M., Claridge, R. F. C., Willard, J. E.: J. Chem. Phys. *47*, 286 (1967).
1998. Shirom, M., Willard, J. E.: J. Amer. Chem. Soc. *90*, 2184 (1968).
1999. Shirom, M., Willard, J. E.: J. Phys. Chem. *72*, 1702 (1968).
2000. Shopov, I.: Vysokomol. Soedin. *B9*, 546 (1967).
2001. Shvartsberg, M. S., Kotlyarevskii, I. L., Andrievskii, V. N., Vasilesvkii, S. F.: Izv. Akad. Nauk SSSR *1966*, 527.

2002. Shvets, A. D., Antipin, A. A., Kirillov, E. I., Stepanov, V. G., Chirkin, G. K.: Cryogenics *6*, 174 (1966).
2003. Sicilio, F., Dousset, M., Florin, R. E., Wall, L. A.: Polym. Prepr. Amer. Chem. Soc. Div. Polym. Chem. *6*, 958 (1963).
2004. Sicilio, F., Florin, R. E., Wall, L. A.: J. Phys. Chem. *70*, 47 (1966).
2005. Siegel, S., Judeiks, H.: J. Chem. Phys. *43*, 343 (1965).
2006. Siegel, S., Hedgpeth, H.: J. Chem. Phys. *46*, 3904 (1967).
2007. Simandi, T. L., Tüdös, F., Mohos, B.: in: Kinetics and Mechanism of Polyreactions. Budapest: Akademiai Kiado 1969, vol. 3, p. 129.
2008. Simionescu, C., Dumitrescu, S.: Rev. Roum. Chem. *12*, 407 (1967).
2009. Simionescu, C., Dumitrescu, S.: in: Kinetics and Mechanism of Polyreactions. Budapest: Akademiai Kiado 1969, vol. 4, p. 183.
2010. Simionescu, C., Dumitrescu, S., Lixandru, T., Dargina, M., Simionescu, B., Vata, M.: Europ. Polym. J. *8*, 719 (1972).
2011. Simionescu, C., Dumitrescu, S., Negulescu, I., Percec, V.: Vysokomol. Soedin. *A16*, 790 (1974).
2012. Simionescu, C., Oprea, K.: Mechanochemistry of Highmolecular Compounds. Moscow: Mir 1970.
2013. Simionescu, C., Vasiliu-Oprea, C., Neguleanu, C.: Europ. Polym. J. *10*, 61 (1974).
2014. Simons, J. P.: Photochemistry and Spectroscopy. New York: Wiley-Interscience 1971.
2015. Singer, L. S.: J. Appl. Phys. *30*, 1463 (1959).
2016. Singer, L. S., Wagoner, G.: J. Chem. Phys. *37*, 1812 (1962).
2017. Singh, S., Hinojosa, O., Arthur, J. C., Jr.: J. Appl. Polym. Sci. *14*, 1591 (1970).
2018. Singh, S., Hinojosa, O., Arthur, J. C., Jr.: J. Appl. Polym. Sci. *14*, 1591 (1970).
2019. Skvortsov, V. G., Milinchuk, V. K., Pshezhetskii, S. Ya.: Radiobiol. *12*, 587 (1972).
2020. Skvortsov, V. G., Zhdanov, G. S., Milinchuk, V. K.: Khim. Vysokh. Energ. *5*, 466 (1971).
2021. Skvortsov, V. G., Zhdanov, G. S., Milinchuk, V. K., Pshezhetskii, S. Ya.: Radiobiol. *14*, 8 (1974).
2022. Slangen, H. J. M.: J. Phys. Educ. *3*, 775 (1970).
2023. Slawinska, D., Slawinski, J., Sarna, T.: Photochem. Photobiol. *21*, 393 (1975).
2024. Slichter, C. P.: Principles of Magnetic Resonances. New York: Herper-Row 1963.
2025. Slonimskii, G. L., Kargin, V. A., Bujko, G. N., Reztsova, E. V., Lewis-Riera, M.: Dokl. Akad. Nauk SSSR *93*, 523 (1953).
2026. Slovokhotova, N. A., Koritskii, A. T., Kargin, V. A., Buben, N. Ya., Illicheva, Z. F.: Vysokomol. Soedin. *5*, 575 (1963).
2027. Smaller, B., Matheson, M. S.: J. Chem. Phys. *28*, 1169 (1958).
2028. Smetankina, N. P., Oprya, V. Ya., Chernaya, N. S., Kuznetsova, V. P.: Khim. Prom. Ukr. *1969*, 17.
2029. Smirnov, B. R., Korolev, G. V., Berlin, A. A.: Kinet. Kataliz. *7*, 990 (1966).
2030. Smith, D. R., Okenka, F., Pieroni, J. J.: Can. J. Chem. *45*, 833 (1967).
2031. Smith, D. R., Pieroni, J. J.: Can. J. Chem. *43*, 876 (1965).
2032. Smith, D. R., Pieroni, J. J.: Can. J. Chem. *43*, 2141 (1965).
2033. Smith, D. R., Pieroni, J. J.: Can. J. Chem. *45*, 2723 (1967).
2034. Smith, P.: in: Biological Application of Electron Spin Resonance (Ed. Swartz, H. M., Bolton, J. R., Borg, D. C.). New York: Wiley-Interscience 1972, p. 483.
2035. Smith, P.: in: ESR Application to Polymer Research (Ed. Kinell, P. O., Rånby, B., Runnström-Reio, V.). Stockholm: Almqvist-Wiksell 1973, p. 29.
2036. Smith, P., Fox, W. M.: Can. J. Chem. *47*, 2217 (1969).
2037. Smith, P., Gilman, L. B., DeLorentzo, R. A.: J. Magnetic Resonance *10*, 179 (1973).
2038. Smith, P., House, D. W., Gilman, L. B.: J. Chem. Phys. *77*, 2249 (1973).
2039. Smith, P., Kaba, R. A., Pearson, J. T.: J. Magnetic Resonance *17* (1975).
2040. Smith, P., Kaba, R. A., Wood, P. B.: J. Phys. Chem. *78*, 117 (1974).
2041. Smith, P., Pearson, J. T., Wood, P. B., Smith, T. C.: J. Chem. Phys. *43*, 1535 (1965).
2042. Smith, P., Stevens, R. D.: J. Phys. Chem. *76*, 3141 (1972).
2043. Smith, P., Wood, P. B.: Can. J. Chem. *45*, 649 (1967).

2044. Smith, R. C., Wyard, S. J.: Nature *186*, 226 (1960).
2045. Smith, W. V., Jacobs, B. E.: J. Chem. Phys. *37*, 141 (1962).
2046. Snipes, W.: Electron Spin Resonance and the Effects on Radiation on Biological Systems. Washington: Natl. Acad. Sci. Natl. Res. Council 1966.
2047. Sobue, J., Tabata, Y., Hiraoka, M.: Kogyo Kagaku Zasshi *64*, 372 (1961).
2048. Sohma, J.: Kobunshi *17*, 1066 (1969).
2049. Sohma, J.: Kobunshi Kagaku *27*, 289 (1970).
2050. Sohma, J., Kashiwabara, H., Komatsu, T., Seguchi, T.: Rep. Fac. Eng. Hokkaido Univ. *35*, 451 (1964).
2051. Sohma, J., Kawashima, T., Shimada, S., Kashiwabara, H., Sakaguchi, M.: in: ESR Application to Polymer Research (Ed. Kinell, P. O., Rånby, B., Runnström-Reio, V.). Stockholm: Almqvist-Wiksell 1973, p. 225.
2052. Sohma, J., Komatsu, T., Kanda, Y.: Japan J. Appl. Phys. *7*, 298 (1968).
2053. Sohma, J., Komatsu, T., Kanda, Y.: Mem. Fac. Eng. Hokkaido Univ. *12*, 319 (1969).
2054. Sohma, J., Komatsu, T., Kashiwabara, H.: J. Polym. Sci. B, *3*, 287 (1965).
2055. Sohma, J., Komatsu, T., Kashiwabara, H., Seguchi, T.: Rep. Assoc. Prom. Technol. Asahi Glass Co. Ltd. *10*, 197 (1964).
2056. Sohma, J., Komatsu, T., Kashiwabara, H., Seguchi, T.: Kogyo Kagaku Zasshi *68*, 1535 (1965).
2057. Sohma, J., Sakaguchi, M.: Adv. Polym. Sci. *20*, 1 (1976).
2058. Solovev, B. V., Tarasov, B. G., Aseyeva, R. M.: Vysokomol. Soedin. *A15*, 2523 (1973).
2059. Sosin, S. L., Korshak, V. V.: Izv. Nauk SSSR *1964*, 354.
2060. Spackman, J. W. C.: Nature *195*, 764 (1962).
2061. Spackman, J. W. C., Charlesby, A.: Proc. Rubber Technology Conf., 4th, London, 1962, p. 24.
2062. Spinks, J. W. T.: An Introduction to Radiation Chemistry. New York: Wiley 1964.
2063. Spitz, R.: J. Catalys. *35*, 335 (1974).
2064. Spitz, R., Revillon, A., Guyot, A.: J. Catalys. *35*, 335 (1974).
2065. Spitz, R., Vaillaume, G., Revillon, A., Guyot, A.: J. Macromol. Sci. Chem. A, *6*, 153 (1972).
2066. Squires, T. L.: Introduction to Microwave Spectroscopy. London: Newnes 1963.
2067. Squires, T. L.: An Introduction to Electron Spin Resonance. New York: Academic Press 1964.
2068. Stamires, D. N., Turkevich, J.: J. Amer. Chem. Soc. *86*, 757 (1964).
2069. Stannett, V.: J. Macromol. Sci. Chem. A, *4*, 1177 (1970).
2070. Stankowsky, J.: Postepy Fizyki *16*, 325 (1965).
2071. Stelnik, C.: Geochim. Cosmochim. Acta *28*, 1615 (1964).
2072. Stevens, J. R., Ward, J. C.: Proc. XIIth Intern. Conf. on: Coordination Chemistry. Sydney, Australia, 1969, p. 185.
2073. Stone, E. W., Maki, A. H.: J. Chem. Phys. *36*, 1944 (1962).
2074. Stone, E. W., Maki, A. H.: J. Chem. Phys. *38*, 1999 (1963).
2075. Stone, E. W., Maki, A. H.: J. Amer. Chem. Soc. *87*, 454 (1965).
2076. Stone, T. J., Buckman, T., Nordio, P. L., McConnell, H. M.: Proc. Natl. Acad. US. *54*, 1010 (1965).
2077. Stone, T. J., Waters, W. A.: Proc. Chem. Soc. *1962*, 253 (1962).
2078. Stoyathenko, I. L., Georgiev, G. S., Golubev, V. B., Zubov, V. I., Kabanov, V. A.: Vysokomol. Soedin. *A15*, 1899 (1973).
2079. Strelko, V. V., Vysotskii, Z. Z.: Sintez i Fiz-Khim. Polim. Akad. Nauk Ukr. SSR, Inst. Khim. Vysokomol. Soedin., Sb. statiei *1964*, 66.
2080. Strelko, V. V., Vysotskii, Z. Z.: Sintez i Fiz-Khim. Polim. Akad. Nauk Ukr. SSR, Inst. Khim. Vysokomol. Soedin., Sb. statiei *1964*, 80.
2081. Strelko, V. V., Vysotskii, Z. Z., Gnyuk, L. N.: Vysokomol. Soedin., Khim. Svoistva i Modyfikatsiya Polimerov, Sb. statiei *1964*, 19.
2082. Stryukov, V. B., Dubovitskii, A. V., Rozenberg, B. A., Enikolopyan, N. S.: Dokl. Akad. Nauk SSSR *190*, 642 (1970).
2083. Stryukov, V. B., Karimov, Yu. S., Rozantsev, E. G.: Vysokomol. Soedin. *B9*, 493 (1967).

2084. Stryukov, V. B., Korolev, G. V.: Vysokomol. Soedin. *A11*, 419 (1969).
2085. Stryukov, V. B., Rozantsev, E. G.: Vysokomol. Soedin. *A10*, 626 (1968).
2086. Stryukov, V. B., Sosina, T. V., Kraitsberg, A. M.: Vysokomol. Soedin. *A15*, 1397 (1973).
2087. Sugden, T. M.: Proc. Roy. Soc. *A302*, 309 (1968).
2088. Sugimoto, S., Ohnishi, S., Nitta, I.: Ann. Rep. Assoc. Res. Radiation Polymerization, Japan *8*, 177 (1966–1967).
2089. Sukhorov, B. I.: Stud. Biophys. *40*, 33 (1973).
2090. Sukhorov, B. I., Kuzmenko, V. A., Blumenfeld, L. A.: Vysokomol. Soedin. Geterotsepnye Vysokomol. Soedin. *1964*, 145.
2091. Sukhov, V. A., Nikitina, L. A., Baturina, A. A., Lukovnikov, A. F., Enikolopyan, N. S.: Vysokomol. Soedin. *A11*, 808 (1969).
2092. Sullivan, P. D.: Magn. Res. Rev. *3*, 251 (1974).
2093. Sullivan, P. J., Koski, W. S.: J. Amer. Chem. Soc. *85*, 384 (1963).
2094. Sumegi, L., Tüdös, G., Kende, I.: in: Kinetics and Mechanism of Polyreactions. Budapest: Ed. Akademiai Kiado 1969, vol. 3, p. 97.
2095. Sumegi, L., Tüdös, F., Kende, I.: Acta Chim. Acad. Sci. Hungaricae *68*, 75 (1971).
2096. Suntsov, E. V., Votinov, M. P.: Opt. Spektr. *16*, 543 (1964).
2097. Suzuki, A., Takahashi, M., Shiomi, K.: Bull. Chem. Soc. Japan *36*, 644 (1963).
2098. Suzuki, A., Takahashi, M., Shiomi, K.: Bull. Chem. Soc. Japan *36*, 998 (1963).
2099. Suzuki, K.: J. Polym. Sci. *11*, 2377 (1973).
2100. Svejda, P., Volman, D. H.: J. Phys. Chem. *73*, 4417 (1969).
2101. Swalen, J. D., Gladney, H. M.: IBM J. Res. Develop. *8*, 515 (1964).
2102. Swallow, A. J.: Radiation Chemistry of Organic Compounds. London: Pergamon Press 1960.
2103. Swartz, H. M., Bolton, J. R., Borg, D. C.: Biological Applications of Electron Spin Resonance. New York: Wiley 1972.
2104. Szöcs, F.: Proc. Third Tihany Symp. on Radiation Chemistry. Budapest: Akademiai Kiado 1972, p. 933.
2105. Szöcs, F.: J. Appl. Polym. Sci. *14*, 2629 (1970).
2106. Szöcs, F.: Chem. Zvesti *26*, 27 (1972).
2107. Szöcs, F., Becht, J., Fischer, H.: Europ. Polym. J. *7*, 173 (1971).
2108. Szöcs, F., Borsig, E., Plaček, J.: J. Polym. Sci. *11*, 185 (1973).
2109. Szöcs, F., Borsig, E., Plaček, J., Klimova, M.: Chem. Zvesti *28*, 581 (1974).
2110. Szöcs, F., Lazar, M.: Europ. Polym. J. Suppl. *1969*, p. 337.
2111. Szöcs, F., Plaček, J.: Europ. Polym. J. *8*, 525 (1972).
2112. Szöcs, F., Plaček, J.: J. Polym. Sci.: A2, *13*, 1789 (1975).
2113. Szöcs, F., Plaček, J., Borsig, E.: J. Polym. Sci. *9*, 753 (1971).
2114. Szöcs, F., Rostašová, O.: J. Appl. Polym. Sci. *18*, 2529 (1974).
2115. Szöcs, F., Rostašová, O.: Europ. Polym. J. *11*, 559 (1975).
2116. Szöcs, F., Rostašová, O., Tino, J., Plaček, J.: Europ. Polym. J. *10*, 725 (1974).
2117. Szöcs, F., Tino, J., Plaček, J.: Europ. Polym. J. *9*, 251 (1973).
2118. Szöcs, F., Ulbert, K.: J. Polym. Sci. *5*, 671 (1967).
2119. Szwarc, M.: Carbanions, Living Polymers and Electron Transfer Processes. New York: Wiley 1968.
2120. Szwarc, M.: Accounts Chem. Res. *2*, 87 (1969).
2121. Szwarc, M.: in: ESR Application to Polymer Research (Ed. Kinell, P. O., Rånby, B., Runnström-Reio, V.). Stockholm: Almqvist-Wiksell 1973, p. 291.
2122. Szwarc, M., Shimada, K.: Polym. Symp. No. 46, 193 (1974).
2123. Symons, M. C. R.: J. Chem. Soc. *1959*, 277.
2124. Symons, M. C. R.: Adv. Phys. Org. Chem. *1963*, 307.
2125. Symons, M. C. R.: J. Chem. Soc. *1963*, 1186.
2126. Symons, M. C. R.: J. Phys. Chem. *71*, 172 (1967).
2127. Symons, M. C. R., Townsend, M. G.: J. Chem. Soc. *1959*, 263.
2128. Symons, M. C. R., Yandell, J. K.: J. Chem. Soc. *970*, 1995.
2129. Tabata, Y., Fujikawa, J.: J. Macromol. Sci. Chem. A, *5*, 821 (1971).

References

2130. Tabata, Y., Fujikawa, J., Oshima, K.: J. Macromol. Sci. Chem. A, 5, 831 (1971).
2131. Tabata, Y., Ito, Y., Mori, K., Oshima, K.: Proc. Third Tihany Symp. on Radiation Chemistry. Budapest: Akademiai Kiado 1972, p. 473.
2132. Takakura, K., Kawa, E., Hayashi, K., Okamura, S.: Ann. Rep. Assoc. Res. Radiation Polymerization, Japan 6, 205 (1964–65).
2133. Takakura, K., Rånby, B.: J. Polym. Sci. B, 5, 83 (1967).
2134. Takakura, K., Rånby, B.: J. Phys. Chem. 72, 164 (1968).
2135. Takakura, K., Rånby, B.: J. Polym. Sci. C, (22), 939 (1969).
2136. Takakura, K., Rånby, B.: Adv. Chem. Ser. No. 91, 125 (1969).
2137. Takakura, K., Rånby, B.: J. Polym. Sci. A1, 8, 77 (1970).
2138. Takamatsu, T., Kaibara, M.: Rep. Inst. Phys. Chem. Res. Tokyo 41, 203 (1965).
2139. Takeda, K., Yoshida, H., Hayashi, K., Okamura, S.: Mem. Fac. Eng. Kyoto Univ. 28, 213 (1966).
2140. Takeda, K., Yoshida, H., Hayashi, K., Okamura, S.: J. Polym. Sci. A1, 4, 2710 (1966).
2141. Takeda, M., Imiura, K., Nozawa, N., Hisatome, M.: J. Polym. Sci. C, (23), 741 (1968).
2142. Takegami, Y., Suzuki, T., Okazaki, T.: Bull. Chem. Soc. Japan 42, 1060 (1969).
2143. Takeshita, T., Hirota, N.: J. Chem. Phys. 51, 2146 (1969).
2144. Takeshita, T., Tsuji, K., Seiki, T.: Rep. Progr. Polym. Phys. Japan 15, 563 (1972).
2145. Takeshita, T., Tsuji, K., Seiki, T.: J. Polym. Sci. A1, 10, 2315 (1972).
2146. Takui, T., Waka, Y., Kawakami, H., Itoh, K.: Chem. Phys. Lett. 35, 465 (1975).
2147. Tamura, N.: J. Phys. Soc. Japan 15, 943 (1960).
2148. Tamura, N.: J. Phys. Soc. Japan 16, 3838 (1961).
2149. Tamura, N.: J. Chem. Phys. 37, 479 (1962).
2150. Tamura, N.: J. Polym. Sci. 60, S5 (1962).
2151. Tamura, N., Nara, S., Kashiwabara, H., Komatsu, T., Sohma, J.: Rep. Fac. Eng. Hokkaido Univ. 35, 443 (1964).
2152. Tamura, N., Oshima, Y., Yotsumoto, K., Sunaga, H.: Japanese J. Appl. Phys. 9, 1148 (1970).
2153. Tamura, N., Shinohara, K.: Proc. Intern. Symp. on Molecular and Structure Spectroscopy. Tokyo, Japan, 1962, p. D204.
2154. Tamura, N., Shinohara, K.: Rep. Progr. Polym. Phys. Japan 6, 261 (1963).
2155. Tamura, N., Shinohara, K.: Rep. Progr. Polym. Phys. Japan 6, 265 (1963).
2156. Tamura, N., Shinohara, K.: Rep. Progr. Polym. Phys. Japan 7, 347 (1964).
2157. Tamura, N., Tachibana, H., Shinohara, K.: Rep. Progr. Polym. Phys. Japan 7, 343 (1964).
2158. Tamura, N., Tachibana, H., Takamatsu, T., Shinohara, K.: Rep. Progr. Polym. Phys. 6, 269 (1963).
2159. Tamura, N., Tachibana, H., Takamatsu, T., Shinohara, K.: Rep. Progr. Polym. Phys. Japan 6, 273 (1963).
2160. Tanaka, H., Matsumoto, A., Goto, N.: Bull. Chem. Soc. Japan 37, 1128 (1964).
2161. Tanei, T.: Bull. Chem. Soc. Japan 40, 2456 (1967).
2162. Taranukha, O. M., Vonsyatskii, V. A., Lebedev, Ya. S.: Khim. Vysokh. Energ. 2, 476 (1968).
2163. Tarasova, Z. N., Elkina, I. A., Kozlov, V. T., Zagumennaya, T. I., Kaplunov, M. Ya.: Vysokomol. Soedin. A14, 1782 (1972).
2164. Tarasova, Z. N., Fogelson, M. S., Kozlov, V. T., Kashlinskii, A. I., Kaplunov, M. Ya., Dogadkin, B. A.: Vysokomol. Soedin. 4, 1204 (1962).
2165. Tarasova, Z. N., Kozlov, V. T., Dogadkin, B. A.: Proc. Tihany Symp. on Radiation Chemistry. Budapest: Akademiai Kiado 1964, p. 287.
2166. Tarvin, R. F., Aoki, S., Stille, J. K.: Macromolecules 5, 663 (1972).
2167. Taylor, B. N., Parker, W. H., Langenberg, D. N.: Rev. Mod. Phys. 41, 375 (1969).
2168. Teleman, A. S., McEvily, A. J.: Fracture of Structural Materials. New York: Wiley 1967.
2169. Teleshov, E. N., Pravednikov, A. N., Medvedev, S. S.: Dokl. Akad. Nauk SSSR 156, 1395 (1964).
2170. Teleshov, E. N., Sharptyi, V. A., Pravednikov, A. N., Medvedev, S. S.: Zh. Strukt. Khim. 5, 62 (1964).
2171. Ten Bosch, J. J., Verhelst, W. F., Chadwick, K. H.: J. Polym. Sci. A1, 10, 1679 (1972).

2172. Tennant, W. C.: N. Z. Dep. Sci. Ind. Res. Chem. Div. Rep. No. CD-2153, 1972.
2173. Terabe, S., Konaka, R.: J. Amer. Chem. Soc. *93*, 4306 (1971).
2174. Terpugova, M. P., Kotlyarevskii, I. L., Mityushova, A. A., Bashkirov, M. F.: Izv. Akad. Nauk SSSR *1967*, 662.
2175. Theodorescu, M.: Stud. Ceret. Fiz. *25*, 831 (1973).
2176. Theodorescu, M.: Stud. Ceret. Fiz. *26*, 229 (1974).
2177. Thomas, J. K.: in: Creation and Detection of the Excited State (Ed. Lamola, A. A.). New York: Dekker 1971, p. 481.
2178. Thomas, J. R.: J. Amer. Chem. Soc. *88*, 2064 (1966).
2179. Thomson, C.: Electron Spin Resonance *1*, 1 (1972).
2180. Thomson, C.: Electron Spin Resonance *2*, 1 (1974).
2181. Thryion, F. C., Baijal, M. D.: J. Polym. Sci. A1, *6*, 505 (1968).
2182. Tiedman, G., Ingalls, R. B.: J. Phys. Chem. *71*, 3092 (1967).
2183. Tikhomirov, L. A., Buben, N. Ya.: Vysokomol. Soedin. *8*, 1881 (1966).
2184. Tikhomirova, N. N., Lukin, B. V., Razmunova, L. L., Voevodskii, V. V.: Dokl. Akad. Nauk SSSR *122*, 264 (1958).
2185. Timm, D., Willard, J. E.: J. Phys. Chem. *73*, 2403 (1969).
2186. Tinkham, M., Strandberg, M. W. P.: Phys. Rev. *97*, 951 (1955).
2187. Tino, J., Čapla, M., Szöcs, F.: Europ. Polym. J. *6*, 397 (1970).
2188. Tino, J., Plaček, J.: Europ. Polym. J. *7*, 1615 (1971).
2189. Tino, J., Plaček, J., Pavlinec, J.: Europ. Polym. J. *10*, 291 (1974).
2190. Tino, J., Plaček, J., Szöcs, F.: Europ. Polym. J. *11*, 609 (1975).
2191. Tino, J., Szöcs, F.: Polym. J. Japan *4*, 120 (1973).
2192. Tino, J., Szöcs, F.: J. Polym. Sci. B, *11*, 323 (1974).
2193. Tino, J., Szöcs, F., Plaček, J.: Chem. Zvesti *28*, 577 (1974).
2194. Tischer, R.: Z. Naturforsch. *22a*, 1711 (1967).
2195. Tkač, A.: Coll. Czechoslov. Chem. Comm. *33*, 1629 (1968).
2196. Tkač, A.: Coll. Czechoslov. Chem. Comm. *33*, 2004 (1968).
2197. Tkač, A.: Coll. Czechoslov. Chem. Comm. *33*, 3001 (1968).
2198. Tkač, A., Adamčik, V.: Coll. Czechoslov. Chem. Comm. *38*, 1346 (1973).
2199. Tkač, A., Frait, Z., Ondris, M.: Coll. Czechoslov. Chem. Comm. *31*, 252 (1966).
2200. Tkač, A., Kresta, J.: Chem. Zvesti *24*, 189 (1970).
2201. Tkač, A., Kresta, J.: Chem. Zvesti *25*, 3 (1971).
2202. Tkač, A., Kresta, J.: Chem. Zvesti *25*, 104 (1971).
2203. Tkač, A., Omelka, L.: Polym. Symp. No. 40, 119 (1973).
2204. Tkač, A., Omelka, L, Holičik, J.: Polym. Symp. No. 40, 105 (1973).
2205. Tkač, A., Přikril, R., Malik, L., Kresta, J.: Chem. Zvesti *25*, 97 (1971).
2206. Tkač, A., Staško, A.: Coll. Czechoslov. Chem. Commun. *37*, 573 (1972).
2207. Tkač, A., Staško, A.: Coll. Czechoslov. Chem. Commun. *37*, 1006 (1972).
2208. Tkač, A., Veselý, K., Omelka, L.: J. Phys. Chem. *75*, 2575 (1971).
2209. Tochin, V. A., Nikolskii, V. G., Buben, N. J.: Dokl. Akad. Nauk SSSR *168*, 360 (1966).
2210. Tolkachev, V. A., Molin, Yu. N., Tchkheidze, I. I., Buben, N. Ya., Voevodskii, V. V.: Dokl. Akad. Nauk SSSR *141*, 911 (1961).
2211. Tolparov, Yu. N.: Prib. Tekh. Eksp. *1975*, 260.
2212. Tomashevskii, E. E., Pavlova, I. N., Savotsin, A. Ya.: Fiz. Tverd. Tela *2*, 485 (1965).
2213. Topchiev, D. A., Kabanov, V. A., Kargin, V. A.: Vysokomol. Soedin. *6*, 1814 (1964).
2214. Toptygin, D. Ya., Pariskii, G. B., Davydov, E. Ya., Ledneva, O. A., Mikheev, Yu. A.: Vysokomol. Soedin. *A14*, 1534 (1972).
2215. Torikai, A., Asai, T., Suzuki, T., Kuri, Z.: J. Polym. Sci. A1, *13*, 797 (1975).
2216. Toriyama, K., Iwasaki, M.: J. Phys. Chem. *73*, 2919 (1969).
2217. Törmälä, P.: Dissert. Univ. Helsinki, Finland, 1973.
2218. Törmälä, P.: Polymer *15*, 124 (1974).
2219. Törmälä, P.: Angew. Makromol. Chem. *37*, 135 (1974).
2220. Törmälä, P.: Europ. Polym. J. *10*, 519 (1974).
2221. Törmälä, P., Brotherus, J.: Finn. Chem. Lett. *1974*, 127.

References

2222. Törmälä, P., Lättilä, H., Lindberg, J. J.: in: ESR Application to Polymer Research (ed. Kinell, P. O., Rånby, B., Runnström-Reio, V.). Stockholm: Almqvist-Wiksell 1973, p. 267.
2223. Törmälä, P., Lättilä, H., Lindberg, J. J.: Polymer *14*, 481 (1973).
2224. Törmälä, P., Lindberg, J. J.: in: Structural Studies of Macromolecules by Spectroscopic Methods. (Ed. Ivin, K. J.). Wiley (in press).
2225. Törmälä, P., Lindberg, J. J., Koivu, L.: Paperi Puu *1972*, 1.
2226. Törmälä, P., Lindberg, J. J., Lehtinen, S.: Paperi Puu *1975*, 1.
2227. Törmälä, P., Lindberg, J. J., Lehtinen, S.: Polymer (in press).
2228. Törmälä, P., Mannila, L., Löfgren, B.: Tutkimus ja Tekniikka *1975*, 43.
2229. Törmälä, P., Martinmaa, J., Silvennionen, K., Vaahtera, K.: Acta Chem. Scand. *24*, 3066 (1970).
2230. Törmälä, P., Salvolainen, A.: Acta Chem. Scand. *24*, 1430 (1973).
2231. Törmälä, P., Silvennionen, K., Lindberg, J. J.: Acta Chem. Scand. *25*, 2659 (1971).
2232. Törmälä, P., Sundholm, F.: Kemia-Kemi *1*, 203 (1974).
2233. Törmälä, P., Tulikoura, J.: Polymer *15*, 248 (1974).
2234. Toussaint, J., Declerck, C.: Bull. Soc. Roy. Sci. Liege *35*, 93 (1966).
2235. Towens, C. H., Turkevich, J.: Phys. Rev. *77*, 147 (1950).
2236. Toy, M. S., Newman, J. M.: J. Polym. Sci. A1, *7*, 2333 (1969).
2237. Tropinin, V. N., Zhdanova, S. V., Rempel, S. I.: Sovrem. Metody Issled. Khim. Lignina *1970*, 15.
2238. Truby, F. K., Storey, W. H.: J. Chem. Phys. *31*, 857 (1959).
2239. Tsuchida, E. Shih, C. N., Shinohara, I., Kambara, S.: J. Polym. Sci. A1, *2*, 3347 (1964).
2240. Tsuchihashi, N., Shimada, S., Kashiwabara, H., Sohma, J.: Rep. Progr. Polym. Phys. Japan *12*, 461 (1969).
2241. Tsuji, K.: Rep. Progr. Polym. Phys. Japan *15*, 555 (1972).
2242. Tsuji, K.: Rep. Progr. Polym. Phys. Japan *15*, 559 (1972).
2243. Tsuji, K.: Rep. Progr. Polym. Phys. Japan *15*, 569 (1972).
2244. Tsuji, K.: Rep. Progr. Polym. Phys. Japan *16*, 565 (1973).
2245. Tsuji, K.: Rep. Progr. Polym. Phys. Japan *16*, 571 (1973).
2246. Tsuji, K.: Adv. Polym. Sci. *12*, 131 (1973).
2247. Tsuji, K.: J. Polym. Sci. A1, *11*, 467 (1973).
2248. Tsuji, K.: J. Polym. Sci. A1, *11*, 1407 (1973).
2249. Tsuji, K.: J. Polym. Sci. A1, *11*, 2069 (1973).
2250. Tsuji, K.: J. Polym. Sci. B, *11*, 351 (1973).
2251. Tsuji, K.: Rep. Progr. Polym. Phys. Japan *17*, 553 (1974).
2252. Tsuji, K.: Rep. Progr. Polym. Phys. Japan *17*, 555 (1974).
2253. Tsuji, K.: Rep. Progr. Polym. Phys. Japan *17*, 557 (1974).
2254. Tsuji, K., Hayashi, K., Okamura, S.: Ann. Rep. Assoc. Res. Radiation Polymerization, Japan *6*, 155 (1964/65).
2255. Tsuji, K., Hayashi, K., Okamura, S.: Bull. Chem. Soc. Japan *43*, 572 (1970).
2256. Tsuji, K., Hayashi, K., Okamura, S.: J. Polym. Sci. A1, *8*, 583 (1970).
2257. Tsuji, K., Imanishi, Y., Hayashi, K., Okamura, S.: J. Polym. Sci. B, *5*, 449 (1967).
2258. Tsuji, K., Iwamoto, T., Hayashi, K., Yoshida, H.: J. Polym. Sci. A1, *5*, 265 (1967).
2259. Tsuji, K., Iwamoto, T., Yoshida, H., Hayashi, K., Okamura, S.: Ann. Rep. Assoc. Res. Radiation Polymerization, Japan *7*, 3340 (1965/66).
2260. Tsuji, K., Iwamoto, T., Yoshida, H., Hayashi, K., Okamura, S.: Mem. Fac. Eng. Kyoto Univ. *31*, 268 (1969).
2261. Tsuji, K., Kondo, T., Takeshita, T., Hirooka, M.: J. Polym. Sci. B, *10*, 189 (1972).
2262. Tsuji, K., Nagata, H.: Rep. Progr. Polym. Phys. Japan *15*, 567 (1972).
2263. Tsuji, K., Nagata, H.: J. Polym. Sci. A1, *11*, 897 (1973).
2264. Tsuji, K., Okamura, S.: Bull. Chem. Soc. Japan *42*, 2827 (1969).
2265. Tsuji, K., Okamura, S.: in: ESR Application to Polymer Research (Ed. Kinell, P. O., Rånby, B., Runnström-Reio, V.). Stockholm: Almqvist-Wiksell 1973, p. 187.
2266. Tsuji, K., Okaya, T., Hayashi, K., Okamura, S.: Ann. Rep. Assoc. Res. Radiation Polymerization *7*, 127 (1965/66).

398

2267. Tsuji, K., Seiki, T.: J. Polym. Sci. B, 7, 839 (1969).
2268. Tsuji, K., Seiki, T.: Polymer J. 1, 133 (1970).
2269. Tsuji, K., Seiki, T.: Rep. Progr. Polym. Phys. Japan 13, 507 (1970).
2270. Tsuji, K., Seiki, T.: J. Polym. Sci. B, 8, 817 (1970).
2271. Tsuji, K., Seiki, T.: Polymer J. 2, 606 (1971).
2272. Tsuji, K., Seiki, T.: J. Polym. Sci. A1, 9, 3063 (1971).
2273. Tsuji, K., Seiki, T.: Rep. Progr. Polym. Phys. Japan 14, 577 (1971).
2274. Tsuji, K., Seiki, T.: Rep. Progr. Polym. Phys. Japan 14, 581 (1971).
2275. Tsuji, K., Seiki, T.: Rep. Progr. Polym. Phys. Japan 14, 585 (1971).
2276. Tsuji, K., Seiki, T.: Rep. Progr. Polym. Phys. Japan 14, 589 (1971).
2277. Tsuji, K., Seiki, T.: J. Polym. Sci. A1, 10, 123 (1972).
2278. Tsuji, K., Seiki, T.: J. Polym. Sci. B, 10, 139 (1972).
2279. Tsuji, K., Seiki, T.: Rep. Progr. Polym. Phys. Japan 15, 573 (1972).
2280. Tsuji, K., Seiki, T.: Polymer J. 4, 589 (1973).
2281. Tsuji, K., Seiki, T., Takeshita, T.: J. Polym. Sci. A1, 10, 3119 (1972).
2282. Tsuji, K., Takakura, K., Nishii, M., Hayashi, K., Okamura, S.: Ann. Rep. Assoc. Res. Radiation Polymerization 6, 179 (1964/65).
2283. Tsuji, K., Takakura, K., Nishii, M., Hayashi, K., Okamura, S.: J. Polym. Sci. A1, 4, 2028 (1966).
2284. Tsuji, K., Takeshita, T.: J. Polym. Sci. B, 10, 185 (1972).
2285. Tsuji, K., Takeshita, T.: Rep. Progr. Polym. Phys. Japan 15, 553 (1972).
2286. Tsuji, K., Takeshita, T.: J. Polym. Sci. B, 11, 491 (1973).
2287. Tsuji, K., Takeshita, T.: Rep. Progr. Polym. Phys. Japan 16, 569 (1973).
2288. Tsuji, K., Tazuke, S., Hayashi, K., Okamura, S.: J. Phys. Chem. 73, 2345 (1969).
2289. Tsuji, K., Williams, Ff.: J. Amer. Chem. Soc. 89, 1526 (1967).
2290. Tsuji, K., Williams, Ff.: J. Phys. Chem. 72, 3884 (1968).
2291. Tsuji, K., Williams, Ff.: Int. J. Radiat. Phys. Chem. 1, 383 (1969).
2292. Tsuji, K., Williams, Ff.: Trans. Faraday Soc. 65, 1718 (1969).
2293. Tsuji, K., Yamaoka, H., Hayashi, K., Kamiyama, H., Yoshida, H.: J. Polym. Sci. B, 4, 629 (1966).
2294. Tsuji, K., Yoshida, H., Hayashi, K.: J. Chem. Phys. 45, 2894 (1966).
2295. Tsuji, K., Yoshida, H., Hayashi, K.: J. Chem. Phys. 46, 810 (1967).
2296. Tsuji, K., Yoshida, H., Hayashi, K.: J. Chem. Phys. 46, 2808 (1967).
2297. Tsuji, K., Yoshida, H., Hayashi, K., Okamura, S.: Ann. Rep. Assoc. Res. Radiation Polymerization 6, 163 (1964/65).
2298. Tsuji, K., Yoshida, H., Hayashi, K., Okamura, S.: Ann. Rep. Assoc. Res. Radiation Polymerization 7, 141 (1965/66).
2299. Tsuji, K., Yoshida, H., Hayashi, K., Okamura, S.: Ann. Rep. Assoc. Res. Radiation Polymerization 8, 103 (1966/67).
2300. Tsuji, K., Yoshida, H., Hayashi, K., Okamura, S.: J. Polym. Sci. B, 5, 313 (1967).
2301. Tsuji, K., Yoshida, H., Hayashi, K., Okamura, S.: Chem. High Polym. Japan 25, 31 (1968).
2302. Tsuji, K., Yoshida, H., Okamura, S.: Ann. Rep. Assoc. Res. Radiation Polymerization 7, 119 (1965/66).
2303. Tsuruya, S., Kawamura, T., Yonezawa, T.: J. Polym. Sci. A1, 9, 1659 (1971).
2304. Tsvetkov, Yu. D., Bubonov, N. N., Makulskii, M. A., Lazurkin, Yu. S., Voevodskii, V. V.: Dokl. Akad. Nauk SSSR 122, 1053 (1958).
2305. Tsvetkov, Yu. D., Lebedev, Ya. S., Voevodskii, V. V.: Vysokomol. Soedin. 1, 1519 (1959).
2306. Tsvetkov, Yu. D., Lebedev, Ya. S., Voevodskii, V. V.: Vysokomol. Soedin. 1, 1634 (1959).
2307. Tsvetkov, Yu. D., Molin, Yu. N., Voevodskii, V. V.: Vysokomol. Soedin. 1, 1805 (1959).
2308. Tsvetanov, Ch. B., Panayotov, I. M., Barzashka, T. D.: Izv. Akad. Nauk Bulgaria 2, 829 (1969).
2309. Tüdös, F., Azori, M., Varsanyi, G., Holly, S.: Act. Chim. Acad. Sci. Hungaricae 33, 433 (1962).
2310. Tüdös, F., Heidt, J., Erö, J.: Acta Chim. Hung. 33, 433 (1962).
2311. Tüdös, F., Kende, I., Berezhnykh, T., Solodovnikov, S. P., Voevodskii, V. V.: Magy. Kem. Folyoirat 69, 371 (1963).

References

2312. Tüdös, F., Kende, I., Berezhnykh, T., Solodovnikov, S. P., Voevodskii, V. V.: Kinet. Kataliz. 6, 203 (1965).
2313. Tupikov, V. I., Malkova, A. I., Sorokin, Yu. A., Pshezhetskii, S. Ya.: Khim. Vysokh. Energ. 2, 352 (1968).
2314. Tupikov, V. I., Pshezhetskii, S. Ya.: 38, 1310 (1964).
2315. Turkevich, J., Fujita, Y.: Phys. Today 18, 26 (1965).
2316. Turro, N. J.: Molecular Photochemistry. New York: Benjamin 1965.
2317. Turro, N. J.: in: Technique of Organic Chemistry (Ed. Leermakers, P. A., Weissberger, A.), Wiley-Interscience, 1969, vol. 14, p. 133.
2318. Turscanyi, B., Koszterszitz, G., Tüdös, F.: Acta Chim. Acad. Sci. Hungaricae 50, 293 (1966).
2319. Tuttle, T. R., Weissman, S. I.: J. Amer. Chem. Soc. 80, 5342 (1960).
2320. Tynan, E. C., Yen Teh Fu.: J. Magn. Resonance 3, 327 (1970).
2321. Uebersfeld, J.: Ann. phys. 1, 395 (1956).
2322. Ueda, H.: J. Phys. Chem. 67, 2185 (1963).
2323. Ueda, H.: J. Polym. Sci. A1, 2, 2207 (1964).
2324. Ueda, H., Kuri, Z.: J. Appl. Polym. Sci. 5, 478 (1961).
2325. Ueda, H., Kuri, Z.: J. Polym. Sci. 61, 333 (1962).
2326. Ueda, H., Kuri, Z., Shida, S.: J. Chem. Phys. 35, 2145 (1961).
2327. Ueda, H., Kuri, Z., Shida, S.: J. Appl. Polym. Sci. 5, 478 (1961).
2328. Ueda, H., Kuri, Z., Shida, S.: J. Polym. Sci. 56, 251 (1962).
2329. Ueda, H., Kuri, Z., Shida, S.: J. Polym. Sci. A1, 1, 3537 (1963).
2330. Ueda, H., Shida, S.: Kogyo Kagaku Zasshi 69, 1527 (1966).
2331. Ueda, H., Shida, S., Kuri, Z.: Shinku Kagaku 5, 3 (1967).
2332. Ueda, H., Yashiro, D.: Bull. Chem. Soc. Japan 44, 595 (1971).
2333. Ueno, T., Yamaguchi, T., Imamura, R.: Kami Pa Gikyoshi 25, 242 (1971).
2334. Ueno, T., Yamaguchi, T., Imamura, R.: Kami Pa Gikyoshi 25, 661 (1971).
2335. Uetsuki, M., Fujiwara, S.: Bunseki Kagaku Shinpo Sosetsu 1966, 104R.
2336. Ulbert, K.: Coll. Czech. Chem. Commun. 30, 3285 (1965).
2337. Ulbert, K.: Anal. Fys. Metody Vyzk. Plastu Pryskiryc 3, 45 (1971).
2338. Umezu, K., Shimada, S., Kashiwahara, H., Sohma, J.: Rep. Progr. Polym. Phys. Japan 11, 475 (1968).
2339. Ungar, I. S., Gager, W. B., Leininger, R. I.: J. Polym. Sci. 44, 295 (1960).
2340. Ursu, I.: Electron Spin Resonance. Bucharest: Acad. Republ. Socialiste Romania 1965 (in Roman.), Paris, Dunod 1968 (in French).
2341. Usmanov, K. U., Azizov, U. A., Milinchuk, V. K., Khamidov, D. S.: J. Polym. Sci. C, (42), 1607 (1973).
2342. Usmanov, K. U., Vakhidov, N.: Vysokomol. Soedin. A11, 1501 (1969).
2343. Vakulskaya, T. I., Kryazhev, Yu. G.: Vysokomol. Soedin. A15, 1783 (1973).
2344. Vakulskaya, T. I., Kryazhev, Yu. G.: Izv. Akad. Nauk SSSR 1974, 1532.
2345. Valitova, F. G., Illyasov, A. V.: Dokl. Akad. Nauk SSSR 144, 600 (1962).
2346. Van Gerven, L., Talpe, J., van Itterbeek, A.: Physica 33, 207 (1967).
2347. Van Gerven, L., van Itterbeek, A., de Laet, L.: in: Paramagnetic Resonance (ed. Low, W.). New York: Academic Press 1963, vol. 2.
2348. Van Reijen, L. L., Cossee, P.: Discuss. Faraday Soc. 41, 277 (1966).
2349. Van de Vorst, A.: Int. J. Radiat. Biol. 24, 605 (1973).
2350. Vargin, V. V., Zakrevskii, V. A., Ivanov, V. A., Yudin, D. M.: Zh. Prikl. Khim. 47, 992 (1974).
2351. Vasserman, A. M., Antisiferova, L. I., Osipova, E. S., Buchachenko, A. L.: Dokl. Akad. Nauk SSSR 222, 384 (1975).
2352. Vasserman, A. M., Buchachenko, A. L.: Zh. Strukt. Khim. 7, 673 (1966).
2353. Vasserman, A. M., Buchachenko, A. L., Kovarskii, A. L., Beiman, M. B.: Vysokomol. Soedin. A10, 1930 (1968).
2354. Vedrine, J. C., Naccache, C.: J. Phys. Chem. 77, 1606 (1974).
2355. Venkataraman, B., Fraenkel, G. K.: J. Chem. Phys. 23, 588 (1955).
2356. Venkataraman, B., Fraenkel, G. K.: J. Amer. Chem. Soc. 77, 2707 (1955).
2357. Venkataraman, B., Segel, B. G., Fraenkel, G. K.: J. Chem. Phys. 30, 1006 (1959).

2358. Verma, G. P. S., Peterlin, A.: J. Polym. Sci. B, 7, 587 (1969).
2359. Verma, G. P. S., Peterlin, A.: J. Macromol. Chem. B, 4, 589 (1970).
2360. Verma, G. P. S., Peterlin, A.: Kolloid-Z. u. Z-Polym. 236, 111 (1970).
2361. Vincent, P. I.: Physical Basis of Yield and Fracture, Conf. Proc. Oxford, Ed. Inst. Physic and Physical Soc., London, 1966, p. 155.
2362. Vinogradov, G. V., Titkova, L. V.: Vysokomol. Soedin. A11, 951 (1969).
2363. Vinogradova, V. G., Shelimov, B. N., Fok, N. V.: Khim. Vysokh. Energ. 2, 107 (1968).
2364. Virnik, R. B., Ershov, Yu. A.: Khim. Tekhn. Proizv. Tsellul. 1971, 333.
2365. Virnik, R. B., Ershov, Yu. A., Frunze, N. K., Livshnitz, R. M., Berlin, A. A.: Vysokomol. Soedin. A12, 1388 (1970).
2366. Virnik, R. B., Ershov, Yu. A., Frunze, N. K., Livshnitz, R. M., Nerlin, A. A.: Vysokomol. Soedin. A12, 2279 (1970).
2367. Vishnevskaya, G. P., Safin, R. Sh., Lipinov, I. N., Kazantsev, E. I., Gorozhanin, V. A.: Teor. Eksp. Khim. 10, 514 (1974).
2368. Vlasova, R. M., Gasparyan, S. N., Kargin, V. A., Rosenthstein, L. D., Kholmogorov, V. E.: Dokl. Akad. Nauk SSSR 171, 132 (1966).
2369. Voevodskii, V. V., Molin, Yu. N.: Radiation Res. 17, 366 (1962).
2370. Voevodskii, V. V.: Proc. Tihany Symp., on Radiation Chemistry. Budapest: Akademiai Kiado 1964, p. 112.
2371. Vonsyatskii, V. A., Mamunya, E. P., Lebedev, Ya. S.: Khim. Vysokh. Energ. 8, 369 (1974).
2372. Vonsyatskii, V. A., Taranukha, O. M., Kalnichenko, A. M., Egorov, Yu. P.: Teor. Eksp. Khim. Kad. Nauk Ukr. SSSR 2, 384 (1966).
2373. Voskerchyan, G. P., Lebedev, Ya. S.: Izv. Akad. Nauk SSSR 1975, 310.
2374. Waggoner, A. S., Griffith, O. H., Christensen, C. R.: Proc. Roy. Acad. Sci. US 57, 1198 (1967).
2375. Wall, L. A.: Symp. on: Materials in Nuclear Application. Special Technical Publ., No. 276, Amer. Soc. for Testing Materials, 1959.
2376. Wall, L. A.: Polym. Prepr. Amer. Chem. Soc., Div. Polym. Chem. 3, 62 (1962).
2377. Wall, L. A.: J. Chem. Phys. 41, 1112 (1964).
2378. Wall, L. A., Brown, D. W., Florin, R. E.: J. Chem. Phys. 30, 602 (1959).
2379. Wall, L. A., Brown, D. W., Florin, R. E.: J. Phys. Chem. 63, 1702 (1959).
2380. Wall, L. A., Fetters, L. J., Straus, S.: J. Polym. Sci. B, 5, 721 (1967).
2381. Wall, L. A., Ingalls, R. B.: J. Polym. Sci. 62, S5 (1962).
2382. Wall, L. A., Ingalls, R. B.: J. Chem. Phys. 41, 1112 (1964).
2383. Walton, A. G., Blackwell, J.: Biopolymers. New York: Academic Press 1973.
2384. Ward, H. R.: in: Free Radicals (Ed. Kochi, J. K.). New York: Wiley 1973, vol. 1, p. 239.
2385. Ward, R. L., Weissman, S. I.: J. Amer. Chem. Soc. 79, 2086 (1957).
2386. Ward, T. C., Books, J. T.: Macromolecules 7, 207 (1974).
2387. Warden, J. T., Bolton, J. R.: Accounts Chem. Res. 7, 189 (1974).
2388. Wartewig, S., Welter, M., Windsh, W.: Exp. Tech. Physik. 12, 354 (1964).
2389. Wasserman, E., Kuck, V. J., Delevan, W. M., Yager, W. A.: J. Amer. Chem. Soc. 91, 1040 (1968).
2390. Wasserman, E., Murray, R. W., Kaplan, M. L., Yager, W. A.: J. Amer. Chem. Soc.
2391. Waterman, D. C., Dole, M.: J. Phys. Chem. 74, 1906 (1970).
2392. Waterman, D. C., Dole, M.: J. Phys. Chem. 74, 1913 (1970).
2393. Waters, W. A.: Proc. Roy. Soc. A302, 287 (1968).
2394. Watson, W. H., Jr., McMordie, W. C., Lands, L. G.: J. Polym. Sci. 55, 137 (1961).
2395. Watt, G. W., Baye, L. J., Drummond, F. O.: J. Amer. Chem. Soc. 88, 1138 (1966).
2396. Wayne, R. P.: Photochemistry. London: Butterworths 1970.
2397. Weiner, S. A., Hammond, G. S.: J. Amer. Chem. Soc. 90, 1659 (1968).
2398. Weiner, S. A., Hammond, G. S,: J. Amer. Chem. Soc. 91, 986 (1969).
2399. Wells, C. H., Jr.: Introduction to Molecular Photochemistry. London: Chapman 1972.
2400. Wells, J. W.: J. Chem. Phys. 52, 4062 (1970).
2401. Wen, W. Y., Johnson, D. R., Dole, M.: Macromolecules 7, 199 (1974).
2402. Wen, W. Y., Johnson, D. R., Dole, M.: J. Phys. Chem. 78, 1798 (1974).
2403. Wertz, J. E.: Chem. Rev. 55, 829 (1955).

References

2404. Wertz, J. E.: in: Kirk-Othmer, Encyclopedia of Chemical Technology. New York: Wiley-Interscience 1965, vol. 7, p. 874.
2405. Wertz, J. E., Bolton, J. R.: Electron Spin Resonance, Elementary Theory and Practical Application. New York: McGraw-Hill 1972.
2406. Westenberg, A. A., de Haas, N.: J. Chem. Phys. 40, 3087 (1964).
2407. Westfahl, J. C.: Rubber Chem. Technol. 46, 1134 (1973).
2408. Westfahl, J. C., Carman, J. C., Layer, R. W.: Rubb. Chem. Technol. 45, 402 (1972).
2409. Whelen, D. J., Pinkerton, D. M.: Austral. J. Chem. 23, 391 (1970).
2410. Willard, J. E.: in: Fundamental Processes in Radiation Chemistry (Ed. Ausloos, P.). New York: Interscience-Wiley 1968, p. 599.
2411. Williams, D., Schmidt, B., Wolfrom, M. L., Michelakis, A., McCabe, L. J.: Proc. Natl. Acad. Sci. 45, 1744 (1959).
2412. Williams, D. J.: Polym. Prepr. Amer. Chem. Soc., Div. Polym. Chem. 14, 830 (1973).
2413. Williams, Ff.: in: Fundamental Process in Radiation Chemistry (Ed. Ausloos, P.). New York: Interscience-Wiley 1968, p. 515.
2414. Williams, Ff.: in: The Radiation Chemistry of Macromolecules (Ed. Dole, M.). New York: Academic Press 1972, vol. 1, p. 7.
2415. Williams, Ff., Hayashi, K.: Nature 212, 281 (1967).
2416. Williams, M. L., DeVries, K. L.: Surfaces and Intersurfaces. Syracuse, USA: Syracuse University Press, vol. 2, p. 139.
2417. Wilmhurst, T. H.: Electron Spin Resonance Spectrometers. London: Hilger 1967.
2418. Wilmhurst, T. H., Bennet, T. J., Smith, R. C.: Proc. Roy. Soc. A302, 305 (1968).
2419. Wilske, J., Heusinger, H.: Radiochem. Acta 11, 187 (1967).
2420. Wilske, J., Heusinger, H.: J. Polym. Sci. A1, 7, 995 (1969).
2421. Wilske, J., Heusinger, H.: Int. J. Radiat. Phys. Chem. 2, 131 (1970).
2422. Wincow, G., Fraenkel, G. K.: J. Chem. Phys. 34, 1333 (1961).
2423. Windle, J. J.: Textile Res. J. 32, 963 (1962).
2424. Windle, J. J., Thurston, W. H.: J. Chem. Phys., J. Chem. Phys. 27, 1429 (1957).
2425. Windle, J. J., Wiersema, A. K.: J. Appl. Polym. Sci. 8, 1531 (1964).
2426. Winslow, F. H., Baker, W. O., Yager, W. A.: J. Amer. Chem. Soc. 77, 4751 (1955).
2427. Wong, P. K.: Radiat. Effects 19, 87 (1973).
2428. Wong, P. K.: Polymer 15, 60 (1974).
2429. Wonn, S. K., Fabes, L., Green, W. J., Wan, J. K. S.: J. Chem. Soc. Faraday Trans. 68, 2211 (1972).
2430. Wood, R. M.: J. Sci. Technol. 33, 50 (1966).
2431. Worthington, K., Baugh, P. J.: Cellul. Chem. Technol. 5, 23 (1971).
2432. Wyard, S. J.: J. Sci. Instrum. 42, 769 (1965).
2433. Wyard, S. J., Smith, R. C., Adrian, F. J.: J. Chem. Phys. 49, 2780 (1968).
2434. Yaffe, L.: Nuclear Chemistry. New York: Academic Press 1968, vol. 1 and vol. 2.
2435. Yagi, K., Toda, F., Iwakura, Y.: J. Polym. Sci. B, 10, 113 (1972).
2436. Yamaguchi, K., Minoura, Y.: J. Polym. Sci. A1, 10, 2875 (1972).
2437. Yamaguchi, K., Yoshida, T., Minoura, Y.: J. Polym. Sci. A1, 10, 2501 (1972).
2438. Yamaguchi, K., Yoshida, T., Minoura, Y.: Polymer (in press).
2439. Yamaguchi, S., Kaneko, M., Furuichi, J.: J. Phys. Soc. Japan 22, 348 (1967).
2440. Yamamoto, M., Sakaguchi, M., Shiotani, M., Sohma, J.: Rep. Progr. Polym. Phys. Japan 16, 549 (1973).
2441. Yamamoto, M., Yano, M., Nishijima, Y.: Rep. Progr. Polym. Phys. Japan 11, 495 (1968).
2442. Yamamoto, M., Fujiwara, S.: Bunseki Kagaku 19, 715 (1970).
2443. Yamamoto, Y., Kanda, S., Kusabayashi, S., Nogaito, T., Ito, K., Mikawa, H.: Bull. Chem. Soc. Japan 38, 2015 (1965).
2444. Yamanouchi, S., Kashiwabara, H., Sohma, J.: Rep. Progr. Polym. Phys. Japan 13, 473 (1970).
2445. Yamaoka, H., Obama, I., Hayashi, K., Okamura, S.: J. Polym. Sci. A1, 8, 495 (1970).
2446. Yamaoka, H., Shiga, T., Hayashi, K., Okamura, S., Sugiura, T.: J. Polym. Sci. B, 5, 329 (1967).
2447. Yamazuki, I.: Tampakushitsu Kakusan Koso 8, 623 (1963).

2448. Yang, J. Y., Ingalls, R. B.: J. Amer. Chem. Soc. *85*, 3920 (1963).
2449. Yashiro, D., Ueda, H.: Kogyo Kagaku Zasshi *73*, 2720 (1970).
2450. Yasukawa, T., Matsuzaki, K.: Tohoku Diagaku Hisuiyoeki Kagaku Kenkyusho Hokoku *19*, 257 (1969).
2451. Yasukawa, T., Matsuzaki, K., Yamagishi, M.: Makromol. Chem. *131*, 305 (1970).
2452. Yasukawa, T., Takahashi, T., Murakami, K., Araki, K., Sasuga, T., Ohmichi, H.: J. Polym. Sci. A1, *10*, 259 (1972).
2453. Yatsu, T., Moriuchi, S., Fujii, H.: Polym. Prepr. Amer. Chem. Soc., Div. Polym. Chem. *16*, 373 (1975).
2454. Yodis, P. P., Koski, W. S.: J. Chem. Phys. *38*, 1313 (1963).
2455. Yokohata, A., Harakon, K., Takimoto, K., Tsuda, S.: Bull. Chem. Soc. Japan *48*, 1629 (1975).
2456. Yonezawa, T., Mirishima, I.: Bunseki Kagaku *20*, 1058 (1971).
2457. Yoshida, H., Hashimoto, S., Iwamoto, T., Okamura, S.: Bull. Inst. Chem. Res. Kyoto Univ. *47*, 1 (1969).
2458. Yoshida, H., Hayashi, K.: Adv. Polym. Ser. *6*, 401 (1969).
2459. Yoshida, H., Hayashi, K., Okamura, S.: Arkiv Kemi *23*, 177 (1964).
2460. Yoshida, H., Hayashi, K., Warashina, T.: Bull. Chem. Soc. Japan *45*, 3515 (1972).
2461. Yoshida, H., Higashimura, T.: Can. J. Chem. *48*, 504 (1970).
2462. Yoshida, H., Irie, M., Hayashi, K.: in: ESR Application to Polymer Research (Ed. Kinell, P. O., Rånby, B., Runnström-Reio, V.). Stockholm: Almqvist-Wiksell 1973, p. 129.
2463. Yoshida, H., Kambara, Y., Rånby, B.: Bull. Chem. Soc. Japan *47*, 2599 (1974).
2464. Yoshida, H., Kodaira, T., Tsuji, K., Hayashi, K., Okamura, S.: Bull. Chem. Soc. Japan *37*, 1531 (1964).
2465. Yoshida, H., Noda, M.: Polym. J. *2*, 359 (1971).
2466. Yoshida, H., Rånby, B.: J. Polym. Sci. *2*, 1155 (1964).
2467. Yoshida, H., Rånby, B.: Acta Chem. Scand. *19*, 72 (1965).
2468. Yoshida, H., Rånby, B.: Acta Chem. Scand. *19*, 1495 (1965).
2469. Yoshida, H., Rånby, B.: J. Polym. Sci. A1, *3*, 2289 (1965).
2470. Yoshida, H., Rånby, B.: J. Polym. Sci. C, (16), 1333 (1967).
2471. Yoshida, H., Tsuji, K., Hayashi, K., Okamura, S.: Bull. Inst. Chem. Res. Kyoto Univ. *41*, 39 (1963).
2472. Yoshida, H., Warashina, T.: Bull. Chem. Soc. Japan *44*, 2950 (1971).
2472. Zachmann, H. G.: Kolloid-Z. u. Z-Polym. *251*, 951 (1973).
2473. Zakrevskii, V. A.: Vysokomol. Soedin. *B13*, 105 (1971).
2474. Zakrevskii, V. A., Baptizmanskii, V. V., Tomashevskii, E. E.: Fiz. Tverd. Tela *10*, 1699 (1969).
2475. Zakrevskii, V. A., Korsukov, V. A.: Vysokomol. Soedin. *A14*, 955 (1972).
2476. Zakrevskii, V. A., Korsukov, V. A.: Plaste Kautsch. *19*, 92 (1972).
2477. Zakrevskii, V. A., Kuksenko, V. S., Savostin, A. Ya., Slutsker, A. I., Tomashevskii, E. E.: Fiz. Tverd. Tela *11*, 1940 (1969).
2478. Zakrevskii, V. A., Tomashevskii, E. E.: Vysokomol. Soedin. *8*, 1295 (1966).
2479. Zakrevskii, V. A., Tomashevskii, E. E., Baptizmanskii, V. V.: Fiz. Tverd. Tela *9*, 1434 (1967).
2480. Zakrevskii, V. A., Tomashevskii, E. E., Baptizmanskii, V. V.: Vysokomol. Soedin. *B10*, 193 (1968).
2481. Zakrevskii, V. A., Tomashevskii, E. E., Baptizmanskii, V. V.: Vysokomol. Soedin. *A12*, 419 (1970).
2482. Zandra, P. J., Weissman, S. I.: J. Amer. Chem. Soc. *84*, 4408 (1962).
2483. Zanchetta, J., Marchand, A., Pacault, A.: C. R. Acad. Sci. *258*, 1496 (1964).
2484. Zaporozhskaya, O. A., Kovarskii, A. L., Pudov, V. S., Vasserman, A. M., Buchachenko, A. L.: Vysokomol. Soedin. *B12*, 702 (1970).
2485. Zefirova, A. K., Tikhomirova, N. N., Shilov, A. E.: Dokl. Akad. Nauk SSSR *132*, 1082 (1960).
2486. Zelenskaya, T. V., Dubinskaya, A. M., Aseeva, R. N., Berlin, A. A.: Izv. Akad. Nauk SSSR *1972*, 1965.

2487. Zhdanov, G. S., Milinchuk, V. K.: Vysokomol. Soedin. *A14*, 2405 (1972).
2488. Zhdanov, G. S., Milinchuk, V. K., Pshezhetskii, S. Ya.: Khim. Vysokh. Energ. *7*, 190 (1973).
2489. Zhimidorov, G. M., Lebedev, Ya. S., Dobryakov, S. N.: Interpretation of Complex ESR Spectra. Moscow: Nauka 1975 (in Russ.).
2490. Zhimidorov, G. M., Lebedev, Ya. S., Tsvetkov, Yu. D.: Zh. Strukt. Khim. *3*, 541 (1962).
2491. Zhimidorov, G. M., Tsvetkov, Yu. D., Lebedev, Ya. S.: Zh. Strukt. Khim. *2*, 696 (1961).
2492. Zimbrick, J., Hoecker, F., Kevan, L.: J. Phys. Chem. *72*, 3277 (1968).
2493. Zhuravleva, T. S., Lebedev, Ya. S., Shuvavlov, V. F.: Zh. Strukt. Khim. *5*, 786 (1964).
2494. Zhurkov, S. N.: Int. J. Fracture Mech. *1*, 311 (1965).
2495. Zhurkov, S. N.: Izv. Akad. Nauk SSSR *1967*, 1765.
2496. Zhurkov, S. N., Kuksenko, V. S.: Mekh. Polim. *5*, 792 (1974).
2497. Zhurkov, S. N., Kuksenko, V. S., Slutsker, A. I.: in: International Congress on Fracture. Proc. Mater. Brighton, 1969, p. 531.
2498. Zhurkov, S. N., Kuksenko, V. S., Slutsker, A. I.: Fiz. Tverd. Tela *11*, 296 (1969).
2499. Zhurkov, S. N., Savostin, A. Yu., Tomashevskii, E. E.: Dokl. Akad. Nauk SSSR *159*, 303 (1964).
2500. Zhurkov, S. N., Savostin, A. Ya., Tomashevskii, E. E.: Soviet Phys. Doklady *9*, 986 (1965).
2501. Zhurkov, S. N., Savostin, A. Yu., Tomashevskii, E. E.: Dokl. Akad. Nauk SSSR *195*, 707 (1969).
2502. Zhurkov, S. N., Tomashevskii, E. E.: in: Physical Basis of Yield and Fracture. Conf. Proc., Oxford, 1966, p. 200.
2503. Zhurkov, S. N., Tomashevskii, E. E., Zakrevskii, V. A.: Fiz. Tverd. Tela *3*, 2841 (1961).
2504. Zhurkov, S. N., Zakrevskii, V. A., Korsukov, V. E., Kuksenko, V. S.: J. Polym. Sci. A2, *10*, 1509 (1972).
2505. Zhurkov, S. N., Zakrevskii, V. A., Korsukov, V. E., Kuksenko, V. S.: Fiz. Tverd. Tela *12*, 1680 (1970).
2506. Zhurkov, S. N., Zakrevskii, V. A., Tomashevskii, E. E.: Fiz. Tverd. Tela *6*, 1912 (1964).
2507. Zhurkov, S. N., Vettegren, V. I., Korsukov, V. E., Novak, I. I.: Fiz. Tverd. Tela *11*, 290 (1969).
2508. Zhurkov, S. N., Vettegren, V. I., Korsukov, V. E., Novak, I. I.: in: International Congress on Fracture, Prov. Mater., Brighton, 1969, p. 545.
2509. Zhurkov, S. N., Vettegren, V. I., Novak, I. I., Kashinseva, K. N.: Dokl. Akad. Nauk SSSR *176*, 623 (1967).
2510. Zhuzgov, E. I., Akundova, L. A., Voevodskii, V. V.: Kinet. Kataliz. *6*, 56 (1965).
2511. Zhuzgov, E. I., Akundova, L. A., Voevodskii, V. V.: Kinet. Kataliz. *9*, 773 (1968).
2512. Zhuzgov, E. I., Bubonov, N. N., Voevodskii, V. V.: Radiats. Khim. Polim. Mater. Simp. Moscow, 1964, p. 224.
2513. Zhuzgov, E. I., Bubonov, N. N., Voevodskii, V. V.: Kinet. Spektrosk. *6*, 56 (1965).
2514. Zigmunt, M. M., Shapiro, B. L., Kuzminskii, A. S.: Vysokomol. Soedin. *A15*, 2361 (1973).
2515. Zott, H., Heusinger, H.: Macromolecules *8*, 182 (1975).
2516. Zubkov, A. V.: Dokl. Akad. Nauk SSSR *261*, 1095 (1974).
2517. Zubov, V. P., Kabanov, V. A.: Vysokomol. Soedin. *A13*, 1305 (1974).
2518. Zubov, V. P., Lachinov, M. B., Golubev, V. B., Kulikova, V. F., Kabanov, V. A., Polak, L. S., Kargin, V. A.: J. Polym. Sci. C, (23), 147 (1968).
2519. Zutty, N. L., Wilson, C. W., Potter, G. H., Priest, D. C., Whitworth, C. J.: J. Polym. Sci. A1, *3*, 2781 (1965).

Index

407

Index